The Quantum-Like Revolution

Arkady Plotnitsky · Emmanuel Haven
Editors

The Quantum-Like Revolution

A Festschrift for Andrei Khrennikov

Springer

Editors
Arkady Plotnitsky
Theory and Cultural Studies Prog
Purdue University
West Lafayette, IN, USA

Emmanuel Haven
Faculty of Business Administration
Memorial University
St. John's, NL, Canada

ISBN 978-3-031-12985-8 ISBN 978-3-031-12986-5 (eBook)
https://doi.org/10.1007/978-3-031-12986-5

This Springer imprint is published by the registered company Springer Nature Switzerland AG
The registered company address is: Gewerbestrasse 11, 6330 Cham, Switzerland

For Professor Plotnitsky: "To Paula"

For Professor Haven: "To 'Popke', Natty, Braunty, Hazel-Noisette and Sammy"

Foreword

It is a great pleasure for me to see the present beautiful collection of articles honoring Andrei Khrennikov. The scientific community, and certainly myself, highly appreciate Andrei's contribution to fundamental questions in quantum mechanics. His papers cover a very broad spectrum of important fundamental issues. As examples, I would mention his numerous papers on foundations of probability theory, on generalized probability models, and on contextual probability and related topics. He also worked on a number of fundamental mathematical issues, including infinite dimensional analysis, systems with an infinite number of degrees of freedom, path integral quantization, and his work on subquantum models. Most remarkable is his extension of quantum methods to biological systems, and even to fields outside physics in the social, political, and economical sciences, independent of whether one agrees or does not agree with some of these positions. Andrei's open spirit and audacity to go beyond the common should be highly appreciated.

Very important for the scientific community is his organizing the series of quantum theory conferences in Växjö. These have become a focus of pilgrimage for a very diverse set of participants. Andrei was able to motivate a very broad spectrum all the way from fundamental and mathematical issues to recent experiments on the edge of developments of the field. At some of these conferences, I remember significant discussions, which occasionally even became stormy, between adherents of different points of view of quantum mechanics. These disputes certainly helped to clarify some of the important issues at stake, and actually were helpful in creating deeper and more consistent points of view for a number of participants. It is a great compliment to Andrei that some people from different fields complain that there are not enough representatives from their own area of research.

I wish Andrei many more fruitful years in his scientific research. I look forward to meeting him again at his Växjö conferences.

Anton Zeilinger
Physics Nobel Laureate 2022
Austrian Academy of Sciences
University of Vienna
VCQ Vienna
Vienna, Austria

Acknowledgments

We would, first of all, like to thank all participants of this volume for their contributions. We also want to express our gratitude to them and many others as participants at the Växjö conferences at Linnaeus University, previously Växjö University, which gave this conference its name, the longest running conference ever on quantum foundations, and which in itself is one of Andrei Khrennikov's great contributions to quantum foundations and quantum-like theories beyond physics. These conferences had a shaping influence on our own thinking concerning these subjects and on conceiving and putting together this volume, as a tribute to both these fields and Khrennikov's contributions to them. We are most grateful to Professor Anton Zeilinger for writing a foreword to this volume.

We are grateful to Springer Nature for publishing the book and would like to thank those with whom we worked there, especially our editor, Angela Lahee, and project coordinator, Chandra Sekaran, whose support of the project and help in bringing it to its final form were invaluable.

Contents

Introduction 1
Arkady Plotnitsky and Emmanuel Haven

On Wheeler's Quantum Circuit 25
Gregg Jaeger

Letters for Andrei: QBism and the Unfinished Nature of Nature 61
Christopher A. Fuchs

Contra Multos Verbos: On Scandals of Quantum Mechanics 91
Theodorus Maria Nieuwenhuizen

Quantum Holism 125
Giacomo M. D'Ariano

Shannon Theory for Quantum Systems and Beyond: Information Compression for Fermions 135
Paolo Perinotti, Alessandro Tosini, and Leonardo Vaglini

Conjugate Logic 157
Niklas Johansson, Felix Huber, and Jan-Åke Larsson

Revisiting Quantum Mysteries 181
Philippe Grangier

Quantum Measurement Theory for Systems with Finite Dimensional State Spaces 191
Masanao Ozawa

Probability Distributions as Alternatives of the Density Operator and Wave Function for Describing Quantum States 215
Margarita A. Man'ko and Vladimir I. Man'ko

Alternative Quantum Mechanics? 235
Dagomir Kaszlikowski and Paweł Kurzyński

Compositionality as We See It, Everywhere Around Us 247
Bob Coecke

Measurement Models in Quantum Cognition 269
Jerome R. Busemeyer

Quantum Cognition: Quo Vadis? 281
Emmanuel M. Pothos and James M. Yearsley

Entanglement in Cognition Violating Bell Inequalities Beyond Cirel'son's Bound 299
Diederik Aerts, Jonito Aerts Arguëlles, Lester Beltran, Suzette Geriente, and Sandro Sozzo

Thinking and Deciding in a Complex World 327
Polina Khrennikova

Contextuality and Random Variables 339
Ehtibar Dzhafarov

Can Quantum Non-identity Exist in Social Phenomena? 361
J. Acacio de Barros

Index 379

Editors and Contributors

About the Editors

Arkady Plotnitsky is a Distinguished Professor in Literature, Theory, and Culture Program, and Philosophy and Literature Program at Purdue University. His research interests include the philosophy and theory of mathematics and science, especially quantum foundations and the role of mathematics in physics, continental philosophy, and the relationships among philosophy, literature, and mathematics and science. He is the author of nine books, several edited and co-edited collections and journal issues, and over 200 hundred articles and book chapters on these subjects. *His latest books are Reality without Realism:Matter, Thought, and Technology in Quantum Physics (Springer 2021) and Logos and Alogon: Thinkable and the Unthinkable in Mathematics, from the Pythagoreans to the Moderns (Spinger 2022).*

Emmanuel Haven holds the Alex Faseruk Chair in Financial Management at Memorial University in Canada. He has research interests in using formalisms from physics to social science. He co-authored several books and published in a variety of fields ranging from Biology, Finance, Economics to Operations Research and Physics. With Andrei Khrennikov he published *Quantum Social Science* (Cambridge University Press 2013) and *The Palgrave Handbook of Quantum Models in Social Science: Applications and Grand Challenges* (Palgrave MacMillan Springer Nature, 2017).

Contributors

J. Acacio de Barros School of Humanities and Liberal Studies, San Francisco State University, San Francisco, CA, USA

Diederik Aerts Center Leo Apostel for Interdisciplinary Studies, Free University of Brussels (VUB), Brussels, Belgium

Jonito Aerts Arguëlles Center Leo Apostel for Interdisciplinary Studies, Free University of Brussels (VUB), Brussels, Belgium

Lester Beltran Center Leo Apostel for Interdisciplinary Studies, Free University of Brussels (VUB), Brussels, Belgium

Jerome R. Busemeyer Psychological and Brain Sciences, Indiana University, Bloomington, IN, USA

Bob Coecke Compositional Intelligence team, Quantinuum, Oxford, UK

Giacomo M. D'Ariano Dipartimento di Fisica dell'Universita di Pavia, Pavia, Italy;
Istituto Lombardo Accademia di Scienze e Lettere and INFN, Gruppo IV, Sezione di Pavia, Pavia, Italy

Ehtibar Dzhafarov Psychological Sciences, Purdue University, West Lafayette, IN, USA

Christopher A. Fuchs Department of Physics, University of Massachusetts Boston, Boston, MA, USA

Suzette Geriente Center Leo Apostel for Interdisciplinary Studies, Free University of Brussels (VUB), Brussels, Belgium

Philippe Grangier Laboratoire Charles Fabry, Institut d'Optique Graduate School, CNRS, Université Paris Saclay, Palaiseau, France

Emmanuel Haven Faculty of Business Administration and CQSCS, Memorial University, St. John's, Canada

Felix Huber Atomic Optics Department, Jagiellonian University, Kraków, Poland

Gregg Jaeger Quantum Communication and Measurement Laboratory, Department of Electrical and Computer Engineering, and Division of Natural Science and Mathematics, Boston University, Boston, MA, USA

Niklas Johansson Department of Electrical Engineering, Linköping University, Linköping, Sweden

Dagomir Kaszlikowski Centre for Quantum Technologies and Department of Physics, National University of Singapore, Singapore, Singapore

Polina Khrennikova Faculty of Behavioural, Management and Social Sciences, Financial Engineering Section, University of Twente, Enschede, The Netherlands

Paweł Kurzyński Faculty of Physics, Adam Mickiewicz University, Poznań, Poland

Jan-Åke Larsson Department of Electrical Engineering, Linköping University, Linköping, Sweden

Margarita A. Man'ko Lebedev Physical Institute, Moscow, Russia

Vladimir I. Man'ko Lebedev Physical Institute, Moscow, Russia;
Moscow Institute of Physics and Technology, Dolgoprudnyi, Moscow, Russia

Theodorus Maria Nieuwenhuizen Institute for Theoretical Physics, University of Amsterdam, Amsterdam, The Netherlands

Masanao Ozawa Center for Mathematical Science and Artificial Intelligence, Chubu University Academy of Emerging Sciences, Chubu University, Kasugai, Japan;
Graduate School of Informatics, Nagoya University, Chikusa-ku, Nagoya, Japan

Paolo Perinotti QUIT group, Physics Department, Pavia University, and INFN Sezione di Pavia, Pavia, Italy

Arkady Plotnitsky Literature, Theory, and Cultural Studies Program, Philosophy and Literature Program, Purdue University, West Lafayette, USA

Emmanuel M. Pothos Department of Psychology, City, University of London, London, UK

Sandro Sozzo Department of Humanities and Cultural Heritage (DIUM) and Centre CQSCS, University of Udine, Udine, Italy

Alessandro Tosini QUIT group, Physics Department, Pavia University, and INFN Sezione di Pavia, Pavia, Italy

Leonardo Vaglini QUIT group, Physics Department, Pavia University, and INFN Sezione di Pavia, Pavia, Italy

James M. Yearsley Department of Psychology, City, University of London, London, UK

Introduction

Arkady Plotnitsky and Emmanuel Haven

It is a great honor for the present authors to edit this collection, dedicated to Andrei Khrennikov's work. Both of us benefited immeasurably from our friendship with Andrei, numerous discussions with him concerning quantum foundations and quantum-like modeling, and physics, mathematics, and philosophy in general, and actual collaborations with him (especially extensive in the case of one of us, Emmanuel Haven). Of course, as so many others, we also benefited greatly from our participation in many Växjö conferences, organized by him. Along with the contributors to this collection, we join in celebrating his extraordinary contribution to both quantum foundations and quantum-like modeling beyond physics, and other areas of mathematics and physics. In making this collection itself part of this celebration, we thought that it would be best to do so by making it a collection of original papers that serve the aim of advancing quantum foundations and quantum-like modeling, two main fields to which this collection is devoted. This aim does not preclude some of these articles from offering historical reflections on the development of both fields, either in connection with Khrennikov's work or in general. Any argument, no matter how innovative, has a history and depends on it, and, when innovative, transforms it. Quantum-like theories beyond physics or earlier quantum theory itself are examples of such joint workings of history and innovation. Adopting this perspective in this introduction as well, we would like, first, to reflect on the nature of both fields and Khrennikov's contributions to them, beginning, respecting the order of history, with

A. Plotnitsky (✉)
Literature, Theory, and Cultural Studies Program, Philosophy and Literature Program, Purdue University, West Lafayette, USA
e-mail: plotnits@purdue.edu

E. Haven
Faculty of Business Administration and CQSCS, Memorial University, St. John's, Canada
e-mail: ehaven@mun.ca

A. Plotnitsky and E. Haven (eds.), *The Quantum-Like Revolution*,
https://doi.org/10.1007/978-3-031-12986-5_1

quantum theory in its foundational aspects and then moving to quantum-like modeling beyond physics. We will then offer introductory summaries of contributions to this volume.

We begin, then, with quantum foundations, dealing primarily with quantum mechanics (QM), where most of Khrennikov's work of quantum foundations belongs. The field is vast and multifaceted, and the literature addressing the subject has by now reached a mammoth scale, with thousands of books and hundreds of thousands articles. Accordingly, the present outline is unavoidably partial, focusing on several key junctures that both give this book's broader potential audience a sense of the field of quantum foundations and provide a suitable background for commenting on Khrennikov's contribution and those by the articles assembled in this collection. While we will comment on other pertinent topics, we will consider in particular three subjects crucial to quantum foundations: the first is the question of realism, via the Bohr-Einstein debate; the second is the question of quantum nonlocality and surrounding concerns, such as Bell's theorem, quantum entanglement, and quantum correlations; and the third is the role of quantum information theory in quantum foundations. All three subjects figure prominently in Khrennikov's work and have been equally prominent in the discussions and debates concerning quantum foundations at the Växjö conferences, the longest ever running conference on quantum foundations, and more recently on quantum-like modelling beyond physics. Organizing these conferences was itself an important contribution to both fields on Khrennikov's part.

We consider first the question of reality and realism in quantum physics, via the Bohr-Einstein debate, which has overshadowed the debate concerning quantum foundations for near a century. This debate also figures significantly, and often expressly, in Khrennikov's work. This work oscillates between Albert Einstein's realist and Niels Bohr's nonrealist positions, in Khrennikov's more recent articles exhibiting a greater affinity with Bohr's ideas, in particular, complementarity, Bohr's most famous concept, a persistent reference in these articles (e.g., Khrennikov [17, 18, 20–24]). Arguably, the greatest significance of the Bohr-Einstein debate has been in the fact that at stake in it was not only the viability of QM (and, by implication, quantum field theory, QFT) but also the nature and future of fundamental physics. Einstein conceived of this conjunction as a form of conceptual realism (he was no naïve realist), a pursuit of an idealized representation of physical reality by means of suitable mathematized physical concepts and models, which he considered as imperative for any fundamental physical theory, a theory that deals with the ultimate constitution of nature.[1] This imperative guided Einstein's thinking throughout his life, in particular his work on general relativity, where this guidance proved to be so successful. Bohr, by contrast, was ready to abandon this imperative even (partially) in his 1913 atomic theory and found it to be no longer viable after the discovery of QM by Werner Heisenberg in 1925. It was Einstein's realist imperative that grounded his discontent with QM, at least if the latter was understood, as it was by Einstein, along nonreal-

[1] All modern physics, from Galileo on, only deals with such mathematical idealizations, and the question, posed anew by quantum theory, is in what way it can or cannot use them.

ist lines of Bohr's interpretation or "the Copenhagen spirit of quantum theory", as Heisenberg called it (putting for the moment aside the complexities associated with different positions within this spirit) (Heisenberg [16], p. iv). Einstein never accepted that QM (or QFT) offered a "useful point of departure for future development", as he stated as late as 1949, more than two decades after the introduction of QM and in the immediate wake of major breakthroughs in QED (related to its renormalization), breakthroughs that would, however, have been unlikely to affect his view in any event (Einstein [7], p. 83). Einstein's discontent with QM is most widely known by his pronouncement that "God does not play dice" (e.g., Born [5], p. 88). It is clear, however, that this discontent had more to do with the difficulty of reconciling QM or QFT with his realist imperative. Einstein realized, just as Bohr did, that the probabilistic nature of quantum predictions is implied by the absence of realism, because of the impossibility of assuming that the ultimate nature of the reality responsible for quantum phenomena could be treated deterministically, even ideally and in principle. Einstein's arguably final pronouncement on the subject was this: "That the Lord should play dice, all right; but that He should gamble according to definite rules, this is beyond me" (reported in Wheeler and Zurek [41], p. 8). It is sometimes said that quantum dice are loaded in view of such phenomena as quantum correlations, likely on Einstein's mind here. In fact, however, the language of dice is misleading. The workings of quantum probability or statistics are very different from those of classical physics, to which a throw of the dice belongs. Accordingly, Einstein's statement confirms and even amplifies, rather than tempers, his discontent with QM. Bohr characterized Einstein's position, which was expressed, arguably most clearly, in Einstein's 1936 article "Physics and Reality" [6] and remained essentially the same thereafter, as follows: 'Einstein....[argued] that the quantum-mechanical description is to be considered merely as a means for of accounting for the average behavior of a large number of atomic systems, and his belief that it should offer an exhaustive description of the individual phenomena [was] expressed in the following words: "To believe this is logically possible without contradiction; but it is so very contrary to my [Einstein's] scientific instinct that [he could not] forego the search for a more complete conception"' [Einstein [6], p. 375]. (Bohr [4], v. 2, p. 61). That quantum-mechanical predictions were fully in accord with all available experimental evidence did not deter Einstein, in part because several of his arguments, especially those concerning Einstein-Podolsky-Rosen (EPR) type experiments for the incompleteness, or else nonlocality, of QM, were based on predictions with probability one, ideally possible in these experiments (Einstein et al. [8]). Einstein had continued to believe that QM could not in fact predict all that was possible to establish as real in considering individual quantum systems, unless one allows for Einstein-nonlocality, as an action at a distance, a subject we discuss below. Bohr argued that "there could be no other way to deem a logically consistent mathematical formalism as inadequate than by demonstrating the departure of its consequences from experience or by proving that its predictions did not exhaust the possibilities of observation" (Bohr [4], v. 2, p. 57). According to Bohr, "Einstein's argumentation could be directed to neither of these ends" (Bohr [4], v. 2, p. 57). Bohr's contention also shaped his response to Einstein's view just stated, rejecting the belief in the completeness of quantum mechanics in spite of the logical consistency of this belief. Bohr countered as follows:

> Even if [Einstein's] attitude might seem well balanced in itself, it nevertheless implies a rejection of [rather than offers a counterargument to] the whole argumentation [essentially that of his ultimate RWR-interpretation] that, in quantum mechanics, we are not dealing with an arbitrary renunciation of a more detailed analysis of atomic phenomena but with a recognition that such an analysis is *in principle* excluded. (Bohr [4], v. 2, p. 62)

This is one of Bohr's strongest expressions of his ultimate, nonrealist, interpretation of quantum phenomena and QM. Bohr's statement may suggest the impossibility of a realist alternative in general, rather than only the fact Bohr himself adopts a nonrealist interpretation of quantum phenomena and QM, possibly as more viable than the realist alternative. While Bohr's ultimate position may be debated, we shall, for the present purposes assume, that such statements by Bohr or others could, regardless of the author's intent, only represent a particular interpretation of quantum phenomena and QM or QFT. Be it as it may on that score, this exchange poses, arguably, the biggest question at stake in the Bohr-Einstein debate and beyond, the question whether nature or (which is Bohr's way of looking at the quantum-mechanical situation) our interactions with nature, will ever allow one to fulfil Einstein's desiderata for a fundamental theory of quantum phenomena or whatever future phenomena may replace or delimit them. It might indeed be argued that the debate between Bohr and Einstein was not primarily about what QM could or could not do, on which point their disagreement, while not discountable, was not so crucial. This debate was ultimately about whether nature could allow us to do better. While Einstein thought that it should and while it is possible that Bohr thought that it will not, a more cautious philosophical position which we prefer to adopt here is that it may not, which is not the same as that it never will. In other words, nature or our interaction with it may or may not allow us to do better. Most have been and remain inclined to accept Einstein's view, beginning, famously with Erwin Schrödinger, a co-creator of QM, whose discontent with it became as strong as that of Einstein. Schrödinger, at some point, referred to it "as the doctrine born from distress," even as he showed a very deep understanding of both QM and the doctrine on the same occasion, in his famous cat-paradox paper (Schrödinger [35], p. 154).

The circumstance that our fundamental theories remain incomplete (as there is no quantum theory of gravity and even handling other forces of nature by means of QFT is still not finalized) has often been used to support the Einsteinian hope that physics will return to realism. Einstein prevailed in this part of his debate with Bohr, with most physicists and philosophers taking his side. Bohr was well aware of this situation, in speaking, again, as late as in 1949, "of scepticism [among the physicists themselves] about the necessity of going so far in renouncing demands as regards the explanation of natural phenomena" (Bohr [4], v. 2, p. 63). He also realized, however, that physics itself is a different matter. While human, it does not necessarily obey human imperatives, which have been defeated by physics throughout its history, not the least by Einstein's own relativity theory. Accordingly, while a realist theory of quantum phenomena is possible or might even become necessary in view of new experimental or theoretical findings, it is equally possible one or another form of nonrealist view (there is no single form of nonrealism any more than of realism) will remain viable in grounding our interpretations of QM and QFT. The debate concerning the subject has never subsided and continues with undiminished intensity with, it appears, no end in sight.

Khrennikov's work occupies a complex position in this regard, arguably (we refrain from making a definitive claim) being ultimately inclined toward a realist view, although, especially in his more recent work, with Bohrian inflections, in particular, adding Bohr's complementarity to the mix. The concept is extensively discussed in his recent articles cited above, also in the context on the question of quantum nonlocality, which we discuss below. It would not be possible to consider this aspect of Khrennikov's work in detail here to do it justice. Our aim is to underscore Khrennikov's extensive engagement with this problematic, central to quantum foundations, and in more recent work with Bohr's ideas. It should be added that Khrennikov also brings both Ludwig Boltzmann's and Heinrich Hertz's arguments concerning the nature of physical theories to bear on this problematic (e.g., [19]). This is a potentially fruitful angle on this part of the foundational debate concerning quantum physics, which has rarely been considered. One important and revealing exception is Schrödinger's view of "the physics of models" in (Schrödinger [35], pp. 152–153).

We now turn to the problematic of quantum entanglement, quantum correlations, and quantum nonlocality (all three concepts are essentially related), which is another central, and at this point of history arguably the dominant, problematic of quantum foundations and debate concerning them. This problematic was brought on the center stage of this debate by the now famous 1935 article by Albert Einstein, Boris Podolsky, and Nathan Rosen "Can the Quantum-Mechanical Description of Physical Reality be Considered Complete?", via the question of the completeness of QM [8]. The significance of their argument in this regard was, however, fully realized only in retrospect. It took a few decades, until the 1960s, for the question of nonlocality to take the center stage in the debate concerning quantum foundations, in the wake of Bell's and the Kochen-Specker theorems and related findings. The terms 'correlations' and 'entanglement' were not used in either EPR's article or in Bohr's nearly equally famous reply, under the same title [3], the exchange that reflects the fact that the Bohr-Einstein debate shaped the foundational debate concerning quantum theory in this regard as well. The concept of entanglement was introduced, in German (*Verschränkung*) and English, by Schrödinger in several papers, including his cat-paradox paper, written in response to EPR's paper [35]. The language of correlations came into prominence, in the wake of Bell's and the Kochen-Specker theorems, in the 1960s.

Most of the key findings and arguments involved in recent debates, following Bell's theorem, deal with discrete variables and Bohm's version of the EPR experiment. There are several reasons for this shift. The main one appears to be that the thought-experiment proposed by EPR cannot be performed in a laboratory. Bohm's version of the EPR experiment can and has been performed, confirming the existence of quantum correlations, which can be ascertained experimentally, apart from QM or any quantum theory. Among the best known of the findings that shaped this, more recent, history are those of Daniel M. Greenberger, Michael Horne, Anton Zeilinger, and Lucien Hardy, and, from the experimental side, Alain Aspect's experiment and related experimental work, such as that by Zeilinger and his group (Aspect [1]; Greenberger et al. [11, 12] and Hardy [13]). Aspect and Zeilinger were, along with John Clauser, awarded the 2022 Nobel prize for their contribution. We only cite some key earlier experiments. There have been numerous experiments performed since, some in order to find loopholes in these and subsequent experiments, which is, we might add, one of the subjects of Khrennikov's works, as well as many discussions at the Växjö conferences. The advent of quantum information theory, in part stimulated by these findings, gave this problematic further prominence and significance, and in turn shaped their understanding.

The meaning of these findings, beginning, again, with Bell's and the Kochen-Specker theorems and their implications continue to be debated, as is, for that matter, the EPR experiment and Einstein's and Bohr's arguments concerning it. It would not be possible to address these debates here. The literature dealing with these subjects is nearly as immense as that on interpretations of QM, and most of the latter in recent decades involves and is even defined by these subjects. This preoccupation confirms that, in the wake of Bell's theorem, the main focus of the debate concerning quantum foundations shifted, from the focus on the completeness of QM, to the question of quantum nonlocality (coupled to those of quantum entanglement and correlation) of QM or quantum phenomena themselves. This assessment may, however, depend on one's concept of completeness. This concept may, for example, be realist, as was that of Einstein, representing the behavior of the ultimate constituents of the physical reality considered, moreover, ideally classically causally, so deterministic predictions concerning their behavior would be ideally possible. This form of completeness may be termed Einstein-completeness. QM, in nonrealist interpretations in the spirit of Copenhagen, such as that of Bohr, is manifestly Einstein-incomplete because it places the ultimate constitution of physical reality responsible for quantum phenomena beyond representation or even conception. On the other hand, QM may be argued to be Bohr-complete, which is a broader conception, with Einstein-completeness being a special case, found in classical physics and relativity: QM is as complete as nature allows our theories of quantum phenomena to be, as things stand now, insofar as it predicts all that is possible to predict in accordance with quantum experiments. Even the Bohr-completeness of QM was, however, questioned by EPR. They argued that QM could not predict all elements of reality that could be established as such in considering any individual quantum systems, no matter how elementary, unless one allows for Einstein-nonlocality, an instantaneous transmission of physical influences between spatially separated physical systems, famously called by Einstein "a spooky action at a distance" [*spukhafte Fernwirkung*] (Born [5], p. 155). Bohr contested this argument on both counts, by arguing that QM could in fact predict all that was possible to ascertain as real, that is, all elements of reality that could be unambiguously defined, without violating Einstein-locality.

Accordingly, the relationships among reality, probability, completeness, and non-locality has never left the debate concerning quantum foundations, the primary emphasis on the question of quantum nonlocality in recent decades notwithstanding. The question of probability, central to all quantum theory, which may be, and have been, seen as a form of probability theory (one of the main areas of Khrennikov's research in both quantum foundations and quantum-like modeling) entered into the EPR problematic in an indirect way because the EPR experiment deals with, and EPR's argument depends on, predictions with probability equal to unity. It can be shown, however, that these predictions pose additional complexities in the case of quantum phenomena. These complexities were, Bohr argued, underappreciated by EPR but assisted Bohr in his counterargument in his reply [3].

Einstein's sentence, where the phrase "a spooky action at a distance" occurs, merits a full citation: "I cannot seriously believe in [QM] because the theory cannot be reconciled with the idea that physics should represent a reality in time and space, free from spooky actions at a distance" (Born [5], p. 155). This combination is also known as 'local realism'. Relaxing either requirement was unacceptable to Einstein. For Bohr, by contrast, only Einstein-locality was required, but not a spacetime representation of the ultimate constitution of the reality responsible for quantum phenomena, even in considering the individual behavior of quantum objects, no matter how elementary. Such a representation is required for Einstein-completeness, but is not necessary for Bohr-completeness, although EPR and Einstein in his other arguments concerning EPR-type experiments, again, claimed that QM is Bohr-incomplete as well.

What is known, more generally, as the principle of locality states that no instantaneous transmission of physical influences between spatially separated physical systems ('action at a distance') is allowed or that physical systems can only be physically influenced by their immediate environment. While relativity conforms to locality and gave rise to the concept, locality is independent of relativity theory. For one thing, relativity prohibits a propagation of physical influences not only instantaneously but faster than the (finite) speed of light in a vacuum, a requirement that could, in principle, be violated (as a different speed limit on physical action may be discovered), while still allowing for Einstein-locality. Einstein's "spooky action at a distance," too, refers to an instantaneous action, which is in conflict with relativity, but is not defined by it. Accordingly, while the Einstein locality of QM would imply that it is in compliance with the requirements of relativity, it may be a deeper fact, as Bell's and the Kochen-Specker theorems indicate. The compatibility with relativity does, however, imply locality. In his reply to EPR, Bohr argued for the compatibility of quantum phenomena and QM, at least in his interpretation and hence abandoning realism, for "all exigencies of relativity theory" and thus for its Einstein-locality (Bohr [3], p. 701n). Hence, Bohr argued, the inference that QM or nature itself is Einstein-nonlocal may be avoided, although this inference and the corresponding interpretations of the EPR-type experiments or quantum correlations have been advanced.

Our aim here is not to adjudicate between EPR and Bohr's arguments in those exchanges, which is very difficult to do with much definitiveness and is a subject of interminable debate of its own, but to emphasize the essential relationships between the questions of quantum nonlocality and completeness of QM, the conceptual and historical significance of which is hardly in doubt. In all of his arguments concerning the subject, Einstein only considered locality or nonlocality in the sense of Einstein-locality, as against 'quantum nonlocality,' the term never used by Einstein. It was introduced, along with several definitions of it, in the wake of Bell's theorem. We may, for the present purposes, adopt the following definition, in accord with several works addressing the subject (e.g., Fine [9], Mermin [31], pp. 81–184). There are, experimentally confirmed, statistical correlations between certain, specifically prepared, distant quantum events, correlations properly predicted by QM (cum Born's or a similar rule). In some idealized cases of this type, such as that of the EPR experiment, involving a pair of spatially separated quantum systems, ($S1$, $S2$) one can also predict with certainty certain measurable quantities for one of them, $S2$, on the basis of measurements performed on the other, $S1$. In nonrealist (or even some realist) interpretations, there is no story to be told and no concept to be formed of how these correlations or, in the first place, quantum phenomena come about or why these predictions are possible. So, there is some 'spookiness' or 'mysteriousness' to them in this sense, but not in the sense of any "mysticism incompatible with the true spirit of science" (Bohr [4], v. 2, p. 63). At the same time, it may be argued, as it was by Bohr and others, that these predictions need not entail a spooky action at a distance or Einstein-nonlocality, including in the EPR-type cases, where they are possible with certainty. 'Quantum nonlocality' may then be defined as the existence of such related or correlated distant quantum events and the possibility of predicting, it follows, at a distance, these relations or correlations. Quantum nonlocality is sometimes defined differently, for example, in terms of violations of Bell's or related inequalities, or still by other mathematical features, dealing with the data obtained in corresponding experiments. These features need not be those of QM, given that these inequalities pertain to these data as such. Such definitions, however, leave space for their physical interpretation, and quantum nonlocality, as just defined, provides one such interpretation, among other possible interpretations, some of which, as noted, interpret quantum nonlocality in terms of Einstein-nonlocality. On the other hand, interpretations of quantum phenomena and QM, such as that of Bohr, which only entail quantum nonlocality in the present or similar definitions and avoid Einstein-nonlocality, would suggest that the latter is not necessarily a feature of quantum phenomena or QM, while quantum nonlocality may be. We reiterate, however, that our aim here is not to advocate any particular position, beginning with those of Einstein and Bohr, concerning this issue but to establish this problematic itself as central to quantum foundations.

Khrennikov's recent work addresses this problematic in detail and from several angles, some of which are new. In particular, it is true that predictions at a distance and correlations between distant events are also possible, without violating Einstein-locality, in classical physics. This circumstance may compel one to argue that quantum nonlocality may not be significant for defining the difference between classical and quantum correlations. This is the aim of several recent arguments by Khrennikov under the general heading "eliminating the issue of quantum nonlocality," while contending that the difference between classical and quantum correlations is essentially defined by the discreteness of quantum phenomena and complementarity (Khrennikov [18, 20–24]). These two features are indeed crucial in considering quantum correlations or quantum phenomena, in the first place, and they are not found in classical physics. Khrennikov's work on this subject deserves credit for bringing these futures back on the stage of quantum foundations, including in the context of quantum information theory. They were more prominent at the initial stages of the development of quantum theory and debates concerning it, but somewhat neglected during 'the age of Bell's theorem,' as it may be called. These works by Khrennikov reflect a certain shift, noted earlier, from his earlier work on quantum foundations, towards a greater emphasis, following Bohr, on complementarity, even if without necessarily adopting Bohr's epistemological position. Khrennikov, arguably, still prefers a realist view, at least as concerns the possibility of a realist theory ultimately underlying QM. Khrennikov also gives due attention to Bohr's concept of phenomenon, as applicable in quantum physics, a concept strictly in terms of what is observed in measuring instruments, to which observations, the concept of complementarity was ultimately applied by Bohr. Khrennikov's work makes one think more deeply concerning quantum nonlocality, even if one believes, as against Khrennikov's position, quantum nonlocality to be important and possibly unavoidable, along with discreteness and complementarity, which concepts would be more generally assumed as valid in these debates.

It may be worth to briefly comment on Bohr's concept of complementarity and phenomena, in view of their significance on Khrennikov's recent work and by way of establishing a bridge, via John Archibald Wheeler, to quantum information, the third key aspect of quantum foundations we would like to discuss here. Complementarity is defined by

(a) a mutual exclusivity of certain phenomena, entities, or conceptions, such as and in particular, those of the position and momentum measurements, which can never be performed simultaneously in view of the uncertainty relations; and yet

(b) the possibility of considering each one of them separately at any given point; and

(c) the necessity of considering all of them at different moments of time for a comprehensive account of the totality of phenomena that one must consider in quantum physics.

Complementarity is a reflection of the fact that, in a radical departure from classical physics or relativity, the behavior of quantum objects of the same type, say, electrons, or, again, the ultimate nature of reality responsible for quantum phenomena defined by such objects, is not governed by the same physical law, especially a representational physical law, in all possible contexts, specifically in complementary contexts. This leads to incompatible observable physical effects in complementary contexts. On the other hand, the mathematical formalism of QM offers correct probabilistic or statistical predictions of quantum phenomena in all contexts, in RWR-type interpretations, under the assumption that the ultimate nature of reality responsible for quantum phenomena is beyond representation or even conception. This situation is also responsible for what is known as 'contextuality', a concept introduced by Simon Kochen and Ernst Specker, and developed, following it, by Bell's theorem, and related findings, mentioned above, during the last half century. Speaking of a physical law in this case requires caution, because, if one adopts a nonrealist view, there is no physical law representing this behavior, only laws that predicted it, moreover, probabilistically or statistically. One might speak, with Wheeler, of a "law without law" (Wheeler [39]), which, as we will discuss presently, can be related to another of Wheeler's famous slogans "it without bit," that helped to usher in quantum information theory (Wheeler [40]). Before we do so, we briefly explain Bohr's, proto-quantum informational, concept of phenomena, which in quantum physics applies strictly to the effects of the interactions between quantum objects and measuring instruments. These are these effects and only them that define Bohr's concept of phenomena, in effect making it an informational concept. According to Bohr:

> I advocated the application of the word phenomenon exclusively to refer to the observations obtained under specified circumstances, including an account of the whole experimental arrangement. In such terminology, the observational problem is free of any special intricacy since, in actual experiments, all observations are expressed by unambiguous statements referring, for instance, to the registration of the point at which an electron arrives at a photographic plate. Moreover, speaking in such a way is just suited to emphasize that the appropriate physical interpretation of the symbolic quantum-mechanical formalism amounts only to predictions, of determinate or statistical character, pertaining to individual phenomena appearing under conditions defined by classical physical concepts [describing the observable parts of measuring instruments]. (Bohr [4], v. 2, p. 64; emphasis added)

Complementarity, too, now, only refers to one or another complementarity of phenomena, such as, those related to the position and the moment measurement, in accordance with the uncertainty relations, which are understood by Bohr in terms of the complementary nature of these two measurements. One can at any point decide to perform one or the other and do so ideally exactly, but one can never perform both simultaneously, as exact measurements. Essentially, then, in QM (or QFT), one deals with the information transfer between measuring devices, information that, as manifested in measuring instruments, can be described physically classically, but that, either as concerns each individual effects or, especially, the structure of the totality of the effects (a structure that included such phenomena as complementarity or quantum correlations) cannot be predicted by means of classical physics. This situation became apparent beginning with Heisenberg's thinking leading him to his invention of QM, although it took a while to realize its meaning and implications. It is in fact, this situation that, coupled to the formalism of QM, defined quantum information theory, and QM itself as a form of (quantum) information theory. While it would be a stretch to say that Heisenberg's approach that led him to the invention of QM was, technically, quantum-informational, it could be viewed as quantum-informational in spirit, and conversely, quantum information theory as Heisenbergian in spirit (Plotnitsky [33, 34], pp. 72–73). The reason for this view is that the quantum-mechanical situation, as Heisenberg conceived of it, was defined by:

(A) certain already obtained information, derived from spectral lines (due to the emission of radiation by the electron), observed in measuring instruments; and

(B) certain possible future information, to be obtainable from spectral lines to be observed in measuring instruments and, hopefully, predictable in probabilistic or statistical terms by the mathematical formalism of a quantum theory.

Heisenberg's aim was to develop such a formalism, without assuming that it needed to represent a spatiotemporal process connecting these two sets of information or how each comes about, a representation that Heisenberg said was "scarcely possible" at the time (Heisenberg [15], p. 265). This information is, in each case, determined by what type of experiment one decides to perform, rather than by arbitrarily selecting one or another preexisting property of physical reality.

It is accordingly not surprising that Wheeler, a champion of Bohr, eventually linked his idea of "law without law" to quantum information theory, which he helped to usher in, along with Richard Feynman, his student, along with several other figures, such as, Paul Benioff, and Yuri I. Manin. As just noted, it became apparent already with Heisenberg's thinking, leading him to his discovery of QM, that quantum objects, in their interactions with measuring instruments, create specifically organized, including, in certain circumstances, complementary, structures of information, composed of classical bits, and allow one to use theories, such as QM, to predict this information. It is this situation that defines quantum information: a particular organization of classical information that cannot be brought about by classical physics. The very nature of quantum measurement, is defined by this situation, making the data observed in quantum phenomena generated by a 'quantum computer' created by our interaction with nature. The ultimate constitution of matter is, to adopt Wheeler's famous maxim, "it from bit." According to Wheeler:

> [E]very it - every particle, every field of force, even the spacetime continuum itself - derives its function, its meaning, its very existence entirely - even if in some contexts indirectly - from the apparatus elicited answers to yes or no questions, binary choices..., bits. It from bit symbolizes the idea that every item of the physical world has at bottom - at a very deep bottom, in most instances - an immaterial source and explanation; that what we call reality arises in the last analysis from the posing of yes-no questions and the registering of equipment-evoked responses; in short, that all things physical are information-theoretic in origin and this is a participatory universe. (Wheeler [40], p. 5)

Wheeler's "it-from-bit" manifesto was inspired by Bohr, whom Wheeler cites at the outset (Wheeler [40], p. 4). Wheeler's very definition of "it from bit" may be seen as a variation of Bohr's view of the irreducible role of measuring technology (the "apparatus" invoked by Wheeler) in the constitution of quantum phenomena and information it contains. It is a separate question whether Wheeler's view converges with that of Bohr or diverges from it. One could think of "it from bit" in more than one way, in particular, in the following two ways. The first is that of 'it' as defined by or even created by 'bit',which appears to be closer to Wheeler's own meaning, correlative to his concept of "a participatory universe," to which he refers next in his manifesto as well. He says: "It from bit symbolizes the idea that every item of the physical world has at bottom—at a very deep bottom, in most instances—an immaterial source and explanation; that what we call reality arises in the last analysis from the posing of yes-no questions and the registering of equipment-evoked responses; in short, that all things physical are information-theoretic in origin and this is a participatory universe" (Wheeler [40], p. 5). This is still somewhat ambiguous, because Wheeler might mean that "all things physical are information-theoretic in origin" only insofar as they are constructed by us, even if constructed as unconstructible, which would be true in Bohr's view. It is possible that this may have been Wheeler's meaning, even though he does not elaborate on this point. The main point here is, however, the quantum-informational nature or potential of QM, which was realized early, at least in effect, by Heisenberg and Bohr, and which was brought by Wheeler to the center stage of quantum foundations, thus helping to usher in the quantum information revolution or epoch, an epoch that is our own.

The relationships between quantum information theory and quantum foundations is now an immense area of research, and several articles in this volume represent both foundational and technical aspects of these relationships. The preceding discussion will, we hope, help to conceptually ground and historically contextualize these articles for this volume's reader. The relationships between quantum foundations and quantum information are considered in many of Khrennikov's articles. The subject has also played a major role in Växjö's conferences, beginning with the second one, which was expressly dedicated to quantum information theory and its relations to quantum foundations, and brought together a formidable array of physicists, including some of the founding figures of quantum information theory, as noted in Christopher A. Fuchs's chapter (Fuchs was a co-organizer of the conference). An outline given here is in accord with Khrennikov's attempts to relate quantum information, including practical aspects of quantum computing, to foundational issues, for example, in his article, with a revealing subtitle: "Echoing the Recent Google Successes: Foundational Roots of Quantum Supremacy." To cite the abstract of this article, which defines this agenda in Khrennikov's work, on this occasion, again, coupled with his inquiry into the problematic of quantum nonlocality and entanglement, as discussed above:

> The recent Google's claim on breakthrough in quantum computing is a gong signal for further analysis of foundational roots of (possible) superiority of some quantum algorithms over the corresponding classical algorithms. This note is a step in this direction. We start with critical analysis of rather common reference to entanglement and quantum nonlocality as the basic sources of quantum superiority. We elevate the role of the Bohr's principle of complementarity1 (PCOM) by interpreting the Bell-experiments as statistical tests of this principle. After a brief presentation of PCOM and endowing it with the information interpretation, we analyze its computational counterpart. The main implication of PCOM is that by using the quantum representation of probability, one need not compute the joint probability distribution (jpd) for observables involved in the process of computation. (Khrennikov [20], p. 1).

The preceding discussion only addressed a relatively small, although, we believe, foundationally especially important, part of Khrennikov's work, the scope and significance of which we wanted to show here. Before, however, leaving this part of our introduction, which was devoted to quantum foundations, we would like to mention other areas of quantum theory to which Khrennikov made important contributions, foundational and technical. Among them are Bohmian mechanics, $p-$adic quantum theory, theory of classical random fields, digital modeling of quantum physics, and his investigations into the nature of quantum probability.

It is, we think, fitting to close our discussion of quantum foundations and Khrennikov's contribution to this field by emphasizing his understanding, both in his own work and in his approach to putting together the Växjö conferences, that physics, mathematics, philosophy each have their role to play in the advancement of quantum theory in its foundational aspects, and, it should be added, also realizing, which is uncommon, that not only experimental physics but also the thinking of experimental physicists are indispensable in this advancement. Remarkable for many other things as well, the Växjö meetings have been unique in bringing together these fields as human endeavors advanced by people and communities of working in them and relating them. The Växjö meetings (and in this context, it may indeed be more fitting to call them meetings rather than conferences) were above all human, and making them human was one of Khrennikov's great contributions to mathematics and science, a contribution profoundly ethical in nature.

So far we have offered our readers a picture of Andrei Khrennikov's seminal contributions to the foundations of quantum mechanics. He can, however, equally be seen as one of the key founders of a whole new area of sprawling research which connects (and applies) elements of the formalism of quantum mechanics to areas outside of physics, such as biology, neuroscience, psychological, cognitive science, sociology, economics and finance, and politics [25].

As might be expected, physicists do not always welcome uses of concepts, even mathematical ones, from their discipline to areas that are only remotely connected with physics, at least prima facie, and some skepticism toward the use of quantum mechanics elsewhere has been voiced by quantum physicists. However, the history of science suggests otherwise. Quite a few ideas of classical mechanics have found ways of application, often in a very fruitful fashion, within key areas of social science. For instance, one can think of finance, with its celebrated trillion dollar option market, in mathematical modeling of which the heat equation and the Brownian motion have played an important role. Even the ideas of entropy found their way in economics beginning with the earlier work of such authors as Georgescu-Roegen [10]. It is not surprising either that, within the social sciences and other fields listed above, there have been varying degrees of resistance towards employing the concepts and mathematical formalism of physics, especially quantum theory. The reasons for this resistance vary, from the objective difficulties of the complexities of quantum theory to social and even political ones. If, for example, a Nobel laureate in economics dismisses this type of research, a member of an academic community could be even less inclined to work on the development and expansion of this area of research and could even be institutionally more limited in pursuing such work.

None of those barriers were ever barriers for Andrei. I (Emmanuel Haven) recall Andrei's visit at the University Essex in the United Kingdom about twenty years ago. At his talk, the fundamental ideas he then already had developed, were presented. One zealous audience member became very active and attempted to 'deconstruct' one by one the ideas Andrei had presented. The talk ended with a heartfelt 'thank you'—given by Andrei to that very person who had criticized his argument. I shall never forget the image that this moment left in me: this was heart-felt. It was not a fleeting 'thank you'—but a reflection of the deep open-ness that Andrei possessed towards anything grounded in 'knowledge'. Andrei has an innate reverence for knowledge. Knowledge that supports or critiques, or deconstructs, ideas, including his own, are for Andrei equally valuable as having the status of 'knowledge', and he always shows a deep respect towards it. This may well be one of the main reasons why Andrei could make such important contributions to areas ranging from mathematics and physics to biology and political science, in other words, why he is in essence a very successful polymath.

What inroads, then, did Andrei make into fields which fall outside of mathematics and physics, or extend both to these fields? In sum: for several decades now, his impact has been important and is becoming increasingly so, and various fields mentioned here become, finally, more open to using the mathematics of quantum theory. For example, a headline in the Financial Times, an important and influential newspaper in the field of finance, indicated that Wall Street was getting very interested in quantum finance [38]. Andrei contributed important thoughts to using the quantum formalism in finance (and economics). We will return to this point below.

For now, however, it is important to pinpoint the starting point of those, as they have been called, 'quantum-like' applications. The quantum-like approach is probably best defined in the way Andrei introduced it in the preface of one of the handbooks he co-edited. With reference to the quantum information revolution which was occurring in physics, he said that ([14], p. v): "...it is natural to use the formalism of quantum theory to describe the processing of information ...not only by physical systems, but by systems of any origin, whether they are biological, social, financial (as long as there is behaviour which has [at least some] distinguishing features of quantum systems)." Andrei, again, contributed to all the 'extra'-physics areas he is mentioning here.

Andrei's work on quantum-like modeling beyond physics can be traced back to more than 20 years ago, to a key article he published in *Foundations of Physics*. Already then, in 1999 [26], he developed information mechanics, through so-called information energy. The same paper also introduced the idea of information measurement on individuals changing the information status of the individual. These were original and important insights into extending quantum-like approaches beyond physics.

As we noted earlier in this introduction, Bohmian mechanics was another area in physics itself that Andrei contributed to. He also worked on applying Bohmian mechanics (which is different from the standard quantum mechanics) beyond physics. Thus, considering how Bohmian mechanics, via the extended Newton equation, could exhibit paths that showed such features as non-zero quadratic variation in finance provided important new insights concerning how this approach could be taken to new levels of application. Andrei's interpretations of the different types of potentials was an important head start in that new avenue of research [25].

Much of the quantum-foundations part of this introduction addressed the centrality of quantum probability. Andrei not only contributed significantly to this area of quantum foundations, but was also a key figure in establishing how that type of probability could be of great value elsewhere, especially in the formalisms of decision making. At least two key ideas lie at the foundation of the success of this approach: (i) incompatible observables in decision making do occur naturally; and (ii) consistent heuristics (observed over many decision making experiments) which now can be formalized by using the quantum formalism (most essentially quantum probability but not exclusively so, because this approach extends beyond this aspect of it). As an example, the quantum rendering of sensation-perception by living species, as devised by von Helmholtz as (i) sensation being a signal interpreted by the brain (which now can be seen as a state of the sensation system); whilst (ii) perception being akin to for instance selective attention (perception is the measurement apparatus) ([2], p. 5). As Andrei then indicated, the unitary operator "describes the process of interaction between the sensation and perception states." ([2], p. 5 and see also [27]). We can say that the measurement apparatus creates a measurement context. The unitarity of the operator will require that the system in question can be assumed to be isolated. In more recent work, Andrei discusses the introduction of so called 'probe systems' within the context of the so called indirect measurements which play a basic role in decision making environments exhibiting more advanced phenomena like the response replicability effect. In this context, it became important to define so-called quantum instruments. Essentially the measurement on a system is described by such an instrument, and the emphasis in quantum instrument theory is not on (non)commutativity of operators but rather on state transformations which result "from the action of (self-)observation" ([2], p. 2). In a key paper published in the *Journal of Mathematical Psychology* [32], it is demonstrated that those quantum instruments are of fundamental importance in the description of question order effects and the response replicability effects. Already in 2014 [28] it was shown that both effects could not be jointly described with quantum measurement theory. Via a generalization of quantum measurement theory, Masanao Ozawa and Andrei [32] show that not only both effects can be jointly described, but also that the important QQ equality that Zheng Wang and Jerome Busemeyer proposed can be further generalized. This equality [36, 37] is a powerful result, essentially confirming, on experimental grounds, that the quantum approach in decision making is much more natural than it was previously anticipated. Following James Yearsley [42], for two binary events, denote $p(A, B)$ as the probability of A and then B happening; and likewise define $p(not\ A, not\ B)$ as the probability of $not\ A$ happening followed by $not\ B$ happening. The QQ equality says (and this is experimentally verified): $p(A, B) - p(B, A) + p(not\ A, not\ B) - p(not\ B, not\ A) = 0$.

Andrei's contributions to quantum-like approaches to decision making in psychology did not stop there, however. He also made important inroads in economic theory. He discovered a fundamental result which showed that agents who use quantum logic may invalidate a core theorem in economic theory (which uses classical probability)—the Aumann theorem. Agents may exhibit agreement (but also disagreement) on posterior probabilities [29].

Before proceeding to a brief outline of the papers in this Festschrift, it may be appropriate to come back to our earlier point in this introduction: "The Växjö meetings were above all human, and making them human was one of Andrei's great contributions to mathematics and science, a contribution profoundly ethical in nature." This rejoins what we tried to bring forward a little earlier too: Andrei has a deep reverence for knowledge. The fact that this knowledge is intimately connected to the human experience, confirms our contention that the Växjö conferences were above all human, and were conceived and organized by Andrei as human.

Both editors of this Festschrift can look back to a superb level of collaboration with the author for whom we have brought together this feast of writing. We trust that all the authors in this book have in one way or another been connected with Andrei's humanity.

We now offer a brief outline of the collection's chapters, beginning with those dealing with quantum foundations and quantum information theory, and then moving to chapters addressing quantum-like modelling beyond physics.

1 Quantum Foundations and Quantum Information Theory

Gregg Jaeger's chapter, "**On Wheeler's Quantum Circuit**," considers "the meaning circuit hypothesis" of John Archibald Wheeler, an important visionary figure in the history of quantum foundations and, as discussed above, one of the founding figures of quantum information theory, indeed arguably, a uniquely important one in establishing its physical and philosophical principles. While offering a comprehensive discussion of the "the meaning circuit hypothesis" and several key aspects of Wheeler's thinking concerning quantum foundations, Jaeger concludes that "Wheeler's Meaning Circuit Hypothesis is better understood as an explanation of how quantum theory arose and is validated rather than one of how the physical universe, much less all of physical existence arose." This view, we might observe, need not mean that this argument replaces physics and philosophy in Wheeler's hypothesis with its history as a reflection of the history of quantum theory; instead, it may help us to put this history in the service of physics and philosophy now, in which Wheeler's thinking continues to play a shaping role. On the other hand, this need not mean either that we should accept his, or anyone's thinking, uncritically or should not question it; quite the contrary. For otherwise Wheeler's thinking, or that of any foundational figure, such as Bohr, Wheeler's mentor, loses its capacity to remain vital, or one might say, no longer remains new. Perhaps, then, we are more faithful to Wheeler by moving our thinking both with and beyond him, rather than by accepting his view as an established doxa.

Wheeler, in this case his famous idea, mentioned by us above, of "the participatory universe," is an important reference, admittedly, in this case hidden in a footnote, but, given Wheeler's significance for **Christopher A. Fuchs**'s work, no less important for that, for Fuchs's chapter, "**Letter to Andrei: QBism and the Unfinished Nature of Nature.**" A special significance of this chapter for the present volume is the role of Khrennikov's work—"Never has there been such a friend to the open and liberal discussion of quantum foundations as Andrei Khrennikov," Fuchs says (it is impossible to disagree!)—and the Växjö conferences in the work on quantum foundations, including QBism (Quantum Baeysianism). Fuchs adds: "It is little known, and who would guess it, but those Växjö meetings and Andrei himself played significant roles in the development of QBism." Andrei might even be called a reluctant godfather to QBism." Fuchs's chapter, however, has an additional significance. Just as Jaeger's chapter does, it, too, puts history in the service of physics and philosophy, in this case, more recent history, to which QBism belongs as part of quantum information theory. What is especially suggested by the chapter is that this history, and with it physics and philosophy, is played out and advances our thinking not only in more conventional published work or commentary, but also, less overtly but no less significantly, in our more private exchanges, letters, conversations, improvised remarks and so forth. As a result, the reader of this chapter will be made part of this richer history and insights it leads to. Fuchs closes by saying: "Then again, as the 20 some odd years of meetings in Växjö have demonstrated, the quantum is no easy code to crack. Congratulations again Andrei, both on your work and the community you have built." Perhaps these meetings also demonstrated that without such meetings and such communities we cannot crack the quantum code, or "the quantum enigma", as Einstein called it, at all, just as it would have been impossible to crack the Enigma code without the community of Bletchley park and the exchanges that took place there, even with Alan Turing's genius at work, important as it was.

Yet another angle on the interplay of history, physics, and philosophy, by combining both more recent and longer history, is offered by **Theo Nieuwenhuizen's** Chapter, "**Contra Multos Verbos: On Scandals of Quantum Mechanics.**" Historically, the argument proceeds via Nicolaas van Kampen, an important figure in the history of quantum theory, even though he rarely figures in more recent discussions, quite unfortunately, the chapter suggests. Conceptually, the article offers both a critique of some recently fashionable trends in quantum theory and fundamental physics (also in connection with the role of the social media in dissemination and promoting some of them, sometimes undeservedly, in Nieuwenhuizen's view) and, this is its most innovative part, in presenting the solution of a class of realistic models for the dynamics of quantum measurement by Armen Allahverdyan, Roger Balian, and Nieuwenhuizen himself. The chapter closes on an open question: "We are left with the question: How does Nature get it done?" That is, the question is: How does nature make quantum phenomena possible? We don't have an answer, of course, and what kind of answer or even what way of posing this question is possible or necessary, has, as noted in our introduction, been at stake in the debates concerning quantum theory, beginning with the Bohr-Einstein confrontation. For Nieuwenhuizen, "Einstein's dream of a reality is as vivid as it ever was. The last but not least of the scandals of quantum mechanics is that, after one century, there is still this fundamental question to answer! This is unbearable, so Andrei: solve it yourself, or else: please organize many further Växjö meetings, so that we can attack as a community!"

Mauro G. D'Ariano's chapter, "**Quantum Holism**," while building on some of the author's previous work, is a new step in his rethinking the nature of operational probabilistic theories, through what D'Ariano defines as "mereological holism—the distinctive holism of quantum theory." This rethinking and the nature of this holism are radical because "mereological holism has the dramatic conceptual consequence of making untenable the usual understanding of the 'quantum system' as being a 'physical object,' since composed objects have properties compatible with those of its parts." The argument builds on the insight originally suggested by Schrödinger in 1935 in his analysis of quantum entanglement, that best possible knowledge of the whole does not include best possible knowledge of its parts (Schrödinger [35], p. 166), which Schrödinger, too, saw as the single defining difference between quantum and classical mechanics. D'Ariano's argument, however, gives it new meaning and implications. The distinctive feature of this argument is in giving this concept a precise mathematical expression, thus bringing together physics, mathematics, and philosophy of quantum physics together.

Paolo Perinotti, Alessandro Tossini, and Leonardo Vaglini's chapter, "**Shannon Theory for Quantum Systems and Beyond: Information Compression for Fermions**," begins with a reminder: "the story of quantum information has been strictly connected to the realm of foundations since its very birth," and, reciprocally, that the concept of information is crucial to understanding quantum foundations. The main aim of the chapter is providing a precise definition of the information content of states of Fermionic systems (rather than more standard quantum informational systems, commonly dealing with photons) in terms of what is known as "data compression." Data compression presents additional difficulties in the case of Fermionic systems, which the chapter addresses thoroughly. The main novel feature of the chapter is the analysis of compression in the absence of local tomography. The chapter has important implications for our understanding of the concept of entropy (which has always been associated with information and probability) in view of the fact that informational theories beyond quantum exhibit inequivalent concepts of entropy, which is the main issue one has to face in order to introduce the notion of information content in the general case. The chapter, thus, offers an example of combining both technical and foundational aspects, which may indeed be necessary for developing a deeper understanding of quantum information theory and information theory in general.

Niklas Johansson, Felix Huber, and Jan-Åke Larrson's Chapter, "**Conjugate Logic**," is a reconstruction project, under the title heading of "conjugate logic," that, as several previous such attempts, aims (in departure from the standard forms of quantum logic, from von-Neumann and Birkhoff on) at avoiding using any Hilbert space structure as an initial postulate. The proposal is similar to the more generic concept of epistemic logic: it encodes predictions concerning outcomes of future observations on some systems, in the case of a quantum system, as statements about future outcomes of measurements performed on specific degrees of freedom of the system. The logic proposed by the chapter includes not only propositions and their relations including connectives, but importantly also transformations between propositions on conjugate degrees of freedom of the systems. A key point is that the addition of a transformation that allows to convert propositions about single systems into propositions about correlations between systems. The chapter argues that subtle choices of the properties of the transformations lead to drastically different underlying mathematical models; one choice gives stabilizer quantum mechanics, while another choice gives Spekkens' toy theory. This difference points to a crucial basic property of quantum and quantum-like systems that can be handled within the present conjugate logic by adjusting the mentioned choice. It also enables a discussion on what behaviors are properly quantum or only quantum-like, relating to that choice and how it manifests in the system under scrutiny.

Philippe Grangier's chapter, "**Revisiting Quantum Mysteries**," revisits some of the quantum mysteries, in part, one might argue, following Bohr's argument concerning the irreducible role of measuring instruments in the constitution of quantum phenomena. Bohr would certainly agree with Grangier's main argument that "in quantum mechanics, and in opposition to classical physics, it is impossible to say that an isolated quantum system 'owns' a physical property." Grangier's overall argument is, however, different from that of Bohr. In particular, Grangier's brings in a new concept to bear on his argument, that of "modality," in association with the essential role of "context" in quantum versus classical physics. This approach also allows a subtle proximity to a (more) Einsteinian form of realism, which Grangier wants to restore to quantum mechanics.

Masanao Ozawa's chapter, "**Quantum Measurement Theory for System with Finite-Dimensional State Space**," aims to remedy the insufficiency, or offer a completion, of John von Neumann's axiomatization of quantum mechanics for finite dimensional state spaces, by "augmenting it by the general measurement axiom that describes all physically realizable measurements consistent with other axioms of quantum mechanics." The argument is in part based on the abandonment of von Neumann's repeatability hypothesis: "If an observable A is measured twice in succession in a system S, then we get the same value each time." An important aspect of the theory is that the mathematical tools for describing measurement statistics, such as POVMs and quantum instruments are not assumed at the outset, but are introduced as natural consequences of the axioms proposed.

Margarita A. Man'ko and Vladimir I. Man'ko's chapter, "**Probability Distributions as Alternatives of the Density Operator and Wave Function for Describing Quantum States**," offers an alternative to the more standard representation of quantum states (in the formalism of quantum mechanics) in terms of wave functions or density operators, by replacing it with the probability representation of quantum states. This program follows earlier work along these lines by R. J. Glauber and E. C. G. Sudarshan, and has parallels with some attempts along similar lines in quantum information theory, including QBism, in dealing with discrete variables. It has foundational, as well as practical, significance by virtue of suggesting that while quantum theory may not be able to avoid probabilities, it may dispense with amplitudes. The mathematics of the two approaches is equivalent, which, one might add, suggests that, although one may be able to avoid amplitudes, one still uses mathematics that involves complex numbers. This fact does not prove that a real-number quantum mechanics (which could, accordingly, not be mathematically equivalent to the standard version) is impossible, but it does indicate that it might be difficult to develop it.

Dagomir Kaszlikowski and Paweł Kurzyński's chapter, "**Alternative Quantum Mechanics?**" aims to give the "exiled" subject of "negative probabilities" its, they argue, deserved place in quantum physics. As they say: "quantum states and operations are equivalent to negative probability distributions and stochastic processes via a mathematical mapping called 'frames'. For various reasons, historical forces pushed this alternative out of the mainstream with a few notable exceptions such as Wigner or Hussimi functions. This is most unfortunate because negative probabilities, exploited as a computational tool, can offer insights into quantum theory where the orthodox, Hilbert space approach may be cumbersome. Motivated by this, we study a binary system with a negative probability distribution called negative bit (nebit)." The article also suggests potential implications of their approach (without, they emphasize, replacing the Hilbert space formalism) for quantum computation and the relationships between classical and quantum probabilistic formalisms.

Bob Coecke's chapter, "Compositionality as We See It, Everywhere Around Us," belongs to both quantum foundations and quantum-like approaches beyond physics, and thus bridges two main areas considered in this volume. The chapter aims at developing a new concept of "compositionality," a phenomenon that, the chapter argues, beginning with its title, is found "everywhere around us." This phenomenon is most commonly understood by "a bottom-up flow of meaning: the meaning of the whole can be derived from the meaning of the part, and how these part are structured together." Quantum theory, beginning arguably, with Schrödinger's original argument concerning quantum entanglement, challenges and even precludes the use of this concept, as the readers of this volume would also know from M. G. D'Ariano's contribution, "Quantum Holism." In confronting this situation, the chapter proposes the concepts of Schrödinger, Whitehead, and complete compositionality, in order "to capture the fact that compositionality is at its best when it is 'real', 'non-trivial,' and even more when it is also 'complete'."

2 Quantum-Like Modelling

The chapters in this book which relate to quantum-like modeling refer mostly (but not uniquely) to the quantum cognition project which is, without much doubt, the 'jewel in the crown' of quantum-like applications.

In the chapter "**Measurement Models in Quantum Cognition**", **Jerome Busemeyer and Zheng Wang** come back to the all important question of how to simultaneously tackle the A-A repeat; A-B-A effect; question order, and, the QQ equality. As we remarked above, the QQ equality developed by the same authors provided for a milestone in terms of how 'quantum' cognition really is. Masanao Ozawa and Andrei Khrennikov provided for a generalization of the QQ equality. At the same time, as the chapter also documents, the A-A repeat effect; the repeat A-B-A effect and question order effects can all be tackled with a generalization of quantum measurement theory. In this chapter, Jerome Busemeyer and Zheng Wang provide a more standard quantum approach to solving this important puzzle. They introduce two spaces, i.e. a 'system space' which represents the cognitive system and an 'environment space' representing measurements. Their approach can now also account for a quantum measurement model which can again tackle all three effects: A-A repeat; A-B-A effect; question order, and, the QQ equality.

In the chapter "**Quantum Cognition: Quo Vadis**", **Emmanuel Pothos and James Yearsley**, stress how well the idea of constructive influences (i.e. the situations when a decision alters the mental state which pertains to the decision) mimics the quantum mechanical collapse. They also discuss how the quantum approach looks at similarity through the lens of subspaces and sequential projections. The chapter looks at some upcoming challenges of the use of quantum-like modeling in cognition. They suggest that the use of real world based data (as opposed to lab data) will be essential. They also stress the potential of the use of the quantum-like approach in artificial intelligence.

Real world data is considered in the chapter "**Entanglement in Cognition Violating Bell Inequalities Beyond Cirel'son's Bound**" by **Diederik Aerts, Jonito Aerts Arguëlles, Lester Beltran, Suzette Geriente and Sandro Sozzo**. Their experimental data shows that participants in two tests which target judgments on conceptual combinations, show strong violations of the CHSH inequality. Quantum entanglement is at the basis of the conceptual combinations. The authors note that, in contrast to widespread belief, a quantum model in a Hilbert space can be elaborated in the Bell-type situations considered in the article even when the CHSH violation exceeds the known 'Cirel'son bound'. The authors also suggest that some of their findings may have implications for the foundations of quantum theory itself.

In the chapter "**Thinking and Deciding in a Complex World**", **Polina Khrennikova** documents some key issues which may make the construction of a unified 'quantum-like' edifice challenging. After an overview on how quantum probability models can aid in refining the decision making formalism, the chapter addresses limitations brought in by the use of Hermitian projector-operators (because of violation of double stochasticity) and Hilbert spaces. In indicating possible future directions, Khrennikova notes, for example, that "the richness of contextuality of human beliefs and preferences might often go beyond the axiomatics of standard quantum formalism."

Contextuality is further discussed by **Ehtibar Dzhafarov** in his chapter "**Contextuality and Random Variables**". Contextuality is a property of random variables and the existence of a non-contextual system can be precisely defined. In this chapter the idea of coupling of random variables is introduced. The article is an important contribution to the understanding of contextuality as a property of systems of random variables and, in part correlatively, to our understanding of random variables themselves.

The last chapter, by **Acacio de Barros** entitled "**Can quantum non-identity exist in social phenomena?**" is an exception to the quantum cognition theme. It makes a strong argument that digital money, which is at the base of enormous amounts of financial transactions, has an indistinguishability characteristic which lends itself to applying the quantum-like formalism to financial systems using such money. Acacio de Barros suggests that the financial system's stability in 2007 may well have been dependent on the existence of such digital money via the huge quantitative easing operations central banks orchestrated through most of the financial systems. The argument is offered that, perhaps, the probabilistic theory of indistinguishables could be applied to digital money.

This brings us almost 'full circle' in that, as Acacio argues, this approach of indistinguishability is also used in Andrei's innovative idea of the social laser. That very topic is the subject of Andrei's latest book [30].

The editors hope this Festschrift provides for a celebration of Andrei's contributions to both science and social science. It has been a pleasure to bring the book together and the editors thank all the contributors for their patience in finalizing this book.

Thanks Andrei for what you have brought (and will continue bringing) to science. We all say: Sei Gesunt!

References

1. A. Aspect, J. Dalibard, G. Roger, Experimental test of Bell's inequalities using time varying analyzers. Phys. Rev. Lett. **49**, 1804–1807 (1982)
2. I. Basieva, A. Khrennikov, M. Ozawa, Quantum-like modelling in biology with open quantum systems and instruments. Biosystems **201**, 104328 (2021)
3. N. Bohr, Can quantum-mechanical description of physical reality be considered complete? Phys. Rev. **48**, 696–702 (1935)
4. N. Bohr, *The Philosophical Writings of Niels Bohr*, 3 vols (Ox Bow Press, Woodbridge, CT, USA, 1987)
5. M. Born, *The Einstein-Born Letters* (tr. Born, I.) (Walker, New York, NY, USA, 2005)
6. A. Einstein, Physics and reality. J. Franklin Inst. **221**, 349–382 (1936)
7. A. Einstein, *Autobiographical Notes* (tr. Schilpp, P. A.) (Open Court, La Salle, IL, USA, 1949)
8. A. Einstein, B. Podolsky, N. Rosen, Can quantum-mechanical description of physical reality be considered complete?, in *Quantum Theory and Measurement*, vol. 1983, ed. by J.A. Wheeler, W.H. Zurek (Princeton University Press, Princeton, NJ, USA, 1935), pp.138–141
9. A. Fine, Do correlations need to be explained? in *Philosophical Consequences of Quantum Theory: Reflections on Bell's Theorem* ed. by J.T. Cushing, E. McMullin, (Notre Dame University Press, Notre Dame, IN, USA, 1989), pp. 175–194
10. N. Georgescu-Roegen, *The Entropy Law and the Economic Process* (Harvard University Press, Harvard, 1971)
11. D.M. Greenberger, M.A. Horne, A. Zeilinger, Going beyond Bell's theorem, in *Bell's Theorem, Quantum Theory and Conceptions of the Universe*. ed. by M. Kafatos (Kluwer, Dordrecht, Netherlands, 1989), pp.69–72
12. D.M. Greenberger, M.A. Horne, A. Shimony, A. Zeilinger, Bell's theorem without inequalities. Am. J. Phys. **58**, 1131–1142 (1990)
13. L. Hardy, Nonlocality for two particles without inequalities for almost all entangled states. Phys. Rev. Lett. **71**, 1665–1668 (1993)
14. E. Haven, A. Khrennikov, *The Palgrave Handbook of Quantum Models in Social Science: Applications and Grand Challenges* (Springer, Palgrave MacMillan, 2017)
15. W. Heisenberg, *Quantum-theoretical re-interpretation of kinematical and mechanical relations*, in *Sources of Quantum Mechanics*, ed. by B.L. Van der Waerden (Dover, Reprint 1968, New York, NY, USA, 1925), pp. 261–277
16. W. Heisenberg, *The Physical Principles of the Quantum Theory* (tr. Eckhart, K., Hoyt, F.C.) (Dover, reprint 1949, New York, NY, USA, 1930)
17. A. Khrennikov, Bohr against bell: complementarity versus nonlocality. Open Phys. **15**, 734–738 (2017)

18. A. Khrennikov, Get rid of nonlocality from quantum physics. Entropy **21**(8), 806 (2019a). https://doi.org/10.3390/e21080806
19. A. Khrennikov, Hertz's viewpoint on quantum theory. Act. Nerv. Super. **61**, 24–30 (2019b). https://doi.org/10.1007/s41470-019-00052-1
20. A. Khrennikov, *Echoing the recent Google success: foundational roots of quantum supremacy* (2019c). arXiv:1911.10337v1 [quant-ph] 23 Nov 2019
21. A. Khrennikov, Quantum versus classical entanglement: eliminating the issue of quantum nonlocality. Found. Phys. **50**, 1762–1780 (2020a). https://doi.org/10.1007/s10701-020-00319-7
22. A. Khrennikov, Two faced Janus of quantum nonlocality. Entropy **22**, 303 (2020b)
23. A. Khrennikov, Is the Devil in h? Entropy **632** (2021a). https://doi.org/10.3390/e23050632
24. A. Khrennikov, Quantum postulate versus quantum nonlocality: on the role of the Planck constant in Bell's argument. Found. Phys. **51**, 16 (2021b)
25. A. Khrennikov, *Ubiquitous Quantum Structure* (Springer, From Psychology to Finance, 2010)
26. A. Khrennikov, Classical and quantum mechanics on information spaces with applications to cognitive, psychological, social and anomalous phenomena. Found. Phys. **29**(7), 1065–1098 (1999)
27. A. Khrennikov, A quantum-like model of unconscious-conscious dynamics. Front. Psychol.**6**, 997 (2015)
28. A. Khrennikov, I. Basieva, E. Dzhafarov, J. Busemeyer, Quantum models for psychological measurements: an unsolved problem. PLoS ONE **9**, e110909 (2014)
29. A. Khrennikov, Quantum version of Aumann's approach to common knowledge: sufficient conditions of impossibility to agree on disagree. J. Math. Econ. **60**, 89–104 (2015)
30. A. Khrennikov, *Social Laser: Application of Quantum Information and Field Theories of Social Processes*. (Taylor & Francis Group, Routledge, 2020)
31. N.D. Mermin, *Boojums All the Way Through: Communicating Science in a Prosaic Age* (Cambridge University Press, Cambridge UK, 1990)
32. M. Ozawa, A. Khrennikov, Modelling combination of question order effect, response replicability effect, and QQ-equality with instruments. J. Math. Psychol. **100**, 102491 (2021)
33. A. Plotnitsky, Quantum atomicity and quantum information: bohr, heisenberg, and quantum mechanics as an information theory, in *Quantum Theory: Reexamination of Foundations 2*, ed. by A. Khrennikov, (Växjö University Press, Växjö, Sweden, 2002)
34. A. Plotnitsky, *The Principles of Quantum Theory, from Planck's Quanta to the Higgs Boson: The Nature of Quantum Reality and the Spirit of Copenhagen* (Springer/Nature, New York, NY, USA, 2016)
35. E. Schrödinger, The present situation in quantum mechanics, in *Quantum Theory and Measurement*, ed. by J.A. Wheeler, W.H. Zurek (Princeton University Press, Princeton, NJ, USA, Princeton 1983, 1935a), pp. 152–167
36. Z. Wang, J. Busemeyer, A quantum question order model supported by empirical tests on a priori and precise prediction. Top. Cogn. Sci. **5**, 689–710 (2013)
37. Z. Wang, T. Solloway, R. Shiffrin, J. Busemeyer, Context effects produced by question orders reveal quantum nature of human judgments. Proc. Natl. Acad. Sci. USA (PNAS) **111**(26), 9431–9436 (2014)
38. R. Waters, Wall Street banks ramp up research into quantum finance. Financ. Times (2020), edition of 6 January 2020
39. J.A. Wheeler, Law without law, in *Quantum Theory and Measurement*. ed. by J.A. Wheeler, W.H. Zurek (Princeton University Press, Princeton, NJ, USA, 1983), pp.182–216
40. J.A. Wheeler, Information, physics, quantum: the search for links, in *Complexity, Entropy, and the Physics of Information*. ed. by W.H. Zurek (Addison-Wesley, Redwood City, CA, USA, 1990), pp.3–28
41. J.A. Wheeler, W.H. Zurek (eds.), *Quantum Theory and Measurement* (Princeton University Press, Princeton, NJ, USA, 1983)
42. J. Yearsley, The quantum question equality (Vanderbilt University, 2021)

Arkady Plotnitsky is a Distinguished Professor in Literature, Theory, and Cultural Studies Program, and Philosophy and Literature Program at Purdue University. His research interests include the philosophy and theory of mathematics and science, especially quantum foundations and the role of mathematics in physics, continental philosophy, and the relationships among philosophy, literature, and mathematics and science. He is the author of nine books, several edited and co-edited collections and journal issues, and over 200 hundred articles and book chapters on these subjects. *His latest books are Reality without Realism:Matter, Thought, and Technology in Quantum Physics (Springer 2021) and Logos and Alogon: Thinkable and the Unthinkable in Mathematics, from the Pythagoreans to the Moderns (Spinger 2022).*

Emmanuel Haven holds the Alex Faseruk Chair in Financial Management at Memorial University in Canada. He has research interests in using formalisms from physics to social science. He co-authored several books and published in a variety of fields ranging from Biology, Finance, Economics to Operations Research and Physics. With Andrei Khrennikov he published *Quantum Social Science* (Cambridge University Press 2013) and *The Palgrave Handbook of Quantum Models in Social Science: Applications and Grand Challenges* (Palgrave MacMillan Springer Nature, 2017).

On Wheeler's Quantum Circuit

Gregg Jaeger

1 Introduction

The Meaning Circuit Hypothesis (MCH) is a synthesis of ideas providing John Wheeler's outline of ultimate physics, which he fine-tuned over several decades from the 1970s onward. It is a 'working hypothesis' in which 'existence is a 'meaning circuit'" that portrays the world as a "system self-synthesized by quantum networking." It was strongly advocated by him for roughly two decades and since then has had an increasingly strong impact on the approach of many investigators of quantum theory [1–3]; in particular, elements such as the quantum participator and 'it from bit' are now considered by others as candidate components of a foundation for quantum theory in which information is involved essentially; cf., e.g., [4–6]. Therefore, it is worthy of review and critique.[1]

[1] This paper is written on the occasion of the 60th birthday of Andrei Khrennikov, whose scientific life has included the service to the international physics community of organizing and running a continuous series of similarly interdisciplinary yearly conferences on the foundations of quantum theory that has spanned two decades, held in Växjö, Sweden, not far from Copenhagen, Denmark where Wheeler carried out his own postdoctoral studies with Niels Bohr. The Växjö quantum foundations conference series, which began with the new millennium, has been singular in its openness to a broad range of intellectual perspectives ranging from mathematical physics to the philosophy and history of science, all of which relate to this analysis and its fostering connections between quantum information studies and the foundations of quantum theory. It has been a privilege to participate in these meetings during which, as it turns out, my own work has run much of the gamut of areas touched on by the ideas of Wheeler's synthesis.

G. Jaeger (✉)
Quantum Communication and Measurement Laboratory, Department of Electrical and Computer Engineering, and Division of Natural Science and Mathematics, Boston University, Boston, MA 02215, USA
e-mail: jaeger@bu.edu

A. Plotnitsky and E. Haven (eds.), *The Quantum-Like Revolution*,
https://doi.org/10.1007/978-3-031-12986-5_2

Physical theory can be viewed as prescriptive and/or proscriptive, setting out what must happen and/or what is forbidden from happening in the physical world (cf., e.g. [9]), and the more metaphysically ambitious of physicists have used it to venture or support theses about what can, must, or cannot *exist* in the physical world, that is, its ontology which has often been rather imprecisely referred to in the physics literature as simply 'reality', as Wheeler himself did. Physical laws and a small number of mathematically oriented mechanical principles have been held to ground physics since at least the introduction of Newtonian mechanics; in its mature period, physics has considered equations of motion, conservation laws, or symmetries as illuminating the structure of physical world and its behavior. It has been possible for physicists to find, or come close to finding sets of axioms serving as a consistent mathematical basis for fundamental physics in addition to techniques for predicting physical behavior with high precision. However, for quantum theory, some of these accomplishments proved more illusive than in classical theory, leading to a re-evaluation of what had been considered self-evident and basic elements of physics.

The apparent foundational difficulties in quantum mechanics led Niels Bohr, Wheeler's postdoctoral supervisor to introduce new notions into the physics of the atomic and subatomic scale such as *the quantum phenomenon* and *the significance of measurement configuration to the definiteness of physical properties*, in addition to its defining element, the quantum of action, $\hbar$. Wheeler believed that "nobody has had a better picture of what quantum theory is and means" than Bohr [7], p. 59. Wheeler viewed the MCH as building "only a little [further] on the structure of Bohr's thinking" that showed "how information may underlie reality." He characterized his own approach to science as a sort of "radical conservative-ism," the conservative aspect being its building on the ideas of the founders of quantum mechanics including Bohr and its connection to experiment, and its radical aspect being the demotion of objective matter-energy and spacetime as fundamental and viewing them as arising from the processing and communication of information resulting from observation. The development of the MCH took place while quantum information science (cf., e.g. [8]), which relates physics, communication, and computation, was developing into a discernible subject area that has also recently been involved in the ongoing investigation of the foundations of quantum theory [3, 4]. Although it takes information as fundamental, the MCH goes far beyond standard quantum information science in its attempt to identify a common basis for both quantum theory and space-time theory—in Wheeler's view, by providing an answer to the question "Why the quantum?" that he saw as essential for understanding both these areas of physics and, indeed, for offering an ultimate conception of reality itself.

Up at least up until the early 1970s, for Wheeler and the physics community at large, the ontology of quantum theory was thought to be one of particles and/or fields (cf., e.g. [10]): The evolution of Wheeler's thinking preceding the MCH has been described as involving first "a major shift ...concerning the fundamental ontology of the universe, from a world of particles (with fields as convenient fictions) to a world of fields (from which particle-like aspects can be derived). This shift can also be seen to have radically altered Wheeler's creative process" in conceiving new physical theory; cf. [11], Sect. 2. With the MCH, Wheeler attempted to shift from the conventional view, proposing that the quantum physical world picture is one where neither matter nor even space and time are primary but, instead, one where reality is 'built up' through the accumulation 'meaningful' information which, in his view, logically precedes them. This notion was captured by his now famous phrase "it from bit" and explicated in a process which he schematized as a circuit, the "meaning circuit" [3].[2] Following this lead, axiomatic foundations for physics involving information have recently been sought; for a review of these attempts, cf. [4].

Wheeler saw information-centric physics as the latest, "third era" of physics following "Era I—Motion with no explanation of motion: the parabola of Galileo and the ellipse of Kepler" and "Era II—Law with no explanation of law: Newton's laws of motion, Maxwell's thermodynamics, Einstein's geometrodynamics, modern chromodynamics, grand unified field theory, and string theory" ([1], p. 109). The closed nature of the universe, in particular, poses a problem: None of the conserved-quantity requirements can be applied there; "For example, every method to determine the mass-energy of a system goes back to or is in some way equivalent to placing a planet in orbit about that system, measuring the period and size of that orbit and applying Kepler's law,"

$$\text{mass (in cm)} = \left(\frac{2\pi}{\text{period, in cm of light travel time}}\right)^2 (\text{radius, in cm})^3 \quad (1)$$

([2], p. 14). Wheeler argued that physical laws as we have conceived them cannot be stated without the involvement of space and time, but space and time themselves cannot be fundamental, he claims, because they must come into existence at the putative beginning and possible end of the known universe according to observationally confirmed implications of general relativity such as an initial singularity and black holes.

Accordingly, it became Wheeler's view that future physical laws—of what he called "Era III" of the history of physics—cannot be absolute and invoke space and time as fundamental but must, instead, be mutable and, so, follow from deeper notion(s), including that of "the quantum" in terms of which space and time themselves can be understood. Through its explication of ultimate reality, Wheeler believed that the meaning circuit resolves two suggested paradoxes: one, that "The universe exists 'out there' independent of acts of registration, but the universe does not exist out there independent of acts of registration" captured by quantum theory; the other, that "The bounds of time tell us that physics comes to an end. Yet physics

[2] The "It from bit" slogan has been since been taken up by others, for example, Jeffrey Bub arguing "assuming the information-theoretic constraints are in fact satisfied in our world, no mechanical theory of quantum phenomena that includes an account of measurement interactions can be acceptable, and the appropriate aim of physics at the fundamental level then becomes the representation and manipulation of information" [12, 13], something extreme in that most physicists accord with the view of John Bell that measurement should be understood as just another physical process—most processes having nothing obvious to do with information—despite the quantum-*mechanical* measurement problem; cf. [14].

has always meant that which goes on its eternal way despite all surface changes in the appearance of things. Physics goes on, but physics stops; physics stops, but physics goes on" ([15], p. 341). Here, Wheeler's MCH hypothesis for resolving these paradoxes is reviewed and critiqued.

2 Development of the Meaning-Circuit Schema

Let us begin by briefly reviewing Wheeler's 'circuit' model for the conception of physical reality, its development, and a number of his arguments in support of it which are dealt with in greater detail in later sections. In this section, this is done primarily through the statements of Wheeler himself, in no small part because his discussions were often analogical and metaphorical, even poetic. The development of the MCH had begun at least as early as 1971 as evidenced by a presentation in Fribourg, "Beyond the End of Time," where the relationship of measurement and physics was considered. Wheeler began with the following question and answer. "[M]ay the universe in some strange sense be 'brought into being' by the participation of those who participate? ...'Participator' is the incontrovertible new concept given by quantum mechanics: it strikes down the term 'observer' of classical theory, the man who stands behind the thick glass wall and watches what goes on without taking part. It can't be done" ([16], p. 180). This image encapsulates his view of the role of the 'observer' in physics, a notion invoked by Bohr's approach to quantum mechanics but with a reach extended beyond the physics of light and matter to that of space and time; his precise position as to what constitutes an 'observer' was to change over time and the *community* of observers, also mentioned by Bohr, would later be argued to play an even more central role in the meaning circuit model than the 'observer'.

In an article of 1973, "From Relativity to Mutability," Wheeler elaborated the metaphor of observer and glass barrier. "Even to observe so minuscule an object as an electron, he must shatter the glass. He must reach in. He must install his chosen measuring equipment. It is up to him to decide whether he shall measure position or momentum. To install the equipment to measure the one prevents and excludes his installing the equipment to measure the other. Moreover, the measurement changes the state of the electron. The universe will never afterwards be the same. To describe what has happened, one has to cross out that old word 'observer' and put in its place the new word 'participator'. In some strange sense the universe is a participatory universe" ([17], p. 244). The main differences in this approach from that of classical physics are that the physical state and even "the universe" is supposed to change as a result of observations and their communication, not merely that any measurement involves physical interaction and a record of the result. Wheeler would later offer a corresponding experiment as a concrete illustration of such cosmic participation.

In a presentation following within a year's time, "Is Physics Legislated by Cosmogony," Wheeler turned more precisely to the spatial aspect of the universe, "No one see any longer how to defend the view that geometry was created on 'Day One' of creation, and quantized on Day Two.' More reasonable today would appear the contrary view, that 'the advent of the quantum principle marked Day One, and out of the quantum principle geometry and particles were both somehow built on Day Two"' ([18], p. 544). So, it appears to him that "[the quantum principle] promotes observer to participator", one whose participatory observations are enabled on Day Two ([18], p. 560). For Wheeler, the 'quantum principle' also "tells what question it makes sense for the observer to ask" in order to obtain 'meaningful' observations, which are of greater significance to understanding reality than theory is ([18], p. 544). In "Include the Observer in the Wave Function?", a paper prepared for a May, 1974 conference in Strasbourg, Wheeler wrote in section entitled "The Role of the Observer" that "That the 'observer' should have a special place in the scheme of quantum mechanics has often been contested. Bohr himself argued at one point that an 'irreversible amplification process' is all it takes to 'complete' a measurement, only to stress on other occasions that an observation is only complete when there is an observer ...Fission, and by extension the pulse of a Geiger counter and the blackening of a silver halide crystal, are only then guaranteed to be indelible in the relevant sense when the act has registered in the consciousness of the observer"—this conclusion supported by the in principle reversibility of fission with the addition of "'mirrors' set up to give sufficient time delays and sufficient accuracy of return of the outgoing particles and radiation..." to reverse it [19]. Here, Wheeler suggests that it is the character of observation by a conscious observer that is necessary for and precipitates the novel change of physical state involved in quantum theory and so alters existence at a basic level with each "meaningful" measurement.

In a following, 1975 presentation, "Genesis and Observership," Wheeler went on exactly to explore "the working hypothesis that 'observership is the mechanism of genesis"' as an answer to Leibniz' question "'Why is there something rather than nothing"' which "William James translated" into "the more meaningful ...'How comes the world to be here'..." ([2], p. 3), and the position that "Quantum mechanics promotes the mere 'observer of reality' to 'participator in the defining of reality'. It demolishes the view that the universe sits 'out there'." Thus, Wheeler rejects any fully objective view of reality ([2], pp. 5–6), which for him came to be one *constructed* by physics of a kind that is intimately connected with a process of observation yielding information which is considered primary to it.[3] Four ideas, "(1) 'mutability', (2) 'no ultimate underpinning', (3) 'observership as prerequisite for genesis' ...(4) 'observer-participator as definer of reality'," were presented as bearing on the origin of existence. Wheeler argued that to lend coherence to this collection and provide an answer "...demands a central theme and thesis. ...Up to now, however, no pattern suggests itself from the available clues except this, to interpret quantum mechanics as evidence for the tie between genesis and observership" ([2], p. 8). He also recalled

[3] This is in direct opposition to the position held by many physicists, for example, John von Neumann, that physical theories are required to accommodate a physical correlate to subjective perception according to the principle of psycho-physical parallelism: "it is a fundamental requirement of the scientific viewpoint—the so-called principle of psycho-physical parallelism—that it must be possible to describe the extraphysical process of the subjective perception as if it were in reality in the physical world—i.e., to assign to it parts equivalent physical processes in the objective environment, in ordinary space" ([20], pp. 418–419).

"Bohr's words about the Schrödinger state function, 'that we are here dealing with a purely symbolic procedure, the unambiguous physical interpretation of which, in the last resort, requires a reference to a complete experimental arrangement.'" Wheeler adds "'that all departures from common language and ordinary logic are entirely avoided by reserving the word 'phenomenon' solely for reference to unambiguously communicable information'" ([2], p. 24). This already indicates the significance of information theory in Wheeler's ultimate scheme which connects it with Copenhagen approach to interpreting the quantum formalism wherein meaningful information is the sort of information *communicable within a community* of participators.

In a 1977 magnum opus, the lecture "Frontiers of Time" for Course LXXII of the Enrico Fermi School of Physics in Varenna, Wheeler reiterated Bohr's reply to Einstein that the "...conditions [of measurement] constitute an inherent element of any phenomenon to which the term 'physical reality' can be attached.' ...[This requires] a final renunciation of the classical ideal of causality and a radical revision of our attitude towards the problem of physical reality'" ([21], p. 7). Under the MCH, all physical form is to arise through acts of observer-participation, even that of space-time: "...individual events. Events beyond law. Events so numerous and so uncoordinated that, flaunting their freedom from formula, they yet fabricate all physical form ...Nowhere more clearly than in the ending of spacetime are we warned that time is not an ultimate category in the description of nature" ([21], p. 143). To support the claim, Wheeler offered an argument involving a defining characteristic of the physics of his Era III: "Law without law." "The absolute central point would seem to be this: The universe had to have a way to come into being out of nothingness, with no prior laws, no Swiss watchworks, no nucleus of crystallization to help it—as on a more modest level, we believe, life came into being out of lifeless matter with no prior life to guide the process. ..." where "when we speak of nothingness we mean nothingness: neither structure, nor law, nor plan" ([21], p. 16). In the anticipated Era III of physics, he held, it will be held that only chance events take place with collective behavior agrees with the laws of Era II in the "everyday world." The connection to the novelty of quantum theory is made specifically as follows. "The necessity for [the] line of separation [between the observer-participant and the system under view] is the most mysterious feature of the quantum. We take that demarcation as being, if not the central principle, the clue to the central principle in constructing out of nothing everything" ([21], p. 17).

Although Wheeler suggested that Bohr rejected a universe that is simply "out there," he nonetheless pointed out that "For Bohr the central point is not 'consciousness,' not even an 'observer,' but an experimental device—grain of silver bromide, Geiger counter, retina of the eye–capable of an "irreversible act of amplification". This act brings the measuring process to a 'close.' Only then, he emphasized, is a person able "to describe the result of the measurement to another in plain language" ([21], p. 19). For Wheeler, this constraint is understood to extend to the earliest moment of the universe, so that "The universe is a self-excited circuit. As it expands, cools and develops, it gives rise to observer-participancy. Observer-participancy in turn gives what we call 'tangible reality' to the universe" ([21], pp. 22–25). As an example of the manifestation of observer-participancy and its cosmogonic significance, Wheeler considered the measurement of photon properties at large times. "There is no more remarkable feature of this quantum world than the strange coupling it brings about between future and past, between (1) the way one will choose to orient his polarization analyzers two years down the road and (2) what one can then say about the photons that already—as judged by us now—had their genesis 3 years ago. There is a sense in which the polarization of those photons, already on their way, 'was' brought into being by the disposition of the polarization analyzers that the observer has yet to make and will make. Moreover there is no difference in principle whether the time between the emission of a photon and its detection is 5 years; or a nanosecond, as it is when one is looking at an object a foot away; or 10^{10} years, as it is when one is receiving direct or indirect radiative evidence of what went on in the first few seconds after the big bang. In each case what one chooses to ask …forms an inseparable part of a phenomenon that in earlier thinking one would have said had'already happened'. In this sense it is incontrovertible that the observer is participator in genesis" ([2], pp. 24–25).

This 'coupling' raises the question of the definiteness and ontological status of the past in general. "How can an observation made now have any influence whatsoever on what has already happened?" along with the point that in the quantum realm, "'what has already happened' is not so easy to say" ([21], pp. 25–26), pointing to the notion of a delayed-choice experiment, one spanning nearly the duration of the history of the universe. "The photons of the primordial cosmic fireball radiation that enter our telescope today, we customarily assume, already had an existence in the very earliest days of the universe long before life evolved. However, not until we catch a particular one of those photons in a particular state with particular parameters, not until the elementary phenomenon is an observed phenomenon, do we have the right even to call it a phenomenon" and, as for the "unbelievably more numerous relict photons that escape our telescope…their 'reality' is of a much paler and more theoretic hue"; ibid., pp. 26–28. And, while there is some tension in the very way these sentences are formulated, in that Wheeler is talking about "catching photons" that become "phenomena," if one keeps in mind that free photons can be thought of—indeed, at the moment, are standardly thought of—as excitations of quantum fields rather than as individual systems of their own, it is a consistently posed statement that relates to more than the polarization but even the presence of each photon. Accordingly, "an immense labor of imagination and theory" is required to provide a conception of physical reality grounded in intersubjectively communicated observation results; ibid., p. 28.

Turning to space and time themselves, Wheeler argued that spacetime is a "deterministic classical concept, applicable only at the level of approximation theory" in light of complementarity ([21], pp. 36–37) and asks "how can any elementary building process be an elementary process for building existence and law unless it transcends the category of time? To identify an elementary building process that transcends time…The act of observer-participancy in such an experiment, right now, irretrievably alters what we have the right to say about 'the past.' In that sense, that carefully restricted sense, that act is an inescapable part of the actual building of 'the past.'"; [21], pp. 42–43. He held that, given both quantum theory and geometrodynamics, "time is transcended, laws are mutable, and observer-participancy matters. 'Before' and 'after' don't rule everywhere, as witness quantum fluctuations in the geometry of space at the scale of the Planck distance. Therefore 'before' and 'after' cannot legalistically rule anywhere. Even at the classical level, Einstein's standard closed-space cosmology denies all meaning to 'before the big bang' and 'after the big crunch.' Time cannot be an ultimate category in the description of nature. We cannot expect to understand genesis until we rise to an outlook that transcends time ... There never was a law of physics that did not require space and time for its statement. With collapse the framework falls down for everything one ever called a law. The laws of physics were not installed in advance by a Swiss watchmaker, nor can they endure from everlasting to everlasting. They must have come into being" ([21], pp. 44–45), the Planck length being

$$l_P = (\hbar G/c^3)^{1/2} , \tag{2}$$

where G is the gravitational constant, c is the speed of light in vacuum, and $\hbar$ is the reduced Planck constant (the quantum of action), and all constants on the right-hand side are all given the value 1. And so, he argued, determinism must be abandoned: "one has to forgo that view of nature in which every event, past, present, or future, occupies its preordained position in a grand catalog called 'spacetime.' There is no spacetime, there is no time, there is no before, there is no after. The question what happens 'next' is without meaning" ([21], p. 105).

However, Wheeler also recognized that there are apparent limitations to a physics based on participancy; he instructs one, at least for now, as follows. "Don't try to 'take time apart' into elementary quantum acts of observer-participatorship out of which we conceive it—and everything—to be built. Instead, sticking to the solid ground of physics as we know it, identify domains where familiar concepts of time and causality come to the limit of applicability and have to be modified. We have just finished exploring one such frontier. We have seen how both time and spacetime, according to existing theory, lose all application at the Planck distance and the Planck time; but how out of a description that transcends time–out of superspace–we come back in the appropriate correspondence principle limit to familiar views of time" ([21], p. 117), where the Planck time is

$$t_P = (\hbar G/c^5)^{1/2} . \tag{3}$$

Accepting this view of space and time, one must of course contend with the implications for causality, which Wheeler did by considering, as a basic example, the situation of the emission of electromagnetic radiation in a system of point charges coupled with each other by elementary electromagnetic actions-across-a-distance—which are individually time symmetric in the sense that the force exerted on a particle by one particle on the other is given by half the retarded field of the first, as usually calculated, plus half the advanced field—and the particle is suddenly accelerated.

> This links past and future in a maze of backward and forward running light rays. Nowhere can the slightest change be made without altering motions everywhere into the indefinite past and future. we want to establish in this one example a point of more general application:

> The apparent inability of an action taken now to influence the past by no means rules out a direct influence of the present in 'bringing about that which we call the past.' It is in no way suggested here that this is the actual mechanics by which acts of observer-participancy in the present bring [about] that which we call the tangible or communicable reality of the universe at an era when no observers existed. That is a deeper question with which physics is not yet prepared to deal. However, one is open to believe that the kind of considerations that elucidate the one may clarify the other ([21], p. 118).[4]

Similarly, "the irreversibility of the emission process is a phenomenon of statistical mechanics connected with the asymmetry of the initial conditions with respect to time...an order in time, ostensibly causal, can originate from an underlying machinery that is very far from causal. ...we here have a sample law, causality, emerging from a description of nature that contains no such law" ([21], pp. 120–121).

Wheeler continued his investigation of physical law as mutable and the possibility of the non-primacy of space-time in a 1979 article "Beyond the Black Hole."

> 'Physical space-time is not mathematical space-time' is the one lesson of mutability; the other, 'Physical law is not ideal mathematical law.' Law that comes into being at the beginning of time and fades away at the end of time cannot be forever 100% accurate. Moreover, it must have come into being without anything to guide it into being...The laws of physics themselves, coming into being and fading out of existence: in what else can they have their root but billions upon billions of acts of chance? What way is there to build law without law, field without field, substance without substance except 'Individual events. Events beyond law. Events so numerous and so uncoordinated that, flaunting their freedom from formula, they yet fabricate firm form?' ([15], p. 352).

Several of the above ideas were reiterated along with some other contributions that quickly followed in 1980 and 1981 to form a shorter 1983 article "Law without Law" ([23], pp. 182–213). Wheeler had raised a challenge to physics with these speculative arguments, and he was optimistic regarding their significance and ability to explain how 'something may arise from nothing.' "No test of these views looks more like being someday doable, nor more interesting and more instructive, than a derivation of the structure of quantum theory from the requirement that everything have a way to come into being out of nothing. If you would have an epitome of this summary, let it be this: Nothing. No time. The line. Acts. Statistics. Law. Spacetime. Substance. Observer-participant Closed circuit. Test." ([21], p. 46).

Wheeler offered a comprehensive model illustrating the structure of the MCH in a presentation of 1986 entitled "How come the quantum?". In this model, discussed in detail below in Sects. 3 and 5, the establishment of physical reality is explained as follows.

> Physics gives rise, ...to light, pressure, and sound. They provide means of communication, of the importance of which Niels Bohr notes, '. . . every analysis of the conditions of

---

[4] As he was later to describe this, in the mid-1980s, one finds "an unbelievable maze of light-like connections running backward and forward in time. Nevertheless, when the number of particles is sufficiently great to guarantee absorption of all radiation, the maze translates itself into the familiar full-strength pure retarded electromagnetic interactions, plus the familiar radiative reaction. No contradiction with everyday notions of causality is to be seen. No less has to be expected of an information-theoretic account of physics;" [22], p. 31.

> human knowledge must rest on considerations of the character and scope of our means of communication. Physics is also the foundation of chemistry and biology, out of which arise communicators....' From meaning back to physics, the circuit under examination makes its way by a less apparent or underground sequence of linkages. Meaning rests on action. Action forces the choosing between complementary questions and the distinguishing of answers. Distinguishability, in the realm of complementary possibilities, demands for its measurement complex probability amplitudes. The change in the phase of a complex probability amplitude around a closed circuit not only measures the flux of field through that circuit, but can even be regarded as the definition and very essence of that field. Fields, in turn, can be viewed as the building stuff of particles; and fields plus particles generate the world of physics with which the hypothesized meaning circuit began ([3], pp. 304–305).

Here, Wheeler's 'meaning' is "the joint product of all the evidence that is available to those who communicate." (This is the subject of Sect. 5.4 below.) After having presented the elements of the MCH by the mid-1980s, Wheeler continued to clarify his views on its detail and implications. For example, in a 1988 follow-up "World as System Self-Synthesized by Quantum Networking," he presents the MCH "in contrast to the view that the universe is a machine governed by some magic equation," it is the "view that the world is a self-synthesizing system of existences, built on observer-participancy via a network of elementary quantum phenomena" [1].

Wheeler's goal was a conception of the physical world in which space and time are no longer considered primary, incorporating the novelties of quantum mechanics in a way consistent with the known natural science, which had been developed with a methodology similar to that used in discovering classical physics but eschewing the mechanistic approach in favor of a focus on communicable information. He believed that the "elementary quantum phenomenon is the strangest thing in this strange world. It is strange because it has no localization in space or time. It is strange because it has a pure yes-no character—one bit of meaning. It is strange because it is more deeply dyed with an information-theoretic color than anything in all physics" ([1], p. 115). Broadly speaking, with the MCH Wheeler went beyond Bohr's position that science, as a human activity, involves the communication of results of its contributors, to a direct grounding of physical systems and processes in the information that is communicated within science. Nonetheless, his extension to the experimental 'asking questions of nature' remains in the spirit of Bohr's understanding of physics, in particular, in its use of the notion of the complementary nature of physical measurements [24]. "In brief, complementarity symbolizes the necessity to choose a question before we can expect an answer...We once thought, with Einstein, that nature exists 'out there', independent of us. Then we discovered—thanks to Bohr and Heisenberg—that it does not" ([1], p. 113).[5]

Throughout the development of his MCH, Wheeler was either vague or vacillated with regard to the question of whether there is a necessary role for consciousness—something he consistently recognized as itself not well understood by science—in elementary quantum phenomena. Finally, in the 1990s, he expressed an affinity for

[5] One should take care not to accept this statement of Wheeler as a correct understanding of the views of Bohr and Heisenberg themselves, which often differ from those of their interpreters like Wheeler; cf., e.g. [25].

the notion of quantum potentiality, which plays a central role in the Heisenberg's mature understanding of the quantum state as indicating a genuine mode of existence and not mere logical possibility; Heisenberg's later view was that measurement outcomes and observations arise independently of human participation, even if they are brought about only through coincidence with the greater physical world; cf. [26]. In the following section, the central elements of Wheeler's conception of physics as represented via the MCH are considered in succession. These include "question and answer feature of the elementary quantum phenomenon" and the concomitant role of distinguishability in it, why the quantum probability amplitude is complex, the attribution of meaning to measurement results, and the circuit as the alternative to a hierarchical built upon a traditional foundation, taking into consideration the issues with his MC model, the role of 'communication' in physics, and the 'It from bit' thesis. Wheeler himself noted some difficulties with the assertion of aspects of this synthesis, and these considered along with a number of other deep concerns.

3 How Come the Quantum? A Critique

Although information plays a fundamental role according to the MCH, the physical mechanism involved in measurement is nearly as significant: "...the meaning-creating community of observer-participants, past, present, and future, is brought into being by the machinery of the world." Indeed, consistently with his late affinity with the potentiality interpretation of quantum state amplitudes in the 1990s, the "machinery" and that the information obtained using it in measurement can be understood on par with each other, for it is the machinery of the world that provides the signal-records through which information is obtained. This machinery and the information it provides to observers relate to *different*, however intimately related domains—those of existence and human knowledge of that existence—that Wheeler failed to adequately distinguish in his writings of the 1970 and 1980s. The hypothesis "goes on to interpret this very world of past, present, and future, and of space, time, and fields, to be (despite all its apparent continuity, immensity, and independence from us) a construction of imagination and theory" where elementary quantum phenomena form "the iron posts of discrete acts of observer-participancy, the ant-like but magnificent labor of a community stretching from far in the past to even farther in the future" and time is "not primordial, precise, and supplied from outside physics, but secondary, approximate, and derived" [3], p. 313. The observed and communicated bits of information are considered the strongest elements of a reality understood as constructed from these facts and scientific knowledge. Yet, consciousness is not required for the signaling of these bits, only for the existence of knowledge to which they may give rise. And, in the long run, Wheeler relinquished the view that observation and measurement *require* consciousness.

And, despite his strong views as to the significance of measurement and information theory to physics, Wheeler was certainly *not* advocating a form of idealism; rather, like Bohr, he put forward a very subtle form of realism, a 'participatory realism.' Wheeler rhetorically raised the question, "How does quantum mechanics today differ from what Bishop George Berkeley told us two centuries ago, '*Esse est percipi*,' to be is to be perceived? ... Yes, and in an important way. Berkeley—like all of us under everyday circumstances—deals with multiple quantum processes ... Anything macroscopic that happened in the past makes, we know, a rich fallout of consequences in the present. ... the number of quanta that come into play is so enormous that the unseen quantum individuality of the act of observation can hardly be said to influence the event observed"; [27], p. 186. And just this is required for the occurrence of an observed quantum phenomenon capable of yielding the communicable information so important to his meaning-circuit model, its "genesis." Two crucial questions are involved in the process of 'genesis' Wheeler hypothesizes: (i) what the 'density' (variable) of such signposts in reality is at various stages of its history, and (ii) how essential, if at all, to their appearance living or conscious observers (who when the universe existed) are.

Regarding (i), Wheeler commented that "If life wins all, then the number of bits of information being exchanged per second can be expected to rise enormously compared to that number rate today. ... And how great must that future total be—tally as it is of times past—to furnish enough iron posts of observation to bear the smooth plaster which we of today call existence? Bits needed. Bits available. Calculate each. Compare. This double undertaking, if and when it becomes feasible, will mark the passage from clues about existence to testable theory of existence" [1], p. 126. However, this is an essentially epistemic question having to do with the facts serving as the basis of our own theorizing rather than with the existence of the physical world that the resulting physical theories are to describe and explain. While it is true that the degrees of (in)determinacy of properties of subatomic systems relate to measurement, the Heisenberg indeterminacy principle—which is what is most relevant here in the quantum realm—regards the relative definiteness involved in the complementary measurement activities that themselves still require physically distinct and mutually excluding apparatus configurations to yield information to observers.

Regarding (ii), Wheeler proposed that, due to presence of the quantum of action, "The observer is elevated from 'observer' to 'participator.' What philosophy suggested in times past, the central feature of quantum mechanics tells us today with impressive force: In some strange sense this is a participatory universe"; [2], pp. 5–6. Space, time, particles, and fields are all "brought into being" from quantum information-yielding events, and only by a "dethronement from primordial and precise to secondary, approximate and derived can 'time' be reduced, like all the rest of physics, to an information-theoretic foundation" [22], p. 30. Thus, the second question is again one about the process of obtaining information by observers and the manner in which that information is rendered meaningful and communicated, which is an epistemic one. Wheeler argued that for this "we possess no other model [than the Meaning Circuit] that puts in central place the quantum and the question of 'how come the quantum?'" [3]. However, this is not so. One need not assume that the entities of physics are *brought into being* in information-transfer events simply because of the presence of the quantum of action in measurement interactions. Standard quantum measurement theory explains how measurements provide records of the properties of quantum systems in the past whose possible different properties upon measurement are made definite at some moment—much, though of course somewhat differently, as the exact color of Caesar's clothing on a particular day could have been more or less better inferred, even from an extreme temporal distance, by an author whose writings we now possess, and be useful as a fact knowable to historians to be combined with other possibly later-obtained information—providing data to those who might attribute it meaning and communicate it between themselves. Wheeler's MCH is a hypothesis that transcends physics and concerns itself with the manner in which physical knowledge and theory arise, claiming that these are the determinants of existence but without sufficient justification.[6]

To reassure those who might be concerned about that the role given observer in the MCH might threaten the unity of the universe, Wheeler notes that his view of quantum theory does not amount to a "many worlds" view: "What keeps the images of something 'out there' from degenerating into separate and private universes: one observer, one universe, another observer, another universe? That is prevented by the very solidity of those iron posts, the elementary acts of observer-participancy. That is the importance of Bohr's point that no observation is an observation unless we can communicated the results of that observation to others in plain language"; [27], p. 203. Wheeler argued further that the *self* is not independent of others.

> The heart of the matter is the word self. What is to be understood by the word self we are perhaps beginning to understand today as well as some of the ancients did. We know that in the last analysis there is no such thing as self. There is not a word we speak, a concept, we use, a thought we think which does not arise, directly or indirectly, from our membership in the larger community. On that community the mind is as dependent as is the computer. A computer with no programming is no computer. ...programming by parents and community that makes a mind a mind. The heart of mind is programming, and the heart of programming is communication. ([1], p. 127)

Thus, "*Communication* is the essential idea. If I see something, but I'm not sure whether it's a dream or reality, there's hardly a better test than to check whether someone else is aware of it and can confirm my observations" [7], p. 62. This is certainly important for establishing the objectivity of data of the senses but, again, is beside the point for physics itself; it regards the validity of data, not its ultimate origins.

Thus, the MC model portrays physical systems and their behavior between measurements as ultimately (inter)subjective creations of the scientific community based on the physical facts (phenomena) arising through measurement. This model is illustrated by Wheeler as a corn-kernel shaped 'meaning circuit,' sketched metaphorically as running about a slope, with the abstract elements—questions and quantum formalism—underground and the signaling apparatus sub-grouped into "means of communication" and "communicators" above ground: its two large parts are shown as two towers labeled "physics," at base and "meaning," at peak. That "part of circuit...is well known"; [28], p. 404. However, no detail is offered in this model regarding the

[6] Indeed, Wheeler repeatedly admits that it is precisely the nature of cognition located in the meaning circuit which is inadequately understood, as shown below.

important distinction between meaning and information/data; as with consciousness, Wheeler regarded this as a part of science that is only emerging: "New aspects of what we mean by 'meaning' are beginning to emerge from recent studies in computer science and in information theory that put quantum at center stage"; [28], p. 404. What is said regarding meaning is that physics "gives rise to chemistry and biology and, through them, to communica*tors*. The communicators and the communications between them generate meaning. This is the first part of the meaning circuit. It makes meaning the child of physics. Physics itself, however, according to this view, is also the child of meaning. The return portion of the meaning circuit, the connection that leads back from meaning to physics, runs 'underground,' out of sight." [22], p. 25. For Wheeler, the elementary quantum phenomenon, which is connected with the summit of his circuit and is a notion the origins of which are clearly traceable to Bohr, "displays two characteristic features: (1) complementarity in the choice of the question and (2) statistics in the distinguishing of the answer." [3], p. 306. The important pair of successive points *below ground* labeled "ask question" and "distinguish the answer" are the two elements of the extraction of information, the basis-choice/measurement as the decoding of information carried by the state-signal—do clearly appear in it, although the opposite side of the loop from this pair; the referent, and hence 'meaning', of a click depends on the physical measurement basis.

Although communication theory standardly connects information and physics via the notion of signaling and provides methods for the quantification of information that can be encoded in signals, the relation between physics and information suggested by Wheeler is anything but standard: It is strongly reductionist in the opposite sense from what is the standard relation where systems in physical states are understood to provide signals that *could be used to communicate information* via a freely chosen encoding/decoding method; instead, he views physical systems as *made of and constructed from information*. Wheeler referred to elementary quantum phenomena as being of a "meaning-theoretical character" [7], p. 66, indicating a conflation of two matters: (i) recording and collecting data and (ii) giving meaning to data.[7] Data collection, on the one hand, and the interpretation of data arising in physical events in terms of the elements of theory (much less the creation of theories of the world or reality), on the other, are very different. Indeed, the operation of the circuit—its provision of meaningful experimental results—was summarized by Wheeler as follows. "Evidence available to the communicator comes from the asking of a question [via an experiment] and the distinguishing of an answer" by observing the result of it ([3], p. 305).

Wheeler offered very different examples of the putative reduction of an 'it' to a 'bit' via 'it from bit', a relatively exotic example being that of a black hole: He claimed that "The it, the area of the horizon of a black hole, expressed in units of the basic Bekenstein-Hawking area

$$4(\hbar G/c^3)\log_e 2\,, \tag{4}$$

[7] And this would be so even if he were understood to considering primarily cognitive.

is given by the bit count, N, of that black hole. Here N represents the number of bits of information it could have taken to distinguish the initial configuration of particles and fields that fell in to make this particular black hole from the 2^N alternative quantum configurations that would have produced a black hole externally identical to it" ([29], p. 555).[8] A less exotic example he considered repeatedly in his writing is that of the amount of information obtained in the determining the polarization of a photon, one bit.[9] One could ask, considering an even more concrete example, the *spin* of an electron, does one bit actually constitute the *spin* and does this bit, together with the other bits corresponding to distinguishing the other properties of the electron, *constitute* the electron? The important point to note here is that, on this view, a physical system is no longer considered physical in any usual sense but rather is *only the information corresponding to its state.*

Perhaps the most striking of the examples of 'it from bit' Wheeler provides is that of "reducing the 'how much' of the field measurement to the 'many a chance yes-no' of elemental quantum phenomena" [22], p. 29. This is to be accomplished through dichotomous measurement results signaled by detector clicks: to find the magnetic field strength in a region, "it is enough to illuminate the atom [chosen to observe a Zeeman shift] with monochromatic radiation of a precisely chosen frequency…just enough for photoionization when the magnetic field is 'on'. The electron, once freed, is drawn…into a device like a photomultiplier. A pulse is registered on a counter. The elemental quantum phenomenon is brought to a close by this irreversible act of amplification. An item of information is generated which one person can communicate to another 'in plain language'. Out of sufficiently many such items of 'yes, no' information obtained for slightly different values of frequency…we determine more and more precisely the value of the Zeeman level shift and measure—and define—the magnetic field" ([22], p. 29). Perhaps, but this does not mean that the magnetic field has thereby come into existence; rather, its properties merely become *more or less definite and knowable* to those contemplating the corresponding experimental records, even when that record is a flash recorded by the nervous system of a human being and immediately attended to by its conscious mind.[10]

Despite the circuitous character of his view of experimentation, Wheeler took an ontological-reductionist as well as theory-reductionist stance wherein all of reality is 'theory' constructed from and relating such insubstantial information-theoretically quantifiable and elementary quantum phenomena to each other, conditioned on the circumstances of their recording. He reduces matter and space-time to 'meaning' in his conception of physical reality. "Physical space-time is not mathematical space-time" and "Physical law is not ideal mathematical law" ([15], p. 352). The issues noted in this section are delved into in greater detail in Sect. 5, below, after the consideration in the next section of Wheeler's views on quantum phenomena and the world

[8] The event horizon of the black hole is implicitly considered here identical with the area.

[9] Again, however, this is as a matter of fact dependent on the circumstances of preparation and encoding/decoding assumed and would be rather *at most* one bit.

[10] It is noteworthy that Wheeler's views in the 1990s which brought quantum potentiality into consideration, appear to recognize this.

between quantum preparation and measurement events—what Hans Reichenbach called "interphenomena"—as well as their relationship to space and time [30].

4 Quantum "Smokiness'

An important aspect of Wheeler's theoretical methodology is that, for every grand physical proposal he produced, he offered a related, ultimately feasible experiment, as Misner, Thorne, and Zurek commented. "This is his enduring legacy: Do not be afraid to think big, but make sure that in the end you have a blueprint for an experiment" [31]. For Wheeler, thought experimentation and speculation must to be connected with concrete experience. His view of the physical past according to the MCH was that it is indeterminate ('smoky'), in particular, contingent on subsequent or future interventions for its 'existence', something intimately related to the secondary character of time by comparison to quantum phenomena, and he offered the class of 'delayed-choice' experiments designed to lend support to this view [31]. These experiments, designed to illustrate the operation of the 'meaning circuit,' can in principal be performed with a relatively simple physical apparatus often realized in quantum optics.[11]

4.1 Delayed-Choice

The delayed-choice experiment which, from a mathematical point of view, is the simplest of those Wheeler provided in relation to the MCH is one with an apparatus that can be quickly switched between (i) simple light-beam splitting system allowing for measurement along one of two paths, or (ii) a (Mach–Zehnder) interferometric configuration allowing for the complementary measurement of interference visibility [3].[12] This experiment serves as the archetype of what is now commonly called a "delayed-choice experiment," the idea being to defer the specific configuration until a time *after* beam components could, assuming travel at light speed, be considered to have entered the apparatus and be already headed toward the detection suite at the opposite end.

Both configurations are made possible, as follows. In one configuration of the apparatus, input light first strikes (at an angle of 45 °C) an evenly balanced beam-splitter located at one corner of a rectangle which provides for two, orthogonal beam paths (say, along the axes in two-dimensions); next, the resulting paths turn 90 °C (one clockwise along a "high" path and one counter-clockwise along a "low" path) at mir-

[11] It is noteworthy that photons are special in that the have minimal self-interaction and are able to travel exceptionally long distances without disturbance, unlike ordinary matter composed of electromagnetically charged particles such as electrons and protons.

[12] For an analysis of the associated complementarity relation cf., e.g., [32].

rors toward a common locus through which they could freely pass at the diagonally opposite corner of the rectangle; in each of these two orthogonal directions *beyond* that locus, a photodetector capable of single-photon counting is placed. Detection of energy in a given direction suggests that any detected photon traveled—assuming it must travel along a path—along the path leading to straight into the detector located along it. In the alternate configuration, in addition, a second evenly balanced beam-splitter is located precisely at the second locus of path-intersection, i.e. at the far corner of the path rectangle, allowing for path recombination in both of the two directions described above. Due to this overlapping coincidence of the last portions of the two paths, this second configuration plainly renders impossible the retrodictive determination of which path a given photon must have traveled—assuming it must travel along one—in the apparatus and instead allows a relative-phase determination and self-interference for an ensemble of identically prepared photons.

Thus, the experimenter may 'ask' one of a pair of alternative 'questions about the photon's behavior' by configuring the apparatus without or with a second evenly balanced beamsplitter at the far corner of the rectangle. In this way, the corresponding 'question' thus 'asked' will be 'answered' by the obtaining of one of two distinguishable outcomes in each configuration. In this experiment, Wheeler said, "We can ask which path does an arriving photon follow-the high road or the low road? That is one choice of question, and the photon detectors stand ready to answer it. To ask for the phase relation between the two beams is a complementary choice of question"; [3], pp. 306–307.The distinguishing of an answer is to be done on the basis of the counting statistics of repeating the experiment.[13] Because one can consider such an apparatus with a beam-path rectangle sufficiently long that one could implement choice *after* an initially localized photon would putatively have passed the first beamsplitter but not yet reached the opposite side of the rectangle, Wheeler argued, "This circumstance shows how wrong it is to say that we are finding out in the one case 'which route' and in the other case the relation of phases in a 'two-route mode of travel.' The world is built in such a way that it denies us the possibility to speak in any well-defined way of 'what the photon is doing' in its travel from point of entry to point of reception" ([3], p. 307).

The delayed-choice experiment was taken by Wheeler to illustrate the dictum "No elementary quantum phenomenon is a phenomenon until it is a registered phenomenon, brought to a close by an irreversible act of amplification"; one is to note that "Definite as are the point of entry and point of detection of the photon, what it is doing in between is totally smoky. Equally smoky is any concept of the path of an electron through an atom. Termination of the smokiness by an act of detection is as close as we can get to establishing reality at the microscopic level"; ibid.. Wheeler's "smokiness" corresponds to the *non-deterministic character of the properties of quantum systems*; formally, these measurable quantities correspond to non-commuting operators in the theory of sharp quantum measurements. Wheeler argued that this "smokiness" can be extended to universe as a whole and that its very genesis involves its own self-

[13] Broader conclusions aside, what this example does clearly illustrate is that complementary properties are complementary, that is, the more definite is one the less definite is the other.

observation as a result, as a "self-excited circuit," which he symbolized iconically by a letter "U" with an eyeball (of observation at one end) on one of its verticals directed at the opposite vertical at the other (the deep past) ([21], p. 23), a rather swift conclusion given the gross character of the universe, drawn in part from the additional assumption that all existence arises from quantum phenomena.[14]

Midway through his elaboration of the details of the MCH, in 1983, Wheeler along with his student Warner Miller reviewed a broad range of relevant experiments in an article entitled "Delayed-choice Experiments and Bohr's Elementary Quantum Phenomenon," wherein a number of scenarios of delayed-choice experiments were provided where various complementary quantum degrees of freedom were treated: direction/angular momentum, "which slit"/interference, transverse position/transverse momentum, and time-of-emission/energy [33]. More advanced such experiments were later performed by others. Let us consider now one of these, which was subsequently offered by Anton Zeilinger et al., who viewed their experiment as providing yet greater insight into the relationship between this experiment and the universe [34]. Zeilinger introduced these experiments and commented on them as follows."In Wheeler's own words, 'One decides whether the photon should have come by one route or by both routes after it has already done its traveling.' ...we brought Wheeler's thought-experiment into the laboratory and carried it a step further. The idea was to demonstrate that it can be decided after the photon has been registered already whether the phenomenon observed can be understood as a particle or as a wave" [34].

Aside from the question of the value or lack thereof of being concerned about the choice of 'visualization' of the system (as 'particle' or 'wave'), a less contentious matter is more clearly addressed, namely, whether it is momentum or position that becomes (more) definite of a system given the timing of the configuration setting. An innovation of this extension of Wheeler's basic optical delayed-choice experiment is that it goes beyond the measurement of a solitary photon to one where a photon can be considered part of a greater, precisely specified system, one which can enter measurable entangled states [34]; in their experiment, Zeilinger et al. considered photons within pairs of entangled-photon pairs produced via type-I spontaneous parametric down-conversion from an appropriate crystal pumped by a laser; cf., e.g., [35, 36]. These pairs were created entangled in energy-momentum, being constrained only by joint energy-momentum with each having indeterminate values upon their creation, but the values of both are determined upon the direct measurement of one regardless of their separation in space and time. What is tested here is whether a given one of the photons from those produced pairwise at a downconversion crystal has either a determinate path or a determinate momentum upon measurement given that the sort of measurement is arranged after production of the pair, that is, in delayed-choice fashion.

[14] Less dramatically, one can imagine the distant, future detection event at the far side of the rectangle involved to "determine" what happened at the first beamsplitter. However, such retrodiction is not sanctioned by standard quantum mechanics and, notably, it is suggestive of a species of retrocausality in that it suggests that what takes place in quantum measurement determines what has already happened in its past.

To probe momentum entanglement, one photon was sent to a double-slit arrangement with a movable single-photon detector (rather than the usual screen), and the other in an orthogonal direction to a suite containing a Heisenberg lens and a single-photon Heisenberg detector capable of placement anywhere behind it. Then, with the detector located behind at focal distance f, the precise individual momentum eigenstate of the detected photons was determined; the individual positions were, therefore, indeterminate, that is, which of the two slits (i.e. path) the other photon would pass through is indeterminate—as it is thereby also in a determinate momentum eigenstate and a standard two-slit interference pattern in position resulted. When, alternatively, positioning the same detector behind the Heisenberg lens at a distance $2f$, the same detection plane was $2f$ in front of the lens, as measured from the lens through the the crystal to the two slits, the corresponding photon positions were precisely found and no interference pattern was found.

What is novel in this experiment is that the first of the two considered photons was detected after the second photon *had already itself been detected.* This was characterized by these experimenters as follows, "whether we obtain the two-slit pattern or not depends on whether the possible position information carried by the other photon has been irrevocably erased or not" [34]; to them,

> The important conclusion is that, while individual events just happen, their physical interpretation ...might depend on the future; it might particularly depend on decisions we might make in the future concerning the measurement performed at some distant spacetime location in the future. It is also evident that the relative spacetime arrangement of the two observations does not matter at all. ...The experiment just discussed also provides a clear illustration of the role of the experimentalist. By choosing the apparatus the experimentalist determines whether the phenomenon observed can be seen as a wave or as a particle phenomenon and once the observer has made this choice, Nature gives the respective answer and the other possibility is forever lost. Thus, we conclude, by choosing the apparatus the experimentalist can determine which quality can become reality in the experiment. [34]

They conclude along with Wheeler, though in a qualified manner, that "In that sense, the experimentalist's choice is constitutive to reality, yet one should be warned strongly against a subjective interpretation of the role of the experimentalist or of the observer. It is clear that the consciousness of the observer does not influence the particle at all, in contradiction to a widespread but unfortunate interpretation of the quantum situation" [34]. This discussion is predicated on the validity of retrodiction involved in discussions about the past actual properties, that is, inferences about past properties. In particular, the results involved are *statistical* in nature, so that although the original, Copenhagen sense of the significance of measurement context (cf. [37]) is evident, no retro-causation of individual measurement event is demonstrated in this experiment.

Leaving the question of the validity of quantum retrodiction aside, it is again important to note that it is a property of a (subatomic) object that is understood as determined in the experiment, and not the existence of the object itself. When the behavior observed in such an experiment is considered to extend logically to situations involving distances significant at the scale of the universe, is the central issue surrounding the idea of the role of the observer as a cosmological participator.

However, unless quantum state superpositions are possible of truly macroscopic systems, in the statistical mechanical sense, such an extension is *cosmologically* irrelevant, even aside from the question of the validity of retrodiction in general using the quantum calculus that it would require.

4.2 Genesis via Live Participation

The first of the delayed-choice experiments considered in Sect. 4.1 with possible photon paths extended to billions of light-years was considered by Wheeler was given by him the same interpretation, that of demonstrating the role of the choice of configuration on the information obtainable about the system is responsible for determining the photon's past reality. The remarkable implication of the experiment for Wheeler was what it suggested about the time: In his publication, "Genesis and Observership," Wheeler noted that "No consideration argues more forcibly that the 'observer' has nothing to do with the scheme of physics that the disparity in size of 26 powers of 10 between man and the universe; and none argues more strongly that life and consciousness are a rather unimportant development in a faraway and not particularly relevant part of space. Quite another assessment of the situation develops when one turns one's attention from scales of distance to scales of time" ([2], p. 18). Wheeler argued that in

> no way is one led more directly into this altered approach than through the considerations of Dicke (1961). Paraphrased, Dicke asks what possible sense it could make to speak of 'the universe' unless there was someone around to be aware of it. But awareness demands life. Life in turn, however anyone has imagined it, demands heavy elements. ...[which] requires thermonuclear combustion. Thermonuclear combustion in turn needs several times 10^9 years cooking time in the interior of a star. But for the universe to provide several times 10^9 years of time, according to general relativity, it must have a reach in space of the order of several times 10^9 lyr. Why then is the universe as big as it is? Because we are here! ([2], p. 18)

An experiment performed in the present on light naturally prepared by the universe in the distant past was viewed by him as directly affecting what can be communicated about its origin and so that affects the origin itself. For him, this grounds an anthropic requirement on the universe, namely, that any universe must last long enough for life to arise in it via such nuclear and biological processes so as to endow it with reality.

However, although it is certainly the case that a consistent scientific theory of the world must be such that living observers are not precluded from arising according to its laws—given the existence proof via the fact that we are present in it—a physical universe without observers is certainly physically possible according to what we know about the purely physical behavior of the physical world, with or without consideration of its quantum aspect; although a universe without life would not be exactly our particularly universe, in that certain sets of facts might be different, it would still be one that could have been realized within the requirements of our physical laws, even if certain properties at certain places and times might not become definite in the way that they would have been in our universe in which life is present.

Given what we know about nature, there is no a priori reason why life need have arisen at all within it for our physics to operate, as it clearly did before our appearance in it. That our universe is one in which sharp quantum observables do not all take determinate values appears to have very little in itself to do with life.[15]

4.3 Existence and the Quantum

Whatever might be the case with biology as a basis for gathering data, the ultimate proof and answer to 'why the quantum?' for Wheeler was to show that the *quantum* is necessary for *existence itself*. Indeed, he quotes Leibniz in relation to this question. "Leibniz put it in his famous words, 'Why is there something rather than nothing?' William James (1911), translated the 'why' to the more meaningful 'how': 'How comes the world to be here'... We ask today, "How did the universe come into being?" ...in the absence, as here, of some clear indication that the question is meaningless or undecidable, the question must be faced and the relevant evidence sought out" ([2], p. 3). Here, Wheeler considers the implications of general relativity to suggest that the universe, at least time and space, came into being because, in closed-universe solutions of its governing equation, time and space can be finite, which for Wheeler raises the question of what physics could, and for him should, be 'beyond' time; he concluded that a model for the explanation of existence, being capable of accounting for the boundaries of time, should be independent of time in itself.

One must also consider the obvious, simpler alternative explanation for its current existence: that it has always existed and persisted with time and nothing in physics requires it ever to not exist, but only perhaps be finite in physical measure in what is all of the physically well-defined past.

5 Closure of the Circuit: The Seven Questions

By placing a profound emphasis on the role of information in the structuring of reality through the operation of the 'meaning circuit', Wheeler had discarded his previous positions regarding physical ontology: By 1976, he had made "a clean break with this 'everything is fields' curved spacetime era" of physics, which was to be followed by the "everything is information" era ([31], p. 45); he argued that "With particles owing their definition and existence to fields, with fields owing their definition and existence to phases, with phases owing their definition and existence to distinguishability and complementarity, and with these features of nature going back for their origin to the demand for meaning, we have exposed to view (at least in broad outline) the main features of the underground portion of the model of existence as a meaning circuit

[15] And, as seen in Sect. 3 above, Wheeler was ultimately to back away from any necessary role for consciousness in measurement itself.

closed by observer-participancy" ([3], p. 309) reviewed here in Sect. 2. Despite being confident that with the MCH he had found a way of achieving the outlines of the ultimate physics, Wheeler still did recognize a number of issues to be addressed in order to make progress in filling out the picture.

Wheeler formulated seven questions for approaching the issues raised by his model, which for the most part differ in their specificity from those raised in Sects. 3 and 4 above.

> Before this would-be model can ever rise to the status of a proper model and be subject to quantitative analysis, some questions and difficulties require clarification: (1) What reality does the model ascribe to the physical world before the advent of any meaning-making community? (2) In what respect does it differ totally from the familiar anthropic principle on the issue of why the dimensionless constants of nature have the values they do? (3) What are we to understand by such a term as the community character of meaning? (4) What is the status of an elementary quantum phenomenon that is not put to use in the establishment of meaning? (5) How are we to reconcile the continuum of the world of physics with the discrete yes-or-no character of elementary quantum phenomena? (6) How can we ever hope to quantify meaning? Finally, (7) why a meaning circuit? Why any closed loop at all? ([3], p. 309)

Wheeler's answers to most of these questions, in various forms in which he stated them, are now critically considered.

5.1 Reality and the Status of the Past

The first of the above questions, "(1) What reality does the model ascribe to the physical world before the advent of any meaning-making community?" addresses the issue of the extent to which this participatory model of physics requires *subjectivity* (briefly discussed also in Sect. 3 above). This question arises because it is suggested by the meaning circuit model that reality depends essentially on the attribution of meaning by communicating subjects: Indeed, according to the MCH, reality is ultimately a theoretical construct built from currently available records due to the element of "observer participancy" inherent in quantum phenomena: "...through the quantum-level questions we put to nature we are participators in the making of what we call reality" [3].[16] An example offered as demonstration of this was the

[16] Wheeler makes use of the term *reality* in his writings without provide a fixed definition, but he does refer his readers to books of Bernard d'Espagnat for "a survey of many sides of the quantum principle as one knows it today" and "what one already knows from quantum mechanics" about its central notions such as non-separability [18], p. 599; d'Espagnat discusses the relationship of physics to existence and summarizes it a way consistent with Wheeler's usage of the term. In his *In Search of Reality*, d'Espagnat writes "As regards physics, it seems quite clear that its present stage of development is sufficient to justify considering it to be the universal natural science—a science of that very 'nature' that, apparently at least, it seems appropriate to identify with reality. Even though this latter view should be corrected to some extent...it is certainly quite a good working assumption" [40], p. 2. As shown here in previous sections, various statements by Wheeler suggest that his view of the realm of science as reducible to physics, albeit one which is psycho-physical.

delayed-choice experiment discussed in Sect. 4.1, above. Wheeler argued that "the respect in which what we ask, or do, in the present has an inescapable consequence for what we have the right to say about the past—even a past long before life. This is the sense in which (through the quantum-level questions we put to nature) we are participators in the making of what we call reality" [3]. This he indicated with the phrase 'observer-participator as definer of reality' ([2], p. 7), where "the 'observer,' far from being isolated from the surrounding world except insofar as he opens upon it one-way windows of perception, is in a certain sense coextensive with it" ([18], p. 695).[17] In particular, Wheeler suggests a positive answer to the question "Is the architecture of existence such that only through 'observership' does the universe have a way to come into being?" ([2], p. 7).

In "How Come the Quantum" [3], the question of whether the participatory activity described by the MC model endows reality to anything other than the present; in particular, its relation to the past is engaged. "If life and mind and meaning are so important in the scheme of things, then what is the status of the past?" For him, according to a strict adherence to quantum theory, the physical world of earlier eras of the universe—the present local effects of which being what are understood by astronomers as what is perceived at our telescopes—are at least equally capable of evidentiary establishment in the present as some aspects of the present. "Through the photons that reach the telescope, we see more clearly a quasar event of six billion years ago (before there was any life anywhere) than we can perceive the three-encyclopedia-long sequence of bits in our own DNA, in the here and now" ([3], p. 309); according to the MC hypothesis if the past exists, and exists only "in the records of the present ...then the past ranks no lower and no higher than the rest of what we call existence...reality is theory" [3].[18] Thus, according to the MCH the past is constructed in a significant sense by phenomena of the present. "Not until the one or the other of the appropriately paired photodetectors (the yes-counter and the no-counter) clicks have we distinguished an answer. Not until then do we have the right to attribute a polarization to the received photon" ([3], p. 309).

Wheeler argued that "There is no more remarkable feature of this quantum world than the strange coupling it brings about between future and past, between (1) the way one will choose to orient his polarization analyzers two years down the road and (2) what one can then say about the photons that already—as judged by us now—had their genesis 3 years ago. There is a sense in which the polarization of those photons, already on their way, 'was' brought into being by the disposition of the polarization analyzers that the observer has yet to make and will make. Moreover there is no difference in principle whether the time between the emission of a photon and its

[17] The manner in which this understood to take place is discussed in Sect. 3 above. Note that it would have been far less controversial for Wheeler to have written "...definer of physics" or "...definer of our physical world view" rather than "...definer of reality", but it appears that the goal of the MCH is exactly to transcend traditional physics to provide a basis for all of existence so far as it is empirically knowable. Nonetheless, Wheeler also argued that the central role played by the *community* of observer-participators precludes solipsism, as discussed Sect. 5.2 below.

[18] "The past is not really the past until it has been measured. Or put another way, the past has no meaning or existence unless it exists in a record in the present" ([7], pp. 67–68).

detection is 5 years; or a nanosecond, as it is when one is looking at an object a foot away; or 10^{10} years, as it is when one is receiving direct or indirect radiative evidence of what went on in the first few seconds after the big bang. In each case what one chooses to ask (as for example by the setting of the polarization analyzer) forms an inseparable part of a phenomenon that in earlier thinking one would have said had 'already happened'. In this sense it is incontrovertible that the observer is participator in genesis" ([2], pp. 24–25).

The genesis of the polarization value at the moment of a photon's measurement along a polarization basis in which it was not previously prepared surely does constitute the definition of a physical property *value* at that moment, but retrodicting from such a quantum measurement a *determinate value* of that property in the past is questionable. The significance of genuine *observer*-participancy is for *directly observed quantum* events rather than those of which records were created long ago by natural information recording systems, for example, environments creating 'microfossils,' so to speak, that could do work similar to deep-space probes, say, that land on an asteroid made of material that is billions of years old. The only clear sense in quantum theory in which a physical system may be understood to come into existence is for a quantity on the grounds of which its existence is considered to depend to become definite. However, this is not the case for classically describable systems, only those requiring quantum theory to be describable *at all*, for example, field quanta upon a number-operator measurement of the field.

Of course all 'observations of the past' about which direct observation is unavailable are indirect by definition; no matter the physics involved, they require inference from direct observations of records *in the physical "here and now."* Moreover, those observations in quantum theory considered above regard only system properties, not the very existence of systems observed, and involve only the inherent limitations on the co-measurement of properties at physical scales that require quantum theory to be most precisely described. In such circumstances, the evidence regarding new observations that is available at a space-time distance from their preparatory circumstances depends on physical circumstances local to their direct observation, and the relevance of the biological or cognitive aspect to this is entirely speculative.[19] It is important to recognize that, for past measurements—for example, from deep-space probes such as the Voyagers or other, possible deep penetrating probes from other, older civilizations elsewhere in the universe—which have produced robust physical records transmitted away as readily detected classical signals, current quantum considerations are irrelevant, just as books written in Caesar's time about the color of his clothing would have no effect on their color but are only records that allow us, now, to obtain knowledge of it. As argued above, by talking about quantum measurement as he did, Wheeler invoked a reductionist view of macroscopic systems in which the properties physical records are describable on the basis of their micro-composition,

[19] As one commentator on Wheeler's model in relation to biology put it, "Wheeler's dictum, as applied to the biological realm should therefore read more as 'it from bit from it', where lower of levels of matter dictate the informational state of a system which then in turn dictates its future evolution" ([38], p. 264).

something which is not justified for all but perhaps the tiniest fraction of highly unusual composition.[20]

The notion of observer-participation is an even more radical one if the observing subject is required to contribute something non-physical to the phenomena or to the observational evidence beyond the framing of the observation via the physical circumstances of observation and measurement. And, quantum phenomena always arise under clear physical circumstances. Moreover, Wheeler recognizes that physical records can be destroyed without being communicated among a community in order to be used as the ground of its knowledge. Use of the term 'records' suggests that the phenomena involved in *their* observation are not ephemeral but are understood to be the result of some previous real, completed process the facts about which different subjects could, in principle, have agreed had they (incidentally) observed them. It is also notable that the meaning circuit involves the establishment of physical records "above ground" and that this portion of the circuit is continuously revisited as the history of the universe becomes established in these records; cf. [24]. Wheeler appears to question the adequacy of our standard physicalist theories of autonomous, non-cognitive macroscopic systems that will have recorded the past, yet he equivocates with regard to significance of physical records. This equivocation is mirrored by his ambivalent position toward the necessity of consciousness or living beings in observation-participation despite having the physical portion of the circuit 'above ground' in the model being supported 'from below' by the cognitive portion ('below ground'). Unlike Bohr, who consistently emphasized the role of classicality in the interpretation of quantum theory, Wheeler often only obliquely recognizes the need for the independent existence of the world outside the mind, in that it is needed to supplying signals bearing information that if properly processed further could serve as knowledge about the world; how does the universe's first "bit" arise if there is no signal-bearer, one must ask.

In the quantum realm, any observer plays a role as a physical system in the determination only of property *values* not of the existence of measured systems themselves. Although there is indeed no question in quantum theory of the physical pre-determination of future values of quantum-system properties—in the absence of knowledge of previous state preparation—information related to a system's past is to be found only in a stable, classical record of property values obtained in preparatory experimental measurement as well as current measurements; these jointly allow one to obtain information about system properties and their history. In the absence of information as to previous state preparation one must assume a fully mixed state preparation, and a current measurement provides information only about the current state of the measured system. Natural preparatory environments, such as the cosmic microwave background, do exist naturally in deep space and obtaining such classical records provides valuable information that can be used in conjunction with more recent measurements to understand the history of a quantum system. But, equally if not more certainly, the world itself must already be present for any record to be pro-

[20] The reduction of classical physics to quantum physics has proven problematic and has not been accomplished, if it ever will be; cf., e.g. [39].

duced, that is, for signals providing information regarding property values to reach the the physical correspondent of the cognitive apparatus of any observer indicating the determinacy of any property upon proper attention. Thus, observers obtain information about pre-existing physical objects rather than bringing them into existence; state preparation is not a matter of system genesis but rather state determination. That the existence of a beam (say, of a mole of photons) is the immediately result of its detection or, more to the point, the genesis of the (say, electromagnetic) field *itself* by such measurement is more than doubtful. Here, one must ask of *what* is the purported 'genesis'? The answer is property values and information about them, not of physical systems.

Wheeler argued that humanity's theory of existence will be limited so long as the history of the universe is still, so to speak, being constructed from communicated experience: "If space is closed, if—following on the present phase of expansion—the system of galaxies contracts, if temperatures rise, all in line with the best known Friedmann cosmology, and if life wins all, then the number of bits of information being exchanged per second can be expected to rise enormously compared to that number rate today. The total count of bits …And how great must that future total be—tally as it is of times past—to furnish enough iron posts of observation to bear the smooth plaster which we of today call existence?" ([1], p. 126). Note that, in itself, this could instead be understood as simply emphasizing limitations of the ability of the collective evidence of science to determine the theory and the incidental particulars of the history of the world and, in a sense, it is just that; the question of whether the MC model ascribes reality to the physical world *before* the advent of any meaning-making community turns on the sense of reality involved and, so, is ambiguous exactly because the model involves both an epistemic aspect, which is "below ground," and an ontological aspect, which is "above ground," which Wheeler consistently fails to distinguish clearly.

A clearer representation of what is involved in the development of the physical world picture which is still largely in accordance with the quantum aspect of the MCH would be not a simple closed circuit but rather an *ever-expanding spiral*, where the physical world with its physical ontology induces the data—the basis of human knowledge—delivered to the community of scientists via measurement records that continually inform all of the natural sciences. However, this would also swap the 'below ground' with the 'above ground' of Wheeler's model and, thereby, be at odds with his view of information as primary. Wheeler's reductive thesis that "we know that every bit of information, every item of sight or touch or sound, goes back in the last analysis for its transmission to elementary quantum phenomena" is unjustified, in that: (i) at best, it only as regards *information about properties*, not the very existence physical systems, and (ii) such elementary phenomena are insignificant for any system that can be adequately described without the need of quantum theory: a distant quasar will not change in any existential way because it has been seen for the first time by telescope, only our evidence about it will become accessible and might influence our cosmological *theories*.

5.2 Measurement and Knowledge

"Question three: Are there then as many pasts, as many presents, and as many futures as there are observer–participators? A proposal so extravagant overlooks the community character of everything we call knowledge. Already at the quantum level, Niels Bohr warns us ...that no measurement is truly a measurement unless the result can be communicated from one person to another 'in plain language.' In a wider context, we know that meaning itself would be impossible without communication. There is not a word we speak, a concept we possess, or an idea we conceive that is not rooted in the larger community" ([3], p. 310). With this, Wheeler makes clear that, at a minimum, intersubjective comparability of observations and measurements is essential to the MC model. Despite his emphasis on the contingent nature of quantum-physical events, no matter how far back in time their preparatory development may be considered to have been initiated, following Bohr, Wheeler required the communicability of observations and measurement results, placing a clear constraint on the subjectivity of scientific data.

However, Wheeler went beyond requiring intersubjective communicability of data by attributing the knower–scientist the role of direct participant in the actualization of what his/her senses reflect of physical properties; he held these data to be 'fence-posts of reality': "...Curie tells us that physics deals with things only—not people. Today, the quantum forces on us a different outlook. It tells us that existence is not a deterministic machine grinding away out there. It is senseless for the uninvolved to try to speak *in abstracto* of what is happening out there at the microscopic level. Involvement is essential: observer-participancy. Not until a choice has been made and not until one or another complementary question has been posed can there be an answer" [3], p. 311. But, again, the focus is on phenomena for which quantum theory is *necessary* to their description, as indicated by the term 'microscopic' here. Despite any ostensible role of the observer-measurer as participator, there is no reason to believe that any *essentially quantum* events, which would be (eventually) considered associated with the distant past but were *not yet already measured or recorded in nature* to be of any significance for the macroscopic-level description of the physical reality of the past universe.[21]

For Bohr, it was precisely the classicality of measurement records—that they do not require a quantum description—that enables the communication of physical values, that is, that makes it sensible to speak publicly about what is the case with an observed quantum system.[22] Although physics requires quantum theory at the atomic scale and below, it rarely needs it to describe and explain what is large in the various physical senses, certainly not to justify a belief in the existence the universe as a whole and its past. By contrast, before the late stages of consideration of the MC

---

[21] Regarding the meaning of "macroscopic" here, see [44].

[22] This is much like an example from Hilary Putnam: For a key to be understood, there is no need to have a ultimate fine-grained, physical explanation of its actions in the opening of a lock, because its geometry and, I would add its impenetrability, suffice to understand its propriety and ability to function.

model, Wheeler went far beyond requiring the establishment of physical records for the attribution of reality to what is measured: he included (at least proto-)conscious observation and communication between of results between (at least primitive) conscious entities. In a 1985 discussion, he commented that "Although this word conscious is a little tricky here because one can think of animals as having brains that are they are so primitive that they may not be so completely aware as you or I are, and, if some flash of light—some elementary quantum phenomenon—occurs in which they're able to respond to in some way, they meaning has come into being even if it has involved consciousness at a pretty low level. So I would not like to put stress on consciousness even though that is a significant element in this story" and that there is a distinction depending crucially on the difference between living and inanimate objects, with there being "a most difficult question of where we draw the line" between these [7], p. 64. Still, there is a vitalist element in this requirement of the presence of life for the production of a communicable record.

Wheeler finally retreated from the requirement of the presence of (proto-)consciousness by the late 1990s. In discussion of the 'self-excited circuit,' he withdrew any requirement that measurement-observation be performed by living beings. "By 'measurement' I do not mean an observation carried out by a human or a human-designed instrument-or by any extraterrestrial intelligence, or even by an ant or an amoeba. Life is not a necessary part of this equation. A measurement, in this context, is an irreversible act in which uncertainty collapses to certainty. It is the link between the quantum and classical worlds, the point where what might happen-multiple paths, interference patterns, spreading clouds of probability-is replaced by what does happen: some event in the classical world, whether the click of a counter, the activation of an optic nerve in someone's eye, or just the coalescence of a glob of matter triggered by a quantum event. The event that I am calling a measurement is what Niels Bohr called 'registration.' …What the particle might have done-or, in a quantum sense, all the things it is doing simultaneously-is replaced by what it did in fact do. Not all potentiality is converted to actuality in any finite time." ([41], pp. 336–337). It is particularly important to note, in this regard, that potentiality can be understood as a mode of existence that is independent of thought; cf. [26]. In this distancing of himself from the requirement that life and consciousness be present for quantum actualization, Wheeler also invoked Heisenberg's thread of the Copenhagen perspective; cf. [26]. This raises the question of the extent to which the existence of his participatory universe requires participation in the sense described by the truly closed, circuit model of the original MCH.

Early on in its development, Wheeler had offered a nearly vitalist circuit metaphor to describe the role of the 'participator': that it acts like the electricity in an electric motor, the point being that it is essential to the functioning of the universe: "the universe would be nothing without observership as surely as a motor would be dead without electricity" [2], p. 21. Connecting this view with experiment, he argued that the study of quantum observership "has the equivalent of Franklin's kite and key experiment in the Einstein–Podolsky–Rosen experiment" [2], p. 21. It is also notable that electricity is just as physical as the bulk-material motor it causes to move. And the choice of measurement to be made is still tantamount to a physical configuration of measurement apparatus. It is illuminating to consider Wheeler's "U" icon, that is, the relation of this to his final views on cosmogony, "The point is that the universe is a grand synthesis, putting itself together all the time as a whole. Its history is not a history as we usually conceive history. It is not one thing happening after another after another. It is a totality in which what happens 'now' gives reality to what happened 'then,' perhaps even determines what happened 'then'." [41], p. 337. Again, while this may be the case regarding current events, and perhaps inferences regarding the past of a system between preparation and measurement in special circumstances in which state-measurement/preparation basis and measurement basis agree, it regards only quantum-level facts and, even then, involves a suspect mode of retrodiction.

In later discussions, Wheeler further related his model and quantum 'smokiness' to the perspectives of Heisenberg and Bohr and offered a less radical view of *existence* while still holding information as fundamental to it. "Measurement, the act of turning potentiality into actuality, is an act of choice, choice among possible outcomes. After the measurement, there are roads not taken. Before the measurement, all roads are possible-one can even say that all roads are being taken at once. ...The laws of physics tell us only what may happen. Actual measurement tells us what is happening (or what did happen). Despite this difference, it is not unreasonable to imagine that information sits at the core of physics, just as it sits at the core of a computer. ...Niels Bohr wrestled for most of his life with the question of how acts of measurement (or 'registration') may affect reality. It is registration—whether by a person or a device or a piece of mica (anything that can preserve a record)—that changes potentiality into actuality. I build only a little on the structure of Bohr's thinking when I suggest that we may never understand this strange thing, the quantum, until we understand how information may underlie reality. Information may not be just what we learn about the world. It may be what makes the world" ([41], pp. 338–340).

Nonetheless, one must recognize that communicable physical facts are not established only, or even primarily at the microscopic level; the subtle aspect of the universe in which quantum state actualization is significant only where quantum descriptions are necessary, which at large scales are only those relatively rare situations in which such small signal-capable objects have effects are very dramatically amplified. Some have considered a possible "initial quantum fluctuation" that might have been amplified at the very beginning of the universe, but the magnitudes of the quantities such as mass-energy that would have to have been present, assuming the continual validity of conservation laws, would have been extremely large. An extreme amplification—and a 'mechanism' for this amplification, and by *what*—of an ur-signal would be required if such an explanation of the origins of the structure of the universe is to be considered at hand. Even then, it is difficult to see what this mechanism would have to do with cognition in any clearly specifiable manner.

5.3 Measurement and Information

Regarding the relationship between information and meaning, Wheeler posed "Question four: What significance, if any, are we to attribute to an elementary quantum phenomenon that is not put to use to establish evidence and meaning? Despite all that we know about measurement theory in the realm of the quantum, nothing is more puzzling than the linkage between the counter's click and the community that makes meaning" [3] and suggested as an example for consideration "a crystal of zinc sulfide thrown up by nature on the back side of the moon. It stops a cosmic-ray proton. Ten million photons emerge, which is an irreversible act of amplification if there ever was one. The signal, however, dissipates out into space. No use is made of it. ... the amplifying device consists of some 10^{22}–10^{25} particles coupled by a complex of electromagnetic interactions. ... Surely all over the universe, in regions out of sight, interactions are going on all the time and between particles that are stupendous in number compared to the count ... here" ([3], p. 311). This may appear problematic for autonomous physical measurement in that, even if irreversibly generated in the absence of life or consciousness, an amplified signal serving as a physical record of an event can be, practically speaking, unrecoverable and incapable of providing information, and so knowledge to any finite observer. Wheeler argued that, more generally, "There is no irreversible act of amplification that may not later be erased. Not unless it is put to use in the establishment of meaning does the elementary quantum phenomenon win any special status" (ibid.). However, it is not clear why any "special status" should be attributed to the mere occurrence of any given quantum-level physical event relative to any another, even in the Copenhagen understanding of physical observation and measurement, or why the provision of *meaning* to them need be relevant to the reality of anything physical.

Although there is no measurement record that cannot be degraded or ignored in principle, this "degradation" is only epistemically significant; it is not intrinsically *physically* significant. Wheeler did view 'meaningful' observation and the mere creation of a quantum phenomena as distinct: the "vital distinction between the elementary quantum phenomenon ... on the one hand and the putting of that observation, that elementary quantum phenomenon, to use in the establishment of meaning ... and it's not enough for just one observer to put it to use—you need a community" ([7], p. 63). Very well. But this relates to the meaningfulness of the results to the community in question, not to the existence of the phenomena that might influence its theorizing. Moreover, any communicated meaning of any kind can be lost through the dissipation of the brains of the members of the community in question. Does the universe cease to be real if the human race ceases to exist?

Wheeler also argued that "the quantum ... tells us that existence is not a deterministic machine grinding away out there." Although that may be true for phenomena requiring quantum mechanics for their description, it is by no means the case that nature does not establish a manifest, independent history at large scales and over large times in the absence of living observers or communities thereof, only that the past of essentially quantum systems may have a different character from those well described by classical mechanics. The perspective of the renowned information-theorist Rolf Landauer, to whom Wheeler often referred during the development of the MCH, provides a clear view of the relation between quantum measurement and information. Landauer held that "Information, numerical or otherwise, is not an abstraction, but is inevitably tied to a physical representation. There really is no software, in the strict sense of disembodied information, but only inactive and relatively static hardware. Thus, the handling of information is inevitably tied to the physical universe, its content and its laws" [42]; he viewed the primary significance of the elements of Wheeler's circuit to be that they allow for a consistent picture of *information processing* involving physical signals.

> Wheeler has given us ...the "meaning circuit". In this, physics gives rise to all the phenomena whose study has usually occupied us. ...Wheeler reminds us that the discovery of laws is, or involves, quantum mechanical measurement. The results of measurement are not independent of the act of measurement, and the outcome of the measurement was not there, all along, to be revealed in its finally measured form. If I can be presumptuous enough to differ from Wheeler, it is related to his stress on human observers and their posing of questions. Measurement does not require intelligent beings, the environment is continually acting on a system, and making measurements. [42]

Only, it is better said in the above that '...the discovery of laws may involve quantum mechanical measurement', for many valid physical laws have been established without the need for specifically quantum measurements, however much the smallest structures within measuring apparatus would be amenable to quantum mechanical description were they in isolation.

5.4 The Meaning Circuit

The question of the closed nature of the 'meaning circuit' which relates to the issues discussed above is addressed with the final question:

> Question seven: Why speak of a circuit? Why not seek instead for a foundation? There is an old legend about the foundation that supports the world: Our globe rests on the back of a great elephant. The elephant stands on the back of a giant tortoise. ..."And what does it stand on," she inquires. The lecturer's reply is famous: 'Tortoises, madam; on and on, nothing but tortoises.' How different is the account we give today of the foundation of existence? Matter is built on molecules. The molecule is built on atoms. The atom has a nucleus. The nucleus is built on nucleons. The nucleon is built on quarks. The quark is built on fields. The field is built on geometry of one or another dimension. ...Is there ever to be an end? How can there be an end if we ask always for foundation of foundation of foundation. . .? No escape is evident from this view of worlds without end-or tortoises without end-except in a line of influence that closes on itself, that forms a loop, that makes a circuit. No model for such a loop is available to us today except one of information-theoretic character, the model of existence as a meaning circuit. [3]

Wheeler points out here that an approach to physics where theories, even if they are each approximately capable of reduction to another, is unsatisfactory when each new, ostensibly more fundamental theory requires a yet more fundamental one, which he saw as being the case so far in the history of physics. This critique focuses specifically on the seemingly endless search for deeper ontic levels in physics.

Wheeler then argued that a potential logical infinite regress this history suggests can be avoided by adaptation of a circuit-like schema, such as that he offers on the basis of the MCH. Of course, there is an alternative solution which is a natural one given what has become increasingly clear to experimental physics over the past century, namely, that there is an inherent physical limit to the ability to probe for data providing information about any possible "deeper" structure, anyway: to abandoning reductionist absolutism and to accept accurate theories appropriate to each new physical realm.[23] In fact, in the current era of physics, theorizing about the ontology of the smallest spatial scales is proving extremely challenging, this challenge being posed primarily by an apparently inevitable dearth of empirical input. Given that probing small-scale structure requires ever greater energies, our capacity to probe is limited by the availability of local sources of energy to be harnessed, even by the universe itself through astrophysical phenomena. Theorizing may soon find itself largely incapable of experimental verification, something already manifest with regard to string theory; this far more conservative position is at least as satisfactory philosophically if not so to the curiosity of physicists. Indeed, Wheeler's question only arises from the acceptance of a reductionist imperative. Discovering physical structure is scientifically valuable, but demanding always to find deeper levels of it, as opposed to simply understanding it always better, is another matter.

In his discussions of the novel element of his 'meaning circuit,' Wheeler often turned "to the role of observation in defining meaning, that central topic in the philosophical thought of our time, we have these words of Føllesdal (1975) in the great Oxford Lectures on Mind and Language, 'Meaning is the joint product of all the evidence that is available to those who communicate'." ([2], p. 25). But, unlike information as clearly related to physics, the question of meaning goes beyond the scope of the problem at hand, namely, the basic question of the nature of physical reality. And, ultimately, Wheeler even decided to "steer clear of the issues connected with 'consciousness.' The line between the unconscious and the conscious begins to fade in our day as computers evolve and develop—as mathematics has—level upon level upon level of logical structure. We may someday have to enlarge the scope of what we mean by a 'who'. This granted, we continue to accept—as essential part of the concept of it from bit—Føllesdal's guideline" ([43], p. 320).[24]

[23] Consider, for example, Heisenberg's position regarding elementary particles: "Because particle number is not conserved in high-energy interactions it may be meaningless to ask about the constituent parts of elementary particles; perhaps the central question is dynamics" [45]. There are other reasons besides lack of particle-number conservation per se for holding this that are also evident in the results of quantum field theory, including "virtual-particle clouds" [46].

[24] Others have recently continued pushing on with ways of continuing the meaning circuit program beyond physics proper; cf., e.g., [47].

6 Conclusion

The development and content of John Wheeler's Meaning Circuit Hypothesis is discussed above and a number of significant difficulties with it are found. The most suspect of its propositions is that existence itself, not simply the precise determination of physical properties of micro-systems, is contingent on the participation of observer-participators or communities thereof. This is seen to follow from Wheeler's particular, unusual form of tacit reductionism in which quantum theory is universally required and existence itself is reduced to 'meaningful' communicated information corresponding to individual quantum phenomena where, however, the connection to meaning is never actually provided in any specificity.

When considering this grand attempt to find an explanation of the origin of existence in the novel aspect of quantum theory, the omnipresence of the quantum (of action), it is valuable to reconsider Einstein's rhetorical question, 'Is the moon not there when nobody looks?' Physical experience demonstrates that neither the moon nor the history of anything classically describable depends on any community of communicators or their observations to exist; there is now an empirically well-supported history of the four-billion Earth–Moon system which theory—the existence and details of which of course are based on human-collected data and theory from non-quantum mechanical observations of pre-existing physical objects not built of elementary quantum phenomena involving observers in any precisely demonstrable sense. Explaining the presence of the moon when no-one looks involves no quantum theory whatsoever, and the the Moon did not come into existence with the appearance of astronomers of even the most primitive level; to expect that the universe itself came into existence, say, when the cosmic microwave background was discovered, is even less plausible. Based on the evidence and arguments presented in his various writings on the topic, Wheeler's Meaning Circuit Hypothesis is better understood as an explanation of how quantum *theory* arose and is validated rather than one of how the physical *universe* arose.

Better service to physics is done by pursuing a model containing the elements of the MCH that emphasizes the role of information in the development of our scientific understanding of the world, to which the physical theorizing of Wheeler and his students have clearly and definitively contributed, rather than suggesting a reduction of physics to information or knowledge, particularly considering the speculative leap involved in connecting elementary quantum phenomena directly to cognition. The result of this clarification corresponds not to a 'circuit,' but to an ever-expanding spiral of history and knowledge creation and verification of the character of that history, where evidence continually informs all of the natural sciences and more facts about the world of the essentially quantum are accumulated, a spiral that first began in the absence of any living observer with the beginning of time but has now accelerated with the advancement of science ever more deeply and broadly in the quantum realm.

References

1. J.A. Wheeler, World as system self-synthesized by quantum networking, in *Probability in the Sciences*, ed. by E. Agazzi (1988), pp. 103–129
2. J.A. Wheeler, Genesis and observership, in *Foundational Problems in the Special Sciences*, ed. by R.E. Butts, J. Hintikka (Dordrecht, Holland 1977), pp. 3–39
3. J.A. Wheeler, How come the quantum? Ann. NY Acad. Sci. **480**, 304–316 (1986)
4. G. Jaeger, Information and the reconstruction of quantum physics. Ann. Phys. **531**, 1800097 (2018)
5. C. Rovelli, Relational quantum mechanics. Int. J. Theor. Phys. **35**, 1673–1678 (1996)
6. B.C. Van Fraassen, Rovelli's world. Found. Phys. **40**, 390–417 (2010)
7. P.C.W. Davies, J.R. Brown, *The Ghost in the Atom* (Cambridge University Press, Cambridge, UK, 1986)
8. G. Jaeger, *Quantum Information: An Overview* (Springer, New York, USA 2007)
9. G. Jaeger, Uncertainty relations and possible experience. Mathematics **4**, 40 (2016)
10. G. Jaeger, *Quantum Objects* (Springer, Germany, Heidelberg, 2017)
11. D. Rickles, Geon wheeler. Eur. Phys. J. H **43**, 243–265 (2018)
12. J. Bub, Why the quantum? Stud. Hist. Phil. Mod. Phys. **35**, 241–266 (2004)
13. A. Aguirre, A. Foster, Z. Merali (eds.), *It From Bit or Bit From It?* (Springer, 2015)
14. G. Jaeger, Overcoming conceptual imprecision in quantum measurement theory: measurement and macroscopicity, in *Quantum Nonlocality and Reality*. ed. by M. Bell, S. Gao (Cambridge University Press, Cambridge, UK, 2016)
15. J.A. Wheeler, Beyond the black hole, in *Some Strangeness in the Proportion*. ed. by H. Woolf (Addison-Wesley, Reading, MA, 1980), pp.341–375
16. J.A. Wheeler, Beyond the End of Time, Marchon lecture, University of Newcastle-Upon-Tyne, May 18, and Nuffield lecture, Cambridge University, July 19, 1971; updated in chapter of same title, in C.W. Misner, K.S. Thorne, J.A. Wheeler (eds.), Gravitation, vol. 1973 (San Francisco, Freeman and Co., 1971), pp. 1196–1217
17. J.A. Wheeler, From relativity to mutability, in *The Physicist's Conception of Nature* (D. Reidel Publishing Company, Dordrecht, Holland, 1973), pp. 202–247
18. J.A. Wheeler, Is physics legislated by cosmogony? in *Quantum Gravity; Proceedings of the Oxford Symposium, Harwell, Berks, England, February 15, 16, 1974* (Clarendon Press, Oxford, UK, 1975), pp. 538–605
19. J.A. Wheeler, Include the observer in the wave function? in *Quantum Mechanics, A Half Century Later* ed. by J.L. Lopes, M. Paty M. Episteme (A Series in the Foundational, Methodological, Philosophical, Psychological, Sociological and Political Aspects of the Sciences, Pure and Applied), vol 5 (Springer, Dordrecht, 1977)
20. J. Von Neumann, *Mathematical Foundations of Quantum Mechanics* (Princeton University Press, Princeton, NJ, 1955). (English translation of *Mathematische Grundlagen der Quantenmechanik* (Springer, Berlin 1932)
21. J.A. Wheeler, Frontiers of time, in *Problems in the Foundations of Physics* ed. by G.T. Di Francia (North-Holland, Amsterdam, 1979), pp. 395–492
22. J.A. Wheeler, Physics as a meaning circuit: three problems, in *Frontiers of Nonequilibrium Statistical Physics*. ed. by M. Scully (Plenum Press, New York, NY, 1986), pp.25–32
23. J.A. Wheeler, W. Zurek (eds.), *Quantum Theory and Measurement* (Princeton University Press, Princeton, NJ, 1983)
24. G. Jaeger, Quantum contextuality and indeterminacy. Entropy **22**, 867 (2020)
25. A. Plotnitsky, *Epistemology and Probability* (Springer, Heidelberg, 2010)
26. G. Jaeger, Quantum potentiality revisited. Phil. Trans. R. Soc. A **375**, 20160390 (2017)
27. J.A. Wheeler, Law without law, in [23]
28. J.A. Wheeler, On recognizing 'Law without law' Am. J. Phys. **51**, 398–404 (1983)
29. J.A. Wheeler, Sakharov revisited; 'It from bit', in *Moscow 1991, Proceedings, Sakharov memorial lectures in physics*, vol. 2, Contribution to: *The First International A.D. Sakharov Conference on Physics* (1991), pp. 751–769

30. H. Reichenbach, *Philosophic Foundations of Quantum Mechanics Berkeley* (University of California Press, CA, 1944)
31. C.W. Misner, K.S. Thorne, W.H. Zurek, John Wheeler, relativity, and quantum information. Phys. Today (2009). S-0031-9228-0904-030-7
32. G. Jaeger, A. Shimony, L. Vaidman, Two interferometric complementarities. Phys. Rev. A **51**, 54 (1995)
33. W. Miller, J.A. Wheeler, Delayed-choice experiments and bohr's elementary quantum phenomenon, in *Proceedings of the International Symposium Foundations of Quantum Mechanics* (Tokyo, 1984), pp. 140–152
34. A. Zeilinger, Why the Quantum? "It" from "Bit"? A Participatory Universe? Three far-reaching challenges from John Archibald Wheeler and their relation to experiment, in *Science and Ultimate Reality*. ed. by J.D. Barrow et al. (Cambridge University Press, Cambridge, UK, 2004), pp. 201–220
35. N. Boeuf, D. Branning, I. Chaperot, E. Dauler, S. Guerin, G. Jaeger, A. Muller, A.L. Migdall, Calculating characteristics of non-collinear phase matching in uniaxial and biaxial crystals. Opt. Eng. **9**, 1016–1024 (2000)
36. D. Simon, G. Jaeger, A.V. Sergienko, *Quantum Metrology, Imaging, and Communication* (Springer, Heidelberg, Germany, 2017)
37. G. Jaeger, Quantum contextuality in the Copenhagen approach. Phil. Trans. Roy. Soc. A **377**, 20190025 (2019)
38. S.I. Walker, Is life fundamental?, in *Questioning the Foundations of Physics* ed. by Aguirre et al. (Springer, Cham, Switzerland, 2015)
39. A. Bokulich, *Reexamining the Quantum-Classical Relation* (Cambridge University Press, Cambridge, UK, 2008)
40. B. D'Espagnat, *In Search of Reality* (Springer, New York, NY, USA, 1983)
41. J.A. Wheeler, K. Ford, *Geons, Black Holes, and Quantum Foam* (W. W. Norton, New York, 1998)
42. R. Landauer, Computation: a fundamental physical view. Phys. Scr. **35**, 88–95 (1987)
43. J.A. Wheeler, Information, physics, quantum: the search for links, in *3rd International Symposium Foundations of Quantum Mechanics* (Tokyo, 1989), pp. 354–368
44. G. Jaeger, What in the (quantum) world is macroscopic? Am. J. Phys. **82**, 896 (2014)
45. W. Heisenberg, The nature of elementary particles. Phys. Today 32–39 (1976)
46. G. Jaeger, Exchange forces in particle physics. Found. Phys. **51**(13) (2021)
47. A. Narasimhan, D. Chopra, M. Kafatos, The nature of the Heisenberg-von Neumann cut: enhanced orthodox interpretation of quantum mechanics. Act. Nerv. Super. **61**, 12–17 (2019)

Gregg Jaeger holds a Ph.D. in Physics from Boston University and B.Sc. degrees in Mathematics, Philosophy, and Physics from the University of Wisconsin-Madison. He is currently an Associate Professor of Natural Sciences and Mathematics at Boston University. His work focuses primarily on quantum theory. In addition to more than 70 journal articles in theoretical and experimental physics, history of science, philosophy, and mathematics, Prof. Jaeger has published three monographs in physics and the philosophy of science, edited two books and a number of conference proceedings in the foundations of quantum theory, and has written numerous book chapters in these various subjects.

50. H. Reichenbach, *Philosophic Foundations of Quantum Mechanics* (Berkeley: University of California Press, CA, 1944)
51. C.W. Misner, K.S. Thorne, W.H. Zurek, John Wheeler, relativity, and quantum information. Phys. Today [illegible]
52. [illegible]

Letters for Andrei: QBism and the Unfinished Nature of Nature

Christopher A. Fuchs

> The import of the difference between pragmatism and rationalism is now in sight throughout its whole extent. The essential contrast is that *for rationalism reality is ready-made and complete from all eternity, while for pragmatism it is still in the making, and awaits part of its complexion from the future*. On the one side the universe is absolutely secure, on the other it is still pursuing its adventures.
>
> — William James

> Are we not obliged to admit that more or less "measurement-like" processes are going on more or less all the time more or less everywhere?
>
> — John S. Bell, suitably reinterpreted for QBist purposes

1 Dedication

Never has there been such a friend to the open and liberal discussion of quantum foundations as Andrei Khrennikov. At his yearly meeting in Växjö—the longest running series of conferences on the subject in history!—one will find all kinds, from the crackpot to the philosopher to the Nobel prizewinner. Andrei's philosophy has always been that everyone is due respect in the forum: For everyone is united in wanting to crack the code of the quantum. This has built the most wonderful atmosphere, and everyone knows it—it is why so many of us come back from year to

C. A. Fuchs (✉)
Department of Physics, University of Massachusetts Boston, 100 Morrissey Boulevard, Boston, MA 02125, USA
e-mail: Christopher.Fuchs@umb.edu

A. Plotnitsky and E. Haven (eds.), *The Quantum-Like Revolution*,
https://doi.org/10.1007/978-3-031-12986-5_3

year, from the early years when Andrei could fund our flights, to the *many, many more years after that* when he could not! There is something about the whole experience of the meetings: from the long twilights and ubiquitous squawks of the seagulls, to the beautiful surrounding lakes of Småland, the thoughts that flow so spontaneously on the long walks between hotel and university, the deep and surprisingly coherent conversations till late at night in the Bishop's Arms, and the camaraderie of the banquets.

It is little known, and who would guess it, but those Växjö meetings and Andrei himself played significant roles in the development of QBism [1–3]. Andrei might even be called a reluctant godfather to QBism. It all started when he asked me to be a co-organizer for his 2001 meeting. I saw it as a watershed opportunity. Essentially *using* his generosity—"You are really American!," Andrei wrote me in an email—I gathered up quite a number of friends from the quantum-information world (Asher Peres, Danny Greenberger, David Mermin, Herb Bernstein, Carl Caves, Rüdiger Schack, Richard Jozsa, Ben Schumacher, John Smolin) to participate, as I felt that revolution was in the air: In the quantum-foundations world, it was just a matter of bringing some adults into the room—those who already understood that quantum states just are *information*. I figured that would fix everything [5]. But how little I was prepared for my own transformation by the end of the meeting!

What was up was that in the interstices, Caves, Schack, and I used the opportunity to finish writing our first "quantum Bayesian" paper, "Quantum Probabilities as Bayesian Probabilities" [4]. However as we did, I became and more and more suspicious of the consistency of our considerations. In the end, I believe we presented little more than a facade of the Bayesianism we had hoped to inject into quantum theory. In fact, what had gotten in the way of our consistency was the very word "information."

The aftermath was a gut-wrenching experience as I started to understand that the usual conception of information had too much of a "factive character" [6] to be consistent with our hoped-for banishment of quantum states from reality. Ultimately, the corrective was to abandon the road of Jaynes's objective Bayesianism with its foundation resting on Cox's axioms [7] and change course to de Finetti's radically subjective Bayesianism with its foundation of Dutch-book coherence [8]. Only that concept was nimble enough to remove the factive character of probabilities and consequently the factive character of quantum states. Quantum states were not information; they were beliefs [9]! That was the start of the long road to QBism [10, 11].

Here is the way I put the process in the Introduction to my largest email samizdat[1] [12]:

> The remediation of these problems required steps more radical than I ever imagined I could take. For instance, I came to recognize that even probability-1 assignments must be personalist Bayesian probabilities of a cloth with any other Bayesian probability. Moreover, I started to understand that not only quantum states, but the operators used for describing measuring

[1] Samizdat is Russian for an underground publication, but I use the term to denote my posted or published collections of emails. This usage is due to David Mermin.

> devices and all quantum time evolution maps as well, even Hamiltonian evolutions, must be viewed in strictly personalist Bayesian terms. If these steps were not taken, the view of quantum theory we had been striving to construct would simply topple from inconsistency. But it was no easy pill for even my most sympathetic colleagues and collaborators to swallow.

Nearly everyone I encountered thought I had gone off the deep end: *That I had completely given up on the concept of reality*. Certainly the participants at Andrei's meetings did. (For instance, after Ted Hänsch's talk on QBism during the 2015 meeting, Avshalom Elitzur stood up and declared angrily, "You cannot tell me that I am a figment of Chris Fuchs's imagination!")

Interestingly I saw the process in the opposite light: I saw the method of seeking absolute consistency between our interpretations of the quantum symbols and de Finetti's subjective notion of probability as a powerful method for homing in on an *ontology* to go with quantum theory—i.e., a positive statement about the character and contents of the world. That's not giving up on reality, but a way of protecting oneself from being seduced by false prophets—you know, like the prophets of nonlocality or pilot waves or the wavefunction of the universe.

On 9 October 2001, just four short months after my first Växjö meeting, I wrote to David Mermin:

> Collecting it up, it's hard to believe I've written this much in the little time since Växjö. I guess it's been an active time for me. I think there's no doubt I've gone through a phase transition. For all my Bayesian rhetoric in the last years, I simply had not realized the immense implications of holding fast to the view that "probabilities are subjective degrees of belief." Of course, one way to look at this revelation is that it is a *reductio ad absurdum* for the point of view—and that will certainly be the first thing the critics pick up on. But (you wouldn't have guessed less), I'm starting to view it as a godsend. For with this simple train of logic, one can immediately stamp out the potential reality or objectivity of any of a number of terms that might have clouded our vision. With so much dead weight removed, the little part left behind may finally have the strength to support an ontology.

In fact I posted one year's worth (229 pages) of all the painful thinking that went into my new radical stance in a mini-samizdat titled, *Quantum States: What the Hell Are They? The Post-Växjö Phase Transition* [13].

But who in their right mind would publish their emails!? Well, I tell my reasons for it in the Introduction to my first samizdat, *Notes on a Paulian Idea: Foundational, Historical, Anecdotal & Forward-Looking Thoughts on the Quantum* [14], posted as `arXiv:quant-ph/0105039` on the first anniversary of the Cerro Grande Fire that swept through Los Alamos, New Mexico when I was a postdoc there. But that brings us back to Andrei and his wonderful liberal ways.

The story goes that Andrei read the abstract of the posting, thought "this sounds interesting," and hit the print button. He hadn't noticed that the pdf was 504 pages long. When he later entered the printing room, he discovered that the printer couldn't handle the stack and there were disheveled pages all over the floor! Only then did he realize what he had gotten himself into. What happened next I'm not completely sure of, but it appears from my email record that while visiting Alexander Holevo and me at Bell Labs in October 2001, Andrei proposed to use his resources at Växjö University (now Linnaeus University) to print a number of paperback copies of the samizdat. There followed a printing of 100 copies, and then later 100 more [15]. Later still, I successfully approached Cambridge University Press with the idea of a hardbound edition of the book (*Coming of Age with Quantum Information: Notes on a Paulian Idea*, 2010, [16]), but I would have never reaped that reward or the eight excellent reviews the book received [17–24] without the confidence Andrei instilled in me through his original kindness.

Therein lies the paradox of Andrei, or what I called earlier his wonderful liberal ways. Andrei has never yet made the step to thinking of QBism as the most appropriate interpretation for *quantum physics*, but that has never stopped his support for those of us developing the idea. Witness the book mentioned above, but it goes deeper than that, as evidenced at every Växjö meeting since 2001. At Andrei's instigation (he would often call attention to QBism in his opening remarks), the debate would start right away. So many tomatoes were thrown at QBism for sure, but learning how to dodge them sharpened our concepts. Certainly QBism was never ignored. Compare that attitude with something the scholar Travis Norsen recently posted on social media, "Are there any serious people that take QBism seriously? It is so obviously empty and stupid. … [T]here is a real danger, I think, of providing undeserved oxygen to something that is, in fact, stupid and empty, by talking about it too much and too seriously." At most "philosophy of physics" conferences (so called), QBism is simply dead in the water.[2] But I think Andrei has taken a kind of pride in recognizing that there is something genuinely to be developed in QBism, even if it is adjacent to his own concerns. On 14 August 2003, he wrote me, "Hej Chris! I just like science as it is. … You know that many people have been in Växjö, but I really see that you have a huge potential for development."

So, how shall I tribute Andrei in this volume? With an email collection of course![3] But with what theme? This time it ought to be something big. Here goes.

One of the troubles of QBism's ontological program is that it is so sideways to the ways most run-of-the-mill philosophers of physics (say, Norsen, Tim Maudlin, Wayne Myrvold, David Albert, Michael Esfeld, and others of similar ilk) think, they don't even have the tools to parse its sentences. They simply can't see it as having to do with ontology at all. Maybe there is no remedy for this except to wait for the generation to die away. But QBism can and must move forward. To this end, I have tried to select some emails for the current collection that might inspire a young researcher to throw in on the QBist ontological project, to help develop it on its own terms. The road to a proper, detailed QBist ontology is sure to be a hard one, but one has to start somewhere.

---

[2] Thank God however that "standardized" philosophy of physics is not the only game in town. Recently a number of philosophers of a phenomenological bent have started a constructive dialogue with QBism. Witness for instance the forthcoming volume, *Phenomenology and QBism: New Approaches to Quantum Mechanics*, edited by Philipp Berghofer and Harald A. Wiltsche (Routledge, New York, 2022). In the meantime see: [25–36].

[3] It doesn't help that besides the publication of my book, Andrei also allowed me to submit two email collections to his conference proceedings [37, 38]—i.e., the precedent was essentially set by our very honoree!

Thus, please take these pages as loose notes or supplements to the more formal QBist writings touching on the ontological project in the past [1–3, 25, 32, 39–41]. At the very least, I hope these exchanges will give some sense of the ways QBists think in contrast to the ways those philosophers do who are happy enough with a "fake," as Schrödinger put it [42]:

> In an honest search for knowledge you quite often have to abide by ignorance for an indefinite period. Instead of filling a gap by guesswork, genuine science prefers to put up with it; and this, not so much from conscientious scruples about telling lies, as from the consideration that, however irksome the gap may be, its obliteration by a fake removes the urge to seek after a tenable answer. So efficiently may attention be diverted that the answer is missed even when, by good luck, it comes close at hand. The steadfastness in standing up to a *non liquet*, nay in appreciating it as a stimulus and a signpost to further quest, is a natural and indispensable disposition in the mind of a scientist. This in itself is apt to set him at variance with the religious aim of closing the picture, unless each of the two antagonistic attitudes, both legitimate for their respective purposes, is applied with prudence.

So, here is to you Andrei. Thank you for tolerating QBism's many *non liquets* past and never giving up hope that one day one of us will crack the code of the quantum.

2 5 Jan. 2009 *What I Want Out of Pauli-Fierz* (to H. C. von Baeyer & D. M. Appleby)

Thank[4] you both for the great New Year's reading gift of the last couple of days. It's funny how these things work out; it arrived at just the right time for me. For, I was able to put it in juxtaposition with my reading of the latest issue of *William James Studies*. The two together caused me to think very much about how I want to frame the coming year. And it struck me as worthwhile to try to record as clearly as I can at the moment what I am *personally* seeking to get out of a better knowledge of the byways explored by Pauli and Fierz (and potentially Jung). There is no implication in this that this is what you *should* be seeking as well—it is only an effort to tie the strands of my life together a little better.

Here is what it is really all about for me—it is the root of the root and the bud of the bud, as E. E. Cummings. A good vehicle for setting up what I want to say is one of Hans's passages:

> Pauli summarily dismissed two extreme attitudes – total separation of science from religion, and complete surrender to mystical experience. The former approach was advocated in our times by the late paleontologist Stephen Jay Gould, who coined the phrase "nonoverlapping magisteria" for the respectful noninterference of the realms of nature and of morality, of what is and what should be. Discussing a book by a German physicist who also believed in the separateness of science and religion, Pauli had once written to Fierz that he was appalled, and continuing:

[4] This essay is a condensation of [12, pp. 1631–1641].

> "[The book] is a reversion to the 19th century when religion and science lived in separate sections of the human soul – politely exchanging greetings at a distance, while continually reassuring each other that they had nothing to do with each other – and when the soul seemed to reside outside the boundaries of science."

> For Pauli it was obvious that science should be able to deal with the soul, and that the soul in turn can inform science.

in conjunction with one of Marcus's comments thereafter:

> It is certainly true that I attach more weight to the opus than I do to quantum mechanics. That is I am interested in quantum mechanics because I am interested in the opus, rather than the other way round. For Chris it is, perhaps, a little different. Though I don't think it is all that different, actually. It is true that Chris doesn't use the language of souls (at least I can't recall him doing so in my hearing). But he is deeply concerned with, for example, the question of human freedom. He will have to speak for himself, but I believe that for him too his interest in quantum mechanics is secondary to other considerations.

For it is true that I rarely speak of "souls," "religion," or "redemption." These terms are mostly dead terms for me—they don't stir my soul, so to speak—or maybe I simply don't understand them well enough yet to see their ultimate usefulness for what I *do* want to get at. (Much like I have never understood what the search for "elegance" can possibly mean when it comes to forming physical theories, say, as a criterion for string theory: it is a term that is dead to me.) It is maybe in this way, or more carefully, *in this detail*, that I part company from your tentative feelings on the opus.

Nonetheless, there is no doubt that I believe there is a place—a very important place—for humanistic concerns within physics proper. It seems to me it goes to the core of what quantum mechanics is trying to tell us. You'll find the point made over and over in my "Activating Observer" resource-material document, which both of you have versions of. But I thought the point was made very nicely in the *setting of pragmatism more generally* in this article that I was reading at the time your emails arrived—" 'The Many and the One' and the Problem of Two Minds Perceiving the Same Thing" by Mark Moller in Vol. 3 of *William James Studies* [43]:

> Each of these claims about reality is crucial to James's attempt to offer an alternative to the metaphysical theories of the absolute idealists. The importance of the claim that reality is continuous and in flux is that it offers an alternative to the absolute idealists' view that the universe is "known by one [infinite] knower in one act, with every feature preserved, and every relation apprehended." This means that for the absolute idealists, the universe is forever fixed so as to make real change impossible. We make no difference in such a universe. We neither improve upon it through our efforts nor make it worse. James rejects this view completely. His aim is to argue for a conception of the universe that allows real change to occur in it and where our efforts have a role in causing it. He goes on in the passage from *The Many and the One* manuscripts quoted above to make his point:
>
>> This picture of the irremediably pluralistic evolution of things, achieving unity by experimental methods, and getting it in different shapes and degrees and in general only as a last result, is what has made me give to my volume the title of *The Many and the One.*

> According to James, we, as conscious agents in the universe, have an active role in introducing new content and unity into it. Such a view thus aligns his radical empiricism with his meliorism. In earlier essays, eventually published together as *The Will to Believe* (1897), and in lectures that he gave to teachers, eventually published as *Talks to Teachers on Psychology* (1899), James took the position against the absolute idealists that the ultimate fate of the universe has yet to be decided. He insisted that it is an open question as to whether evil will triumph over good or the other way around. This, in turn, led him to claim that our choices and actions do make a difference in the universe, and, in fact, a crucial one. They help to decide how "the everlasting battle of the powers of the light with those of darkness" will turn out. This melioristic attitude only makes sense if the universe is malleable to human action, and, thus, it is one of James's aims in developing his metaphysics to explain how this malleability is possible.

The very phrases "our choices and actions do make a difference in the universe" and "this melioristic attitude only makes sense if the universe is malleable to human action" mean outright that there must be room for a humanistic element of some sort within physics itself.

Here's the way I put it to Lucien Hardy and Joy Christian when I was in a poetic mood last year:

> My view of the universe is that it is many—that it ultimately cannot be unified, for it is alive and changing and creative in a very deep sense. Moreover, that reality is, to some not-yet well-understood extent, plastic: It can be molded by our actions. Thus, though humanity is quite well a Darwinian accident, now that it is here, it is a significant component of the universe that must be reckoned with.

Finally, it seems worthwhile for me to let James say it himself (from his article "Pragmatism and Humanism"). It captures with his beautiful sweep the romantic thought that really keeps me going from day to day:

> In many familiar objects every one will recognize the human element. We conceive a given reality in this way or in that, to suit our purpose, and the reality passively submits to the conception. You can take the number 27 as the cube of 3, or as the product of 3 and 9, or as 26 *plus* 1, or 100 *minus* 73, or in countless other ways, of which one will be just as true as another. You can take a chess-board as black squares on a white ground, or as white squares on a black ground, and neither conception is a false one.
>
> You can treat the adjoined figure as a star, as two big triangles crossing each other, as a hexagon with legs set up on its angles, as six equal triangles hanging together by their tips, etc. All these treatments are true treatments—the sensible *that* upon the paper resists no one of them. You can say of a line that it runs east, or you can say that it runs west, and the line *per se* accepts both descriptions without rebelling at the inconsistency.
>
> We carve out groups of stars in the heavens, and call them constellations, and the stars patiently suffer us to do so,—*though* if they knew what we were doing, some of them might feel much surprised at the partners we had given them. We name the same constellation diversely, as Charles's Wain, the Great Bear, or the Dipper. None of the names will be false, and one will be as true as another, for all are applicable.
>
> In all these cases we humanly make an *addition* to some sensible reality, and that reality tolerates the addition. All the additions 'agree' with the reality; they fit it, while they build it out. No one of them is false. Which may be treated as the *more* true, depends altogether on the human use of it. If the 27 is a number of dollars which I find in a drawer where I had left 28, it is 28 minus 1. If it is the number of inches in a board which I wish to insert as a shelf into a cupboard 26 inches wide, it is 26 plus 1. If I wish to ennoble the heavens by

the constellations I see there, 'Charles's Wain' would be more true than 'Dipper.' My friend Frederick Myers was humorously indignant that that prodigious star-group should remind us Americans of nothing but a culinary utensil.

What shall we call a *thing* anyhow? It seems quite arbitrary, for we carve out everything, just as we carve out constellations, to suit our human purposes. For me, this whole 'audience' is one thing, which grows now restless, now attentive. I have no use at present for its individual units, so I don't consider them. So of an 'army,' of a 'nation.' But in your own eyes, ladies and gentlemen, to call you 'audience' is an accidental way of taking you. The permanently real things for you are your individual persons. To an anatomist, again, those persons are but organisms, and the real things are the organs. Not the organs, so much as their constituent cells, say the histologists; not the cells, but their molecules, say in turn the chemists.

We break the flux of sensible reality into things, then, at our will. We create the subjects of our true as well as of our false propositions.

We create the predicates also. Many of the predicates of things express only the relations of the things to us and to our feelings. Such predicates of course are human additions. Caesar crossed the Rubicon, and was a menace to Rome's freedom. He is also an American schoolroom pest, made into one by the reaction of our schoolboys on his writings. The added predicate is as true of him as the earlier ones.

You see how naturally one comes to the humanistic principle: you can't weed out the human contribution. Our nouns and adjectives are all humanized heirlooms, and in the theories we build them into, the inner order and arrangement is wholly dictated by human considerations, intellectual consistency being one of them. Mathematics and logic themselves are fermenting with human rearrangements; physics, astronomy and biology follow massive cues of preference. We plunge forward into the field of fresh experience with the beliefs our ancestors and we have made already; these determine what we notice; what we notice determines what we do; what we do again determines what we experience; so from one thing to another, altho the stubborn fact remains that there is a sensible flux, what is *true of it* seems from first to last to be largely a matter of our own creation.

We build the flux out inevitably. The great question is: does it, with our additions, *rise or fall in value*? Are the additions *worthy* or *unworthy*? Suppose a universe composed of seven stars, and nothing else but three human witnesses and their critic. One witness names the stars 'Great Bear'; one calls them 'Charles's Wain'; one calls them the 'Dipper.' Which human addition has made the best universe of the given stellar material? If Frederick Myers were the critic, he would have no hesitation in 'turning down' the American witness.

Lotze[5] has in several places made a deep suggestion. We naïvely assume, he says, a relation between reality and our minds which may be just the opposite of the true one. Reality, we naturally think, stands ready-made and complete, and our intellects supervene with the one simple duty of describing it as it is already. But may not our descriptions, Lotze asks, be themselves important additions to reality? And may not previous reality itself be there, far less for the purpose of reappearing unaltered in our knowledge, than for the very purpose of stimulating our minds to such additions as shall enhance the universe's total value. *'Die Erhöhung des vorgefundenen Daseins'* is a phrase used by Professor Eucken[6] somewhere, which reminds one of this suggestion by the great Lotze.

It is identically our pragmatistic conception. In our cognitive as well as in our active life we are creative. We *add*, both to the subject and to the predicate part of reality. The world

[5] Hermann Lotze (1817–1881) was a German philosopher and logician.

[6] Rudolf Christoph Eucken (1846–1926) was a German philosopher who won a Nobel prize in literature in 1908. One of several potential translations of this phrase given to me by Rüdiger Schack was, "The enhancing of that which is pregiven." I am told it sounds much better in German!

> stands really malleable, waiting to receive its final touches at our hands. Like the kingdom of heaven, it suffers human violence willingly. Man *engenders* truths upon it.
>
> No one can deny that such a role would add both to our dignity and to our responsibility as thinkers. To some of us it proves a most inspiring notion. Signore Papini, the leader of Italian pragmatism, grows fairly dithyrambic over the view that it opens of man's divinely-creative functions.[7]
>
> The import of the difference between pragmatism and rationalism is now in sight throughout its whole extent. The essential contrast is that *for rationalism reality is ready-made and complete from all eternity, while for pragmatism it is still in the making, and awaits part of its complexion from the future*. On the one side the universe is absolutely secure, on the other it is still pursuing its adventures. ...
>
> *The alternative between pragmatism and rationalism, in the shape in which we now have it before us, is no longer a question in the theory of knowledge, it concerns the structure of the universe itself.*
>
> On the pragmatist side we have only one edition of the universe, unfinished, growing in all sorts of places, especially in the places where thinking beings are at work.
>
> On the rationalist side we have a universe in many editions, one real one, the infinite folio, or *édition de luxe*, eternally complete; and then the various finite editions, full of false readings, distorted and mutilated each in its own way.

My anti-Platonist tendencies certainly come out in my feeling of agreement with this quote. There is no sympathy in it to the idea that one might be *attuned* with a divine mind privy to "the way" the universe is constructed—which as Marcus, I think, rightly points out, seems to have been Einstein's modus operandi when theorizing. At most, one might be attuned to the content of this other quote of James:

> If you follow the pragmatic method, you cannot look on any [theory] as closing your quest. You must bring out of each [theory] its practical cash-value, set it at work within the stream of your experience. It appears less as a solution, then, than as a program for more work, and more particularly as an indication of the ways in which existing realities may be *changed*.
>
> *Theories thus become instruments, not answers to enigmas, in which we can rest.* We don't lie back upon them, we move forward, and, on occasion, make nature over again by their aid.

Physical theories, by this view, are conceptual means and tools for making change in the world. Can one be attuned to such a thing in the Platonic sense that Marcus explores? Perhaps. But if so, it's not the medieval Platonism Marcus was talking about—it is instead being attuned to how best be an agent of change.

In any case, what's important here is that I genuinely do see a humanistic role in the very maintenance of the universe. And all of this is potentially independent, and certainly broader, than considerations to do with quantum mechanics. The universe is partially powered by the inhuman, to be sure. But it is also partially powered by *belief*—that I do believe full well. And to the extent that religion or religious feeling are sources of belief, they do indeed play a role in the very construction of the universe. However, from this point of view religion is a special case of something much bigger—namely, belief generally.

---

[7] Giovanni Papini (1881–1956) was an Italian essayist and philosopher of whom James had great regard.

I'll quote James one last time, so that you get a precise sense of what I mean here, but then I'll get back to the quantum and Pauli/Fierz and tell you way I'm saying all this. From "The Sentiment of Rationality":

> Now, I wish to show what to my knowledge has never been clearly pointed out, that belief (as measured by action) not only does and must continually outstrip scientific evidence, but that there is a certain class of truths of whose reality belief is a factor as well as a confessor; and that as regards this class of truths faith is not only licit and pertinent, but essential and indispensable. The truths cannot become true till our faith has made them so.
>
> Suppose, for example, that I am climbing in the Alps, and have had the ill-luck to work myself into a position from which the only escape is by a terrible leap. Being without similar experience, I have no evidence of my ability to perform it successfully; but hope and confidence in myself make me sure I shall not miss my aim, and nerve my feet to execute what without those subjective emotions would perhaps have been impossible. But suppose that, on the contrary, the emotions of fear and mistrust preponderate; or suppose that, having just read [W. K. Clifford's] *Ethics of Belief*, I feel it would be sinful to act upon an assumption unverified by previous experience—why, then I shall hesitate so long that at last, exhausted and trembling, and launching myself in a moment of despair, I miss my foothold and roll into the abyss. In this case (and it is one of an immense class) the part of wisdom clearly is to believe what one desires; for the belief is one of the indispensable preliminary conditions of the realization of its object. *There are then cases where faith creates its own verification.* Believe, and you shall be right, for you shall save yourself; doubt, and you shall again be right, for you shall perish. The only difference is that to believe is greatly to your advantage.
>
> The future movements of the stars or the facts of past history are determined now once for all, whether I like them or not. They are given irrespective of my wishes, and in all that concerns truths like these subjective preference should have no part; it can only obscure the judgment. But in every fact into which there enters an element of personal contribution on my part, as soon as this personal contribution demands a certain degree of subjective energy which, in its turn, calls for a certain amount of faith in the result—so that, after all, the future fact is conditioned by my present faith in it—how trebly asinine would it be for me to deny myself the use of the subjective method, the method of belief based on desire!
>
> In every proposition whose bearing is universal (and such are all the propositions of philosophy), the acts of the subject and their consequences throughout eternity should be included in the formula. If M represent the entire world minus the reaction of the thinker upon it, and if $M + x$ represent the absolutely total matter of philosophic propositions (x standing for the thinker's reaction and its results)—what would be a universal truth if the term x were of one complexion, might become egregious error if x altered its character. Let it not be said that x is too infinitesimal a component to change the character of the immense whole in which it lies imbedded. Everything depends on the point of view of the philosophic proposition in question. If we have to define the universe from the point of view of sensibility, the critical material for our judgment lies in the animal kingdom, insignificant as that is, quantitatively considered. The moral definition of the world may depend on phenomena more restricted still in range. In short, many a long phrase may have its sense reversed by the addition of three letters, *n-o-t*; many a monstrous mass have its unstable equilibrium discharged one way or the other by a feather weight that falls.

These, however, are very big things and very big thoughts James is speaking of. And even though they stir my heart, they remain too vague to transform science as a whole (and indeed the world) in the way that I hope we'll one day transform it. I love the sound of what I'm hearing, but at the end of the day, I'm still not completely sure what I *am* hearing. How do I know I'm not simply fooling myself with pretty words? The main point I want to make at this juncture is that these ideas which drive me forward—these still-too-vague ideas—are *supra*-quantum mechanics. It is 19th century philosophy, not physics.

What now is the role of quantum mechanics within this system of thought I'm laying out? It is that it is a miniature version of these general points. *But*, though it is a miniature version, it is an extremely *precise* version! And that's its ace in the hole, as Hans said of Pauli's discussion with Huxley. Here's where I agree with Marcus,

> Quantum mechanics is important because it is as close to a refutation of the classical world picture as one could hope to get. It is very rare that something gets refuted in science as completely and finally as the classical hypothesis, of the world machine, has been refuted.

Thus, quantum mechanics is a precision laboratory for defining and testing these things *we think we see* on the horizon. When we've got it in quantum mechanics, we know we have it.

This finally is where I can express the value for me of learning more about the Pauli–Fierz and the Pauli–Jung correspondence. There is an amazing amount of development of the metaphysics of a malleable world in the writings of William James and F. C. S. Schiller.[8] And lo and behold, in which direction does that development lead? In significant parts at least, it leads in quite the same direction as Pauli's metaphysic. I don't know that I've ever emphasized this to you two. When James, for instance, asks the question, "What are the materials of the universe's composition?," the answer he tries to develop is that they are things *neutral* to the material/mental or physical/psychical distinction. In fact, he saw the material and mental as *complementary*, but exclusive, aspects of the basic stuff. So, you see the similarity.

But James and Schiller knew no quantum mechanics of course. Particularly, they were not privy to this precision development laboratory for their metaphysic that Pauli and Fierz knew so well. Therefore, I need, I crave, to know more of the precise things Pauli and Fierz were talking about for just this reason. The suspicion is that it will inspire us and help us connect the dots between the quantum formalism and the bigger, much richer, and far-from-really-developed idea of a malleable/alchemical world.

Marcus says,

> Quantum mechanics, as it stands now, is little more than a set of calculational procedures. The calculations show that the vision of the world-machine is completely without empirical foundation. However, the mechanical vision of things has not been replaced with any other.

of which I have a minor qualm, but I think he is right in further saying, "The classical vision of things … was deeply wrong about the relation of mind to matter." For I think this neutral stuff that Pauli and James speak of—James called it "pure experience" but I am not completely happy with that term—will indeed be the "shocking" ingredient that will replace the mechanical vision of things. So, it is not that we have not started the process of replacement; it is only that it has been slow going.

---

[8] F.C.S. should not be confused with the *famous* Schiller, namely Friedrich. F.C.S. stands for Ferdinand Canning Scott Schiller (1864–1937), a now mostly forgotten pragmatist, but one for which I have the greatest respect. Of relevance here is his masterful essay "Axioms as Postulates" [44].

Hans, I suspect, has wondered why I place so much emphasis on wanting to know the most, within all of Pauli's non-x-ing writings and correspondence, about the part to do with the idea of detached and nondetached observers (activating observers). It is because, as I see it, a thorough study of what it means for an observer or agent to be nondetached from the phenomena he helps bring about is the very starting point for a development of James and Pauli's neutral-stuff ontology. If we cannot get the activating-observer idea right and understand all its facets, then I feel we have no chance whatsoever of developing a full-blown "neutral monism."[9] It is the very place to start, for it forces the issue of this new ontology in its very statement. What is an observer if not psychical? What is an activator if not physical? How can we really combine these two aspects of the phenomenon via the help of a Paulian neutral language?

Unfortunately in the technical development of this first step, Marcus and I have hit a great snag by not yet understanding the structure (a.k.a. knowing how to prove the existence of) these damned SICs.[10] The reason I say this is because I see no way forward to a more precise definition of the activating-observer concept than by a kind of "generalized Newton's third law" of the style I wrote you about last month. And I see no way to make progress in that intermediate program without the ability to completely rewrite quantum mechanics (and particularly the Born rule) in terms of SICs. For completeness, let me reproduce that little essay:

> Everybody has their favorite speculation about what powers quantum information and computing. Some say it is the superposition principle, some say it is the parallel computation of many worlds, some say it is the mysteries of quantum entanglement, some say it is the exponential growth of computational space due to the tensor product. For my own part though, my favorite speculation is that it is Newton's Third Law: For every action, there is an equal and opposite reaction. Indeed I sometimes wonder if the very essence of quantum mechanics isn't just this principle, only carried through far more consistently than Newton could have envisioned. That is to say, absolutely *nothing* is exempt from it.
>
> What do I mean by this? What might have been exempt from the principle in the first place? To give an answer, let me note an equivalent formulation of old Newton. For every *reaction*, there is an equal and opposite *action*. Strange sounding, but there's nothing wrong with it, and more importantly, this formulation allows for the possibility of an immediate connection to information theory. In particular, we should not forget how information gathering is represented in the Shannon theory. An agent has gathered information … when something in his environment has caused him to *react* by way of revising a prior expectation $p(h)$ (for some phenomenon) to a posterior expectation $p(h|d)$ (for the same phenomenon).
>
> When information is gathered, it is because we are reacting to the stimulation of something external to us. The great lesson of quantum mechanics may just be that information gathering is physical. Even something so seemingly unimportant to the rest of the universe as the reactions that cause the revisions of our expectations are not exempt from Newton's Third Law. When we react to the world's stimulations upon us, it too must react to our stimulations upon it.

[9] Neutral monism is what it is mostly called in the literature, but a much better name for the idea in the Jamesian context would be "neutral pluralism." See [45].

[10] SICs refers to the notion of the symmetric informationally complete quantum measurements [46]. For some updated references about this aspect of the QBist research project, see [47–50].

> The question is, how might we envision a world with this property—i.e., with such a serious accounting of Newton's law—but in a way that does not make a priori use of the information gathering agent himself? If the question can be answered at all, the task of finding an answer will be some tall order. For never before in science have we encountered a situation where the theorizing scientist is so inextricably bound up with what he is trying to theorize about in the first place.
>
> It's almost a paradoxical situation. On the one hand we'd like to step outside the world and get a clear view of what it looks like without the scientist necessarily in the picture. But on the other hand, to even pose the question we have to imagine an information gathering agent set in the middle of it all. You see, neither Shannon nor any of modern information theory has given us a way to talk about the concept of information gain without first introducing the agent-centered concept of an expectation $p(h)$.
>
> So, how to make progress? What we do know is that we actually are in the middle of the world thinking about it. Maybe our strategy ought to be to use that very vantage point to get as close as we can to the goal. That is, though we may not know what the world looks like without the information gathering agent in it, we certainly do know something about what it looks like with him in: We know, for instance, that he ought to use the formal structure of quantum mechanics when thinking about physical systems. Beyond that, we know of an imaginary world where Newton's Third Law was never taken so seriously: It is the standard world of classical physics and Bayesian probability.
>
> Thus, maybe the thing to do first is to look inward, before looking outward. About ourselves, at the very least, we can ask how has the formal structure of our *behavior* changed since moving from what we thought to be a classical Bayesian world to what we now believe to be a quantum world? In that *differential*—the speculation is—we may just find the cleanest statement yet of what the quantum world is all about. For it is in that differential, that the world without us surely rears its head.
>
> To do this, we must first express quantum mechanics in a way that it can be directly compared to classical Bayesian theory, where the information-gathering agent was detached from the world.

So, you see, there's work to be done all the way down to the SICs, and all the way up to a full-blown ontology of neutral pluralism and a malleable/alchemical reality. In the case of the SICs, I'm actually more confident: That problem is going to be solved if I have to recruit all of Hannibal's army. But in the case of the ontology it's just verbal hints we have presently—there was only so much James, Schiller, Dewey, and others could discern from general premises, without detailed empirical input. Clearly we will have to see our way through in that big task. But Pauli and Fierz are the only two quantum physicists I know of who have already waded these waters to any depth. If I could get hold of what little they may have already strained from the rocks below, it would be invaluable for this effort. And that is what the title of this New Year's note is all about: What I Really Want Out of a Pauli/Fierz-Correspondence Study.

Thank you both for not losing faith that there is something really important here.

3 16 Oct. 2009 *The More and the Modest* (to L. Hardy)

Greetings[11] from New Orleans. This note is to say thanks for your remarks at the panel discussion Friday [51]. You did something that none of the others did—you caused me to think and clarify, rather than piss me off from a display of overt closed-mindedness. It amazes me the way Norsen and Valentini preach to me without understanding an ounce of the way I'm trying to see things.

Your point on Hilbert-space dimension on the other hand was a productive one, and thinking about how I should reply to you has helped me realize that my language needs a significant overhaul. With better language, we inevitably get better understanding. Here goes.

I used to ask, "What is real of a quantum system?," and give "Zing" as a placeholder for the answer. What is Zing? The only thing I've been able to comfortably identify for it in the quantum formalism is Hilbert-space dimension. But that certainly can't be enough to get the world going—or at least, I think you would (and did) say something like this. John Sipe once wrote, "the quantum world of this interpretation is a fixed, static thing. It is a frozen, changeless place."

That is, *at face value* there's not enough engine, not enough gears and pinions, in this bare "reality" to account for motion and change and indeterminism and novelty and growth and evolution and ... make the list just as big as you want ... all the things we see around us. What is weird is that I agree. Such a minimalist world doesn't even track with what I myself believe. So it becomes a question of why I have doggedly resisted saying more.

I believe I have tracked down the troublemaker in my mode of expression, and it is this. My trouble is *not* that I believe that there is *nothing* in the world, or perhaps a minimalist number of things (too minimal), as the solipsism-chargers seem so obsessed with accusing me of. It is rather that my point of view admits *too many* things into the world—too many things of an independent and self-sustaining reality, things for which there are no equations; realities of which I am only willing to *point* to and say effectively, "Yeah the world includes that too."

This is because the ontology I imagine is the one of the humanist, rather than the trained or indoctrinated scientist. Take Democritus as an example: for him, the universe literally was "atoms and void." That simple phrase was meant to account for all that is. Similarly, the universe Einstein sought was literally meant to be a differentiable manifold supporting the solution to a very clever (though never found) differential equation—all else, even such as the flow of time, was simply illusion. (Recall Mermin's talk at PIAF [52].) The unifying theme in these two visions is that chairs and thumbs and bricks and all the like from our common experiences are secondary things—effective (operational) descriptions or illusions—having no real, primary existence of their own. For a most relevant example, consider David Wallace's universe. As far as I can tell, it is literally a single quantum state on a Hilbert space. That vector, timelessly unchanging in a rotating frame, is the universe's whole

[11] This essay is a condensation of [12, pp. 1853–1860].

substance. The apparent indeterminism of the quantum world is subordinate to the greater monistic determinism and timelessness of the whole.

But here's the funny thing: It is exactly of Einstein's and Wallace's universes, not my own, that I would say, "the world of this interpretation is a fixed, static thing. It is a frozen, changeless place." To my mind, these are both completely barren visions for the world. I think our distaste for each other's proclivities comes from this: It is a very old philosophical divide, that between the rationalist attitude and the empiricist attitude. It has only become clear to me recently, but at my core I am an extreme empiricist.

Here is the way William James put the distinction when he was in a mild mood:

> By empiricism I mean the tendency which lays most stress on the part, the element, the individual, treats the whole as a collection, and calls the universal an abstraction. By rationalism I mean the tendency to emphasize the universal, and to make the whole prior to the part, in the order both of logic and of being. The temper of rationalism is dogmatic: it willingly claims necessity for its conclusions. Empiricism is more modest, and professes to deal in hypotheses only.

A more full-blooded statement of the issues involved, and just beautiful reading, can be found in a long quote I've plucked out for you from his essay "The Sentiment of Rationality." See [12, pp. 1856–1860]. If you find something worth reading in this note of my own composition, I hope you will take the time to read the longer quote of James. In the end, it has become crucial to my thinking.

Let me give a briefer hint of the issue here. It has to do with the way James put the dangers of rationalism when he was in a fiercer mood:

> Let me give the name of 'vicious abstractionism' to a way of using concepts which may be thus described: We conceive a concrete situation by singling out some salient or important feature in it, and classing it under that; then, instead of adding to its previous characters all the positive consequences which the new way of conceiving it may bring, we proceed to use our concept privatively; reducing the originally rich phenomenon to the naked suggestions of that name abstractly taken, treating it as a case of 'nothing but' that, concept, and acting as if all the other characters from out of which the concept is abstracted were expunged. Abstraction, functioning in this way, becomes a means of arrest far more than a means of advance in thought. It mutilates things; it creates difficulties and finds impossibilities; and more than half the trouble that metaphysicians and logicians give themselves over the paradoxes and dialectic puzzles of the universe may, I am convinced, be traced to this relatively simple source. *The viciously privative employment of abstract characters and class names* is, I am persuaded, one of the great original sins of the rationalistic mind.

The way this is relevant to me is the following. Far from thinking the world is an empty place, a place only with me in it. I think it is full of things, overflowing with things. *All* distinct things, from head to toe. And literally so. It is not a world made of six flavors of quarks glued together in various combinations. It is not a world that maps to a single algorithm running on Rob Spekkens's favorite version of Daniel Dennett's mechanistic cellular automaton. It is a world of heads and toes and doorknobs and dreams and ambitions and every kind of particular. (And that is not a typo: It is a world in which even dreams and ambition have substance.) It is a world in which Vivienne Hardy is a distinct entity, not "constructed" of anything else, but a true-blue crucial piece of the universe as it is today—no less crucial than spacetime itself.

In modern parlance, I am not a reductionist. And when my world is judged as empty (of all but me), I claim it is because I am being interpreted from the standpoint of an (explicit or implicit) reductionist worldview. It is true that my envisioned world may be as cockamamie as James's heads of Borneo:

> Taken as it does appear, our universe is to a large extent chaotic. No one single type of connection runs through all the experiences that compose it. If we take space-relations, they fail to connect minds into any regular system. Causes and purposes obtain only among special series of facts. The self-relation seems extremely limited and does not link two different selves together. Prima facie, if you should liken the universe of absolute idealism to an aquarium, a crystal globe in which goldfish are swimming, you would have to compare the empiricist universe to something more like one of those dried human heads with which the Dyaks of Borneo deck their lodges. The skull forms a solid nucleus; but innumerable feathers, leaves, strings, beads, and loose appendices of every description float and dangle from it, and, save that they terminate in it, seem to have nothing to do with one another. Even so my experiences and yours float and dangle, terminating, it is true, in a nucleus of common perception, but for the most part out of sight and irrelevant and unimaginable to one another.

But it is not an *empty* world.

With this as a background, let me now return to Zing, Hilbert-space dimension, and the needed overhaul of my language. I ask, "What is real of a quantum system?" and answer, "its Hilbert-space dimension," but that is a very funny thing to say. There is *so much* that is real of a quantum system, why would I ever say that? I wouldn't say it of Vivienne Hardy, for instance—I already declared this quite vocally—and she is a perfectly good example of a quantum system. So, what am I really up to?

It is a bit like this: Forget quantum mechanics, and think back to the days when the only physics known was basically Newtonian gravity. Would any Newtonian have ever answered the question, "What is real of a physical system?," with the declaration, "Only its gravitational mass is real of it."? Most definitely not. But that is because Newton merely/boldly taught us but a singular thing: That every body in the universe, that every thing that can be carved out from it, had a previously undisclosed capacity—a numerical capacity to (try to) attract every other body in the universe. On the one hand, I say "merely" because looked at in this way, it really is a very *modest* move—modest at least by the lights of the Steven-Weinberg–Stephen-Hawking-theory-of-everything generation: It was no theory of everything—never pretended to be. On the other hand, I say "boldly" because I certainly agree with my friend Hans Christian von Baeyer who wrote in one of his books [53]:

> Great revolutionaries don't stop at half measures if they can go all the way. For Newton this meant an almost unimaginable widening of the scope of his new-found law. Not only Earth, Sun, and planets attract objects in their vicinity, he conjectured, but all objects, no matter how large or small, attract all other objects, no matter how far distant. It was a proposition of almost reckless boldness, and it changed the way we perceive the world.

And therein lies the key, I think, to how I should modify my language. When I ask myself what have we learned with quantum mechanics, I want to say first and foremost that we have learned how we should more consistently gamble upon the consequences of our interactions with external physical systems. But why this new calculus for gambling? Because quantum mechanics is uncovering that every object in the universe has a previously undisclosed *capacity*.

I would have liked to have said "uncovered," but at this stage of research I must still settle for "is uncovering"—it is an unfinished project to understand the significance of quantum mechanics in these terms. Hilbert-space dimension, like gravitational mass, is representative of some universal capacity. That's the real idea. Dimension is a quality a body possesses in a way that, in a QBayesian understanding, it does not "possess" a quantum state. If we conceptually delete an agent gambling upon the consequences of his interactions with a quantum system—the QBist says—we also conceptually delete its quantum state. But there is no reason to believe the system and the system's capacity are deleted as well.

Capacity for what? That's where the hard part begins. Sometimes I say the capacity for birth and creation. Sometimes I say the capacity to entertain counterfactuals. Sometimes I say it is a capacity that can be harnessed to aid of computation, as in quantum computation. The truth is, I don't yet know what I mean in any precise way. I only know that I have a strong inner tug to thinking that the SIC calculus will help reveal a precise idea.

Anyway, I write all this to put the key idea into perspective and to practice a way of speaking that I had not used to any great extent before. You were indeed right to suspect that Hilbert-space dimension doesn't have enough gears and pinions to get things going, but neither did gravitational mass. Nonetheless, the disclosing of that old capacity—gravitational mass—"changed the way we perceive the world." And so I think of our new capacity.

Thus I end this instalment by saying thanks again for pointing out a great deficiency in my choice of words and for consequently causing me to think.

4 7 Jan. 2010 *What Comes Next?* (to M. Schlosshauer)

Since you and I have been in so much contact recently, I asked Marcus Appleby if I could forward you the note [12, pp. 1926–1927] he sent me last night. I thought it hit the mark, and in an eloquent way: What comes next? That's the only really important question.

I'm feeling doubly good today. I feel I had a very deep insight this morning just as I was getting into the shower. It is the inevitable conclusion to draw from this QBist trek I've been taking, but I only feel that I really saw it with clarity this morning. In a single sentence, it goes like this: Just as quantum theory is not in conflict with probability theory, only an *addition* to it, quantum mechanics is not in conflict with the normal, empirically perceived world of common experience (sometimes called the classical world); it is only an *addition* to it. It is not that the classical world undergirds quantum mechanics, as I read Bohr as saying; it is not that the classical world should be *derivable* from quantum mechanics (a thought that you are very familiar and probably still quite sympathetic with). It is that one is an *addition* to the other, just as quantum theory is an addition to Bayesian probability. It is an overpowering feeling I have in my chest, a feeling like I've only had a few times in my career—my whole body is most definitely reacting to this thought.

5 3 Oct. 2019 *On M & W* (to the Reflection Team)

I send you this note because of our recent discussions on von Fraassen's reflection principle [54]. In defending our 2011 take on quantum decoherence [55] to Ulrich Mohrhoff, I believe I had a thought which had not come so to the front of my mind before. Maybe Rüdiger understood it, but it strikes me only now. It's why I'm sharing this. Ulrich wrote,

Ulrichism 1: *I am well aware of your paper on decoherence [55]. ... When I first read this paper, the statement that "Decoherence does not come conceptually before a 'selection' ... but conceptually after the recognition of the future possibilities" caused me some headache. It seems to be an example of obscuring a point by trying to make it clearer. (I find myself doing this all the time!) While every probability assignment comes after the recognition of all possible sequences of events, decoherence remains associated with a measurement (by the environment) that comes before the measurement to whose outcomes probabilities are being assigned.*

To which I replied:

> But the whole point of the result was to do away with talk of "measurement by the environment" and the contrived Hamiltonians upon which all that way of thinking requires. A measurement in QBism is an experience of an agent full stop, and there is no place for an agent to sit in such a "measurement by the environment" way of thinking. (Kind of like the little guy I made fun of in a box in this old paper [56].) Thus decoherence becomes a law of thought in QBism, like it had to.

The new thought that comes to me is that that little paper was as crucial to the consistency of QBism as was the quantum de Finetti theorem [56] years earlier. Both in fact addressed little guys in a box that had to be exorcised for our own consistency. Now I'm even further ashamed that we did so little with that paper over the years. We must remedy this with a fantabulous new paper.

Here's one more little piece from my replies to Ulrich that I think worth recording. It was my claim to him that, "I'll warn you that I am fairly skeptical of your phrase, 'the incontestable irreversibility of human sensory experience.' All one has to do is witness the decline of a grandmother or aunt with Alzheimer's disease to become skeptical of that. At the moment, I side with Whitehead that experience too lives and dies. Living and dying is what experience does." This is tied up with my (planned) response to Richard Healey over his accusation that QBism hasn't made much progress toward an ontology. It is a question of tying up loose ends. QBism must milk Wigner's friend for all its worth [57] before it can make a leap to a Whiteheadian style ontology. Bell asked, "Are we not obliged to admit that more or less 'measurement-like' processes are going on more or less all the time more or less everywhere?" To that, I want to say, "Yes, we *are* obliged! Thank you, John Bell. But measurement is not at all what you imagined it to be!" First we have to understand what it actually is in those moments of measurement involving agents, and only then we might abstract. To be 'measurement-like' for an event is to have a subject pole and an object pole, like Whitehead's "actual occasions," but to be neither intrinsically. To be "measurement-like" is to be born, live, and die, like Whitehead's "actual occasions." To be "measurement-like" is to be ... This takes careful thought. This takes absolute consistency before carefully moving forward.

6 9 Oct. 2020 *We Are Such Stuff ...* (to the QBuki)

"We are such stuff as quantum measurement is made on." [10, p. 27]

Let me quote something Marcus Appleby wrote last week:

> To convince Nick Herbert and others like him we need to develop the QBist conception of reality in much more detail, and much more explicitly so that it is completely and totally obvious that we are saying something utterly different from instrumentalism. ... I really and deeply and strongly feel that the time has come to make an all out attack on the problem. No more pussy-footing around the edges. What is required is a full-frontal attack.

In this regard, let me also record a quote of Emanuel Swedenborg that I just came across: "All things, at the present day, stand provided and prepared, and await the light. The ship is in the harbour; the sails are swelling; the east wind blows; let us weigh the anchor, and put forth to sea."

7 13 Sept. 2020 *Concerning the Genuine Stuff* (to A. Gefter)

Thanks again for the notes you sent us last Friday and the cute cartoons you threw together quickly before the meeting. I feel there's something in your heart that's very much on the right track, but we are *all* most definitely still confused.

One comment you made in your presentation caused me to wonder whether we're yet on the same page with regard to personalist Bayesianism. It came when you showed the slide with Wigner deriving from his 50-50 assignment that it was possible that Alice would say down and he would frown. But Alice believed that impossible. If I understood you correctly, you were inclined to think that Alice should have the privileged judgment. But in personalist Bayesianism, no one has a privileged judgment, even when they have a 1-0 probability assignment.

A lot of the Wignerstein paper is about establishing just that kind of idea: That neither agent in the Wigner's friend scenario has a privileged position, either with regard to Bayesianism or quantum theory. Časlav Brukner always likes to call Wigner a "super-observer," whereas we say, "Uh uh. That is an error."[12]

In fact, it is not difficult to create examples where two agents start out with some amount of compatibility in their probability assignments, but then after one observation, completely disagree on what they will find next [61, fn. 16]. I.e., one will make a 1-0 probability assignment, whereas the other will make a 0-1 assignment. Suppose they now perform the observation and the 1-0 agent gets exactly what she expected, whereas the 0-1 agent is caught completely off-guard. What does one make of that? Nothing more than that the first agent won the bet, whereas the second agent is probably devastated. Whatever happened in the observation was something that one of the two equal-status agents thought would be absolutely impossible. But even after the fact, no one has a right to claim their probability assignment was metaphysically correct beforehand. The fact that comes about with the second observation does not establish it so. *If new reality arises with each quantum measurement, then new reality arises.*

What I think this helps emphasize is a train of thought that goes all the way back to [9]. When an agent writes down a POVM to describe a measurement he wishes to perform, that is equivalent to giving the potential outcomes of his soon-to-be action *names*. But the actual outcomes are not equivalent to their names. They are something much more, something much deeper than that. [This idea is significantly expanded upon in the section below titled, "Amanda's Hooks."]

So, here's where I definitely have sympathy for what you're aiming for. When Alice and Bob shake hands, no matter what their expectations for the consequences of that handshake (no matter what their expectations for the experience it will elicit), no matter how they each associate a POVM with this action and its consequences, something will happen between the two of them that is very real and will go on to shape the future. But that thing—*the genuine stuff*—is far more than either of the two names that Alice and Bob might name it. The real essence of the "connective tissue," the "semi-permeable membrane" you keep calling our attention to is something just a bit beyond what quantum theory has the resources to describe. This is why I was calling it the pre-theoretic or pre-linguistic outcome on Friday. It's just a name I pulled out of a hat for the occasion, but it was an attempt to express that experiences are part of the stuff of the world, but propositions like "Alice saw the spin was up in the x-direction" are but names—they are further removed from the actual stuff.[13]

I don't yet know where all of this can go, but I do think we're starting to circle around the correct issues, and I really value that you're helping steer us that way.

Oh, let me also point you to the anti-inductivist example I told you about the other day: See "Priors in Quantum Bayesian Inferernce" [58]. A 2009 paper! It seems like all the good things happened in 2009; that was the year QBism got its name [58].

---

[12] This paragraph refers to a lengthy treatment of Wigner's friend tentatively titled "QBism's Account of Wigner's Friend (and the friend's account of Wigner)," which has become so large and patched together as to be like Frankenstein's monster. In case the paper is never completed, see [57].

[13] For a more formal discussion of the contents of this paragraph, see [41].

8 3 Nov. 2020 *Consistency Is Way Overrated* (to A. Gefter)

I don't don't have the bandwidth to be able to reply to you properly over email at the moment, so we'll just have to duke it out in person eventually. But I do want to register that I am in disagreement with some points in your note. Maybe the most important point for the present is this one:

Gefterism 1: *To me, the interesting next question is, what happens when Wigner enters the lab and interacts with the friend? At this point in the story, while still rejecting any 3rd person description, there are, as far as I can tell, only two options:*

1. *we can stick to the story that measurement outcomes are truly private and therefore Wigner and the friend can have contradictory accounts of one and the same interaction in which they both take part, or*
2. *we can allow consistency to be created through the interaction itself (and because this consistency would not extend to anyone not directly interacting with Wigner and friend, it would not be 3rd person consistency, but 2nd)*

If QBism wants to stick to the first option—insisting that outcomes are truly private, and that different observers **(even while interacting)** *live in different spacetimes or different realities—that's fine ...but it is solipsistic.*

I've added boldfacing to a portion of what you wrote because, even in 2012 I was already saying things like this [60, Ans. 5]:

> This is where the Wigner's friend question comes into play. This is a story of two agents with a different physical system in front of each: (1 The *friend*, with say an **electron** in front of himself, and 2) *Wigner*, with the **friend+electron** in front of himself. (Agents are italicized; systems are boldfaced.) Which agent's quantum state assignment for his own system is the correct one? Quantum Bayesianism knows of no agent-independent notion of "correct" here—and this is why we say there is no paradox. The source of each assignment is the agent who makes it, and the *concern* of each assignment is not of what is going on out in the world, but of the uncertain consequences each agent might experience if he takes any actions upon his system. The only glaringly mutual world there is for Wigner and his friend in a QBist analysis is the partial one that might come about if these two bodies were to later take actions upon each other ("interact")—the rest of the story is deep inside each agent's private mesh of experiences, with those having no necessary connection to anything else.
>
> But what a limited story this is: For its concern is only of agents and the systems they take actions upon. What we learn from Wigner and his friend is that we all have truly private worlds in addition to our public worlds. But QBists are not reductionists, and there are many sources of learning to take into account for a total worldview—one such comes from Nicolas Copernicus: That man should not be the center of all things (only some things). Thus QBism is compelled as well: What we have learned of agents and systems ought to be projected onto all that is external to them too. The key lesson is that each part of the universe has plenty that the rest of the universe can say *nothing* about. That which surrounds each of us is more truly a pluriverse.

There's nothing in my writing that is as exclusionary as the way you portray me. Particularly, I'll emphasize the last sentence of my first paragraph: "The only glaringly mutual world there is for Wigner and his friend in a QBist analysis is the partial one that might come about if these two bodies were to later take actions upon each other ...." (This is what Cavalcanti calls the bursting of the "Wigner bubbles" in [62].)

But I should have also boldfaced your phrase "we can allow consistency to be created through the interaction itself." From my point of view, consistency is way overrated: Agents need not have agreement to live in the same world. *The very fact that they are fighting it out (perhaps to the death) is already indication enough that they are in a single world.* Just look at the Democrats and the Trumpists on this very day. What one declares to be true, the other declares to be a lie: They fight it out precisely because they are in one world. In fact, for the Darwin-inspired ontology the classical pragmatists wanted, they nor I would want it any other way: This is one of the mechanisms by which novelty seeps into the pluriverse.

So when you want to find a yet-unearthed mechanism within QBism for how consistency between Wigner and the friend is *enforced* or *necessitated*, I think you are on the wrong track. Wigner and the friend *may* or *may not* agree when they eventually speak to each other about the electron: It is not within the *raw* quantum formalism (the user's guide for individual agents) to say. What Rovelli and everyone previously have left out of their analyses is that the problem is simply way underdetermined. Before our recent paper with Rüdiger [57], I don't believe anyone had ever given a flip about treating Wigner and the friend on the same footing, completely symmetrically. But then once one does that, one sees that Wigner and his friend can disagree just as any two Bayesians can.

This is not at all to say that one could not identify various conditions by which to adorn the friend's and Wigner's uses of the quantum formalism so that they feel confident they will come to agreement upon speaking to each other. But then such an addition would be in the spirit of, for instance, de Finetti's assumption of exchangeability for his representation theorem. I.e., he identifies a condition (a smidgen of agreement they have beforehand) through which two agents believe they will come to complete agreement after enough data. The point I want to make is this: The belief in an ultimate common agreement is not bought from the *raw* formalism of probability theory, but from a certain feature common to the two Bayesians' otherwise distinct priors. The formalism by itself cannot set the priors. QBists expect it to be no different when it comes to quantum theory—in fact, QBism is pretty much built on that very idea!

But I said I don't have so much bandwidth at the moment. There's still a lot more to talk about! So, let me just shut down the note, and in the spirit of Michael Murphy, say "To be continued!"

9 12 Feb. 2021 *Getting Merleau-Pontified* (to J. Pienaar)

Jacques, I felt it a very productive and stimulating discussion today. Thanks again! I am genuinely starting to get Merleau-Pontified by you.

I loved the line of yours about how in M-P's hands phenomenology becomes not an epistemology, but an ontology. It reminded me of this little bit of William James that I always come back to:

> The import of the difference between pragmatism and rationalism is now in sight throughout its whole extent. The essential contrast is that *for rationalism reality is ready-made and complete from all eternity, while for pragmatism it is still in the making, and awaits part of its complexion from the future.* On the one side the universe is absolutely secure, on the other it is still pursuing its adventures. [...]
>
> *The alternative between pragmatism and rationalism, in the shape in which we now have it before us, is no longer a question in the theory of knowledge, it concerns the structure of the universe itself.*

It's not about a theory of knowledge, but about the structure of the universe itself.

By the way, on my bookshelf I have a small book titled *William James and the Reinstatement of the Vague* by William Joseph Gavin, which I skimmed probably 10 years ago. On the back cover, a reviewer writes, "Gavin has taken an often cited but seldom explored text from James and shown its significance for James's overall philosophy. He clearly and insightfully delineates the character and role of 'the vague' in James's metaphysics ..." On page 1, Gavin writes:

> In *The Principles of Psychology*, the American philosopher William James makes what at first sight appears as a strange statement. He tells the reader: "It is, in short, the re-instatement of the vague to its proper place in our mental life which I am so anxious to press on the attention." ... In *Pragmatism*, he tells the reader, "Profusion, not economy, may after all be reality's key-note"; and asks: "May there not after all be a possible ambiguity in truth?" And in *Essays in Radical Empiricism*, he calls "our experiences, taken all together, a quasi-chaos." Now most philosophers, and indeed most people, do not set out to emphasize the vague. Quite the opposite: most have pursued certainty, objectivity, and some form of universal truth. Furthermore, most have assumed that such truth can be captured in language, that is, that a complete linguistic description of reality is at least possible, and certainly desirable. Yet James does not.

It might be useful to do a comparative study between M-P's "partiality" and James's "the vague."

Finally, here are a couple of passages from my paper [2] which I tried unsuccessfully to recall today:

> There is a sense in which this unhinging of the Born Rule from being a "law of nature" in the usual conception—i.e., treating it as a normative statement, rather than a descriptive one—makes the QBist notion of quantum *indeterminism* a far more radical variety than anything proposed in the quantum debate before. It says in the end that nature does what it wants, without a mechanism underneath, and without any "hidden hand" of the likes of Richard von Mises's *Kollective* or Karl Popper's *propensities* or David Lewis's *objective chances*, or indeed any conception that would diminish the autonomy of nature's events. Nature and its parts do what they want, and we as free-willed agents do what we can get away with.

> Quantum theory, on this account, is our best means yet for hitching a ride with the universe's pervasive creativity and doing what we can to contribute to it.
>
> The case QBism makes before the forum is this. What the quantum agent—the protagonist in the drama of any application of quantum theory—is ultimately doing is hitching a ride with a *new kind of ontology or metaphysic*. An ontology of all-pervasive, *pan-creative*[14] experience, not unakin to what William James was thinking of when he wrote his essay "Does 'Consciousness' Exist?"

To the extent that I understand it, I think M-P did well to make a distinction between "perception" and "phenomenon," of which my present understanding is this: The first refers to a particular person's experience, while the second attempts to depersonalize the concept and apply it more widely. To mix metaphors, "perception" refers to those "drops of experience" to which an agent is hitching a ride, and "phenomena" refer to those that might be outside any human's outfit. Not making such a linguistic distinction is potentially one of the things that helped derail James and Dewey's program of an ontology of "pure experience." Here's a passage from Martin Jay's book *Songs of Experience* [64, p. 298]:

> Was it ever really possible, other critics asked, to suppress the subjective connotation of the very word "experience," as both James and Dewey insisted was necessary, and allow it to stand for something deeper than the subject/object split? Could experience, moreover, serve as a foundationless alternative to traditional grounding concepts in philosophy, one based on accepting uncertainty rather than seeking to overcome it, or was it merely another variation of the same fruitless quest? Was it, in the final analysis, really anything more than what one critic damned as "a vague, incantatory expression of Dewey's," which served less to answer hard questions than postpone them?
>
> That Dewey himself came to realize he was fighting an uphill battle is evident in a frequently quoted admission he made in a revised edition of *Experience and Nature*, where he wrote in frustration: "I would abandon the term 'experience' because of my growing realization that the historical obstacles which prevented understanding of my use of 'experience' are, for all practical purposes, insurmountable." In its place, he mused, he would now want to put the word "culture," thus duplicating virtually the same move made by Raymond Williams a generation later (and which would have surely led into the same swamps of meaning that almost drowned his British counterpart). In his final work, jointly written with Arthur Bentley, *Knowing and the Known*, Dewey ruefully acknowledged that "experience" would no longer do the work he had thought it might, preferring "trans-action" instead, even if the basic premises of his thought had not really changed.

Anyway, again, a really stimulating discussion for me today. I'm looking forward to next week's.

---

[14] This terminology is adapted from, but may not be identical to, Michel Weber's designation of Alfred North Whitehead's later ontology as a *pancreativism* [63].

10 3 Mar. 2021 *Amanda's Hooks* (to the QBuki)

It is hard for me keep up with all the responses *I dream of making* to so many of your notes. A case in point was the note Amanda wrote to all of us on February 26:

Gefterism 2: *Let's say, on the one hand, I take for granted all the levels of personalism that Blake was mentioning. But let's say I also take seriously Merleau-Ponty's "reversibility of the flesh" and the notion of a shared interaction, which says that when I interact directly with another agent (I'm Wigner asking my friend what outcome she got after breaking the hermetic seal of the lab), the outcome of my measurement of my friend can also be read in reverse as my friend's outcome of her measurement on me. Can I then say, that after my measurement of my friend, with its outcome personal to me, I update my beliefs/priors, which will then inform future quantum state assignments I give to relevant measurements—and meanwhile my friend updates her beliefs/priors, which will then inform future quantum state assignments she makes on relevant measurements, and that while all of that is happening in our own "island universes," the reversibility of our original interaction means that we should be updating our priors "in the same direction," as it were, and therefore that while our quantum state assignments will never be the same, because we will always have different contexts as a result of our having different bodies with different histories, they can at least begin to converge as a result of our interactions? Or have I violated QBism in some way in that scenario?*

This caused a great excitement in me, and I jotted down very quickly:

> There are some elements I really like in this!! I don't think it is convergence necessarily (though it could be), but a kind of synchronic update (the consequence for each of them) for each agent. A natural mapping between consequences. ... I think there is a road to progress in this thought!

and then a little bit later

> I should have said "between their individual consequences." In Merleau-Ponty speak, these considerations concern all that happens after the reflective individuation occurs. I.e. the level at which the quantum formalism might be made use of. At lower levels of reality, there's no good sense to it. At best it can inform us of the lower (actually *wider*) levels of reality.

But I never *adequately* responded to express all the flurry of thoughts Amanda provoked in me.

The epiphany in me (or more accurately the déjà vu) provoked by Amanda's remarks was this idea: The formal expression that Wigner and his friend in their handshake are two poles of the same "genuine stuff" (as I called it in the section titled "Concerning the Genuine Stuff" above) is that Wigner and friend both use the apparatus of quantum to update from prior to posterior states. The handshake is the "hook" or "pivot" or "spur" upon which they base an update. This alleviates the problem that intersubjective agreement must necessarily find a place either in the individual practices or experimental consequences of quantum theory.

When I wrote to Amanda in the section titled "Consistency Is Way Overrated," "From my point of view, consistency is way overrated: Agents need not have agreement to live in the same world. The very fact that they are fighting it out (perhaps to the death) is already indication enough that they are in a single world," I didn't have quite this clarity in my mind. The key point I want to try to express is that I have always been suspicious of intersubjectivity as a criterion for agents to be said to be living in a common world, and I think Amanda's remarks got me a little closer to clarifying what the real issue is here.

Now, why did I say "more accurately the déjà vu"? In the abstract to my monstrous samizdat [39], I wrote, "More roundly, the document is an attempt to provide an essential ingredient, unavailable anywhere else, for turning QBism into a live option within the vast spectrum of quantum foundational thought." How I should heed my own words and look through it from time to time. Take this passage from 4 November 2002, essentially 18 years before I had ever heard of Merleau-Ponty:

> We generally write a POVM as an indexed set of operators, E_d. Here is how I would denote the referents of those symbols. The index d should be taken to stand for the raw data that can enter our attention when a quantum measurement is performed. The whole object E_d should be construed as the "meaning" we propose to ascribe to that piece of data when/if it comes to our attention. It is important here to recognize the logical distinction between these two roles. The symbol d stands for something beyond our control, something that enters into us from the world outside our head. The ascription of a particular value d is not up to us, by definition. The *function* E_d, however, is of a completely different flavor. It is set by our history, by our education, by whatever incidental factors that have led us to believe whatever it is that we believe when we walk into the laboratory to elicit some data. That is to say, E_d has much the character of a subjective probability assignment. It is a judgment. [...]

[With that in mind,] here's a point of view I'm finding myself more and more attracted to lately.

I think it is safe to say that the following idea is pretty commonplace in quantum mechanical practice. Suppose I measure a single POVM twice—maybe on the same system or two different systems, I don't care—and just happen to get the same outcome in both cases. Namely, a single operator E_d. The common idea, and one I've held onto for years, is that there is an objective sense in which those two events are identical copies of each other. They are like identical atoms ... or something like the spacetime equivalent of atoms. But now I think we have no warrant to think that. Rather, I would say the two outcomes are identical only because we have (subjectively) chosen to ignore almost all of their structure.

That is to say, I now count myself not so far from the opinion of Ulfbeck and Bohr, when they write [65]:

> The click ... is seen to be an event entirely beyond law. ... [I]t is a unique event that never repeats ... The uniqueness of the click, as an integral part of genuine fortuitousness, refers to the click in its entirety .... [T]he very occurrence of laws governing the clicks is contingent on a lowered resolution.

For though I have made a logical distinction between the role of the d's and the E_d's above, one should not forget the very theory-ladenness of the set of possible d's. What I think is going on here is that it takes (a lot of) theory to get us to even recognize the raw data, much less ascribe it some meaning. In Marcus Appleby's terms, all that stuff resides in the "primitive theory" (or perhaps some extension of it), which is a level well below quantum mechanics. What quantum mechanics is about is a little froth on the top of a much deeper sea. Once that deeper sea is set, then it makes sense to make a distinction between the inside and the outside of the agent—i.e., the subjective and the objective—as we did above. For even in this froth on the top of a deeper sea, we still find things we cannot control once our basic beliefs—i.e., our theory—are set.

Without the potential d's we could not even speak of the possibility of experiment. Yet like the cardinality of the set of colors in the rainbow—Newton said seven, Aristotle said three or four—a subjective judgment had to be made (within the wide community) before we could get to that level. If this is so, then it should not strike us as so strange that the raw data d in our quantum mechanical experience will ultimately be ascribed with a meaning E_d that is subjectively given. (I expressed some of this a little better in a note I wrote to David last month; I'll place it below as a supplement.) More particularly, with respect to the EPR example above, it should not strike us as odd that the phenomenon comes about solely because of an interpretive convention we set: All quantum measurement outcomes are unique and incomparable at the ontic level. At least that's the idea I'm toying with.

In a later note from 28 July 2003, I complained to Rüdiger Schack that he didn't have enough ontological inclinations for my tastes:

> In your newest turn, such discussion would be "meaningless"—correct? You would no longer allow yourself to contemplate what the *click* d might be in its own essence. You would no longer say that "successive clicks with a single value d are truly the same," nor would you contemplate that "successive clicks with a single value d may actually be truly different."
>
> That is to say, in this new philosophy you are toying with, if something (i.e., an event, a fact, etc.) is not a "hook" upon which a probability can be conditioned, you are not willing to speak of it at all. It is meaningless—I believe you say. You don't even let yourself conjecture about the stuff that is out there independently of us. (Or at least I don't see how you're going to be able to do this with such a strongly positivistic line.)
>
> As I tried to express today, and as I'm trying harder to articulate now, I guess I don't like that. Pragmatists are not positivists, but more opportunists.

Attached is the full discussion for anyone who cares to look. Also attached is a nice find by Jacques today that's quite relevant to this very discussion [66]. Its abstract reads, "For two ideally rational agents, does learning a finite amount of shared evidence necessitate agreement? No. But does it at least guard against belief polarization, the case in which their opinions get further apart? No. OK, but are rational agents guaranteed to avoid polarization if they have access to an infinite, increasing stream of shared evidence? No."

OK, time for QBuki. One more articulated response down; 458 to go!

11 Conclusion

Indeed, 458 to go, and probably more. Then again, as the 20 some odd years of meetings in Växjö have demonstrated, the quantum is no easy code to crack. Congratulations again Andrei, both on your work and the community you have built.

Acknowledgements I thank Blake Stacey for a number of suggestions about the material included herein and Arkady Plotnitsky and Emmanuel Haven for infinite patience. This work was supported in part by the John E. Fetzer Memorial Trust.

References

1. C.A. Fuchs, B.C. Stacey, QBism: quantum theory as a hero's handbook, in *Proceedings of the International School of Physics "Enrico Fermi" Course 197 – Foundations of Quantum Physics*, ed. by E.M. Rasel, W.P. Schleich, S. Wölk (IOS Press, Amsterdam; Società Italiana di Fisica, Bologna, 2018), pp. 133–202. arXiv:1612.07308
2. C.A. Fuchs, Notwithstanding Bohr, the reasons for QBism, *Mind and Matter*. **15**, 245–300 (2017). arXiv:1705.03483
3. C.A. Fuchs, "On participatory realism," in *Information and Interaction: Eddington, Wheeler, and the Limits of Knowledge*, ed. by I.T. Durham, D. Rickles (Springer, Berlin, 2016), pp. 113–134. arXiv:1601.04360
4. C.M. Caves, C.A. Fuchs, R. Schack, Quantum probabilities as Bayesian probabilities. Phys. Rev. A **65**, 022305 (2002)
5. C.A. Fuchs, Quantum Foundations in the Light of Quantum Information, in *Decoherence and its Implications in Quantum Computation and Information Transfer*. ed. by A. Gonis, P.E.A. Turchi (IOS Press, Amsterdam, 2001), pp.38–82
6. C.G. Timpson, Quantum Bayesianism: A study, *Studies in History and Philosophy of Modern Physics*. **39**, 579–609 (2008). arXiv:0804.2047
7. E.T. Jaynes, *Probability Theory: The Logic of Science* (Press, Cambridge U, 2003)
8. B. de Finetti, *Theory of Probability* (Wiley, New York, 1990)
9. C.A. Fuchs, Quantum mechanics as quantum information (and only a little more), in *Quantum Theory: Reconsideration of Foundations*, ed. by A. Khrennikov (Växjö University Press, Växjö, Sweden, 2002), pp. 463–543. arXiv:quant-ph/0205039
10. C.A. Fuchs, QBism, the perimeter of quantum bayesianism, arXiv:1003.5209
11. B.C. Stacey, Ideas abandoned en route to QBism, arXiv:1911.07386
12. C.A. Fuchs, *My Struggles with the Block Universe: Selected Correspondence, January 2001 – May 2011*, vol. 2, ed. by B.C. Stacey (2014), p. 349. arXiv:1405.2390
13. C.A. Fuchs, Quantum States: What the Hell Are They? The Post-Växjö Phase Transition (2002). [The document should soon be available at https://qbism.org]
14. C.A. Fuchs, *Notes on a Paulian Idea: Foundational, Historical, Anecdotal & Forward-Looking Thoughts on the Quantum*, with foreword by N. David Mermin, p. 504. arXiv:quant-ph/0105039
15. C.A. Fuchs, *Notes on a Paulian Idea: Foundational, Historical, Anecdotal & Forward-Looking Thoughts on the Quantum,* with foreword by N. David Mermin, (Växjö University Press, Växjö, Sweden, 2003), p. 718
16. C.A. Fuchs, *Coming of Age with Quantum Information: Notes on a Paulian Idea*, (Cambridge University Press, Cambridge, UK, 2010), p. 598
17. D.M. Greenberger, Coming of Age with Quantum Information: Notes on a Paulian Idea. Christopher A. Fuchs. Am. J. Phys. **79**, 1083–1084 (2011). (October)
18. E. Cavalcanti, Quantum subversives. Am. Sci. **99**, 500–502 (2011)

19. A. Trabesinger, Inside quantum information. Nat. Phys. **7**, 443–444 (2011). (June)
20. F. Benedictus, Quantum information. Metascience **21**, 595–600 (2012). (November)
21. E. Cavalcanti, Christopher A. Fuchs: Coming of age with quantum information: notes on a Paulian idea. Quantum Inf. Process. **11**, 633–636 (2012). (April)
22. S. Dick, Christopher A. Fuchs. Coming of age with quantum information: Notes on a Paulian idea. Isis **104**, 646–647 (2013). (September)
23. F. Morikoshi, C.A. Fuchs, *Coming of Age with Quantum Information: Notes on a Paulian Idea*," Nihon Butsuri Gakkaishi .**70**(3), 224 (2015), in Japanese
24. T. Bilban, C.A. Fuchs, *Coming of age with quantum information*." Phainomena. **29**(114/115), 223–229 (2020), in Slovenian
25. C.A. Fuchs, "Interview with Physicist Christopher Fuchs," with Robert P. Crease and James Sares, *C*ontinental Philosophy Review, (5 Feb 2021)
26. H.A. Wiltsche, P. Berghofer, Phenomenological Approaches to Physics: Mapping the Field, in *Phenomenological Approaches to Physics, Synthese Library 429*. ed. by H.A. Wiltsche, P. Berghofer (Switzerland, Springer, Cham, 2020), pp.1–47
27. M. Bitbol, A phenomenological ontology for physics: Merleau-ponty and QBism, in *Phenomenological Approaches to Physics, Synthese Library 429*. ed. by H.A. Wiltsche, P. Berghofer (Springer, Cham, Switzerland, 2020), pp.227–242
28. L. de La Tremblaye, QBism from a Phenomenological Point of View: Husserl and QBism, in *Phenomenological Approaches to Physics, Synthese Library 429*. ed. by H.A. Wiltsche, P. Berghofer (Switzerland, Springer, Cham, 2020), pp.243–260
29. M. Bitbol, Is the life-world reduction sufficient in quantum physics?. Cont. Philos. Rev. (2020)
30. P. Berghofer, Scientific Perspectivism in the Phenomenological Tradition. Eur. J. Philos. Sci. **10**, 30 (2020)
31. F. J. Boge, Realism without interphenomena: Reichenbach's cube, Sober's evidential realism, and quantum solipsism. Int. Stud. Phil. Sci. (2021)
32. J. Pienaar, *"Unobservable Entities in QBism and Phenomenology*, forthcoming in Phenomenology and QBism: New Approaches to Quantum Mechanics, ed. by P. Berghofer, H.A, Wiltsche (Routledge, New York, 2022)
33. F.J. Boge, *"Back to Kant! QBism, Phenomenology, and Reality from Invariants,"* forthcoming in Phenomenology and QBism: New Approaches to Quantum Mechanics, ed. by P. Berghofer, H.A, Wiltsche (Routledge, New York, 2022)
34. S. French, Putting some flesh on the participant in participatory realism, forthcoming in Phenomenology and QBism: New Approaches to Quantum Mechanics, ed. by P. Berghofer, H.A. Wiltsche (Routledge, New York, 2022)
35. M. Bitbol, L. de La Tremblaye, *"QBism: An Eco-Phenomenology of Quantum Physics,"* forthcoming in Phenomenology and QBism: New Approaches to Quantum Mechanics, ed. by P. Berghofer, H. A, Wiltsche (Routledge, New York, 2022)
36. H.C. von Baeyer, *"On the Consilience between QBism and Phenomenology*, forthcoming in Phenomenology and QBism: New Approaches to Quantum Mechanics, ed. by P. Berghofer, H.A, Wiltsche (Routledge, New York, 2022)
37. C.A. Fuchs, The anti-Växjö interpretation of quantum mechanics, in *Q*uantum Theory: Reconsideration of Foundations, ed. by A. Khrennikov (Växjö University Press, Växjö, Sweden, 2002), pp. 99–116. arXiv:quant-ph/0204146
38. C.A. Fuchs, Delirium quantum, in *Foundations of Probability and Physics – 4*, ed. by G. Adenier, C.A. Fuchs, A.Yu. Khrennikov, in AIP Conference Proceedings, Vol. 889 (American Institute of Physics, Melville, NY, 2007), pp. 438–462. arXiv:0906.1968
39. C.A. Fuchs, R. Schack, Quantum measurement and the Paulian idea, in *The Pauli-Jung Conjecture and Its Impact Today*, ed. by H. Atmanspacher, C.A. Fuchs (Imprint Academic, Exeter, UK, 2014), pp. 93–107. arXiv:1412.4209
40. J. Pienaar, Extending the agent in QBism. Found. Phys. **50**, 1894–1920 (2020)
41. J. Pienaar, QBism and relational quantum mechanics compared. arXiv:2108.13977
42. E. Schrödinger, *Nature and the Greeks and Science and Humanism* (Cambridge University Press, Cambridge, UK, 2014)

43. M. Moller, 'The many and the one' and the problem of two minds perceiving the same thing. William James Stud.**3**, (2008)
44. F.C.S. Schiller, Axioms as postulates, in *Personal Idealism*, ed. by H.C. Sturt (MacMillan and Co., New York, 1902), pp. 43–133. posted at https://en.wikisource.org/wiki/Personal_Idealism/Axioms_as_PostulatesWikisource
45. R.A. Putnam, Introduction, in *The Cambridge Companion to William James*. ed. by R.A. Putnam (Cambridge University Press, Cambridge, UK, 1997), pp.1–10
46. C.A. Fuchs, M.C. Hoang, B.C. Stacey, The SIC question: History and state of play. Axioms **6**, 21 (2017). arXiv:1703.07901
47. J.B. DeBrota, C.A. Fuchs, J.L. Pienaar, B.C. Stacey, Born's rule as a quantum extension of Bayesian coherence. Phys. Rev. A **104**, 022207 (2021)
48. J.B. DeBrota, C.A. Fuchs, B.C. Stacey, Symmetric informationally complete measurements identify the irreducible difference between classical and quantum systems. Phys. Rev. Res. **2**, 013074 (2020)
49. D.M. Appleby, C.A. Fuchs, B.C. Stacey, H. Zhu, Introducing the Qplex: A novel arena for quantum theory. Eur. Phys. J. D **71**, 197 (2017)
50. D.M. Appleby, C.A. Fuchs, H. Zhu, Group theoretic, Lie algebraic and Jordan algebraic formulations of the SIC existence problem. Quant. Inf. Comp. **15**, 61 (2015)
51. Y. Aharonov, C.A. Fuchs, S. Goldstein, T. Norsen, T. Rudolph, R. W. Spekkens, C.G. Timpson, "PIAF 09 – Panel Discussion," 30 Sept 2009. https://pirsa.org/09090098PIRSA:09090098
52. N.D. Mermin, Confusing Ontic and Epistemic Causes Trouble in Classical Physics Too, talk at PIAF '09 New Perspectives on the Quantum State, 27 Sept 2009
53. H. von Baeyer, *Petite Leçons de Physique dans les Jardins de Paris* (Dunod, Paris, 2009)
54. B.C. van Fraassen, Belief and the will. J. Philos. **81**, 235–256 (1984)
55. C.A. Fuchs, R. Schack, Bayesian conditioning, the reflection principle, and quantum decoherence, in *Probability in Physics*, ed. by Y. Ben-Menahem, M. Hemmo (Springer, Berlin, Frontiers Collection, 2012), pp. 233–247. arXiv:1103.5950
56. C.M. Caves, C.A. Fuchs, R. Schack, Unknown quantum states: The quantum de Finetti representation. J. Math. Phys. **43**, 4537–4559 (2002)
57. J.B. DeBrota, C.A. Fuchs, R. Schack, Respecting one's fellow: QBism's analysis of Wigner's friend, *Foundations of Physics*. **50**, 1859–1874 (2020). arXiv:2008.03572
58. C.A. Fuchs, R. Schack, Priors, in quantum Bayesian inference, in Foundations of probability and physics - 5, Vol. 1101, ed. by L. Accardi et al. A.I.P. Conference Proceedings, (American Institute of Physics, Melville, NY, 2009), pp.255–259
59. C.A. Fuchs, R. Schack, From quantum interference to Bayesian coherence and back round again, in Foundations of Probability and Physics - 5, vol. 1101, ed. by L. Accardi et al. A.I.P. Conf, Proc (American Inst. Phys, Melville, NY, 2009), pp.260–279
60. C.A. Fuchs, Interview with a quantum Bayesian, in *Elegance and Enigma: The Quantum Interviews*, ed. by M. Schlosshauer (Springer, Berlin, 2011)
61. C.A. Fuchs, R. Schack, Quantum-Bayesian coherence. Rev. Mod. Phys.**85**, 1693–1715 (2013)
62. E. Cavalcanti, The view from the Wigner bubble. Found. Phys. **51**, 39 (2021)
63. M. Weber, *Whitehead's Pancreativism: The Basics* (Ontos Verlag, Frankfurt, 2006)
64. M. Jay, *Songs of Experience: Modern American and European Variations of a Universal Theme* (U. California Press, Berkeley, 2005)
65. O. Ulfbeck, A. Bohr, Genuine fortuitousness. Where did that click come from? Found. Phys. **31**, 757–774 (2001)
66. M. Nielsen, R.T. Stewart, Persistent disagreement and polarization in a bayesian setting. Br. J. Philos. Sci. **72**(1), 51–78 (2020)

Contra Multos Verbos: On Scandals of Quantum Mechanics

Theodorus Maria Nieuwenhuizen

Abstract In 2008 Nico van Kampen wrote in his letter *The scandal of quantum mechanics*: "The scandal is that there are still many articles, discussions and textbooks, which advertise various interpretations and philosophical profundities." Not much has changed since then, while social media have given a platform for more of what Nico would term "a scandal". A detailed viewpoint is presented on the status of quantum mechanics, distilled from two decades of work with Armen Allahverdyan and Roger Balian on the dynamical solution of Curie-Weiss models for quantum measurement. It embodies a certain minimal form of the statistical interpretation and stays clear of ontological connections. Along the way, comments on various related subjects, terms and interpretations are given. Connections are made to contributions of Andrei Khrennikov, to whom this essay is dedicated.

1 Introduction

Congrats!
Half way-i,
Andrei!

This essay is dedicated to the 65th birthday of Prof. Andrei Khrennikov. We have shared some history, a warm friendship and deep mutual understanding. We met when Andrei invited me to the 2004 edition of his Växjö conference series on Foundations of Quantum Mechanics (and Probability), and I have participated basically every time since then. It was great to realize that we share our basic view on quantum mechanics (QM) and it was great to find at the Växjö meetings a podium and a specialized audience for works on dynamics of quantum measurement and, from that, the interpretation of QM. The various questions led to further consideration and

T. M. Nieuwenhuizen (✉)
Institute for Theoretical Physics, University of Amsterdam, PO Box 94485, 1090 GL, Amsterdam, The Netherlands
e-mail: t.m.nieuwenhuizen@uva.nl

A. Plotnitsky and E. Haven (eds.), *The Quantum-Like Revolution*,
https://doi.org/10.1007/978-3-031-12986-5_4

Fig. 1 Andrei (left) and the Author during the 2012 Växjö conference, amazed by an asymmetric top. If it is turned clockwise it continues its motion; if turned counterclockwise it stops and starts slowly turning clockwise. Courtesy by Marian Kupczynski

better formulations, resulting in deeper understanding. Another effect, of course, is spreading the good news about steps of progress.

I would like to recall the 2005 Växjö meeting. I was co-organizer and celebrated my 50th birthday during the meeting. On a sunny afternoon we made a cultural tour that ended by boat at an island. Andrei offered me a special gift, one of his books, and I toasted with the participants. Next day, the conference dinner in the splendid Växjö castle, was on my birthday. Figure 1 shows us at another Växjö meeting.

Andrei and I coauthored a paper [1] and we co-organized or participated in various meetings, notably the FQMT series in Prague, and Andrei lectured at summer schools I organized in João Pessoa. So Andrei, thanks for that; I wish you a bright future and hope that the ideas I present here inspire you for further modelling, analysis and discussion.

1.1 Oh, There is a Scandal?

I read the news today, oh boy
And though the news was rather sad
Well, I just had to laugh ...
A Day in the Life
The Beatles

The title of this essay is inspired by the 2008 letter to the editor of the American Journal of Physics with title *The scandal of quantum mechanics* by Nicolaas 'Nico' Godfried van Kampen [2]. Next to Martinus Veltman and Gerardus 't Hooft, Nico was one of my main teachers and influencers at the University of Utrecht in the 1970's. We coauthored a criticism [3], while our roads regularly crossed until his passing in 2013. Nico published on the interpretation of QM in 1958 and 1988. The latter paper, titled *Ten theorems about quantum mechanical measurements* [4], can

Fig. 2 Nicolaas 'Nico' Godfried van Kampen. Courtesy by Gerard 't Hooft

be summarized as: Theorem I: *Quantum mechanics works* and Theorem IV: *Whoever endows ψ with more meaning than is needed for computing observable phenomena is responsible for the consequences.* I remember Nico presenting it at the annual meeting of the Netherlands Physical Society in the early 1990s, a vivid presentation in a fully packed room. Nephew Gerard(us) came in late and had to sit on the floor, as he must have done so often as a boy, trying to grasp as much as possible from his uncle's lessons. For a photograph of Nico, see Fig. 2.

Nico kept on still going strong. In his 2008 letter, Nico writes at age 88: *The scandal is that there are still many articles, discussions and textbooks, which advertise various interpretations and philosophical profundities.* So for Nico nothing changed in the two decades since his 1988 paper. Has something changed since then? Yes, but not only in a positive direction. The good news is that one class of realistic models for the dynamics of quantum measurement has been solved by Armen Allahverdyan, Roger Balian and myself (hereafter: ABN), to be discussed below.

1.2 The Battle at the Social Media

Please don't bother me with your many worlds,
I have already so many problems in this world;
Neither bother me with your multiverse,
I have too many problems in this Universe
Anonymous

In contrast to the situation on which Nico reflected in 2008, now there are also the social media. One may find many clips, podcasts and panel discussions on the interpretation of QM. Next to giving explanation concerning interesting subjects or points of progress, they can promote subjects that would be better off with silence. While support for science has always been based on a motivation for rewards or spin offs, social media emphasize one further component: popularization. Clearly, concerning the ill-informed general audience, this "new axis in Hilbert space" is not immediately–and often not at all–connected to scientific relevance.

Thus, while, for example, Sabine Hossenfelder stays on the safe side of the debate by supporting a version of the statistical interpretation, but would be better off by not speaking about metaphysical notion of "the wave function of the particle", various others, including other respectable colleagues such as Sean Carroll and Max Tegmark, are caught by the spell of the many worlds interpretation, and still others, like Mario Livio, by the multiverse interpretation. In my opinion, Sean, Max and Mario fall in the trap of Nico's above mentioned Theorem IV. The battle to get out of the trap has the smell of the criminal assuring his innocence.

The idea of a multiverse can neither be proven nor disproven; I do not consider this idea to be a part of science, since, certainly for now, there is no way to test it. Scientific guessing is needed, and some of it can become part of testable science, but most of it will not. Quantum mechanics should deal with experiments in the laboratory or even in detectors in satellites, but it should not have to rely on what happens in the remote Universe, let alone in other universes with which no communication is possible. Trying to convince audiences that (the measure problem [63] on) other universes or "unvisited worlds" are relevant to get the proper Born rule for the experiments in a laboratory, seems no less than a scandal to me. Asher Peres stressed that unperformed experiments have no results [5]. There is a good reason for that: during an experiment, a (quantum) force is exerted on the particle or system; this creates a situation different from the original one, a fact of Nature which should not be lost in vague discussion. How this generalizes to "unvisited worlds" or "unvisited universes" is a question solely for science fiction.

So Nico's scandal is certainly not eliminated by the arrival of social media, and, let us admit, neither is it in the literature, not even in the scientific literature.

1.3 Praise of Quantum Theory

O store gud [64]
O store Gud, när jag den verld beskådar
Som du har skapat med ditt allmaktsord,
Written by Carl Boberg
Sung by Sissel with the Tabernacle Choir

Quantum mechanics has allowed us to understand Nature: the solid state, particle physics and even the microwave background of early cosmology. We are now in the era of nanotechnology, mobile phones that do not only allow to phone everybody anywhere on the world, but also have much greater computer power than the big computing halls that were available half a century ago.

There is no doubt about what the quantum formalism embodies, be it that of quantum mechanics or quantum field theory. The structure of Hilbert space and Fock space is well known. There are strict rules for the allowed operators and how to handle them. The Schrödinger equation and the ensuing quantum field theory describe the whole condensed matter. The proof of renormalizability of quantum gauge theories by 't Hooft and Veltman allowed quantum field theory also to describe the whole particle physics. The predictions are unique, given the choice of Hamiltonian or Lagrangian. For a historical overview, see [6].

In laboratories, repeated experiments are performed to obtain statistical accuracy so as to compare to theoretical predictions. At the Large Hadron Collider in Geneva (LHC) the discovery of the Higgs particle was an exciting achievement, finishing off the standard model (SM) of particle physics. But, to the regret of many, no further fundamental physics has been discovered so far. This has been termed "the nightmare scenario". The Dutch language allows to put this in the succinct phrase "De Higgs en verder niks" (The Higgs and further "nicks").

Theoretical predictions for the occurrence rates of many interaction channels has led to an impressive confirmation of the SM, with mild question marks for muon related channels, that may still end up as statistical flukes. But the point to stress here is that there is no doubt about the applicability of quantum field theory in physics. Moreover, its ideas have been applied, among others, by Andrei and his pals in cognitive science, psychology, genetics, economy, finances, and game theory alike [7]; for now, the only question is whether the SM Lagrangian is still good enough, or that lepton universality is absent.

1.4 Is Quantum Theory the Formalism Plus the Postulates?

Eén plus één is twee [65]
Dutch expression

The (Copenhagen) measurement postulates are so well known and rightfully so often taught in freshmen courses, that they may leave the impression that quantum theory consists of two parts: first the quantum formalism (wave functions, Hilbert space, Schrödinger equation, Fock space, operator formalism, ...) and second, the postulates. This gives the postulates a higher status than they deserve. They are merely shortcuts.

In a laboratory one may observe many objects: devices, machines, cables, coolers, ventilators, next to students and the supervisor. But *one does not observe postulates*. The postulates are an abstraction for what happens in a realistic measurement. The proper physics is to model the experimental setup ("context") and the ensuing dynamics that constitutes the measurement. That deals with the only point of contact between the quantum formalism and the reality in the lab, so interpretation of the formalism should be based on that, not on creative (wishful) thinking.

To describe a realistic measurement setup is a hopelessly complicated task, and that has hampered progress in quantum foundations. The best to strive for are idealized models, which are still rich enough to capture the essential physics, but can be handled theoretically, preferably be solved analytically. Even this was a long road, the various models have been reviewed by ABN [8]. The Curie-Weiss model for quantum measurement, to be discussed in Sect. 4, serves this purpose. Its solution is rich, with various aspects of how the dynamics of the measurement is proceeding. But often one is not interested in the apparatus, only in its indications of the pointer (measurement outcomes). To a good but incomplete extent, the essentials of the indications are captured by the postulates.

The principles for a general physical theory are formulated by Darrigol [9]. They involve 4 steps:

- A mathematical formalism, here the Hilbert and Fock spaces, symmetry groups, gauge structure, etc.
- Physical laws, here the Schrödinger and Liouville-von Neumann equations, shortly: Hamiltonian dynamics
- Interpretation, understood as a way to interpret some idealized experiments
- Approximations. They are needed to discard irrelevant aspects so as to come to 'clean' statements

In quantum physics, where an experiment influences the system, one also needs its setting (context) [10],

- A module: the principle or postulate that individual (idealized) experiments yield individual pointer values in a given experimental context.

1.5 *The Ontology of Nature Needs Not Be Encoded in Quantum Mechanics*

Das Leben is hart,
aber ungerecht [66]
German saying

Individual experiments yield individual outcomes. One may ask the question: what happened such that this outcome arose? High energy cosmic ray particles hit the atmosphere now and then, and can be detected by large collective instruments such as IceCube and Auger. A standard question is: how did this particle arrive here? The standard answer is: it got ejected long ago in some remote stellar explosion, and travelled all the years with one goal, namely to hit our detector. All of this is an ontology, with the question: *how* did it precisely happen? Much effort has been devoted to this question, see, e.g., Andrei's works [11, 12].

Since quantum mechanics works so well in practice, it fulfils its primary task of offering an epistemology for the experiments, which is good enough: quantum mechanics works. If it would offer more, such as an ontology, that would be pleasant, but if not, so be it. There is no reason for a theory to be final or to deal with all aspects. Newton's theory does not deal with the atomic structure of the planets and apples it considers, but it is still of great value in its domain of application. Nobody will dismiss Newton theory and, likewise, there is no reason to abandon QM. Still, there are other physical questions, that may once be described in a theoretical framework, such as: in a double slit experiment, through which slit went the first electron that hit the screen?

Bohmian mechanics offers an ontology, but we consider it as inapt, as we discuss in section 3.6.

1.6 *How I Got Here*

Get here, if you can
Written by Brenda Gordon
Sung by Oleta Adams

Let me recall coffee time at the Institute for Theoretical Physics at the University of Utrecht during my Ph.D. time in the early 1980s. All the staff came for common coffee and discussion, and the informal Friday seminar. The coffee was prepared by the youngest student. At some moment Nico started to discuss the foundations of quantum mechanics. When Nico spoke, everybody listened, including professors Johnny Tjon in nuclear physics, Tiny Veltman and Gerard 't Hooft in particle physics, the latter was my master thesis supervisor, and further my most honorable Ph.D. supervisor Theo Ruijgrok in stochastic and quantum physics, and my later postdoctoral supervisor Matthieu Ernst in statistical physics. But busy with their own subjects, they all left one by one, when no clear conclusion was reached. When the topic got stretched out over weeks and even months, it ended with Nico testing his ideas solely on his colleague Ben Nijboer, statistical physicist, deep thinker and every inch a gentleman. And I was the only other one left; I was the audience for this play without script...

Ben exposed a firm opinion about doing a Ph.D. on quantum foundations: A student should first learn how quantum mechanics works; if he has written some 20 papers, he has worked enough with it to have acquired a solid basis for the foundations if he is interested; before that, it can not lead to a serious contribution. I followed Ben's advise, for two decades I just "had no time" for foundations. I carried out research on other subjects, mostly in statistical physics. In fact, I liked Nico's courses on mathematical physics so much, that my first research decade was devoted to toy models of statistical physics [13, 14]. For my field-theoretic approach to compact random walks with relatively few distinct sites visited, published in 1989 [15] and confirmed numerically in $d = 2$ [16] and $d = 3$ [17], I expected it for mathematicians to take 30 years to prove the result firmly, but it appears that they need more.

It was my move to the experimental group on multiple light scattering of Ad Lagendijk in Amsterdam that taught me to look with the eyes of an experimenter [18, 19], and this profoundly changed my view on what is most relevant in physics, to solve various questions in this field [20–22]. Working also on spin glasses, I found the first model where the Parisi function $x(q)$ for the overlaps between states is solved analytically [23]. Remarks by Giorgio Parisi about the breakdown of the Ehrenfest relations for glasses led me to solve this classical paradox [24] and next to formulate the thermodynamics of glasses [25], which had been a hot topic in the 1950s, 60s and 70s. With Ph.D. student Luca Leuzzi we wrote a book on the subject [26], Giorgio was so kind to write the preface.

Next, I wondered what subjects to work on in the next decade. Just around that time in early 1999, Armen Allahverdyan joined my group and we went over many subjects of common interest, including quantum thermodynamics and interpretation of QM. Not long thereafter, we were joined by Roger Balian, forming the "ABN" team for our research on quantum thermodynamics; I coined the by now well known term "ergotropy" for the maximal work extractable from a quantum system [27]. A second theme was our ABN decades long research line on quantum measurement, the main theme of the present essay. What I present below is my view on our achievements and related issues.

1.7 Disclosure

I fully take no responsibility for this
Anonymous

While my own thinking is inspired by my teachers, Nico in particular, and even the more so by my coworkers Roger Balian and Armen Allahverdyan, this does not imply that I follow anyone of them blindly, and, perhaps unsurprisingly, this happens to be completely mutual. They bear no responsibility for the present viewpoints, but would likely do well in agreeing with all of them. Discussions at the various Växjö conferences were very illuminating and stimulating, for which I thank the various participants and Andrei in particular.

In Sect. 2 we discuss some issues of the density matrix. In Sect. 3 we discuss some interpretations of QM. In Sect. 4 we analyze a model for ideal quantum measurement. In Sect. 5 we stress the need of models for quantum measurement for the interpretation of the quantum formalism. Finally, in Sect. 6 we stress the need for an ontology.

2 Meaning of the Density Matrix

Say you don't mind
Written by Denny Laine
Sung by Colin Blunstone

The interpretation of "the wave function", or, more generally, the density matrix, has been subject of fierce debates in last century.

2.1 The least biased interpretation of the density matrix

The least biased interpretation of the density matrix in quantum mechanics is that it codes, in an abstract manner, our best knowledge about the ensemble of identically prepared systems. The ensemble can be a beam of neutrons during one day, or the collisions in the proton beams at the Large Hadron Collider at the CERN during one season. But it may also refer to one ion in a Pauli trap, which is repeatedly excited by a lasar beam and repeatedly emits a photon. Finally, for macroscopic systems, one may imagine a Gedanken ensemble of similar systems.

Like Gibbs state in classical statistical mechanics, the density matrix of an ensemble of quantum systems is a catalog of incomplete knowledge. In both cases, the entropy is a measure for the lack of knowledge. So let me recall the modern view on entropy.

2.2 Entropy in Classical Statistical Physics

Increasing knowledge,
decreasing satisfaction
Anonymous

In Newtonian dynamics of a large set of particles, the dynamics can be followed for some time. It can be said that the entropy is zero then (or, at least, conserved in time). Indeed, entropy is a measure for the lack of information, but in that early time window we know the positions and speeds of the particles, so no information is lost. But at larger times an exponential divergence between orbits starting close to each other, set by the Lyapunov rate (unfortunately called "Lyapunov entropy"), is unavoidable due to chaotic behavior over long times. Weather forecasts, however, bear more upon chaos theory (since the 1970s) than upon classical statistical physics. It is the very reason why reliable forecasts are still limited to a few days despite the enormous increase in computer power in the last decades, with only a modest improvement in accuracy.

In classical statistical physics the modern view is that entropy codes our lack of knowledge of or the missing information about the physics at the microscopic scale [28–31].

Actually, there are many ways of coarse graining and there are many entropies. The founding fathers of thermodynamics already introduced the Clausius entropy, the thermodynamic entropy and the Boltzmann entropy. They considered macroscopic systems, so that the leading term in the volume coincides, which allows to speak about "the entropy". However, if the system is small, while both the bath and the work force are large, one enters the field of Quantum Thermodynamics. Then there are again many entropies and they are all different. Each of them relates in principle to a measurement which could be performed on the system. Hence entropy is a state of knowledge, defined in an experimental context, true or in Gedanken.

The "state of knowledge" role is exemplified by the Gibbs paradox [32]. Imagine a pool table with many red and blue billiard balls in some random configuration. An observer will associate a certain entropy to it. But a color blind observer will add an additional "mixing entropy", $k_B \ln 2$ per ball. Gibbs himself considered two identical compartments with a gas; removing the wall between them, also leads to this mixing entropy. Taking the Gibbs paradox into quantum thermodynamics, Armen and I showed that the difference in entropy can be connected to the possibility of extracting work from the setting difference, which depends on the abilities ("knowledge") of the experimenter ("observer") [33].

Somewhat related to this is the demonstration that independent (tensor product of) quantum states of identical particles acquire their symmetrization dynamically in their collision [34]. So the symmetrized wave functions always used in the solid state need not be used when considering electrons some at the Earth and some at the Moon.

2.3 The Density Matrix for Microscopic Systems

Veel weten van een klein beetje [67]
Dutch expression

A similar observer-dependence applies to the meaning of the quantum wave function or, more generally, the density matrix. The density matrix is an abstract mathematical object that describes our best knowledge of the considered ensemble of identically prepared systems. Notice that it describes *our* best knowledge, to be employed for prediction of outcomes of future experiments.

Concerning "best knowledge", the following example was considered in footnote 29 of our "Opusculo" [35]: A spin $\frac{1}{2}$ is measured along a certain axis. Observer A knows the axis direction and the results, so he can construct the collapsed post-measurement state. Observer B knows that the measurement has taken place, but not the direction of the measurement axis. The best he can do, is to average the final density matrix over all directions of the axis. We may add to this Observer C, who is unaware of this measurement and will continue with the original density matrix; he has least (namely: no) knowledge of the measurement. The best statistical predictions for future experiments can be made by Observer A, the second-best by Observer B and the least-best by Observer C.

2.4 Contra "the Wave Function of the Particle"

The wave function of the system
is a misconception;
The wave function of the Universe
is a cosmic misconception
Anonymous

A person is not a wave, also not when taking part in a wave in a sport stadium. The wave is a description of the collection (ensemble) of persons. A given person can participate in many different waves, so it is quite a stretch to proclaim that the wave is some type of a property of a person in it. Moreover, the behavior of any person will depend on the environment he/she is in; I find it amusing that this sociological property has an analogon in quantum physics, or perhaps even its root in it.

In my opinion the very same holds for ensembles of particles, like the beam of protons in the LHC beam during some period. The same protons could have been a member of a different beam, at another moment, hence to me "the wave function of the particle" is a horribility. Setting apart my opinion, this concept is not needed anyhow, and actually only a burden. Employment of Ockham's razor allows a cleaner scientific situation.

2.5 Contra "the Particle is Here and There at the Same Time"

The electrons in my pinkie are mine!
Anonymous

In colloquial presentations, often statements are made like "in a Stern–Gerlach experiment, the electron is at two places at the same time". This is an absurdity, a nonsense, a scandal. No measurement has ever been done to support that claim, and, actually, none can ever be done. It is a colloquial way of trying to express some of the wonders of quantum mechanics, in particular the wave nature. But Einstein said already: one must simplify as much as possible, but not further. In the end, quantum mechanics will remain a mysterious theory.

The culmination of this reasoning was proposed by John Wheeler in a telephone call to Richard Feynman in the spring of 1940, "all electrons are the same, actually there is only one electron popping up here and there" [36]. Needless to say that this absurdity (scandal) causes questions in electromagnetism and gravitation. To stress that a sound approach is possible, we recall the treatment of Allahverdyan et al, in which electrons at a certain location in the initial state are treated as really different from the ones at other locations by the (non-symmetrized) tensor product [34].

2.6 Pure State as a Purified Limit of a Mixed State

Dat is geen zuivere koffie [68]
Dutch saying

A pure state $|\psi_n\rangle$ corresponds to a density matrix $|\psi_n\rangle\langle\psi_n|$, while a mixed state involves a mixture of them, $\sum_n p_n|\psi_n\rangle\langle\psi_n|$, with $p_n \geq 0$ and $\sum_n p_n = 1$. It is often thought that a pure state has a special meaning, like being attachable to a particle, but we oppose this point of view. In a statistical approach, a pure state is a limit of a mixed state, where all $p_n \to 0$, except for one, which goes to 1. No special meaning is gained in this limit: also the resulting pure state stands for an ensemble of identically prepared particles or systems.

Since the general case is a mixed state, it has to be purified to reach a pure state. To purify one variable, one performs a measurement, selects the members with the desired outcome, and discards the other members. Then a second variable, commuting with the first, can be purified, by a similar procedure; next a third one, commuting with both previous ones, can be purified, and so on. However, in practice only a handful (or perhaps a dozen) of purifications can be carried out. For one reason, because of the difficulty to do any experiment, for another, because one would have to discard nearly all members of the initial ensemble. This is uneconomic, to say the least, and an over focus on very special cases.

2.7 Schrödinger's Ambiguity

In psychology,
the opposite is also true
A. Ross

The quantum formalism has an unexpected property. Suppose we have one purified beam of neutrons with spins in the $+z$ direction and a similar one with spins in the $-z$ direction. Combining them yields the mixed density matrix

$$|\uparrow\rangle\langle\uparrow| + |\downarrow\rangle\langle\downarrow| = \begin{pmatrix} 1 & 0 \\ 0 & 1 \end{pmatrix}. \tag{1}$$

However, the same result is obtained when combining similar beams in the $\pm x$ directions, viz.

$$|\rightarrow\rangle\langle\rightarrow| + |\leftarrow\rangle\langle\leftarrow| = \begin{pmatrix} 1 & 0 \\ 0 & 1 \end{pmatrix}, \tag{2}$$

or by combining similar beams polarized in arbitrary $\pm\hat{n}$ direction. This implies that after the merging, no test can be performed to establish which beams were combined. This simple example shows that a mixed density matrix can in general not be related to individual systems. It is a form of the measurement problem, that is, the inability of QM to describe individual events, i.e., the lack of an ontology of quantum theory. Since in our view a pure state is a limit of a mixed state (obtained after purification by a set of measurements), it does not obtain a special meaning in this (never really reached) limit, so it can neither be connected to individual particles.

2.8 Individuality of Macroscopic Systems

Me, myself and I
Expression

Microscopic systems, like atoms or cosmic ray particles, exist, but there is presently no tool to describe them other than classically. Quantum theory captures their behavior in a statistical way.

Classical systems are macroscopic, they typically have some 10^{25} atoms and degrees of freedom. One can identify a few of them as "system properties", like point of gravity, the speed, the temperature... . From the abstract quantum formalism point of view, it is natural to postulate that when a density matrix contains a mixture of such macroscopic states, it can be decomposed in *individual* macroscopic states. Along this line of reasoning, individuality is a phenomenon emerging in the thermodynamic limit, and overcoming Schrödingers ambiguity. Still, this emergence likely only holds, and is only needed, for the "classical" system variables: "what is quantum should remain quantum".

While needed for interpreting the final state of a quantum measurement, this principle is much more general. Imagine two dynamical descriptions of an Ising magnet in zero field below the thermodynamic phase transition temperature. In the first approach, one may work with classical spins ± 1 and employ Glauber dynamics, which, because of the classicality, deals with individual systems. Alternatively, by directly employing the quantum approach with its quantum dynamics, one may follow Darrigol and idealize the situation [9] by neglecting the anyhow tiny "cat" terms, after which the classicality postulate allows to connect to individual systems.

In the alternative view that the quantum formalism is just a technical framework to describe outcomes of measurements, one starts from the pointer readings on a macroscopic apparatus, and the postulate is evidently not needed [10, 37–39]. Another treatment of the context is by Andrei [40].

2.9 Classical Systems are Mostly Quantum and Thermodynamic

Je kunt niet verder springen
dan je polsstok lang is [69]
Dutch saying

The notion of a "classical apparatus", so often discussed in connection to quantum measurement, is an oversimplification. What the founding fathers like Heisenberg and Bohr were aiming at, was, of course, the pointer of the apparatus. In order that it can be read off or processed electronically, the pointer has to be macroscopic and thus have classical features, but its dynamics will involve, in principle, all the microscopic degrees of freedom.

Like every system in Nature, an apparatus consists of atoms, and many of them, say, 10^{25}, and it has even more degrees of freedom. This holds also for any "classical" system. We can speak about classicality of a few of its macroscopic properties, like the position, momentum, angular momentum and temperature. In a fluid one often considers local quantities like density, pressure, temperature, average speed (wind) and so on, which are averages in a mesoscopic cell, still containing some 10^{12} particles or so. But even here the mentioned ones are only a handful out of the 10^{12} degrees of freedom. Nearly 100% of the degrees of freedom remain quantum and never become classical.

As an example, in section 10.1.2 of their "Opus Magnum" [8], ABN consider a small metallic grain. Its center of mass can be treated as a classical variable, but at low and moderate temperatures its specific heat must be described by Debye theory, that is to say, by quantum mechanics. So Heisenberg's celebrated classical-quantum cut depends not only on the considered system or variable but even on the condition; the classical regime of the specific heat of the grain pertains to high temperature.

2.10 No Apparatus is Microscopic

An elephant cannot easily hide itself
Expression

A "microscopic apparatus" is a contradictio in terminis; it is not an apparatus, but just another quantum system. To say that it "measures" something is misleading (it is a scandal); like any other quantum system, a "microscopic apparatus" just undergoes some quantum dynamics, that can be tested by employing a macroscopic apparatus, of which the outcomes can be read off or processed.

It is often said that the Stern-Gerlach magnet acts as an apparatus. This is misleading (thus another scandal), the apparatus is the screen that collects the electrons in the Stern-Gerlach setup. The also popular term "pre-measurement" does more justice to the action of the magnet [70].

One often sees the apparatus as having both quantum and classical parts with the outcome of the interaction (the registration of the measurement), manifested in the classical part; however, as is evident from ABN's solution of the Curie-Weiss model, this "classical part" has a dynamics that stems purely from the quantum nature of its constituents.

In fact, quantum objects can be macroscopic, too (Josephson junctions) even though their quantum nature is defined by their microscopic constitutions and can only be detected by a suitable apparatus.

Not every system can act as an apparatus. Obviously, a broken apparatus, or even an apparatus not under the right conditions, does not act as a proper apparatus, let alone other objects, such as rocks. These trivial examples corroborate that a working apparatus is a special macroscopic system under special conditions, adept for the task under consideration.

2.11 No Macroscopic System, Cat nor Universe, can be in a Pure State

Next time I see a Schrödinger's cat,
I won't be able to contain myself
Anonymous

The idea that a cat or an apparatus, or for the sake of the matter, any macroscopic system, can start in a pure state (on the eigenbasis of some operator), is just another scandal. A state is a mathematical description of the knowledge about the ensemble of identically prepared systems, in reality or in Gedanken. For a macroscopic system, only a relatively very limited amount of information can be known and described.

Starting from a complete description, going to a reduced one amounts technically to tracing out all other degrees of freedom. In the Liouville-von Neumann equation of motion for the density matrix ρ, viz. $\mathrm{d}\rho/\mathrm{d}t = (i/\hbar)[H, \rho]$, the tracing is easily carried out on the left hand side. The right hand side involves the trace of the commutator. Except for extremely fine-tuned Hamiltonians, no simplification occurs, so that the traced equation is not closed. It can not be solved and at best be used for very special purposes or in approximation. The physical reason herefore is, of course, that the multitude of microscopic degrees of freedom can not be neglected; they are all relevant, and in principle they all influence the dynamics of reduced system parts such as the pointer.

Adherents of many worlds theories or spontaneous collapse models omit this check, so that they effectively work with extremely fine tuned but unspecified Hamiltonians, or, more often, abandon modelling at all, and restrict themselves to arguing with linguistic concepts, not guided nor constrained by detailed analysis of the physics.

Hence a *cat* can not be in a pure state. The statement "the cat is in a superposition of a state dead and a state alive" is a poetic construct with no bearing on reality. The conclusions drawn from it already convinced Schrödinger that this line of reasoning leads to absurdity (to scandals). We side with him.

An *apparatus* has to be macroscopic, so it can not be in a pure state. This invalidates the von Neumann-Wigner theory for quantum measurements and exposes the Many Worlds Interpretation as inapt. Instead, in a quantum measurement with the apparatus necessarily in a mixed state, the total system of tested system plus apparatus always starts out in a mixed state, also when the tested system starts out in a pure state. Consequently, there is no unitarity paradox: the system itself can go from pure to mixed, since during the measurement it is an open system coupled to the apparatus. More precisely, the unitarity paradox is solved by discarding the off-diagonal blocks, after they have become irrelevant for all practical purposes, in line with step 4 in Darrigol's general scheme of Sect. 1.4.

A *black hole* can not be in a pure state, certainly not an astrophysical one that starts out from an imploding star, but neither one that arises by the collapse of some chunk of matter in the early Universe. Hence there is no black hole unitarity paradox; Hawking evaporation will not cause a transition from a pure state to a mixed one because considering a pure state is a mathematical or philosophical exercise unrelated to a physical black hole. The fine grained entropy (von Neumann entropy) is conserved under unitary motion, also when Hawking radiation occurs. It is impossible to measure the fine grained entropy in the solid state, say when burning coal, let alone by a Dyson sphere of detectors around the black hole. The coarse grained entropy of a closed system increases according to the second law of thermodynamics. The increase of coarse grained entropy in Hawking radiation is compatible with the decrease the coarse grained entropy of the black hole, as only the sum has to increase. It can be summarized as "the information goes out with the Hawking radiation", and "finally the black hole has very small entropy".

The *Universe* can not start out in a pure quantum state. Quantum theory describes our probabilistic knowledge about the Gedanken ensemble we imagine it to be a member of. No set of apparatuses for measuring commuting variables can be imagined which would purify the mixed state describing our incomplete knowledge about an ensemble of identically prepared Universes. And it seems difficult to give a meaning to discarding the very, very nearly 100% of the candidate universes with wrong measurement outcomes. Assuming to have the absolute knowledge about (the ensemble to which) the whole Universe (belongs), is making oneself equal to God. I would beg: then please make next step, and also create, for instance, another Universe. Unitarity considerations (to go from a pure to a mixed state in a measurement) based on "the wave function of the Universe" are mental exercises that have no bearing on our Universe; they can be called "poetry". Even so, this is not to claim that applications of e.g. the Hartle-Hawking wave function will always lead to false results, but to stress that its use should be motivated.

2.12 Many Mathematical Issues are Devoid of Physical Relevance

Much ado about nothing
Shakespeare

The von Neumann entropy is conserved in time and so is the trace of any power of the density matrix. These quantities express the unitary character of the evolution, but they can not be measured and there is nothing physical that can be deduced from them.

It is possible to consider the wave function of a cat or of the Universe, thereby imagining that the mixed density matrix of such entities has been purified completely. However, there is no sound way of even imagining that the purification can be carried out in any reality. Hence in my opinion there is no sense in considering such concepts, rather they are a reliable source for paradoxes. Not a complete ridicule: a more physical and urgent question seems to me: how many angels can dance on the tip of a needle?

Whether or in how far the landscape of string theory and the many universes in the multiverse fall in this category, is an interesting point for reflection.

3 On Some Interpretations of Quantum Mechanics

For progress in this field
less study is needed
Anonymous

The interpretation of QM has eluded the community for long. Erwin Schrödinger originally supposed that his wave function describes the charge density, but Max Born brought the connection to probabilities, which is so deep, that now, about a century later, it is still debated. Historical papers on the subject are collected by Wheeler and Zurek [41].

3.1 *Interpretatio Copenhagenensis*

Most often employed is the so-called Copenhagen interpretation. It serves in introductory courses on QM, since a short cut for interpretation will work for the many applications. It is also often employed in practice where one needs to compare measured frequencies with theoretical probabilities, it is David Mermin's minimalistic and empirical "shut up and calculate" approach without bothering about the pitfalls of the foundations.

While there is a broad literature on its various formulations, we recall the main points: QM applies to individual objects in a probabilistic way; a particle has its wave function (wave packet); observation leads to reduction of the wave packet (collapse of the wave function); the Born rule holds for probabilities of outcomes; and the quantum description is independent of the mind of the experimenter.

As seen in Sect. 5, ABN basically support this interpretation, except for their more modest interpretation of the wave function or density matrix.

3.2 *Contra Multos Mundos*

To excel in explaining concepts
that cannot be understood
Anonymous

After Bohr and Heisenberg presented initial ideas, von Neumann presented a first model for quantum measurement, so it became possible to connect interpretation to models of measurement. The "many worlds" interpretation (MWI) starts out with supposing that the system S and the apparatus A both begin in a pure state. By the coupling, A is said to evolve to a pure state correlated to S. But this is a magic dynamics, if not a Kindergarten dynamics. In Nico's terms: it is a scandal that it is still presented. The MWI has never been backed up by a solvable model and, let us admit, it is counterintuitive and not economical. The bishop of Ockham would find a worlds-wide if not heavens-wide task here for employing his razor.

The MWI is sometimes termed the "many *words* interpretation", and there is some truth in this bizarre and even ridiculous term, which also entered the title of this work. Indeed, many papers have been written where authors try to explain various issues in sharp words, and the internet is flooded with discussions about it. That something can not be explained, as is claimed, has indeed a spell for many people. While society could choose to go to the next level and decide to any answer scientific question by a popularity poll rather than by fundamental research, that would obviously be a dead end street for the development of new materials and devices. "QAnon" physics may play a political card, but it can not bring long term progress and prosperity for society. Nature is just as it is. If it is to be liked, we better adjust our taste to it. Personally, I have a sweet spot for things that exist and mechanisms that work.

The problem with approaches without stringent mathematics is that there are far more concepts than words. Not only for the Bible can one argue for hours about the meaning of a certain phrase, it can and has also be done for statements by physicist and philosophers alike, including famous people like Einstein, Bohr and Feynman. Not intuitive reasoning, but only a stringent handling of the well determined mathematical structure of QM can offer the hope of reaching a sensible and, hopefully, unique interpretation of QM. The unease of this field is summarized in David Mermin's mantra "shut up and calculate", don't waste your time on this philosophical issue, but work on verifiable predictions. After all, "the" Copenhagen interpretation works quite well in practice.

After physicists gave up the hope of "solving the measurement problem", it was left for philosophers. They did not solve it either. In hindsight, this is not surprising since for long the physicists had no sufficiently good models to direct them. Working from the postulates is just a black-box treatment of measurements, which bypasses a lot of physics that is covered in the dynamical solution of an ideal measurement. But in a laboratory, a measurement is the bread and butter of the experimenter, rather than a black box. Philosophy cannot bypass a physical process, it should, instead, closely follow what happens in the laboratory.

In the many-worlds interpretation, reduction of the density matrix is even denied, and regarded as a delusion due to the limitations of the human mind. Yes, it is often proclaimed that our minds are not evolved to understand certain abstract notions. But that smells like another scandal, "a proof by complete intimidation", to cover up the fact that the proper framework or tool for the question under consideration has just not been discovered.

Below we discuss that the truncation of the density matrix on the measurement basis does occur in a fundamental analysis of dynamics of a quantum measurement. Hence the MWI proponents should now take up the challenge of demonstrating that the MWI dynamics is also derivable in a serious microscopic model, to avoid scandal. Or else, to give up on it.

The failure of MWI can be understood as follows. Often, the approach is to reason from imagining to follow a particle along its path in time, during which measurements are performed. The assumption–even by many though not all adherents of the Copenhagen interpretation–that the particle has its own wave function is particularly unfortunate here. At each measurement, its wave function is said to split up so as to create one new world; this idea is as mind boggling as science fiction and equally absurd; it is a scandalous story told about the particle based on no more than wishful thinking, more precisely, on the hope that QM embodies an ontology. Which, alas, it does not.

3.3 *Contra Theoriam Bohmi*

The double slit experiment can be carried out at such a low intensity, that the probability for two electrons being present at the same time, is very small. While the resulting many-electron interference pattern on the photographic plate will not be different from setups with higher intensity where the electrons may overlap in time, this clarifies one issue: electron-electron interaction does not lie at the heart of quantum interference and the Bohm potential.

The de Broglie-Bohm theory (Bohmian mechanics) arises from the map of the complex wave function of a particle to its amplitude and phase, after which the amplitude is connected to the probability density of the particle and the phase to its velocity [42, 43]. It is supposed to be an ontology of the particles involved. They move in a potential which is the sum of the external potential and a "quantum" potential, set by the shape of the wave function. But the low-intensity experiment dashes the hope that Bohm's "quantum potential" can be the cause of quantum interference.

While it is intuitively clear to view the Bohm trajectories as real particle trajectories, there is another problem: measurement. One can not just read off the coordinate of the particle and be allowed to say that the particle is at that point. Since Bohm's theory makes the same predictions as standard QM, the dynamics in the Curie-Weiss model for quantum measurement, should be expressed in Bohm's framework. Hence the already complicated dynamics must be written in a much more complicated way, with much pain and without any gain.

3.4 *Contra Verbos Zureki*

Another issue is decoherence, important for describing the classical world from quantum mechanics [44, 45]. But it is a far stretch to hold decoherence as the main actor in measurement. Although the ideas of decoherence and pointer states is sensible, it is by far not enough to describe quantum measurements.

In a rather abstract approach, Zurek introduces "pointer states of the environment", which have a long lifetime, and are claimed to explain quantum measurements [46]. But this linguistic stretch is a scandal. While it is true that the environment plays some role, this can not be connected to the complete measurement; in fact, the environment is by assumption that part of our full system that we can not control. The "pointer states of the environment" can not be read off, and this name is only creating confusion. An analogy: The air that surrounds me measures my size, but this cannot be called a measurement; no one can use it to know my size. Otherwise, zillions of solar neutrinos pass through our bodies, but they measure nothing; would they, then they would destroy life itself.

Despite their name, these "pointer states" have no bearing on realistic pointers that are necessarily macroscopic. No experimenter will count on uncontrollable resonances in the environment of his setup; to the opposite, he would discard events where they occur as flukes. Rather, he needs a controllable setup where he can start and finish the measurement at his planning, and he needs a pointer that can be read off his pointer in a time window of his choice. To achieve this, the pointer must be macroscopic, so he needs a macroscopic apparatus consisting of very many (up to 10^{26}) atoms and an even larger number of degrees of freedom. When the pointer has a small number of stable states (2 in the Curie-Weiss model), one may speak of pointer states; they are thermodynamic states of the apparatus in various models for ideal measurement, not pure states of the environment.

In the Curie-Weiss model then bath (environment) plays two roles. In the off-diagonal elements it leads to decoherence after the dephasing has hidden the signal. In the diagonal elements the thermal bath acts as usual, and absorbs the free energy dumped during the thermodynamic phase transition from paramagnet to one of the ferromagnetic states; more precisely, it determines the form of the evolution of the magnetization distribution $P(m, t)$. The essential result, a magnetized final state that can be read off, is universal, however, for all good apparatuses.

4 Basics of the Dynamics of an Ideal Quantum Measurement

The locomotion
Written by King and Goffin
Sung by Little Eva

When an experimenter employs an apparatus, he has been sure that it has a pointer, so that he can read off the result of each measurement. Clearly, the pointer has to be macroscopic and with it the apparatus. The operator to be measured should enter the system-apparatus coupling linearly. In an ideal measurement, it will not change during the measurement, and one will only "measure what is there".

The apparatus has to start in a metastable "ready" state, a thermodynamic state triggered by the coupling to the tested system to go to one of its thermodynamically stable states. This is also what happens in night vision: a handful of photons trigger an avalanche of electrons in the retina, and the generated electric current is processed in the visual cortex. Recharging leads to a short period of blind vision.

4.1 Introducing the Curie—Weiss Model for Quantum Measurement

Listen, people,
to what, I say
Written by Graham Gouldman
Performed by Herman's Hermits

The so called Curie-Weiss model for quantum measurement was introduced and partly solved by ABN in 2003 [47, 48] and 2004 [49]. Some progress reports were published as contributed papers to the Växjö meetings [50, 51]. In 2013 we present an overview of such dynamical models and then go on to solve many aspects of the Curie-Weiss model in over 550 equations [71] [8]. This allows to base the interpretation on

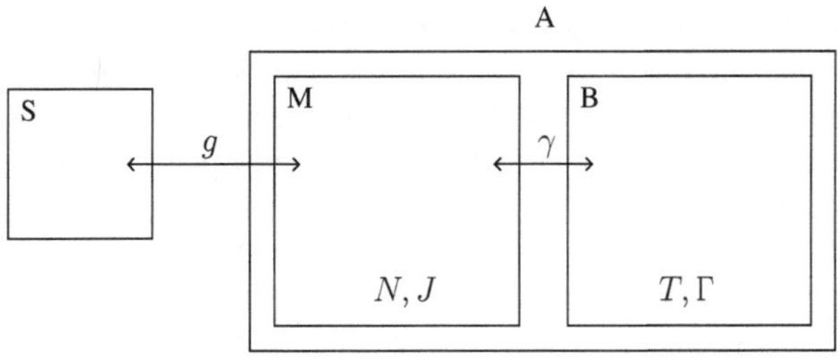

Fig. 3 Schematic presentation of the Curie-Weiss measurement model and its parameters. The system S is a spin $\hat{\mathbf{s}}$. The apparatus A includes a magnet M and a bath B. The magnet, which acts as a pointer, consists of N spins coupled in quartets through an Ising interaction J. The phonon bath B is characterized by its temperature T and a Debye cutoff Γ. It interacts with M through a spin-boson coupling γ. The measurement is started by establishing a finite interaction g between the observable $\hat{s}_z$ and the magnetization $\hat{M}$ of the pointer; it is ended by putting $g \to 0$ after the dynamics is stabilized

a solid basis. In 2017 ABN extend the approach to a class of models and pay further attention to interpretation [35]. Lecture notes are presented in 2014 [52]. A paper entitled "How to teach quantum measurement" is underway.

The tested system S is just a single spin $\frac{1}{2}$ and its tested operator $\hat{S}$ is $\hat{s}_z$, taking eigenvalues ± 1 in units of $\frac{1}{2}\hbar$. The apparatus A consists of a magnet M and a thermal bath B. M consists of N spins $\frac{1}{2}$, denoted as $\hat{\sigma}^{(1)}, \cdots, \hat{\sigma}^{(N)}$. In the simplest model, quartets of spins have a ferromagnetic coupling energy J. B is a phonon bath coupled to each spin component $\hat{\sigma}_a^{(j)}$ for $a = x, y, z$ and $j = 1, 2, \cdots, N$; it is characterized by its temperature T and Debeye cutoff frequency Γ. For simplicity, the bath is assumed to act independently for each j and a. Its coupling to M is weak and characterized by the small, dimensionless coupling constant γ. The setup is schematized in Fig. 3. The extension of the model to higher spins like $s = 1$ is worked out in ref. [53].

Let the tested system S have density matrix $\hat{r}$, the apparatus A have a density matrix $\hat{\mathcal{R}}$, and the total system a density matrix $\hat{\mathcal{D}}$. Initially, S and A are uncorrelated, so $\hat{\mathcal{D}}(0) = \hat{r}(0) \otimes \hat{\mathcal{R}}(0)$. After coupling S to A at time $t = 0$, they become correlated. At later times, the density matrix of S is defined by $\hat{r}(t) = \mathrm{tr}_A \hat{\mathcal{D}}(t)$ and the one of A by $\hat{\mathcal{R}}(t) = \mathrm{tr}_S \hat{\mathcal{D}}(t)$. The Hamiltonian of the total system has terms from system and apparatus alone, and from their coupling, $H = H_S + H_A + H_{SA}$. When measuring a system operator $\hat{S}$, it has to be coupled to a property $\hat{M} = N\hat{m}$ of the apparatus. In the Curie–Weiss model $\hat{S} = \hat{s}_z$ is the z-component of a single spin and $\hat{M} = \sum_{j=1}^{N} \hat{\sigma}_z^{(j)}$ the magnetization of the apparatus in the z-direction, N is the number of particles in the magnet and $\hat{m}$ the magnetization per particle in the z-direction. The interaction Hamiltonian $H_{SA} = -g\hat{S}\hat{M}$ evidently aims to produce effects dependent on the state of S. The S-A coupling g vanishes at $t < 0$, is turned on to a finite value at $t = 0$ and is put back to zero at the final time t_f in order to decouple S from A.

4.2 Born Probabilities from a Conservation Law in the System-Apparatus Dynamics

Saved by the bell
Robin Gibb

The greatest predictive power of quantum theory lies in the Born rule. In models it is often derived by posing appropriate assumptions, see, e.g., [54]. But in a microscopic modelling of system and apparatus, it automatically emerges from a conservation law in the dynamics [47], as we now discuss.

In an ideal measurement $\hat{S}$ will not change by itself during the measurement, so that $[\hat{S}, \hat{H}_S] = 0$, or, at least, it vanishes effectively. Also $[\hat{S}, \hat{H}_A] = 0$ because for an unbiased measurement, the apparatus is not correlated to the system. Hence $\hat{S}$ commutes with the total Hamiltonian and is conserved in the course of time. In the density matrix of S,

$$\hat{r} = \begin{pmatrix} r_{\uparrow\uparrow} & r_{\uparrow\downarrow} \\ r_{\downarrow\uparrow} & r_{\downarrow\downarrow} \end{pmatrix}, \quad r_{\uparrow\uparrow} + r_{\downarrow\downarrow} = 1, \quad r_{\downarrow\uparrow} = r^*_{\uparrow\downarrow}, \tag{3}$$

the diagonal elements $r_{\uparrow\uparrow}$ and $r_{\downarrow\downarrow}$ therefore keep their initial values, so that at the end of the measurement they are given by the values before the measurement, which is the Born rule.

In reality experiments are not ideal, and actually even generally not in models. In the ABN "Opus Magnum" [8] various non-idealities are considered. For instance, a small transversal field will cause dynamics of the tested spin during the measurement process, and cause pointer reading statistics deviating from Born's rule. Another case is the situation where the number of spins in the apparatus is large but not very large. Then the pointer may spontaneously "jump" between the upwards and downwards magnetized state on a timescale not very much larger than the read-out time. Also then non-idealities in the pointer readings occur, here even despite the conservation of the diagonal elements of the density matrix of the tested system.

4.3 Truncation of the Density Matrix of the System by Dephasing

At very short times, the dynamics is solely driven by the interaction Hamiltonian [8]. Since the SA coupling $-g\hat{s}_z \sum_j \hat{\sigma}_z^{(j)}$ acts in the $|\uparrow\rangle$ or $s_z = +1$ sector as a magnetic field $h = +g$ on each of the apparatus spins and as a field $-g$ in the $|\downarrow\rangle$ or $s_z = -1$ sector, this induces a Larmor rotation around the z-axis, and, consequently, dephasing. For the elements $r^*_{\uparrow\downarrow}$ and $r_{\downarrow\uparrow}$ of $\hat{r}$, the density matrix of the tested spin, each apparatus spin brings a phase factor $\cos 2gt/\hbar$. For the combination of the N spins this brings $\cos^N(2gt/\hbar)$. This factor decays quickly, as $\exp(-t^2/\tau^2)$ with

$\tau = \hbar/g\sqrt{2N}$ for large N. Unlike what is often stated, this is a *dephasing* effect, not a decoherence effect. There is an analogy with NMR or MRI physics and with T_2 relaxation, both being dephasing effects.

Decoherence by coupling to the environment, on the other hand, is the cause that the important phase information in MRI after putting the spins in the same direction by a strong magnetic pulse, gets washed out if one waits too long before giving a π pulse to restore the spins to their $t = 0$ directions and record the accumulative effect.

Correlations between the tested spin and $k = 1, 2, 3 \ldots$ spins of the magnet, $\langle(\hat{s}_x \pm i\hat{s}_y)\hat{\sigma}_z^{(1)} \cdots \hat{\sigma}_z^{(k)}\rangle$ start at zero because S and A are uncorrelated at $t = 0$, but grow a bit and then decay to zero [8]. This demonstrates that the truncation $r_{\uparrow\downarrow}$, $r_{\downarrow\uparrow} \to 0$ occurs in a cascade with multiparticle correlations between S and M: the information in S about $\hat{s}_x$, $\hat{s}_y$ "leaks away into M".

4.4 *Truncation of the Density Matrix of the Apparatus by Decoherence*

The previous cosine has undesired recurrences at $t_n = n\pi\hbar/g$. They are suppressed when the bath B sets in at a time t_B well before t_1; this effect referred to as *decoherence* means that the information on S leaks away in the bath (environment). Then the contribution of each of the cosines is suppressed, so that the elements $\hat{\mathcal{D}}_{\uparrow\downarrow}$ and $\hat{\mathcal{D}}_{\downarrow\uparrow}$ (defined in terms of $\hat{\mathcal{D}}$ as in (3)) themselves are suppressed, and with them their traces over A that yield $r_{\uparrow\downarrow}$ and $r_{\downarrow\uparrow}$, respectively. This is termed "truncation of the (off-diagonal elements of the) density matrix" [55], an essential ingredient for the later collapse interpretation of the density matrix, i.e., the description of selecting individual systems with identical measurement outcome.

4.5 *Registration of the Measurement*

Wie schrijft, die blijft [72]
Dutch saying

In the Curie–Weiss model, in the $s_z = \pm 1$ sector, the coupling acts as an external magnetic field of strength g in the $\pm z$ direction on each of the apparatus spins. The coupling strength g has to be large enough to overcome the free energy barrier between the initial paramagnetic state and one of the thermodynamically stable states, so that a final magnetized state is reached. If g is too small, there is still a truncation but no registration, since after decoupling the system from the apparatus, the magnet will go back to its paramagnetic state. These options are schematized in Fig. 4.

The quantum dynamics can be expressed in $P(m, t)$, the distribution of the magnetization per magnet spin in the z-direction at time t. At $t = 0$ this is a narrow

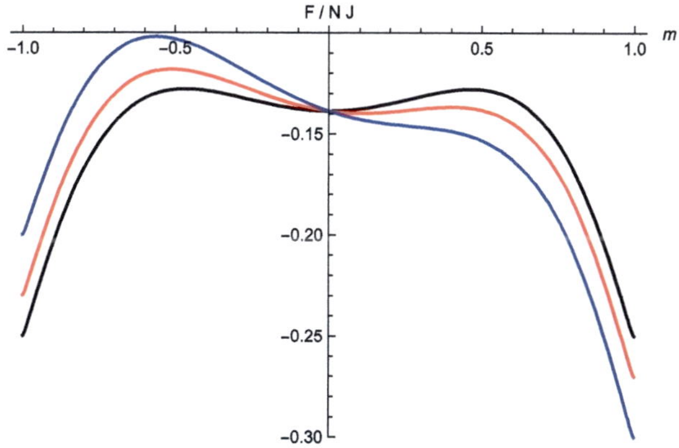

Fig. 4 The free energy of the magnet in the sector $s_z = +1$ at $T = 0.2J$ as function of normalized magnetization m. For $t < 0$ the coupling g of S to A vanishes (black, symmetric curve). The magnet lies in its metastable paramagnetic state near $m = 0$, while there are stable states at $\pm m_F \approx \pm 1$. The barrier towards one of the stable states (in this case, the one at $+m_F$) is suppressed by a coupling $g > 0.0357J$ so as to achieve registration of the measurement; depicted is the case $g = 0.05J$ (blue curve). For smaller g, like for $g = 0.02J$ (red curve), no registration will be achieved, only truncation

peak around $m = 0$ reflecting the initial paramagnetic state. In the course of time, the peak moves, widens and finally narrows around the magnetized state $+m_F$ in $r_{\uparrow\uparrow}$ and $-m_F$ in $r_{\downarrow\downarrow}$, respectively (Fig. 4).

4.6 *Post—Measurement State*

After all was said and done,
more was said than done
Aesop

At the end of the measurement, and putting $g \to 0$ to decouple S from A, we end up with

$$\hat{\mathcal{D}}(t_f) = p_\uparrow |\uparrow\rangle\langle\uparrow| \otimes \mathcal{R}_\Uparrow + p_\downarrow |\downarrow\rangle\langle\downarrow| \otimes \mathcal{R}_\Downarrow, \qquad p_\uparrow = r_{\uparrow\uparrow}(0), \quad p_\downarrow = r_{\downarrow\downarrow}(0). \tag{4}$$

As discussed, the "cat terms, viz, the $\uparrow\downarrow$ and $\downarrow\uparrow$ elements, have effectively vanished. The shape is stable. Any composition of $\hat{\mathcal{D}}(t_f)$ in two terms with the sum equal to (4) can be shown, in a microcanonical approach but conserving the value of m, to relax, for each of the terms, to a shape like (4), and hence for their combination. This is due to small additional terms in the Curie–Weiss Hamiltonian, which cause a weak relaxational effect relevant only at this final stage. The mechanism has been termed subensemble relaxation in [8] and polymicrocanonical relaxation in [35].

4.7 A Run of Experiments Needs Repeated Resetting of the Apparatus

Keep on running,
keep on hiding
Written by Jackie Edwards
Performed by the Spencer Davis Group

In an experiment the macroscopic pointer goes from its metastable "ready" state to one of its stable states. The difference in free energy is dumped in the bath and lost. To reset the apparatus for a next measurement, it has to be brought back to its initial state, so one has to do work on it, which costs energy. This is the dominant reason why experiments are costly (and energy bills are a worry).

In theoretical considerations of quantum foundations starting from the postulates, this physical aspect is not accounted for, showing how remote they are from what goes on in reality.

4.8 On the Frequency Interpretation of Quantum Mechanics

Already in classical physics, the notion of probability has many interpretations, as discussed, for instance, in one of Andrei's books [56]. This dichotomy will not be solved–but rather be deepened–when going to the quantum level.

In their "Opusculo" [35], ABN consider the split of a general final density matrix into physical sub-ensembles with post-measurement density matrices $\hat{\mathcal{D}}_i$, viz. $\hat{\mathcal{D}}(t_f) = \sum_i q_i \hat{\mathcal{D}}_i$; in the example (4) the sum over i pertains to its two possible measurement outcomes. When joining two sub-ensembles with density matrices $\hat{\mathcal{D}}^{(1,2)}$ and $N_{1,2}$ members, respectively, the new density matrix is

$$\hat{\mathcal{D}} = \lambda \hat{\mathcal{D}}^{(1)} + (1-\lambda)\hat{\mathcal{D}}^{(2)}, \qquad \lambda = \frac{N_1}{N_1 + N_2}. \tag{5}$$

The weights q_i of the respective splits obey the rule

$$q_i = \lambda q_i^{(1)} + (1-\lambda) q_i^{(2)}. \tag{6}$$

This is the combination rule for frequencies in probability theory, so quantum theory directly embodies the frequency interpretation, the only one that experimenters can reasonably employ. The Bayesian interpretation of QM (QBism) adds priors of expectations to the quantum formalism [57]; it is then capable to give (additional) uncertainties in otherwise sharply defined theoretical values, such as employed at the LHC.

4.9 Useless Measurements are Often Still Measurements

Asher Peres says: unperformed measurements have no outcome. They are not only useless, they are meaningless. Their repeated use in foundational considerations is a scandal, based on the misinterpretation of the postulates, something that can not occur when modelling the physical apparatus and solving its dynamics.

But also performed measurements can be "useless". Imagine a Curie–Weiss apparatus–to be discussed in detail in Sect. 4 below–which is turned off after the truncation process but well before the registration of the measurement has been been achieved. The density matrix will have been truncated, the information coded in the off-diagonal elements will have been scrambled on its very short timescale. But the magnet will return to its paramagnetic initial state, and no registration will have taken place. For practical purposes, this is a useless experiment. But (in the ensemble of measurements of this type,) forces have acted on the (the ensemble of) system(s) and have altered the density matrix, being the best coding of knowledge about the ensemble.

Another case of such a "useless" measurement is when the system-apparatus coupling (g in the Curie-Weiss model below) is not strong enough to overcome the free energy barrier between the paramagnetic state and the stable, magnetized states. After decoupling the apparatus from the system, the former will again go back to its paramagnetic state, and no sensible pointer reading can be achieved. Also this is a "useless" experiment, in the sense of useless for testing elements of the system's density matrix. But it is not useless if one merely wants to truncate it.

The best case of really useless experiments involves solar neutrinos. Each second our bodies are bombarded by hundreds of billions of them. Their coupling to atoms is so weak, that they they do not even achieve the truncation mechanism and can be left out of modelling in QM. It is allowed to invoke them in idealized theoretical considerations, but they are of no help since they alter nothing. Thank heavens, if they would couple stronger and really be able to interact with the atoms in our bodies, they would not only destroy life, but even have prevented it.

4.10 Measuring a "Classical" System

In order to understand the classical limit of quantum systems, it is instructive to consider the case where the also tested system itself is large, consisting of $n \gg 1$ spins $\hat{\mathbf{s}}^{(i)}$ for $i = 1, 2, \cdots, n$, coupled to some low temperature thermal bath, so that we consider an ensemble of systems magnetized upwards or downwards. Taking now as observable $\hat{S} = (1/n)\sum_{i=1}^{n} \hat{s}_z^{(i)}$ in the interaction Hamiltonian $-g\hat{S}\hat{M}$, the previous $n = 1$ theory can be generalized directly.

In a first measurement, a fast truncation occurs over a time $(\hbar n/g)\sqrt{2/N}$, are after which registration produces an equivalent of (4) on the time scale $\hbar J/\gamma$ provided $n \ll g\sqrt{N}/\gamma J$. The would-be recurrences at $t_k = 2nk\hbar/g$ come too late to influence the pointer readings. The decoherence should set in between them. Upon repeating the measurement an arbitrary number of times in each of the up-up and down-down sectors, the previous outcome gets reproduced. The standard repeatability has for measuring a macroscopic system the smell of a classical system property: the moon is observed to be there "again" if you look again. This supports the individuality postulate of macroscopic systems in Sect. 2.8.

4.11 Simultaneously Measuring Non-Commuting Variables: A Model Setup

Je kunt niet op twee bruiloften tegelijk dansen [73]
Dutch saying

It is often said that non-commuting (incompatible) variables cannot be measured simultaneously; however, this statement is *incorrect*. The proper statement is: *when measuring incompatible variables simultaneously, the Copenhagen postulates do not apply.*

In practice, such measurements are not impossible though difficult, as they involve at least two apparatuses, which will influence each other. An idealized case has been modeled where both $\hat{s}_z$ and $\hat{s}_x$ of S, a spin $\frac{1}{2}$, are measured by two Curie–Weiss magnets: one, to be called M_z, having stable magnetized states in the $\pm z$-direction, and the other, to be called M_x, having them in the $\pm x$-direction [8, 58]. The apparatuses will influence each other via their coupling to the common tested spin. The distribution of outcomes depends on many if not all details of the apparatuses; if M_x is coupled to S before (or after) the M_z-apparatus is coupled, it also depends on the time delay between the coupling instances, relaxing to the standard results in the limit of infinite time delay.

The case of zero time-delay between the coupling of the apparatuses to the spin, the exactly simultaneous measurement, has been worked out in detail [59]. The probabilities for the four possible outcomes: magnetization of M_z up or down, and magnetization of M_x left or right, are explicit, depend on many system parameters and have no bearing on any Born rule that one could imagine for this case. Moreover, there is no clear analog of the truncation of the density matrix, or "collapse of the wavefunction". But the measurement is still informative so as to allow reconstruction of the elements $\langle\hat{s}_x\rangle$ and $\langle\hat{s}_z\rangle$ of the initial density matrix of S from the outcomes of repeated measurements in this setup [59]. In the standard setup, this would require 2 sets of repeated measurements, for measuring along the x and z directions, to determine $\langle\hat{s}_x\rangle$ and $\langle\hat{s}_z\rangle$ separately.

With a third apparatus, such that $\hat{s}_x$, $\hat{s}_y$ and $\hat{s}_z$ are all measured simultaneously, one should be able to determine all elements of the density matrix of S from one (repeated) measurement. This task has so far not been worked out.

5 Realistic Quantum Measurement as Sine-Qua-Non Basis for Interpretation

I will liken him unto a wise man,
which built his house upon a rock
Matthew VII:24

The project of interpretation of QM of course goes all the way back to the introduction of this theory. Classical papers on interpretation of QM and quantum measurement have been collected in [41]. Lacking a convincing solvable model, the opinions remained divergent.

Like the proof of the pudding is in the eating, the only point of contact between quantum theory and the reality in the laboratory lies in quantum measurements. Hence interpretation should be based on analyzing measurement in a specified setup (context), which can be an optical table with beam splitters, mirrors, phase delayers, photon counters, and so on. Together they form the context for the measurement and interpretation needs only refer to a given context; all further attempts can not offer a way to test them. Grangier and Auffèves stress this by introducing the term of the "modality" for the quantum system in its specified context [10]. Other interpretations like Wigner's friend [41, 60] and relational QM [61] try to grasp the contextuality, bounded more by the imagination of the human mind than by the stringent structure of the quantum formalism.

Following Bohr, we consider QM as a theory that makes predictions in a well defined context. The role of context in ordinary life is well known. For example, a joke which does well in one context, can be sour in another. Even the same word in the same language can have a different meaning in a different context.

With the solution of the Curie-Weiss model at hand, ABN generalize this for a class of similar models and interpret the results in a minimalistic fashion. The main points are:

- The natural interpretation of QM is a specified version of the statistical interpretation. QM describes ensembles of identically prepared systems or particles, tested in an ensemble of measurements in a specified context.
- The density matrix is an abstract mathematical object, which codes our best knowledge about the ensemble. It needs no interpretation (no "wave function of the particle"). It leads to an epistemology without connection to an ontology, it describes statistics but is silent about the individual events.
- Results from measurements in different contexts should not be combined, for instance because apparatuses can not stand in each others way.
- Individual events do exist, but quantum theory can only treat them statistically.
- Truncation of the density matrix of the tested system (disappearance of off-diagonal "cat" terms terms) occurs by dephasing followed by decoherence. The latter also erases the related terms for the apparatus.

- The registration of the measurement is related to a dynamical phase transition from the initial "ready" metastable thermodynamic phase to one of the stable thermodynamic phases. This transition amplifies the small quantum signal and causes finally a macroscopic effect in the apparatus, which can be read off or processed electronically.
- The system-apparatus coupling has to be large enough to overcome the free energy barrier between the initial metastable and final stable state. The excess free energy is dumped into the bath. Resetting the apparatus costs energy.
- Ideal measurements constitute the mathematical limit of a series of more and more idealized physical measurements. Interpretation of the quantum formalism follows from considering suitable ideal measurements.
- The Born rule follows from the quantum formalism itself, namely from a dynamical conservation law in ideal measurements.
- Born probabilities can be related to the macroscopic pointer readings, with which the tested microscopic system is correlated [74]. Non-idealities lead to false readings.
- Noncommuting variables can be measured simultaneously. The outcome probabilities are not described by the Born rule, but involve the rich physics of the apparatuses, which get coupled to each other via the commonly tested system.

6 The Elephant in the Room: No Theory for Individual Events

This emptiness deep inside
Expression

There is no working model for the shape of an individual electron. Quantum electrodynamics states it to be a point particle with a quantum cloud of virtual particles and antiparticles around it. But point particles do not exist, they are an abstraction for a description of some small entity, so we still dream of a better modelling.

In our view, particles do exist (virtual particles live shortly). Clear evidences are solid state materials and high energy cosmic rays. QM provides the Born probabilities for ensembles of measurements on ensembles of particles. But what about the individual measurements? What happens in them? It is our thesis that we have no theory for it, no ontology! This is "a problem as big as an elephant" or, said otherwise, the proverbial elephant in the room.

6.1 The EPR Setup

Wanting to beat Einstein
equates to
Being alive and kicking
Anonymous

In an Einstein-Podolsky-Rosen (EPR) experiment for a pair of, say, electrons in an $S = 0$ Bell state, measurement at A along an arbitrary unit vector **n**, will give an outcome $s_A = \pm 1$ (in units of $\frac{1}{2}\hbar$). Measurement at the opposite station B along the same axis **n** will result in $s_B = \mp 1$, so that the sum leads to $s_A + s_B = 0$. This is a manifestation of the conservation law of total spin in this sector, $\mathbf{S} \cdot \mathbf{n} = 0$. Whether the experiment at A is performed before or after the one at B is set by the distance from the source to the respective detectors. They can be such that the second experiment is far enough that it gets performed before a light signal from the first could arrive.

In lack of a subquantum theory that would clarify the situation, it has been proposed that there has to be faster-than-light communication between the particles. This poses some questions. How is that compatible with the relativity principle? Would the information be sent when the first particle is measured, to the second, carried by it till that also gets measured? Or sent to the second apparatus to be employed when its particle arrives? No detailed model is known, and no definite answer for this conundrum is agreed upon.

We are left with the question: How does Nature get it done? A simple model is that the information for the measurement outcomes travels with the particles, so that they can instruct the detectors. A role is then likely played by the clouds of virtual quantum particles. For this much thinking will be needed, but the goal is assured by the existence of the individual events.

6.2 Wanted Living or Alive: Subquantum Mechanics as a Less Statistical Theory

One may consider that the quantum uncertainty arrives from averaging over some very fast underlying dynamics, with quantum theory representing the behavior at the longer timescales, the standard case of time scale separation in glasses [26]. A consistent research line based on cellular automata is developed by Gerard 't Hooft [62]. For the time being one can be amazed that quantum mechanics captures the essentials of this subquantum physics, partly as conservation laws, partly in a statistical way.

To me Einstein's dream of a reality is as vivid as it ever was. The last but not least of the scandals of quantum mechanics is that, after one century, there is still this most fundamental question to answer! This is unbearable, so Andrei: solve it soon. Or else: please organize many further Växjö meetings, so that we can attack it as a community!

References

1. A.E. Allahverdyan, A. Khrennikov, T.M. Nieuwenhuizen, Brownian entanglement. Phys. Rev. A. **72**(3), 032102 (2005)
2. N. Van Kampen, The scandal of quantum mechanics. Am. J. Phys. **76**(11), 989 (2008)
3. T.M. Nieuwenhuizen, D. Frenkel, N. Van Kampen, Objections to handel's quantum theory of 1/f noise. Phys. Rev. A. **35**(6), 2750 (1987)
4. N. Van Kampen, Ten theorems about quantum mechanical measurements. Phys. A: Stat. Mech. Appl. **153**(1), 97 (1988)
5. A. Peres, Unperformed experiments have no results. Am. J. Phys. **46**(7), 745 (1978)
6. A. Plotnitsky, On the reasonable and unreasonable effectiveness of mathematics in classical and quantum physics. Found. Phys. **41**(3), 466 (2011)
7. A.Y. Khrennikov, *Ubiquitous Quantum Structure*,(Springer, 2014)
8. A.E. Allahverdyan, R. Balian, T.M. Nieuwenhuizen, Understanding quantum measurement from the solution of dynamical models. Phys. Rep. **525**(1), 1 (2013)
9. O. Darrigol, Why Some Physical Theories Should Never Die, *Évora Studies in the Philosophy and History of Science. In memoriam Hermínio Martins*, (2015), pp. 319–368
10. A. Auffèves, P. Grangier, Contexts, systems and modalities: a new ontology for quantum mechanics. Found. Phys. **46**(2), 121 (2016)
11. A. Khrennikov, *Beyond Quantum*, (CRC Press, 2014)
12. A. Plotnitsky, A. Khrennikov, Reality without realism: on the ontological and epistemological architecture of quantum mechanics. Found. Phys. **45**(10), 1269 (2015)
13. T.M. Nieuwenhuizen, Exact electronic spectra and inverse localization lengths in one-dimensional random systems: I. random alloy, liquid metal and liquid alloy. Phys. A: Stat. Mech. Appl. **120**(3), 468 (1983)
14. C. de Calan, J.-M. Luck, T.M. Nieuwenhuizen, D. Petritis, On the distribution of a random variable occurring in 1d disordered systems. J. Phys. A: Math. Gen. **18**(3), 501 (1985)
15. T.M. Nieuwenhuizen, Trapping and lifshitz tails in random media, self-attracting polymers, and the number of distinct sites visited: a renormalized instanton approach in three dimensions. Phys. Rev. Lett. **62**(4), 357 (1989)
16. M. van Rossum, T.M. Nieuwenhuizen, E. Hofstetter, M. Schreiber, Band tails in a disordered system, in *Photonic Band Gaps and Localization*. ed. by C. Soukoulis (Plenum Press, New York, 1993), pp.509–513
17. M. Van Rossum, T.M. Nieuwenhuizen, E. Hofstetter, M. Schreiber, Density of states of disordered systems. Phys. Rev. B **49**(19), 13377 (1994)
18. P. Den Outer, T.M. Nieuwenhuizen, A. Lagendijk, Location of objects in multiple-scattering media. J. Opt. Soc. Am. A. **10**(6), 1209 (1993)
19. J.F. de Boer, M. Van Rossum, M.P. van Albada, T.M. Nieuwenhuizen, A. Lagendijk, Probability distribution of multiple scattered light measured in total transmission. Phys. Rev. Lett. **73**(19), 2567 (1994)
20. T.M. Nieuwenhuizen, J. Luck, Skin layer of diffusive media. Phys. Rev. E. **48**(1), 569 (1993)
21. T.M. Nieuwenhuizen, M. Van Rossum, Intensity distributions of waves transmitted through a multiple scattering medium. Phys. Rev. Lett. **74**(14), 2674 (1995)
22. M. van Rossum, T.M. Nieuwenhuizen, Multiple scattering of classical waves: microscopy, mesoscopy, and diffusion. Rev. Mod. Phys. **71**(1), 313 (1999)
23. T.M. Nieuwenhuizen, Exactly solvable model of a quantum spin glass. Phys. Rev. Lett. **74**(21), 4289 (1995)
24. T.M. Nieuwenhuizen, Ehrenfest relations at the glass transition: solution to an old paradox. Phys. Rev. Lett. **79**(7), 1317 (1997)
25. T.M. Nieuwenhuizen, Thermodynamics of the glassy state: effective temperature as an additional system parameter. Phys. Rev. Lett. **80**(25), 5580 (1998)
26. L. Leuzzi, T.M. Nieuwenhuizen, *Thermodynamics of the Glassy State*,(CRC Press, 2007)
27. A.E. Allahverdyan, R. Balian, T.M. Nieuwenhuizen, Maximal work extraction from finite quantum systems. Eur. Lett. **67**(4), 565 (2004)

28. E.T. Jaynes, *Probability Theory: The Logic of Science*, (Cambridge University Press, 2003)
29. R. Balian, Entropy, a protean concept, progress in mathematical. Phys. **38**, 119 (2004)
30. R. Balian, Information in statistical physics. Stud. Hist. Philos. Sci. Part B: Stud. Hist. Philos. Mod. Phys. **36**(2), 323 (2005)
31. L. Brillouin, *Scientific Uncertainty and Information*, (Academic Press, 2014)
32. J.W. Gibbs, *Scientific Papers of J. Willard Gibbs, in Two Volumes*, vol. 1, (Longmans, Green, 1906)
33. A.E. Allahverdyan, T.M. Nieuwenhuizen, Explanation of the gibbs paradox within the framework of quantum thermodynamics. Phys. Rev. E. **73**(6), 066119 (2006)
34. A.E. Allahverdyan, K.V. Hovhannisyan, D. Petrosyan, Dynamical symmetrization of the state of identical particles. Proc. R. Soc. A. **477**(2249), 20200911 (2021)
35. A.E. Allahverdyan, R. Balian, T.M. Nieuwenhuizen, A sub-ensemble theory of ideal quantum measurement processes. Ann. Phys. **376**, 324 (2017)
36. wikipedia, *One-electron universe*(2021), https://en.wikipedia.org/wiki/One-electron_universe. Accessed on Dec 1
37. H. De Raedt, M.I. Katsnelson, K. Michielsen, Quantum theory as the most robust description of reproducible experiments. Ann. Phys. **347**, 45 (2014)
38. H. De Raedt, M.I. Katsnelson, H.C. Donker, K. Michielsen, Quantum theory as a description of robust experiments: Derivation of the pauli equation. Ann. Phys. **359**, 166 (2015)
39. A. Auffèves, P. Grangier, Deriving born's rule from an inference to the best explanation. Found. Phys. **50**(12), 1781 (2020)
40. A. Khrennikov, Contextual viewpoint to quantum stochastics. J. Math. Phys. **44**(6), 2471 (2003)
41. J.A. Wheeler, W.H. Zurek, *Quantum Theory and Measurement*, vol. 53, (Princeton University Press, 2014)
42. D. Bohm, *A Suggested Interpretation of the Quantum Theory in Terms of Hidden Variables. i.* Phys. Rev. **85**(2), 166 (1952)
43. P.R. Holland, *The Quantum Theory of Motion: An Account of the de Broglie-Bohm Causal Interpretation of Quantum Mechanics*, (Cambridge university press, 1995)
44. H.D. Zeh, On the interpretation of measurement in quantum theory. Found. Phys. **1**(1), 69 (1970)
45. E. Joos, H.D. Zeh, C. Kiefer, D.J. Giulini, J. Kupsch, I.-O. Stamatescu, *Decoherence and the Appearance of a Classical world in Quantum Theory*, (Springer Science & Business Media, 2013)
46. W.H. Zurek, Decoherence, einselection, and the quantum origins of the classical. Rev. Mod. Phys. **75**(3), 715 (2003)
47. A.E. Allahverdyan, R. Balian, T.M. Nieuwenhuizen, Curie-weiss model of the quantum measurement process. Eur. Lett. **61**(4), 452 (2003)
48. A.E. Allahverdyan, R. Balian, T.M. Nieuwenhuizen, The quantum measurement process: an exactly solvable model, vol. An. LVIII. (2003), https://arxiv.org/abs/cond-mat/0309188
49. A.E. Allahverdyan, R. Balian, T.M. Nieuwenhuizen, Dynamics of quantum measurements. (2004), https://arxiv.org/pdf/quant-ph/0412045.pdf
50. A.E. Allahverdyan, R. Balian, T.M. Nieuwenhuizen, The quantum measurement process in an exactly solvable model, in *AIP Conference Proceedings*, vol. 750 (American Institute of Physics, 2005), pp. 26–34
51. A.E. Allahverdyan, R. Balian, T.M. Nieuwenhuizen, Phase transitions and quantum measurements, in *AIP Conference Proceedings*, vol. 810 (American Institute of Physics, 2006), pp. 47–58
52. T.M. Nieuwenhuizen, M. Perarnau-Llobet, R. Balian, Lectures on dynamical models for quantum measurements. Int. J. Mod. Phys. B. **28**(21), 1430014 (2014)
53. T.M. Nieuwenhuizen, Models for quantum measurement of particles with higher spin. Entropy **24**(12), 1746 (2002)
54. A. Khrennikov, Born's rule from classical random fields. Phys. Lett. A. **372**(44), 6588 (2008)
55. A.E. Allahverdyan, R. Balian, T.M. Nieuwenhuizen, Dynamics of a quantum measurement. Phys. E: Low-Dimens. Syst. Nanostructures **29**(1–2), 261 (2005)

56. A. Khrennikov, *Interpretations of Probability*, (de Gruyter, 2009)
57. C.A. Fuchs, N.D. Mermin, R. Schack, An introduction to qbism with an application to the locality of quantum mechanics. Am. J. Phys. **82**(8), 749 (2014)
58. A.E. Allahverdyan, R. Balian, T.M. Nieuwenhuizen, Simultaneous measurement of non-commuting observables. Phys. E: Low-Dimens. Syst. Nanostructures **42**(3), 339 (2010)
59. M. Perarnau-Llobet, T.M. Nieuwenhuizen, Simultaneous measurement of two non-commuting quantum variables: solution of a dynamical model. Phys. Rev. A. **95**(5), 052129 (2017)
60. F. Laloë, *Do we Really Understand Quantum Mechanics?*, (Cambridge University Press, 2019)
61. C. Rovelli, Relational quantum mechanics. Int. J. Theor. Phys. **35**(8), 1637 (1996)
62. G.t. Hooft, *The Cellular Automaton Interpretation of Quantum Mechanics*,(Springer Nature, 2016)
63. One can express the measure problem in the multiverse as: When I do a measurement, there are fortunately other universes in which little green men with their little green women carry out the same experiment, helping to get my statistics right
64. How great Thou art
65. One plus one is two
66. Life is hard, but unjust
67. Knowing much of a little bit
68. That is no pure coffee
69. You can't jump further than your vaulting pole allows
70. Since the divergence of the magnetic field is zero, a gradient in B_z has to be accompanied by a gradient in B_x and or B_y, which cause non-idealities. Hence no Stern-Gerlach setup is ideal; an ideal Stern-Gerlach setup is the mathematical limit of a series of setups with ever larger Stern-Gerlach magnets, which involve ever smaller gradients in the magnetic field
71. I mention that number of equations to contrast this fundamental study to typical works in the eld of quantum measurement with, at best, a few equations, often even only posing expressions of a supposed dynamics
72. Who writes, stays
73. You can't dance at two weddings simultaneously
74. This "view from the top" is opposite to the Copenhagen interpretation, which focusses on the tested system and treats the apparatus in an abstract way by a postulate

Theodorus Maria Nieuwenhuizen The author studied theoretical physics in Utrecht, the Netherlands. After a master thesis in quantum field theory under professor Gerard 't Hooft, he did PhD research under professor Theo W. Ruijgrok, presenting exact solutions for the spectral density of random mass and random electronic chains. After a postdoc under professor Matthieu H. Ernst, he first moved to Saclay, France, and then to Aachen, Germany. Since 1990 he is member of the University of Amsterdam. There he worked on multiple light scattering and on spin glasses, presenting a model with exact Parisi function. He solved the problem of thermodynamics for glassy systems, in a coauthored book with foreword by Giorgio Parisi. In the last decades, he worked on quantum thermodynamics, introducing the terrm "Ergotropy" and on EPR-Bell experiments, introducing the term "Contextuality Loophole". He also worked on the dynamics of quantum measurement and the and the statistical interpretations of quantum mechanics he and his coauthors developed, on mass models for galaxy clusters, stochastic electrodynamics, and on exact solutions for the interior of black holes within standard model physics. He is a co-organizer of the FQMT conference series in Prague.

Quantum Holism

Giacomo M. D'Ariano

Abstract A composite quantum system has properties that are incompatible with every property of its parts. The existence of such global properties incompatible with all local properties constitutes what I call *mereological holism*–the distinctive holism of quantum theory. Mereological holism has the dramatic conceptual consequence of making untenable the usual understanding of the "quantum system" as being a "physical object", since composed objects have properties compatible with those of its parts. The notion of "property" can be extended in a unique way to the whole class of operational probabilistic theories (shortly OPTs), of which the most relevant cases are quantum theory and classical theory. Whereas classical theory is not mereologically holistic, we can now search for other OPTs that are so. Within the OPT framework the role of the "system" is that of an input-output connection between two objective events. In non holistic theories, such as classical theory, the system can still be regarded as an "object". By contrast, in holistic theories interpreting "system" as "object" constitutes an hypostatization of a theoretical notion.

1 Introduction

What is a "physical object?"[1] An object without properties is inconceivable: it is not an object. Also, implicit in the notion of object is its mereological aspect, i. e.

[1] W.V.O.Quine in his *Whither Physical Objects?* [1] made a thorough attempt to arrive at a comprehensive meaning of the word "object", but he ended up with a progressive evaporation of the concept–starting from that of "body", toward "space-time region", ending with a mere "set of numerical coordinates".

This work was made possible through the support of the Elvia and Federico Faggin Foundation, Grant 2020-214365.

G. M. D'Ariano (✉)
Dipartimento di Fisica dell'Universita di Pavia, via Bassi 6, PO Box 27100, Pavia, Italy

Istituto Lombardo Accademia di Scienze e Lettere and INFN, Gruppo IV, Sezione di Pavia, Pavia, Italy
e-mail: dariano@unipv.it

A. Plotnitsky and E. Haven (eds.), *The Quantum-Like Revolution*,
https://doi.org/10.1007/978-3-031-12986-5_5

its compositional nature by virtue of which putting objects together results in a new object, whose properties are made with the properties of its parts.

In this paper it is shown that the common position of regarding the quantum system as an "object" is untenable, since a quantum system has properties that are incompatible with any property of its parts (the "property" being described by an orthogonal projector (hereafter "projector") over a subspaces of the system Hilbert space, as we learned from quantum logic [2]). Such part-whole incompatibility is ubiquitous in quantum theory, as entanglement is.

The present result strongly supports the Copenhagen interpretation, according to which the quantum system (i. e. the "particle" in the nonrelativistic theory) is never directly observable, but only through its objective manifestations.[2]

The part-whole incompatibility considered here corresponds to a special notion of holism: the *mereological holism*. This is the distinctive holism of quantum theory. The theory *is not* holistic in the only sense thus far defined in literature, namely the *local-discriminability* holism [4, 5]. A relevant conceptual consequence of mereological holism of quantum theory, is that regarding the "system" as an "object" constitutes an hypostatization of a notion that is strictly theoretical, and whose role is just that of *a theoretical input-output connection between objective events*, as we learnt from OPTs.[3] The "object" connotation of "system "can be recovered only in theories that are not mereologically holistic, such as classical theory.

After a short review of quantum properties in Sect. 2, the mereological holism of quantum theory is derived in Sect. 3. In Sect. 4 it is shown that the mereological holism is the distinctive quantum holism, and its reconciliation with local-discriminability is discussed. In Sect. 5 the notion of "property" is generalized to OPTs for further analysis of mereological holism in such theories. Section 6 summarizes the main results and considers open problems and possibilities for further analysis.

2 Properties

Let's consider a quantum system A with associated Hilbert space $\mathscr{H}_{\mathrm{A}}$. To a proper subspace $\mathscr{S} \subsetneq \mathscr{H}_{\mathrm{A}}$, or, equivalently, to the corresponding projector $P_{\mathscr{S}}$,[4] we associate a *property* of the system A. Precisely, we say that

---

[2] Bohr, as an anti-realist, brought to his side many physicists, including Werner Heisenberg, Wolfgang Pauli, and Max Born. The Copenhagen interpretation had taken hold by the 1930s, and many textbooks state that Bohr's view 'won'. Although this view has been challenges in recent decades, the present article offers a support for it.

[3] For an introductory review on OPTs, see Refs. [5, 6].

[4] There is trivially a one–to-one correspondence between Hilbert subspaces $\mathscr{S} \subseteq \mathscr{H}_{\mathrm{A}}$ and projectors $P_{\mathscr{S}}$ on $\mathscr{S}$.

Definition 1 The[5] system A in a state $\rho \in \mathrm{St}(\mathrm{A})$ with $\mathsf{Supp}\,\rho \subseteq \mathscr{S}$ has property $\mathscr{S}$. If, instead, $\mathsf{Supp}\,\rho \subseteq \mathscr{S}^{\perp}$, then A it has not the property $\mathscr{S}$. When neither of the two cases apply the property $\mathscr{S}$ is meaningless for state ρ.

Example 1 Consider a two-dimensional system A in the state corresponding to the vector $|\uparrow\rangle \in \mathbb{C}^2$. We say that the system has the property of being "up", and it has not the property of being "down"–the latter corresponding to state vector $|\downarrow\rangle$, with $\langle\downarrow|\uparrow\rangle = 0$. On the other hand, for state $|\rightarrow\rangle = \frac{1}{\sqrt{2}}(|\uparrow\rangle + |\downarrow\rangle)$ properties "up" or "down" are meaningless.

Example 2 Consider a four dimensional system A with canonical basis $\{|n\rangle\}_{n=0}^{4}$. For state $\rho = \frac{1}{4}|0\rangle\langle 0| + \frac{3}{4}|2\rangle\langle 2|$ we say that the system has the property of being *even*. The same property holds if e. g. $\rho = \frac{1}{2}(|0\rangle + |2\rangle)(\langle 0| + \langle 2|)$.

Thanks to the identifications *subspace* ≡ *projector* ≡ *property*, with a little abuse of language, we will use the word *property* for all the three notions.

Example 3 (Property) The state $\rho{=}|\uparrow\uparrow\rangle\langle\uparrow\uparrow| \in \mathrm{St}(\mathrm{AB})$ has the property of being "symmetric", the property corresponding to the Hilbert subspace

$$(\mathscr{H}_{\mathrm{AB}})_{+} \subset \mathscr{H}_{\mathrm{AB}} \equiv \mathscr{H}_{\mathrm{A}} \otimes \mathscr{H}_{\mathrm{B}},$$

$(\mathscr{H}_{\mathrm{AB}})_{+}$ being spanned by vectors invariant under the *swap* operator E, which acts on $\mathscr{H}_{\mathrm{AB}}$ as follows

$$E(|a\rangle \otimes |b\rangle) = |b\rangle \otimes |a\rangle. \tag{1}$$

Definition 2 A property P is nontrivial iff $0 < P < I$.

Resorting to the usual notion of *compatibility* for observables, we say:

Definition 3 Two properties are compatible if and only if they commute.

Example 4 (Incompatible properties) Properties *up* and *left* are incompatible (see also Example 1).

Corollary 1 *1. Two properties are compatible iff their product is a property.*
2. Two properties are mutually exclusive when $PQ = QP = 0$.
3. $\dim(\mathscr{H}_{\mathrm{A}})$ *is the maximum number of mutually exclusive properties of A.*

---

[5] This notion of property has been introduced by John von Neumann in his celebrated treatise [7]. Von Neumann noted that projections on a Hilbert space can be viewed as propositions about physical observables. The idea led to the research line of *quantum logic*, the first major attempt to derive quantum theory from first principles.

3 The Quantum System is not an Object!

In the following we will consider a bipartite system, and we will use the following definitions.

Definition 4 *(Property of the parts)* We call *property of the parts* any property of the composite system AB of the form $P_{\rm AB} = P_{\rm A} \otimes P_{\rm B}$, meaning that system A has the property $P_{\rm A}$, and system B has the property $P_{\rm B}$.

Definition 5 *(Property of the whole)* We call *property of the whole* any joint property $P_{\rm AB}$ of the composite system AB which is not a property of the parts.

Generalization of the above definitions to multipartite systems are straightforward.

Theorem 1 (Quantum Holism properties) *There exist properties of the whole that are incompatible with any nontrivial property of the parts.*

Proof For the sake of proving this existence, we can restrict the proof to the simplest nontrivial case the pair of systems, A and B, with $\dim(\mathscr{H}_{\rm A}) = \dim(\mathscr{H}_{\rm B}) = d \leq \infty$ (extensions to more general cases will be provided after the present proof.)

A nontrivial property of the parts is represented by either a factorized projector of the form

$$P_{parts} = P \otimes Q, \qquad (P, Q) \in \Pi_d^2 \tag{2}$$

where Π_d^2 are the couples of projectors in dimension d, of which at least one is non trivial. For the property of the whole we consider the following one [6]

$$P_{whole} = |\Gamma\rangle\!\rangle\langle\!\langle\Gamma|, \qquad \Gamma \in \mathrm{HS}(\mathbb{C}^d) \text{ invertible}, \ \mathrm{Tr}(\Gamma^\dagger\Gamma) = 1.$$

Let's consider the case of Eq. (2). We want to prove that there exist a Γ such that

$$\Big[P \otimes Q, |\Gamma\rangle\!\rangle\langle\!\langle\Gamma|\Big] \neq 0, \qquad \forall P, Q \in \Pi_d^2.$$

[6] Here we are using the *double-ket* notation [8] (for a thorough treatment see [5].) Once it is chosen the orthonormal factorized canonical basis $\{|i\rangle \otimes |j\rangle\}$ for $\mathscr{H} \otimes \mathscr{H}$, we have the one-to-one correspondence between vectors in $\mathscr{H} \otimes \mathscr{H}$ and operators on $\mathscr{H}$

$$|\Psi\rangle\!\rangle := \sum_{ij} \Psi_{ij} |i\rangle \otimes |j\rangle \longleftrightarrow \Psi = \sum_{ij} \Psi_{ij} |i\rangle\langle j| \in \mathrm{HS}(\mathbb{C}^d),$$

where $\mathrm{HS}(\mathbb{C}^d)$ denotes Hilbert-Schmidt operators in dimensions d. One then can veryify the following identity

$$(A \otimes B)|C\rangle\!\rangle = |ACB\rangle\!\rangle,$$

B denoting the *transposed operator* of B, the operator that has the transposed matrix w.r.t. the canonical basis. Notice that e. g. $(|\phi\rangle\langle\psi|) = |\psi^*\rangle\langle\phi^*|$ where $|\psi^*\rangle$ is the vector $|\psi\rangle$ with complex-conjugated coefficients w.r.t. the canonical basis.

Using the following identity for self-adjoint operators A and B

$$[A, B] = 2i\Im(AB), \tag{3}$$

one obtains

$$\Big[P \otimes Q, |\Gamma\rangle\!\rangle\langle\!\langle\Gamma|\Big] = 2i\Im[(P \otimes Q)|\Gamma\rangle\!\rangle\langle\!\langle\Gamma|].$$

One has

$$(P \otimes Q)|\Gamma\rangle\!\rangle\langle\!\langle\Gamma| = |P\Gamma Q\rangle\!\rangle\langle\!\langle\Gamma|$$

whose imaginary part is zero if and only if

$$|P\Gamma Q\rangle\!\rangle\langle\!\langle\Gamma| = |\Gamma\rangle\!\rangle\langle\!\langle P\Gamma Q|,$$

namely

$$|P\Gamma Q\rangle\!\rangle = |\Gamma\rangle\!\rangle,$$

i. e.

$$P\Gamma Q = \Gamma, \tag{4}$$

which implies $P = Q = I$, since Γ is invertible, thus contradicting the hypothesis that at least one of the two projectors is non trivial. ■

The above theorem can be easily generalized to two quantum systems with different dimensions, by considering Γ only to be left or right-invertible, depending on which of the two systems has larger dimension. Upon considering one or both of the two systems A and B as composite, say $\mathrm{A} = \mathrm{A}_1\mathrm{A}_2$, the theorem can be easily generalized to three, or to any larger number of subsystems, since incompatibility with all properties of A implies incompatibility with any factorized property of A.

We have established the existence of properties of the whole that are incompatible with all properties of its parts, for every set of unbounded dimension and finite partitioning.[7] In the following we will call such properties *holistic properties*.

The holistic properties that we have found are rank-one. One may proceed to constructing higher-rank holistic properties e. g. by taking any operator Λ orthogonal to Γ (i. e. with $\langle\!\langle\Gamma|\Lambda\rangle\!\rangle = 0$) using a Gram-Schmidt procedure, thus building a full lattice of compatible holistic properties. Then by unitary transformation of the holistic properties one can obtain a set of holistic properties that is presumably dense within the set of compatible properties. One then may argue that, similarly to what happens for non separable states, the holistic properties are ubiquitous in quantum theory (i. e. they make a set dense within the set of properties), as it happens for state-entanglement. It would also be interesting to see if quantum discord may give rise to mereological holism (see e. g. Refs. [10, 11]).

We then reach the following conclusion.

[7] For infinite partitioning, one needs to move to a von Neumann algebra setting [9].

No object interpretation: the "quantum system" lacks the "object" interpretation, since it misses its mereological connotation of having parts compatible with the whole.

We end this section by observing that a kind of Schrödinger-picture version of the above statement is the customary argument that by taking the marginal of a pure entangled state one gets a mixed state. This means that, quantifying the "knowledge of the state" in terms of its von Neumann entropy, one has

Proposition 1 *The knowledge of the whole does not imply the knowledge of the parts.*

4 Mereological Holism is the Distinctively Quantum Holism

Mereological holism is the distinctly quantum holism in the sense that there are properties of the whole that are incompatible with all possible properties of the parts, thus violating the mereological connotation of the notion of "object" as being made of parts compatible with the whole. This has led us to the conclusion that the interpretation of the quantum "system" as "object" is untenable, in contrast with the classical case, where all properties are compatible.

In the arena of OPTs we have already encountered another kind of holism [4], related to the principle of local discriminability [5]. Different theories can have different degrees of such holism, depending on being local, bilocal, and generally n-local–the larger n, the more holistic the theory is. We will term such holism *local-discriminability holism* (shortly: *ld-holism*), to distinguish it from the kind of holism considered here, to which we will refer as *mereological holism*.

We should note that just as classical theory, quantum theory is not ld-holistic. On the other hand, in contrast to classical theory, quantum theory is mereologically holistic. Indeed, the fact that quantum theory is not ld-holistic guarantees its *compatibility with the scientific reductionistic approach* based on local observations in the presence of holism [5, 12], and this kind of holism is precisely the mereological holism established here.

5 Notion of "Property" and Mereological Holism in OPTs

Whereas the ld-holism of OPTs has been analyzed in the literature [4, 5, 12–15], the same has still to be done for the mereological one. Many questions can be raised. The first that comes to mind is: Does mereological holism hold for all theories with entanglement, e. g. symplectic theories with

Thc above considerations suggest to confront different OPTs in regards of mereological holism, also for deepening our understanding of the notions of *system* in

OPTs. In order to do so we need to generalize the notion of *property* to the OPT level.

At first sight this looks unfeasible, due to the quantum nature of the quantum-logic notion of "property", in one-to-one correspondence with Hilbert subspaces of the system Hilbert space. However, we will see now that the quantum notion of property is in one-to-one correspondence with a genuine OPT notion: the *repeatable atomic transformation*,[8] namely an atomic transformation $\mathscr{P} \in \mathrm{Trn}(\mathrm{A})$ satisfying the identity

$$\mathscr{P}\mathscr{P} = \mathscr{P}. \tag{5}$$

We recall that by the same definition of "transformation" within the OPT framework [5], Eq. (5) holds independently on the test to which $\mathscr{P}$ belongs, and independently on the circuit in which the test is embedded, namely independently on $\mathscr{P}$ input state, and on what is performed at the output of $\mathscr{P}$, thus holding also for non causal OPTs. It is easy to see that there is a one-to-one correspondence between quantum repeatable atomic transformations $\mathscr{P}$ and its relative projector P, since they are mapped to each other as follows

$$\mathscr{P} = P \cdot P, \quad P = \mathrm{Tr}_1[(\mathscr{P} \otimes \mathscr{I})(E)], \tag{6}$$

E denoting the swap operator in Eq. (1). The second identity in Eq. (6) is derived as follows

$$\begin{aligned}\mathrm{Tr}_1[\mathscr{P} \otimes \mathscr{I})(E)] &= \mathrm{Tr}_1[(P \otimes I)E(P \otimes I)] = \mathrm{Tr}_1[(E(P \otimes P)] \\ &= \mathrm{Tr}_1[E(P \otimes I)]P = \mathrm{Tr}_1[(I \otimes P)E]P = P\,\mathrm{Tr}_1[E]P = P.\end{aligned}$$

Thanks to bijection (6) we can now identify a property with the corresponding repeatable atomic transformation $\mathscr{P}$. We can then define the OPT notion of property in terms of a repeatable atomic transformation, and we can test mereological holism on various OPTs. Particularly interesting would be the cases of quantum-like theories (e. g. Fermionic) and of simplicial theories with entanglement [14, 15]. It would also be interesting to explore the interplay of mereological holism with OPT notion of complementarity [16] and the relation with incompatibility of properties, also in in connection with the principle of no-information without disturbance.

6 Conclusions

This paper introduced the notion of the mereological holism, and it was shown that it is the distinctive holism of quantum theory. The mereological holism corresponds to the existence of global properties of systems that are incompatible with all local properties, where "properties" are identified with projectors. It has also been shown

[8] We remind that a trasformation $\mathscr{T}$ is atomic if it is not the coarse-graining of other transformations, namely it cannot be written as the sum of two transformations e.g. as $\mathscr{T} = \mathscr{T}_1 + \mathscr{T}_2$.

how to extend uniquely the notion of property to the whole class of OPTs. to establish which other OPTs are mereologically holistic. Because of mereological holism, the usual interpretation of *quantum system* as a *physical object* with "properties" is untenable. Therefore, regarding the "system" as an "object" constitutes a hypostatization of the strictly-theoretical concept of "system", which in OPTs represents only a theoretical connection between objective events.

These conclusions suggest testing mereological holism on different OPTs, based on a unique extension of the notion of "property" to a repeatable atomic transformation, as suggested here. The concept of system can recover its "object" connotation only in theories which are not mereologically holistic, such as classical theory, whereas it is still an open problem weather this also holds for classical theories with entanglement, such as those in Refs. [14, 15].

Acknowledgements I thank Paolo Perinotti for noticing the relevance of atomicity of the OPT version of "property".

References

1. W. Quine, Whither physical objects?. In: R.S. Cohen, P.K. Feyerabend, M.W. Wartofsky (eds.), Essays in Memory of Imre Lakatos. Boston Studies in the Philosophy of Science, vol 39. (Springer, Dordrecht, 1976). https://doi.org/10.1007/978-94-010-1451-9_29
2. G. Birkhoff, J. von Neumann, Math. Ann. **37**, 823 (1936)
3. Merriam-Webster Object, https://www.merriam-webster.com/dictionary/object
4. L. Hardy, W.K. Wootters, Found. Phys. **42**, 454–473 (2012)
5. G.M. D'Ariano, G. Chiribella, P. Perinotti, *Quantum Theory from First Principles* (Cambridge University Press, Cambridge, 2017)
6. G. Chiribella, G.M. D'Ariano, P. Perinotti, *Quantum from Principles*, (Springer Netherlands, Dordrecht, 2016), pp. 171–221
7. J. von Neumann, *Mathematical Foundations of Quantum Mechanics* (Princeton University Press, Princeton, 1996)
8. G.M. D'Ariano, P. Lo Presti, M. Sacchi, Phys. Lett. A. **272**, 32 (2000)
9. G. Murphy, *C*-Algebras and Operator Theory*, (Academic Press, Boston, 1990), ISBN 978-0125113601
10. P. Perinotti, Phys. Rev. Lett. **108**(12), 120502 (2012)
11. A. Ferraro, L. Aolita, D. Cavalcanti, F.M. Cucchietti, A. Acín, Phys. Rev. A. **81**(5), 052318 (2010)
12. G.M. D'Ariano, Quantum Mechanics, How to derive the hilbert-space formulation of quantum mechanics from purely operational axioms. in *AIP Conference Proceedings*, vol. 844, ed. by A. Bassi, D. Duerr, T. Weber, N. Zanghi (American institute of Physics, USA, 2006)
13. G.M. D'Ariano, F. Manessi, P. Perinotti, A. Tosini, Eur. Lett. **107**, 20009 (2014)
14. G.M. D'Ariano, M. Erba, P. Perinotti, Phys. Rev. A. **102**(5), 052216 (2020)
15. G.M. D'Ariano, M. Erba, P. Perinotti, Phys. Rev. A. **101**(4), 042118 (2020)
16. G.M. D'Ariano, P. Perinotti, A. Tosini, 2021 in preparation

Giacomo Mauro D'Ariano is a professor of theoretical physics at the University of Pavia, where he is the leader of the QUIT (Quantum Information Theory) group. He is a member of the Center of Photonic Communication and Computing at Northwestern University, a member of the Instituto Lombardo Accademia di Scienze e Lettere, and a member of the Foundational Questions Institute. His primary areas of research are Quantum information theory, the mathematical structure of quantum theory, and foundational problems of contemporary physics. As one of the pioneers of Quantum Information Theory, he has made major contributions to the informational-theoretical derivation of Quantum Theory and Quantum Field Theory. He has published numerous articles in these areas and co-authored (with Giulio Chiribella and Paolo Perinotti) the book Quantum Theory from First Principles: An Informational Approach (Cambridge University Press, 2017). Giacomo Mauro D'Ariano received the International Quantum Award 2022 "for outstanding achievements in quantum science research".

Shannon Theory for Quantum Systems and Beyond: Information Compression for Fermions

Paolo Perinotti, Alessandro Tosini, and Leonardo Vaglini

Abstract We address the task of compression of fermionic quantum information. Due to the parity superselection rule, differently from the case of encoding of quantum information in qubit states, part of the information carried by fermionic systems is encoded in their delocalised correlations. As a consequence, reliability of a compression protocol must be assessed in a way that necessarily accounts also for the preservation of correlations. This implies that input/output fidelity is not a satisfactory figure of merit for fermionic compression schemes. We then discuss various aspects of the assessment of reliability of an encoding scheme, and show that entanglement fidelity in the fermionic case is capable of evaluating the preservation of correlations, thus revealing itself strictly stronger than input/output fidelity, unlike the qubit case. We then introduce a fermionic version of the source coding theorem showing that, as in the quantum case, the von Neumann entropy is the minimal rate for which a fermionic compression scheme exists, that is reliable according to the entanglement fidelity criterion.

1 Introduction

The story of quantum information has been strictly connected to the realm of foundations since its very birth. The development of this discipline into a growingly technical field of research did not weaken its bound with foundations, on the contrary, the experience of quantum information theory provided new perspectives and tools to face fundamental questions about the nature of quantum mechanics. Andrei Khrennikov was surely a witness of this process, and contributed to it actively by

P. Perinotti (✉) · A. Tosini · L. Vaglini
QUIT group, Physics Department, Pavia University, and INFN Sezione di Pavia, via Bassi 6, 27100 Pavia, Italy
e-mail: paolo.perinotti@unipv.it

A. Tosini
e-mail: alessandro.tosini@unipv.it

L. Vaglini
e-mail: leonardo.vaglini01@universitadipavia.it

A. Plotnitsky and E. Haven (eds.), *The Quantum-Like Revolution*,
https://doi.org/10.1007/978-3-031-12986-5_6

running for many years a renowned series of conferences in Växjö, which were a fertile ground for the development of new ideas. As one can easily check by reviewing the titles of the conferences, the genuine interest of Andrei for foundations [1, 2] was always accompanied by attention to the most recent developments of applied quantum theory, and especially to the pioneering view that quantum theory might be brought out of the realm of physics to approach the description of other phenomena than microscopic physical processes.

The community of scientists that apply ideas and techniques from quantum information to quantum foundations, to which the authors belong, thus largely benefitted from the lively and open environment that Andrei Khrennikov contributed to construct. This community looks at information as a primitive notion, from which the mechanical semantics of the quantum formalism should better be derived than postulated in the first place. The importance of the notion of information is thus crucial, and the present contribution aims at providing a precise definition of the information content of states of Fermionic systems, in terms of the minimum number of modes where many copies of the state can be mapped reversibly—a task commonly known as *data compression*. This is a standard notion in the literature, but the quantification of the compression ratio in the Fermionic case presents with some technical difficulties that we face here thoroughly.

As mentioned above, the task of data compression addresses the primary question in information theory as to how redundant is the information contained in a message and to what extent the message can then be compressed.

In classical information theory this question is answered by the source coding theorem [3], which establishes the fundamental role of Shannon entropy in information theory and its operational interpretation. The coding theorem recognizes the Shannon entropy as the fundamental limit for the compression rate in the i.i.d. setting. This means that if one compresses at a rate above the Shannon entropy, then the compressed data can be recovered perfectly in the asymptotic limit of infinitely long messages, while this is not possible for compression rate below the Shannon entropy. As a result the Shannon entropy, which can be intuitively thought of as the uncertainty about the outcome of an experiment that we are going to perform on a classical system, quantifies in a stringent way the amount of "non-redundant" information that is encoded in the state of the classical system, what one would definitely call *information content*.

In quantum information theory the Shannon entropy is replaced by the von Neumann entropy. In particular, the quantum source coding theorem [4] identifies von Neumann entropy as the rate at which quantum compression can be reliably achieved. Consider a quantum information source described by a system A and density operator $\rho \in \mathsf{St}(\mathrm{A})$, with $\mathsf{St}(\mathrm{A})$ the set of states of system A. The density operator describes the preparation of a state σ_i from any ensemble $\{p_i\sigma_i\}$, with probabilities p_i, such that $\sum_i p_i\sigma_i = \rho$. A quantum message of N letters can now be understood in terms of N uses of the quantum source, that output a sequence of N states $p_{i_j}\sigma_{i_j}$, with $1 \leq j \leq N$, drawn independently. One instance of this preparation protocol thus produces $\sigma_{\mathbf{i}} := \bigotimes_{j=1}^{N} \sigma_{i_j}$, with probability $p_{\mathbf{i}} := \prod_j p_{i_j}$. Each of the N systems has density operator ρ, and the density operator of the entire message is then given by

$\rho^{\otimes N}$. A compression scheme for messages from the above described source consists of two steps. *Encoding*: Alice encodes the system $A^{\otimes N}$ according to a compression map given by a channel $\mathscr{E} : \mathsf{St}(A^{\otimes N}) \to \mathsf{St}(B)$, where B is generally a system with dimension $d_B(N)$ smaller then $A^{\otimes N}$. The compression rate is defined as the asymptotic quantity $R = \lim_{N\to\infty} \log_2 d_B(N)/N$. Typically, one estimates the "size" of the compressed message through the capacity of system B given in terms of $\log_2 d_B(N)$, namely the number of qubits that are needed to simulate B. Alice then sends the system B to Bob using NR noiseless qubit channels. *Decoding:* Finally, Bob applies a decompression map $\mathscr{D} : \mathsf{St}(B) \to \mathsf{St}(A^{\otimes N})$ to the message encoded in system B, with the purpose of recovering the original message as reliably as possible.

As one might expect, the above scheme generally introduces an error in the decoding: we now discuss the figure of merit by which we estimate the error introduced by the compression scheme. In order to understand the operational meaning of the figure of merit, think of a referee (Charlie) who prepares the states σ_i with probability p_i, and receives the final states $\mathscr{D}\mathscr{E}(\sigma_i)$. The figure of merit that we use corresponds to the probability that, after receiving Bob's final state, Charlie is able to distinguish it from the input one. For a single instance, this probability is a linear function of the trace-norm distance $\|\sigma_i - \mathscr{D}\mathscr{E}(\sigma_i)\|_1$. The probability of successful discrimination is thus evaluated to $\sum_i p_i \|\sigma_i - \mathscr{D}\mathscr{E}(\sigma_i)\|_1 = \|\rho^{\otimes N} - \mathscr{D}\mathscr{E}(\rho^{\otimes N})\|_1$. The protocol has then error ϵ if the compressed and decompressed states $\mathscr{D}\mathscr{E}(\sigma_i)$ are ϵ-close to the original states σ_i, in the trace-norm distance. In the case of qubits the above quantity is equivalent to fidelity, thanks to the Fuchs-van der Graaf inequalities [5]. The optimal quantum encoding will then make the error arbitrarily small for N large enough, with rate R as small as possible. Schumacher's quantum source coding theorem shows that the optimal rate is equal to the von Neumann entropy $S(\rho)$ of the state ρ.

Another way to evaluate the error for a compression scheme is the following: Charlie prepares a purification of the density operator $\rho^{\otimes N}$ and sends the N copies of system A to Alice. Alice then sends her share of the pure state to Bob, sending as few qubits to Bob as possible. After decompressing the received qubits, Bob shares an entangled state with Charlie. The quality of the compression scheme can then be evaluated considering how well Charlie can distinguish the initial state from the final one, after receiving Bob's N systems. The probability that Charlie detects a compression error can be evaluated through the input/output fidelity. Again, Schumacher's theorem states that Alice can transfer her share of the pure state to Bob by sending $NS(\rho)$ qubits and achieving arbitrarily good fidelity, increasing the length N of the message. This second perspective answers the question whether the compression protocol preserves the correlations that system $A^{\otimes N}$ has with a remote system C.

Equivalence of the two approaches in assessing the quality of a compression scheme shows that the ability to send quantum superpositions is equivalent to the ability to send entanglement. In other terms, the amount of quantum information preserved by a compression scheme represents the dimension of the largest Hilbert space whose superpositions can be reliably compressed, or equivalently the amount of entanglement that can be reliably compressed.

According to the above discussion, a crucial point in the compression protocol is to quantify the reliability of the compression map $\mathscr{C} := \mathscr{D}\mathscr{E}$, which in the asymptotic limit of $N \to \infty$ must coincide with the identity map. In quantum theory checking the reliability of the compression map looking only at its local action, namely via the fidelity between states $\mathscr{C}(\rho^{\otimes N})$ and $\rho^{\otimes N}$, or at the effects on correlations, namely via entanglement fidelity, is equivalent. This follows from *local process tomography* of quantum theory where, given a map $\mathscr{C}$ on system A one has

$$\begin{aligned} (\mathscr{C} \otimes \mathscr{I}_{\mathrm{C}})(\Psi) =& \Psi \qquad \forall \Psi \in \mathsf{St}(\mathrm{AC}) \\ &\Leftrightarrow \\ \mathscr{C}(\rho) =& \rho \qquad \forall \rho \in \mathsf{St}(\mathrm{A}). \end{aligned} \tag{1}$$

This equivalence is due to *local discriminability* [6–8] of quantum theory, where the discrimination of bipartite quantum states can always be performed using local measurements only (this property is equivalent to the one known in the literature as local tomography, or tomographic locality [9–12]). However, in the absence of local discriminability, a map preserving local states still can affect correlations with remote systems [13]. This raises a crucial issue if one aims at studying the compression task beyond quantum theory, where the reliability of a protocol generally needs to be verified on extended systems. Indeed, in general, testing a compression scheme using ancillary systems is strictly stronger than testing them with local schemes.

As a first step in the direction of generalizing the compression protocol to an arbitrary information theory, in this paper we consider the case of fermionic systems as carriers of information. Fermionic computation has been proposed in Ref. [14] and later studied in several works [15–19]. Differently from quantum systems, fermions obey the *parity superselection rule*. As a consequence, fermionic information theory does not satisfy local discriminability, thus providing a physically relevant example of a theory where the task of compression is not straightforward. Indeed, in the case we consider, a map $\mathscr{C}$ that acts as the identity on local states $\rho^{\otimes N}$ could still destroy the correlations with remote systems, and then be mistakenly considered as a reliable compression map.

After reviewing the structure of fermionic quantum information, we prove that the entanglement fidelity is a valid criterion for the reliability of a fermionic compression map. We then show an analogous of the quantum source coding theorem in the fermionic scenario, showing that the minimal compression rate for which a reliable compression scheme exists is the von Neumann entropy of the fermionic state. We conclude therefore that the von Neumann entropy provides the informational content of the state also in the case of fermionic theory, namely in the presence of parity superselection. The above result, however, is not a straightforward consequence of Schumacher's source coding theorem.

2 Fermionic Information Theory

We now briefly review fermionic information theory. The systems of the theory are made by local fermionic modes (LFMs). A LFM is the counterpart of the qubit in quantum theory, and can be thought of as a cell that can be either empty or occupied by a fermionic excitation. An L-LFMs system, denoted $\mathrm{L_F}$, is described by L fermionic fields φ_i, satisfying the canonical anticommutation rule (CAR) $\{\varphi_i, \varphi_j^\dagger\} = \delta_{ij} I, \{\varphi_i, \varphi_j\} = 0$ where $i, j = 1, \ldots, L$. With these fields one constructs the occupation number operators $\varphi_i^\dagger \varphi_i$, which can be easily proved to have only eigenvalues 0 and 1. The common eigenvector $|\Omega\rangle$ of the operators $\varphi_i^\dagger \varphi_i$, $i = 1, \ldots, L$ with eigenvalue 0 defines the vacuum state $|\Omega\rangle\langle\Omega|$ of $\mathrm{L_F}$, representing the state in which all the modes are not excited. The fermionic vacuum state in terms of the field operators is given by $|\Omega\rangle\langle\Omega| = \prod_{i=1}^{L} \varphi_i \varphi_i^\dagger$. By applying the operators $\varphi_i^\dagger$ to $|\Omega\rangle$ the corresponding i-th mode is excited and, by raising $|\Omega\rangle$ in all possible ways, we get the 2^L orthonormal vectors forming the Fock basis in the occupation number representation: a generic element of this basis is

$$|n_1, \ldots, n_L\rangle := (\varphi_1^\dagger)^{n_1} \ldots (\varphi_L^\dagger)^{n_L} |\Omega\rangle, \tag{2}$$

with $n_i = \{0, 1\}$ corresponding to the occupation number at the ith site. The linear span of these vectors corresponds to the antisymmetric Fock space $\mathcal{F}_L$ of dimension $d_{\mathcal{F}_L} = 2^L$. Notice that the Fock space $\mathcal{F}_L$ is isomorphic to the Hilbert space $\mathcal{H}_L$ of L qubits, by the trivial identification of the occupation number basis with the qubit computational basis. This correspondence lies at the basis of the Jordan-Wigner isomorphism [20–22] typically used in the literature to map fermionic systems to qubits systems and vice-versa. We recall here the definition of the Jordan-Wigner map

$$\begin{aligned} J_L(\varphi_i) &= \left(\bigotimes_{l=1}^{i-1} \sigma_l^z\right) \otimes \sigma_i^- \otimes \left(\bigotimes_{k=i+1}^{L} I_k\right), \\ J_L(\varphi_i^\dagger) &= J_L(\varphi_i)^\dagger, \\ J_L(XY) &= J_L(X) J_L(Y), \\ J_L(aX + bY) &= a J_L(X) + b J_L(Y), \end{aligned} \tag{3}$$

with X, Y linear combinations of products of field operators on the L-LFMs, and where we used the standard notation for Pauli sigma operators. In the following we will drop the dependence on the number of LFMs in the Jordan-Wigner map, namely we will write $J(X)$ in place of $J_L(X)$, when it will be clear from the context. Notice that the Jordan-Wigner isomorphism is implicitly defined in Eq. (2), and, as such, it depends on the arbitrary ordering of the modes. All such representations are unitarily equivalent.

Differently from standard qubits, fermionic systems satisfy the parity superselection rule [18, 23–26]. One can decompose the Fock space $\mathcal{F}_L$ of system $\mathrm{L_F}$ in the direct sum $\mathcal{F}_L = \mathcal{F}_L^e \oplus \mathcal{F}_L^o$, with $\mathcal{F}_L^e$ and $\mathcal{F}_L^o$ the spaces generated by vectors with even and odd total occupation number, respectively. The convex set of states $\mathsf{St}(\mathrm{L_F})$ corresponds, in the Jordan-Wigner representation, to the set of density matrices on $\mathcal{F}_L$ of the form $\rho = \rho_e + \rho_o$, with $\rho_e, \rho_o \geq 0$, $\mathrm{Tr}[\rho_o] + \mathrm{Tr}[\rho_e] \leq 1$ and with ρ_e and ρ_o having support on $\mathcal{F}_L^e$ and $\mathcal{F}_L^o$, respectively, and pure states are represented by rank one density operators. Moreover, the density matrices representing the states represent linear combinations of products of an even number of field operators (see Appendix 1 and [26] for further details). Conversely, every linear combination of products of an even number of field operators that is represented by a density matrix is an admissible state. Analogously, effects in the set $\mathsf{Eff}(\mathrm{L_F})$ are represented by positive operators on $\mathrm{L_F}$ of the form $a = a_e + a_o$, with a_e and a_o having support on $\mathcal{F}_L^e$ and $\mathcal{F}_L^o$, respectively. Notice that set of states and effects of system $\mathrm{L_F}$ have dimension $d_{\mathcal{F}_L}^2/2 = 2^{2L-1}$, compared to the quantum case where the set of states and effects associated to the Hilbert space $\mathcal{H}_L$ of L qubits has dimension $d_{\mathcal{H}_L}^2 = 2^{2L}$.

Given a state $\rho \in \mathsf{St}(\mathrm{L_F})$ we define the refinement set of ρ as $\mathsf{Ref}(\rho) := \{\sigma \in \mathsf{St}(\mathrm{L_F}) | \exists \tau \in \mathsf{St}(\mathrm{L_F}) : \rho = \sigma + \tau\}$, and a state is pure when all the elements in the refinement are proportional to the state itself. In the following we will denote by $\mathsf{PurSt}(\mathrm{L_F})$ and $\mathsf{St}_1(\mathrm{L_F})$ the set of pure states and the set of normalized states (of trace one) of system $\mathrm{L_F}$, respectively.

Given two fermionic systems L_F and M_F, we introduce the composition of the two as the system made of $K \equiv L + M$ LFMs, denoted with the symbol $\mathrm{K_F} := \mathrm{L_F} \boxtimes \mathrm{M_F}$, or simply $\mathrm{K_F} := \mathrm{L_F M_F}$. We use the symbol $\boxtimes$ to distinguish the fermionic parallel composition rule from the quantum one, corresponding to the tensor product $\otimes$.

Given a state $\Psi \in \mathsf{St}(\mathrm{L_F M_F})$, one can discard the subsystem $\mathrm{M_F}$ and consider the marginal state, which we denote by $\sigma := \mathrm{Tr}^f_{\mathrm{M_F}}(\Psi)$. We use the symbol $\mathrm{Tr}^f_{\mathrm{M_F}}$ to denote the fermionic partial trace on the subsystem $\mathrm{M_F}$. This is computed by performing the following steps (see Ref. [26] for further details): (i) drop all those terms in Ψ containing an odd number of field operators in any of the LFMs in $\mathrm{M_F}$; (ii) remove all the field operators corresponding to the LFMs in $\mathrm{M_F}$ from the remaining terms. The fermionic trace $\mathrm{Tr}^f(\rho)$ of a state $\rho \in \mathsf{St}(\mathrm{M_F})$ is then defined as a special case of the partial one, corresponding to the case in which $L = 0$.

Finally, the set of transformations from $\mathrm{L_F}$ to $\mathrm{M_F}$, denoted by $\mathsf{Tr}(\mathrm{L_F} \to \mathrm{M_F})$, is given by completely positive maps from $\mathsf{St}(\mathrm{L_F})$ to $\mathsf{St}(\mathrm{M_F})$ in the Jordan-Wigner representation. Moreover, we denote by $\mathsf{Tr}_1(\mathrm{L_F} \to \mathrm{M_F})$ the set of deterministic transformations, also called *channels*, from $\mathrm{L_F}$ to $\mathrm{M_F}$, corresponding to trace-preserving completely positive maps. Like in quantum theory, any fermionic transformation $\mathscr{C} \in \mathsf{Tr}(\mathrm{L_F} \to \mathrm{M_F})$ can be expressed in Kraus form $\mathscr{C}(\rho) = \sum_i C_i \rho C_i^\dagger$, with deterministic transformations having Kraus operators $\{C_i\}$ such that $J(\sum_i C_i^\dagger C_i) = I_{\mathcal{H}_L}$, $I_{\mathcal{H}_L}$ denoting the identity operator on $\mathcal{H}_L$. For a map $\mathscr{C} \in \mathsf{Tr}(\mathrm{L_F} \to \mathrm{M_F})$ with Kraus operators $\{C_i\}$, we define its Jordan-Wigner representative $J(\mathscr{C})$ as the quantum map with Kraus operators $\{J(C_i)\}$. Now, given two transformations $\mathscr{C} \in \mathsf{Tr}(\mathrm{L_F} \to \mathrm{M_F})$ and $\mathscr{D} \in \mathsf{Tr}(\mathrm{K_F} \to \mathrm{N_F})$, we denote by $\mathscr{C} \boxtimes \mathscr{D} \in \mathsf{Tr}(\mathrm{L_F K_F} \to \mathrm{M_F N_F})$ the *parallel composition* of $\mathscr{C}$ and $\mathscr{D}$, with Kraus operators $\{C_i D_j\}$, where $\{C_i\}$ are Kraus operators for $\mathscr{C}$ and $\{D_j\}$ for $\mathscr{D}$. We observe that in the Jordan-Wigner representation one generally has $J_{L+K}(C_iD_j) \neq J_L(C_i) \otimes J_K(D_j)$, and $J_{L+K}(\mathscr{C} \boxtimes \mathscr{D}) \neq J_L(\mathscr{C}) \otimes J_K(\mathscr{D})$. If $\mathscr{C}$ is a transformation in $\mathsf{Tr}(\mathrm{L_F} \to \mathrm{M_F})$, its extension to a composite system $\mathrm{L_F N_F}$, is given by $\mathscr{C} \boxtimes \mathscr{I}$, with $\mathscr{I}$ the identity map of system $\mathrm{N_F}$—whose Jordan-Wigner representative is the quantum identity map—and its Kraus operators involve field operators on the $\mathrm{L_F}$ modes only. It is worth noting that, despite the Jordan-Wigner representative of this map is not necessarily of the form $J_L(\mathscr{C}) \otimes \mathscr{I}$, upon suitable choice of the ordering of the LFMs that defines the representation, one can always reduce to the case where, actually, $J_{L+N}(\mathscr{C} \boxtimes \mathscr{I}) = J_L(\mathscr{C}) \otimes \mathscr{I}$.

As a special case of the above composition rule, one can define $\rho \boxtimes \sigma := \rho\sigma \in \mathsf{St}(\mathrm{L_F M_F})$ for the parallel composition of states $\rho \in \mathsf{St}(\mathrm{L_F})$ and $\sigma \in \mathsf{St}(\mathrm{M_F})$, and similarly $a \boxtimes b := ab \in \mathsf{Eff}(\mathrm{L_F M_F})$ for the parallel composition of effects $a \in \mathsf{Eff}(\mathrm{L_F})$ and $b \in \mathsf{Eff}(\mathrm{M_F})$.

A useful characterization of fermionic maps in $\mathsf{Tr}(\mathrm{L_F} \to \mathrm{L_F})$, proved in Ref. [27], is the following:

Proposition 1 (Fermionic transformations) *All the transformations in* $\mathsf{Tr}(\mathrm{L_F} \to \mathrm{L_F})$ *with Kraus operators being linear combinations of products of either an even number or an odd number of field operators are admissible fermionic transformations. Conversely, each admissible fermionic transformation in* $\mathsf{Tr}(\mathrm{L_F} \to \mathrm{L_F})$ *has Kraus operators being superpositions of products of either an even number or an odd number of field operators.*

Corollary 1 (Fermionic effects) *Fermionic effects are positive operators bounded by the identity operator that are linear combinations of products of an even number of field operators. Conversely, every linear combination of products of an even number of field operators that is represented by a positive operator bounded by the identity is a fermionic effect.*

The corollary follows immediately from Proposition 1, since an effect A is obtained as a fermionic transformation $\mathscr{A}$ followed by the discard map, i.e. the trace. Thus

$$\mathrm{Tr}[\rho A] = \mathrm{Tr}[\mathscr{A}(\rho)] = \sum_i \mathrm{Tr}[K_i \rho K_i^\dagger] = \mathrm{Tr}[\rho \sum_i K_i^\dagger K_i],$$

namely $A = \sum_i K_i^\dagger K_i$. Having the polynomial K_i a definite parity (though not necessarily the same for every i), A is an even polynomial.

In the following we denote by $\mathcal{L}(\mathcal{H}_L)$ the set of linear operators on the Hilbert space $\mathcal{H}_L$ of L-qubits and by $\mathcal{L}(\mathcal{H}_L, \mathcal{H}_M)$ the set of linear operators from $\mathcal{H}_L$ to $\mathcal{H}_M$. It is useful to introduce the isomorphism between operators X in $\mathcal{L}(\mathcal{H}_L, \mathcal{H}_M)$ and vectors $|X\rangle\rangle$ in $\mathcal{H}_M \otimes \mathcal{H}_L$ given by

$$|X\rangle\rangle = (X \otimes I_{\mathcal{H}_L})|I_{\mathcal{H}_L}\rangle\rangle = (I_{\mathcal{H}_M} \otimes X^T)|I_{\mathcal{H}_M}\rangle\rangle, \tag{4}$$

where $I_{\mathcal{H}_L}$ is the identity operator in $\mathcal{H}_L$, $|I_{\mathcal{H}_L}\rangle\rangle \in \mathcal{H}_L^{\otimes 2}$ is the maximally entangled vector $|I_{\mathcal{H}_L}\rangle\rangle = \sum_l |l\rangle|l\rangle$ (with $\{|l\rangle\}$ a fixed orthonormal basis for $\mathcal{H}_L$), and $X^T \in$

$\mathcal{L}(\mathcal{H}_M, \mathcal{H}_L)$ is the transpose of X with respect to the two fixed bases chosen in $\mathcal{H}_L$ and $\mathcal{H}_M$. Notice also the useful identity

$$Y \otimes Z|X\rangle\rangle = |YXZ^T\rangle\rangle, \tag{5}$$

where $X \in \mathcal{L}(\mathcal{H}_L, \mathcal{H}_M)$, $Y \in \mathcal{L}(\mathcal{H}_M, \mathcal{H}_N)$ and $Z \in \mathcal{L}(\mathcal{H}_L, \mathcal{H}_K)$. Moreover, for $X, Y \in \mathcal{L}(\mathcal{H}_L, \mathcal{H}_M)$, one has $\mathrm{Tr}_{\mathcal{H}_L}[|X\rangle\rangle\langle\langle Y|] = XY^\dagger$, and $\mathrm{Tr}_{\mathcal{H}_M}[|X\rangle\rangle\langle\langle Y|] = X^T Y^*$. We remark that, in the above paragraph, we are dealing with abstract linear operators on an Hilbert space, disregarding their possible interpretation as Jordan-Wigner representatives of some fermionic operator.

A notion that will be used in the following is that of states dilation.

Definition 1 (*Dilation set of a state* ρ) For any $\rho \in \mathsf{St}(\mathrm{L_F})$, we say that $\Psi_\rho \in \mathsf{St}(\mathrm{L_F M_F})$ for some system $\mathrm{M_F}$, is a dilation of ρ if $\rho = \mathrm{Tr}^f_{\mathrm{M_F}}[\Psi_\rho]$. We denote by D_ρ the set of all possible dilations of ρ. A pure dilation $\Psi_\rho \in \mathsf{PurSt}(\mathrm{L_F M_F})$ of ρ is called a purification.

Naturally, any purification of ρ belongs to D_ρ, more precisely the set of purifications of ρ is the subset of D_ρ containing pure states. A main feature of quantum theory that is valid also for fermionic systems is the existence of a purification of any state, that is unique modulo channels on the purifying system.

Proposition 2 (Purification of states) *For every* $\rho \in \mathsf{St}(\mathrm{L_F})$, *there exists a purification* $\Psi_\rho \in \mathsf{PurSt}(\mathrm{L_F M_F})$ *of* ρ *for some system* $\mathrm{M_F}$. *Moreover, the purification is unique up to channels on the purifying system: if* $\Psi_\rho \in \mathsf{PurSt}(\mathrm{L_F M_F})$ *and* $\Phi_\rho \in \mathsf{PurSt}(\mathrm{L_F K_F})$ *are two purifications of* ρ *then there exists a channel* $\mathscr{V} \in \mathsf{Tr}_1(\mathrm{M_F} \to \mathrm{K_F})$ *such that* $(\mathscr{I}_{\mathrm{L_F}} \boxtimes \mathscr{V})(\Psi_\rho) = \Phi_\rho$.

Proof It can be easily verified that every purification of $\rho \in \mathsf{St}(\mathrm{L_F})$, having even part ρ_e and odd part ρ_o, can be obtained in terms of the minimal one $J^{-1}(|F\rangle\rangle\langle\langle F|) \in \mathsf{PurSt}(\mathrm{L_F M_F})$, with $F = J(\rho)^{\frac{1}{2}}$, $M = \lceil \log_2 2r \rceil$ and $r = \max(\mathrm{rank}(\rho_e), \mathrm{rank}(\rho_o))$. Now, let $\Psi_\rho \in \mathsf{PurSt}(\mathrm{L_F M_F})$ and $\Phi_\rho \in \mathsf{PurSt}(\mathrm{L_F K_F})$ be two purifications of ρ. If $M = K$, let us choose the ordering defining the Jordan-Wigner isomorphism of Eq. (2) in such a way that the modes in the purifying systems $\mathrm{M_F}$ precede the modes of $\mathrm{L_F}$. Then, using the quantum purification theorem, we know that there exists a reversible map $\mathscr{U}$ with unitary Kraus U such that $|F_\rho\rangle\rangle = (U \otimes I)|P_\rho\rangle\rangle$, where

$$|F_\rho\rangle\rangle\langle\langle F_\rho| = J(\Phi_\rho), \quad |P_\rho\rangle\rangle\langle\langle P_\rho| = J(\Psi_\rho).$$

The unitary U can be chosen in such a way that $J^{-1}(\mathscr{U})$ is an admissible fermionic map, namely in such a way that it respects the parity superselection rule. Moreover, due to Lemma 2, $J^{-1}(U \otimes I)$ cannot contain field operators on the modes in $\mathrm{L_F}$, and is then local on the purifying system $\mathrm{K_F}$. Now, let $K > M$. Then, we can consider a pure state ω on the $K - M$ modes and take the parallel composition $\Psi_\rho \boxtimes \omega$. This is still a purification of ρ, and by the previous argument, there exists a reversible channel $\mathscr{U} \in \mathsf{Tr}_1(\mathrm{K_F} \to \mathrm{K_F})$ such that $\Phi_\rho = (\mathscr{I}_{\mathrm{L_F}} \boxtimes \mathscr{U})(\Psi_\rho \boxtimes \omega) = (\mathscr{I}_{\mathrm{L_F}} \boxtimes \mathscr{V})(\Psi_\rho)$ where $\mathscr{V}$ is the channel defined by $\mathscr{V} = \mathscr{U}(\mathscr{I} \boxtimes \omega)$. If $K < M$, we consider $\Phi_\rho \boxtimes \omega$, where ω is any pure state on $N = M - K$ modes system, and we have $\Phi_\rho \boxtimes \omega = (\mathscr{I}_{\mathrm{L_F}} \boxtimes \mathscr{U})(\Psi_\rho)$. Now we discard the additional modes, and the channel connecting the purifications is the sequential composition of $\mathscr{U}$ and the discarding map: $\mathscr{V} := (\mathscr{I}_{\mathrm{K_F}} \boxtimes \mathrm{Tr}^f_{\mathrm{N_F}})\mathscr{U}$. □

The main difference between fermionic and quantum information lies in the notion of which Kraus operators correspond to local maps. While in the case of qubit systems local maps acting on the i-th qubit of a composite system have Kraus operators that can be factorized as a non trivial operator on the i-th tensor factor $\mathbb{C}^2$ of the total Hilbert space, in the case of the fermionic Fock space $\mathcal{F}_L$ a local transformation on the i-th mode can be represented in the Jordan-Wigner isomorphism by operators that act non trivially on factors $\mathbb{C}^2$ different from the i-th one. This fact is the source of all the differences between the theory of qubits and fermionic theory, including superselection and features that it affects, such as the notion of entanglement [27] and local states discrimination protocols [28, 29]. Due to parity superselection, fermionic theory does not satisfy *local process tomography*, namely the property stating that two transformations $\mathscr{C}_1, \mathscr{C}_2 \in \mathsf{Tr}(\mathrm{L_F} \to \mathrm{M_F})$ are equal iff they act in the same way on the local states in $\mathsf{St}(\mathrm{L_F})$, namely $\mathscr{C}_1(\rho) = \mathscr{C}_2(\rho)$ for every $\rho \in \mathsf{St}(\mathrm{L_F})$ (see for example Eq. (1) in the introduction on the equality between the compression map $\mathscr{C}$ and the identity map). As a consequence, fermionic theory also violates *local tomography*. A typical example of a transformation that is locally equivalent to the identity but differs from it when extended to multipartite systems is the parity transformation, as shown in the following. Let us consider a single fermionic mode system 1_F, whose possible states are constrained to be of the form $J(\rho) = q_0 \left|0\right\rangle\left\langle 0\right| + q_1 \left|1\right\rangle\left\langle 1\right|$ by the parity superselection rule. Let P_0 and P_1 be the projectors on $\left|0\right\rangle$ and $\left|1\right\rangle$ respectively, namely on the even and odd sector of the Fock space. The parity transformation $\mathscr{P}$, that in the Jordan-Wigner representation $J(\mathscr{P})$ has Kraus operators P_0 and P_1, acts as the identity $\mathscr{I}_{1_\mathrm{F}}$ when applied to states in $\mathsf{St}(1_\mathrm{F})$. However, taking the system 2_F and considering the extended transformation $\mathscr{P} \boxtimes \mathscr{I}_{1_\mathrm{F}}$ on $\mathsf{St}(2_\mathrm{F})$ one notices that $\mathscr{P}$ differs from the identity map $\mathscr{I}_{1_\mathrm{F}}$. Indeed, the state $J^{-1}(\left|\Psi\right\rangle\left\langle\Psi\right|)$, with $\left|\Psi\right\rangle = \frac{1}{\sqrt{2}}(\left|00\right\rangle + \left|11\right\rangle)$ is a legitimate fermionic state in $\mathsf{St}(2_\mathrm{F})$, and one can straightforwardly verify that

$$
\begin{aligned}
(\mathscr{P} \boxtimes \mathscr{I}_{1_\mathrm{F}})[J^{-1}(\left|\Psi\right\rangle\left\langle\Psi\right|)] &= \frac{1}{2} J^{-1}(\left|00\right\rangle\left\langle 00\right| + \left|11\right\rangle\left\langle 11\right|) \\
&\neq J^{-1}(\left|\Psi\right\rangle\left\langle\Psi\right|).
\end{aligned}
$$

2.1 *Identical Channels Upon-Input of ρ*

In the following we will be interested in quantitatively assessing how closely a channel (the compression map) resembles another one (the identity map), provided that we know that the input state corresponds to a given ρ. To this end we introduce the notion of identical fermionic channels upon-input of ρ.

Given two fermionic channels $\mathscr{C}_1, \mathscr{C}_2 \in \mathsf{Tr}_1(\mathrm{L_F} \to \mathrm{M_F})$ and a state $\rho \in \mathsf{St}(\mathrm{L_F})$, we say that $\mathscr{C}_1$ and $\mathscr{C}_2$ are equal upon-input of ρ if

$$(\mathscr{C}_1 \boxtimes \mathscr{I})(\Sigma) = (\mathscr{C}_2 \boxtimes \mathscr{I})(\Sigma) \qquad \forall \Sigma \in \mathsf{Ref}(D_\rho). \tag{6}$$

Operationally, this means that one cannot discriminate between $\mathscr{C}_1$ and $\mathscr{C}_2$ when applied to any dilation Ψ_ρ of the state ρ, independently of how Ψ_ρ has been prepared. Suppose that $\Psi_\rho \in D_\rho$ was prepared as $\Psi_\rho = \sum_i \Sigma_i$, for some refinement of Ψ_ρ. Even using the knowledge of the preparation, one cannot distinguish between $\mathscr{C}_1$ and $\mathscr{C}_2$. Notice that, differently from the quantum case here it is necessary to check the identity between channels on bipartite systems.

Following the above definition, one can quantify how close two channels are. One has that $\mathscr{C}_1$ and $\mathscr{C}_2$ are ε-close upon-input of ρ if

$$\sum_i \|[(\mathscr{C}_1 - \mathscr{C}_2) \boxtimes \mathscr{I}](\Sigma_i)\|_1 \leq \varepsilon \quad \forall \{\Sigma_i\} : \sum_i \Sigma_i \in D_\rho, \tag{7}$$

where $\|X\|_1$ is the 1-norm of $J(X)$. One can straightforwardly prove that the trace distance $d(\rho, \sigma) := \frac{1}{2}\|\rho - \sigma\|_1$ has a clear operational interpretation in terms of the maximum success probability of discrimination between the two states ρ and σ. Equation (7) provides then an upper bound for the probability of discriminating between $\mathscr{C}_1$ and $\mathscr{C}_2$ when applied to the dilations of ρ, including their refinements: $\mathscr{C}_1$ and $\mathscr{C}_2$ cannot be distinguished with a succes probability bigger than $\frac{1}{2} + \frac{1}{4}\varepsilon$. Accordingly, a sequence of channels $\mathscr{C}_N \in \mathsf{Tr}_1(\mathrm{L_F} \to \mathrm{M_F})$ converges to the channel $\mathscr{C} \in \mathsf{Tr}_1(\mathrm{L_F} \to \mathrm{M_F})$ upon-input of ρ if

$$\lim_{N \to \infty} \|[(\mathscr{C}_N - \mathscr{C}) \boxtimes \mathscr{I}](\Sigma)\|_1 = 0 \quad \forall \Sigma \in \mathsf{Ref}(D_\rho).$$

3 Fermionic Compression

Consider now a system $\mathrm{L_F}$ and let $\rho \in \mathsf{St}(\mathrm{L_F})$ be the generic state of the system. As usual the source of fermionic information is supposed to emit N independent copies of the state ρ . A fermionic compression scheme $(\mathscr{E}_N, \mathscr{D}_N)$ consists of the following two steps:

1. Encoding: Alice encodes the system $\mathrm{L_F}^{\boxtimes N}$ via a channel $\mathscr{E}_N : \mathsf{St}(\mathrm{L_F}^{\boxtimes N}) \to \mathsf{St}(\mathrm{M_F})$, where the target system is generally a system of M-LFMs. The map $\mathscr{E}_N$ produces a fermionic state $\mathscr{E}(\rho^{\boxtimes N})$ with support $\mathsf{Supp}(\mathscr{E}(\rho^{\boxtimes N}))$ on a Fock space $\mathcal{F}_M$ of dimension $d_{\mathcal{F}_M}(N)$ smaller than the one of the original state $\rho^{\boxtimes N}$. The compression rate is defined as the quantity

$$R = \log_2 d_{\mathcal{F}_M}(N)/N.$$

Alice sends the system $\mathrm{M_F}$ to Bob using $N\lceil R \rceil$ noiseless fermionic channels.

2. Decoding: Finally Bob sends the system $\mathrm{M_F}$ through a decompression channel $\mathscr{D}_N : \mathsf{St}(\mathrm{M_F}) \to \mathsf{St}(\mathrm{L_F}^{\boxtimes N})$.

The scheme $(\mathscr{E}_N, \mathscr{D}_N)$ overall transforms the $L^{\boxtimes N}$ LFMs, with a compression map $\mathscr{C}_N := \mathscr{D}_N \mathscr{E}_N$. The latter can be more or less "good" (in a sense that will be precisely defined) in preserving the information which is contained in the system, depending on ρ itself. The goal now is to define the notion of reliable compression scheme once we are provided with an information source ρ.

3.1 Reliable Compression Scheme

The aim of a compression scheme, besides reducing the amount of information carriers used, is to preserve all the information that is possibly encoded in a given state ρ. What we actually mean is that the aim is not only to preserve the input state and keep track of the correlations of our system with an arbitrary ancilla, but also to preserve all information involved, for any procedure by which the input system and its correlations have been prepared. In other words, even the agent that prepared the system along with possible ancillary systems, must have a small success probability in detecting the effects of compression on the original preparation. This amounts to require that the compression channel $\mathscr{C}_N$ must be approximately equal to the identity channel upon-input of ρ, and more precisely that in the limit of $N \to \infty$ the two channels must coincide upon-input of ρ.

In Sect. 2.1 we introduced the notion of ε-close channels upon-input of ρ. This notion can now be used to quantify the error, say ε, introduced by the map $\mathscr{C}_N$ in a compression protocol given the source ρ. According to Eq. (7) we have indeed the following definition of a reliable compression scheme

Definition 2 (*Reliable compression scheme*) Given a state $\rho \in \mathsf{St}(\mathrm{L_F})$, a compression scheme $(\mathscr{E}_N, \mathscr{D}_N)$ is ε-reliable if $\sum_i \|(\mathscr{C}_N \boxtimes \mathscr{I})(\Sigma_i) - \Sigma_i\| < \varepsilon$ for every $\{\Sigma_i\}$ such that $\sum_i \Sigma_i \in D_{\rho^{\boxtimes N}}$, where $\mathscr{C}_N := \mathscr{D}_N \mathscr{E}_N$.

It is clear from the definition that in order to check the reliability of a fermionic compression map one should test it on states of an arbitrary large system, since the dilation set $D_{\rho^{\boxtimes N}}$ includes dilations on any possible ancillary system. It is then necessary to find a simpler criterion to characterize the reliability of a compression scheme. Let us start with a preliminary definition.

Definition 3 Let $\rho \in \mathsf{St}(\mathrm{L_F})$. We define its *square root* $\rho^{\frac{1}{2}}$ as follows

$$\rho^{\frac{1}{2}} := J^{-1}[J(\rho)^{\frac{1}{2}}]. \tag{8}$$

One can easily prove that the square root of a fermionic state is well defined, i.e. it does not depend on the particular Jordan-Wigner representation J chosen (see Appendix 3). In the following we show that a useful criterion for reliability can be expressed via *entanglement fidelity*:

Definition 4 (*Entanglement fidelity*) Let $\rho \in \mathsf{St}_1(\mathrm{L_F})$, $\mathscr{C} \in \mathsf{Tr}_1(\mathrm{L_F} \to \mathrm{M_F})$ and $\Phi_\rho \in \mathsf{PurSt}(\mathrm{L_F K_F})$ be any purification of ρ. The entanglement fidelity is defined as and $F(\rho, \mathscr{C}) = F(\Phi_\rho, (\mathscr{C} \boxtimes \mathscr{I})(\Phi_\rho))^2$, where $F(\rho, \sigma) := \mathrm{Tr}[J(\rho^{1/2}\sigma\rho^{1/2})^{1/2}]$ denotes the Uhlmann's fidelity between states $\rho, \sigma \in \mathsf{St}_1(\mathrm{L_F})$.

We notice that the Uhlmann fidelity of fermionic states is well defined, namely it is independent of the ordering of the fermionic modes (see Appendix 3). As a consequence also the Entanglement fidelity, given in terms of the Uhlmann one, must be well defined.

Since by definition the Uhlmann fidelity of fermionic states coincides with the one of their Jordan-Wigner representatives and the same for their trace-norm distance, given $\rho, \sigma \in \mathsf{St}(\mathrm{L_F})$, the Fuchs-van der Graaf inequalities [5] hold as a trivial consequence of their quantum counterparts

$$1 - F(\rho, \sigma) \leq \frac{1}{2}\|\rho - \sigma\|_1 \leq \sqrt{1 - F(\rho, \sigma)^2}. \tag{9}$$

The following proposition summarizes the main properties of fermionic entanglement fidelity that will be used in the remainder.

Proposition 3 *Let $\rho \in \mathsf{St}_1(\mathrm{L_F})$, $\mathscr{C} \in \mathsf{Tr}_1(\mathrm{L_F} \to \mathrm{L_F})$ and $\Phi_\rho \in \mathsf{PurSt}(\mathrm{L_F K_F})$ be any purification of ρ. Entanglement fidelity has the following properties.*

1. *$F(\rho, \mathscr{C})$ is independent of the particular choice for the purification Φ_ρ.*
2. *If the ordering is chosen in such a way that the L modes are all before the purifying ones, the following identity holds:*

$$F(\rho, \mathscr{C}) = \sum_i |\mathrm{Tr}[J(\rho)C_i]|^2 \tag{10}$$

for arbitrary Kraus decomposition $J(\mathscr{C}) = \sum_i C_i \cdot C_i^\dagger$ of the Jordan-Wigner representative $J(\mathscr{C})$. From the second inequality in (9) it follows that, if $F(\rho, \mathscr{C}) \geq 1 - \delta$, one has

$$\|(\mathscr{C} \boxtimes \mathscr{I}_\mathrm{C})(\Phi_\rho) - \Phi_\rho\|_1 \leq 2\sqrt{\delta} \tag{11}$$

for every purification Φ_ρ of ρ.

Proof Let $\Phi_\rho \in \mathsf{PurSt}(\mathrm{L_F M_F})$ be a purification of ρ. If we choose the trivial ordering for the LFMs, the Kraus of $J(\mathscr{C} \boxtimes \mathscr{I})$ are of the form $C_i \otimes I$. Moreover, since the minimal purification $|F\rangle\!\rangle\langle\!\langle F|$ (introduced in the proof of proposition 2) and $J(\Phi_\rho)$ both purify the same quantum state, they are connected through an isometry V.

Recalling that for quantum states $|\psi\rangle\langle\psi|$ and σ the quantum Uhlmann fidelity is given by $F(|\psi\rangle\langle\psi|, \sigma) = \langle\psi|\sigma|\psi\rangle^{1/2}$, we find

$$\begin{aligned} F(\rho, \mathscr{C}) &= \sum_i \mathrm{Tr}(|FV^T\rangle\!\rangle\langle\!\langle FV^T|C_i FV^T\rangle\!\rangle\langle\!\langle C_i FV^T|) \\ &= \sum_i |\mathrm{Tr}(|C_i FV^T\rangle\!\rangle\langle\!\langle FV^T|)|^2 \\ &= \sum_i |\mathrm{Tr}[J(\rho)C_i]|^2, \end{aligned}$$

namely, the claimed formula in (10). Since Φ_ρ is arbitrary, this also implies independence from the choice of the purification. □

In quantum theory a compression scheme $(\mathscr{E}_N, \mathscr{D}_N)$ is reliable when the entanglement fidelity $F(\rho^{\boxtimes N}, \mathscr{C}_N)$, with $\mathscr{C}_N := \mathscr{D}_N\mathscr{E}_N$, approaches 1 as $N \to \infty$. Here we prove an analogous reliability criterion for the fermionic case.

We can now prove the following proposition and the subsequent corollary providing a simple reliability criterion for fermionic compression.

Proposition 4 *Given a state $\rho \in \mathsf{St}_1(\mathrm{L_F})$ and a channel $\mathscr{C} \in \mathsf{Tr}_1(\mathrm{L_F} \to \mathrm{L_F})$, $\forall\varepsilon > 0$ there exists $\delta > 0$ such that if $F(\rho, \mathscr{C}) > 1 - \delta$ then $\sum_i \|[(\mathscr{C} - \mathscr{I}) \boxtimes \mathscr{I}](\Sigma_i)\|_1 \leq \varepsilon$ for every $\{\Sigma_i\}$ such that $\sum_i \Sigma_i \in D_\rho$.*

Proof Firstly we observe that, given a set of states $\{\Sigma_i\}$ such that $\sum_i \Sigma_i \in D_\rho$, considering any purification $\Psi_\rho \in \mathsf{PurSt}(\mathrm{L_F K_F N_F})$ of $\Sigma := \sum_i \Sigma_i$, one can find a POVM $\{b_i\} \in \mathsf{Eff}(\mathrm{N_F})$ such that $\Sigma_i = \mathrm{Tr}_{\mathrm{N_F}}[(I_{\mathrm{L_F K_F}} \boxtimes b_i)\Psi_\rho]$. As a consequence, we have

$$\begin{aligned} &\sum_i \|(\mathscr{C} - \mathscr{I})\Sigma_i\|_1 \\ &\quad= \sum_i \|\mathrm{Tr}_{\mathrm{N_F}}[\{(\mathscr{C} - \mathscr{I}) \boxtimes \mathscr{I}_{\mathrm{N_F}}\}(\Psi_\rho)(I_{\mathrm{L_F K_F}} \boxtimes b_i)]\|_1 \\ &\quad\leq \|\{(\mathscr{C} - \mathscr{I}) \boxtimes \mathscr{I}_{\mathrm{N_F}}\}(\Psi_\rho)\|_1, \end{aligned}$$

where the last inequality follows from the equivalent definition of the 1-norm for $X \in \mathsf{St}_{\mathbb{R}}(\mathrm{L_F})$

$$\|X\|_1 = \max_{b \in \mathsf{Eff}(\mathrm{L_F})} \mathrm{Tr}[Xb],$$

and from the fact that, for $\{a_i\} \subseteq \mathsf{Eff}(\mathrm{L_F K_F})$ such that

$$\begin{aligned} &\|\mathrm{Tr}_{\mathrm{N_F}}[\{(\mathscr{C} - \mathscr{I}) \boxtimes \mathscr{I}_{\mathrm{N_F}}\}(\Psi_\rho)(I_{\mathrm{L_F K_F}} \boxtimes b_i)]\|_1 \\ &\quad= \mathrm{Tr}_{\mathrm{N_F}}[\{(\mathscr{C} - \mathscr{I}) \boxtimes \mathscr{I}_{\mathrm{N_F}}\}(\Psi_\rho)(a_i \boxtimes b_i)], \end{aligned}$$

one can write

$$\begin{aligned}\sum_i \|(\mathscr{C}-\mathscr{I})\Sigma_i\|_1 &= \sum_i \mathrm{Tr}_{\mathrm{N_F}}[\{(\mathscr{C}-\mathscr{I})\boxtimes\mathscr{I}_{\mathrm{N_F}}\}(\Psi_\rho)(a_i\boxtimes b_i)]\\ &= \mathrm{Tr}_{\mathrm{N_F}}[\{(\mathscr{C}-\mathscr{I})\boxtimes\mathscr{I}_{\mathrm{N_F}}\}(\Psi_\rho)A],\end{aligned}$$

where $A := \sum_i (a_i \boxtimes b_i)$. Now, by the Fuchs-van der Graaf inequalities, if $F(\rho,\mathscr{C}) \geq 1-\delta$, then

$$\|\{(\mathscr{C}-\mathscr{I})\boxtimes\mathscr{I}_{\mathrm{N_F}}\}(\Psi_\rho)\|_1 \leq 2\sqrt{1-F(\rho,\mathscr{C})} \leq 2\sqrt{\delta}.$$

The thesis is then obtained just taking $\delta \leq \varepsilon^2/4$. □

Corollary 2 (Reliable compression scheme) *Given a state $\rho \in \mathsf{St}(\mathrm{L_F})$, a compression scheme $(\mathscr{E}^N, \mathscr{D}^N)$ is ϵ-reliable if one has $F(\rho^{\boxtimes N}, \mathscr{C}_N) > 1-\delta$, where $\delta = \epsilon^2/4$, and $\mathscr{C}_N := \mathscr{D}^N\mathscr{E}^N$.*

3.2 Fermionic Typical Subspace

At the basis of the quantum source coding theorem lies the notion of typical subspace, that in turn generalizes to the quantum case that of typical sequences and typical sets of classical information. We now introduce the notion of typical subspace also for fermionic systems and use it to show that, like in quantum theory, the von Neumann entropy of a fermionic state is the rate that separates the region of rates for which a reliable compression scheme exists from that of unachievble rates. In order to do this we have to verify that the compression map given in terms of the projection on the typical subspace represents an admissible fermionic map.

We start by defining the notion of logarithm of a fermionic state

Definition 5 Let ρ be a fermionic state. We define its logarithm as

$$\log_2 \rho = J^{-1}[\log_2 J(\rho)]. \tag{12}$$

Then we define the von Neumann entropy of a fermionic state via its Jordan-Wigner representative.

Definition 6 Given a fermionic state ρ, its von-Neumann entropy is defined as

$$S_f(\rho) := S(J(\rho)) = -\mathrm{Tr}(J(\rho)\log_2 J(\rho)). \tag{13}$$

These definitions are independent of the particular Jordan-Wigner transform corresponding to a given ordering of the modes (see Appendix 3).

When we use the orthonormal decomposition for $J(\rho) = \sum_{x_i} p_i |x_i\rangle\langle x_i|$, this reduces to the Shannon entropy of the classical random variable X that takes values in $\mathsf{Rng}(X) = \{x_1, x_2, \dots x_n\}$, called range of X, with probability distribution $(p_1, p_2, \dots, p_n)$: $S_f(\rho) = H(X) = -\sum_i p_i \log_2 p_i$. We remind that N i.i.d. copies of the state ρ are represented as $J(\rho^{\boxtimes N}) = J(\rho)^{\otimes N} = \sum_{x_{\mathbf{i}} \in \mathsf{Rng}(X)^N} p_{\mathbf{i}} |x_{\mathbf{i}}\rangle\langle x_{\mathbf{i}}|,$. With $\mathsf{T}_{N,\varepsilon}(\rho)$ we will denote the typical set of the random variable X.

Definition 7 (*Typical subspace*) Let $\rho \in \mathsf{St}(\mathrm{L_F})$ with orthonormal decomposition $J(\rho) = \sum_{x_i \in \mathsf{Rng}(X)} p_i |x_i\rangle\langle x_i|$. The ε-*typical* subspace $\mathsf{F}_{N,\varepsilon}(\rho)$ of $\mathcal{H}_L^{\otimes N}$ is defined as

$$\mathsf{F}_{N,\varepsilon}(\rho) := \mathsf{Span}\{|x_{\mathbf{i}}\rangle \mid x_{\mathbf{i}} \in \mathsf{T}_{N,\varepsilon}(X)\}, \tag{14}$$

where $|x_{\mathbf{i}}\rangle := |x_{i_1}\rangle |x_{i_2}\rangle \dots |x_{i_N}\rangle$, and X is the random variable with $\mathsf{Rng}(X) = \{x_i\}$ and $\mathbb{P}_X(x_i) := p_i$.

It is an immediate consequence of the definition of typical subspace that

$$\mathsf{F}_{N,\varepsilon}(\rho) := \mathsf{Span}\left\{|x_{\mathbf{i}}\rangle \;\middle|\; \left|\frac{1}{N}\log_2 \frac{1}{\mathbb{P}_{X^N}(x_{\mathbf{i}})} - S_f(\rho)\right| \le \varepsilon\right\}.$$

We will denote the projector on the typical subspace as

$$\begin{aligned} P_{N,\varepsilon}(\rho) &:= \sum_{x_{\mathbf{i}} \in \mathsf{T}_{N,\varepsilon}(X)} |x_{\mathbf{i}}\rangle\langle x_{\mathbf{i}}| \\ &= \sum_{x_{\mathbf{i}} \in \mathsf{T}_{N,\varepsilon}(X)} |x_{i_1}\rangle\langle x_{i_1}| \otimes \dots \otimes |x_{i_N}\rangle\langle x_{i_N}|, \end{aligned} \tag{15}$$

and we have that $\dim(\mathsf{F}_{N,\varepsilon}(\rho)) = \mathrm{Tr}[P_{N,\varepsilon}(\rho)] = |\mathsf{T}_{N,\varepsilon}(X)|$.

Notice that some of the superpositions of vectors in the typical subspace might not be legitimate fermionic pure states, as their parity might be different. However, up to now, we only defined the typical subspace as a mathematical tool, and it does not need a consistent physical interpretation. We will come back to this point later (see Lemma 1), when we will discuss the physical meaning of the projection $P_{N,\varepsilon}(\rho)$. Now, it immediately follows that

$$\begin{aligned} &\mathrm{Tr}[P_{N,\varepsilon}(\rho) J(\rho)^{\otimes N}] \\ &\quad = \sum_{x_{\mathbf{i}} \in \mathsf{T}_{N,\varepsilon}(\mathrm{X})} \mathbb{P}_{X^N}(x_{\mathbf{i}}) = \mathbb{P}_{X^N}[x_{\mathbf{i}} \in \mathsf{T}_{N,\varepsilon}(X)]. \end{aligned} \tag{16}$$

As in quantum theory, also the fermionic typical subspace has the following features:

Proposition 5 (Typical subspace) *Let* $\rho \in \mathsf{St}(\mathrm{L_F})$. *The following statements hold:*

1. *For every* $\varepsilon > 0$ *and* $\delta > 0$ *there exists* N_0 *such that for every* $N \ge N_0$

$$\mathrm{Tr}[P_{N,\varepsilon}(\rho) J(\rho)^{\otimes N}] \ge 1 - \delta. \tag{17}$$

2. *For every* $\epsilon > 0$ *and* $\delta > 0$ *there exists* N_0 *such that for every* $N \geq N_0$ *the dimension of the typical subspace* $\mathsf{F}_{N,\varepsilon}(\rho)$ *is bounded as*

$$(1-\delta)2^{N(S_f(\rho)-\varepsilon)} \leq \dim(\mathsf{F}_{N,\varepsilon}(\rho)) \leq 2^{N(S_f(\rho)+\varepsilon)} \tag{18}$$

3. *For given* N*, let* S_N *denote an arbitrary orthogonal projection on a subspace of* $\mathcal{F}_L^{\otimes N}$ *with dimension* $\mathrm{Tr}(S_N) < 2^{NR}$*, with* $R < S_f(\rho)$ *fixed. Then for every* $\delta > 0$ *there exists* N_0 *such that for every* $N \geq N_0$ *and every choice of* S_N

$$\mathrm{Tr}[S_N J(\rho)^{\otimes N}] \leq \delta. \tag{19}$$

The proof of the above properties is exactly the same as the one of quantum theory (see for instance [30]). However, order to exploit the same scheme proposed by Schumacher for the quantum case, one has to check that the encoding and decoding channels given in the constructive part of the proof are admissible fermionic maps. In particular, the encoding channel makes use of the projector $P_{N,\varepsilon}(\rho)$ as a Kraus operator, therefore, we have to show that it is a legitimate Kraus for a fermionic map. This is proved in the following lemma based on characterization of fermionic transformations of Proposition 1.

Lemma 1 *Let* ρ *be a fermionic state. The projector* $P_{N,\varepsilon}(\rho)$ *of Eq. 15 is the Kraus operator of an admissible fermionic transformation.*

Proof By Proposition 1 the projector on the typical subspace $P_{N,\varepsilon}(\rho)$ is a legitimate fermionic Kraus if it is the sum of products of either an even or an odd number of fermionic fields. Let us consider the single projection $|x_\mathbf{i}\rangle\langle x_\mathbf{i}|$. This is given by the tensor product $|x_{i_1}\rangle\langle x_{i_1}| \otimes \cdots \otimes |x_{i_N}\rangle\langle x_{i_N}|$, where each $|x_{i_k}\rangle$ is an eigenvector of the density matrix $J(\rho)$ representing the fermionic state ρ, and, as such, it has a definite parity. Thus, each factor in the above expression of $|x_\mathbf{i}\rangle\langle x_\mathbf{i}|$ is the Jordan-Wigner representative of an even polynomial, and also the projection $|x_\mathbf{i}\rangle\langle x_\mathbf{i}|$ is thus the representative of an even polynomial for every $\mathbf{i}$, which is given, in detail, by the product $J^{-1}(|x_\mathbf{i}\rangle\langle x_\mathbf{i}|) = \prod_{j=1}^N J^{-1}(|x_{i_j}\rangle\langle x_{i_j}|)$. Now, by Proposition 1, $P_{N,\varepsilon}(\rho)$ is the Jordan-Wigner representative of a legitimate fermionic Kraus operator. □

3.3 Fermionic Source Coding Theorem

We can now prove the source coding theorem for fermionic information theory.

Theorem 1 (Fermionic source coding) *Let* $\rho \in \mathsf{St}_1(\mathrm{L_F})$ *be a state of system* $\mathrm{L_F}$*. Then for every* $\delta > 0$ *and* $R > S_f(\rho)$ *there exists* N_0 *such that for every* $N \geq N_0$ *one has a compression scheme* $\{\mathscr{E}_N, \mathscr{D}_N\}$ *with rate* R*, and* $F(\rho^{\boxtimes N}, \mathscr{D}_N\mathscr{E}_N) \geq 1-\delta$*. Conversely, for every* $R < S_f(\rho)$ *there is* $\delta \geq 0$ *such that for every compression scheme* $\{\mathscr{E}_N, \mathscr{D}_N\}$ *with rate* R *one has* $F(\rho^{\boxtimes N}, \mathscr{D}_N\mathscr{E}_N) \leq \delta$*.*

The proof follows exactly the lines of the original proof for standard quantum compression, that can be found e.g. in Ref. [30]. As the direct proof is constructive, we only need to take care of the legitimacy of the compression protocol as a fermionic map. To this end, we recapitulate the construction here.

1. Encoding: Perform the measurement $\{P_{N,\varepsilon}(\rho), I - P_{N,\varepsilon}(\rho)\}$. If the outcome corresponding to $P_{N,\varepsilon}(\rho)$ occurs, then leave the state unchanged. Otherwise, if the outcome corresponding to $I - P_{N,\varepsilon}(\rho)$ occurs, replace the state by a standard state $|S\rangle\langle S|$, with $|S\rangle \in \mathsf{F}_{N,\varepsilon}(\rho)$. Such an map is described by the channel $\mathscr{M}_N : \mathrm{L}_\mathrm{F}^{\boxtimes N} \to \mathrm{L}_\mathrm{F}^{\boxtimes N}$ given by

$$J(\mathscr{M}_N)(\sigma) := \\ P_{N,\varepsilon}(\rho)\sigma P_{N,\varepsilon}(\rho) + \mathrm{Tr}[(I - P_{N,\varepsilon}(\rho))\sigma]\,|S\rangle\,\langle S|$$

   Notice that this is a well defined transformation since by Lemma 1 the projector on the typical subspace is a legitimate fermionic Kraus operator. The second term is a measure and prepare channel, which is also a legitimate fermionic transformation. Then consider a system M_F made of $M := N\lceil R\rceil$ LFMs and the (partial) isometric embedding $V : \mathsf{F}_{N,\varepsilon}(\rho) \to \mathcal{H}_{N\lceil R\rceil}$ such that $V^\dagger V = I_{\mathsf{F}_{N,\varepsilon}(\rho)}$. Since the first stage of the protocol never produces states in the complement of $\mathsf{F}_{N,\varepsilon}(\rho)$, we can complete the map $V \cdot V^\dagger$ to a fermionic channel $\mathscr{V}_N$. The encoding is then given by the composite map $\mathscr{E}_N := \mathscr{V}_N \mathscr{M}_N$.
2. Decoding: For the decoding channel, we simply choose the co-isometry $V^\dagger$, which inverts V on $\mathsf{F}_{N,\varepsilon}(\rho)$.

As for the converse statement, the proof for quantum compression is based on item 3, which we proved for fermionic theory as well. Thus, the quantum proof applies to the fermionic case.

4 Discussion

We have studied information compression for fermionic systems, showing the fermionic counterpart of the quantum source coding theorem. In spite of parity superselection rule and the non locality of the Jordan-Wigner representation of fermionic operators, the von Neumann entropy of fermionic states can still be interpreted as their information content, providing the minimal rate for which a reliable compression is achievable.

The novelty in this paper is the analysis of compression in the absence of local tomography. Here, the properties of a map, and in the specific case of study of the compression map, cannot be accessed locally. This poses stronger constraints on the set of reliable compression maps.

Despite the significant differences between fermionic and quantum information [27], the source coding theorem holds also for fermions. We can now wonder which are the minimal features of a theory that lie behind the coding theorem. As we learn from classical, quantum, and now also fermionic information theory, the task of information compression is intimately related to the notion of entropy. However, it is known that information theories beyond quantum exhibit inequivalent notions of entropy [31–33]. This is the main issue one has to face in order to introduce the notion of information content in the general case. On one side one has to provide a definition of information content including a broad class of probabilistic theories. On the other side one can compare such a notion with the different notions of entropy, identifying the one that plays the same role of Shannon entropy in the compression task.

Acknowledgements A.T. acknowledges financial support from the Elvia and Federico Faggin Foundation through the Silicon Valley Community Foundation, Grant No. 2020–214365. This work was supported by MIUR Dipartimenti di Eccellenza 2018–2022 project F11I18000680001.

Appendix 1: Fermionic States

In a L-LFM system, fermionic states in $\mathsf{St}(\mathrm{L_F})$ are represented by density matrices on the antisymmetric Fock space $\mathcal{F}_L$ satisfying the parity superselection rule. As such they can be written as combinations of products of field operators. Indeed, a fermionic state ρ can be split in its even and odd part as follows

$$\rho = \sum_e E_e \left|\Omega\right\rangle \left\langle\Omega\right| E_e^\dagger + \sum_o O_o \left|\Omega\right\rangle \left\langle\Omega\right| O_o^\dagger,$$

where E_e and O_o are linear combinations of products of even and odd number of field operators respectively. By recalling that $\left|\Omega\right\rangle \left\langle\Omega\right| = \prod_{i=1}^L \varphi_i \varphi_i^\dagger$ one can easily realize that ρ can be written as combination of products of even number of field operators. Moreover, by using the CAR, the generic state can be written as follows

$$\rho = \sum_{\underline{s},\underline{t}} \rho_{\underline{st}} \prod_{i=1}^{L} \varphi_i^{\dagger s_i} \varphi_i \varphi_i^\dagger \varphi_i^{t_i},$$

where $\underline{s}, \underline{t} \in \{0, 1\}^L$ and $\rho_{\underline{st}} \in \mathbb{C}$.

Appendix 2: Technical Lemma

Here we prove a lemma that is used in the main text.

Lemma 2 *Let* $\mathrm{N_F} := \mathrm{L_F K_F}$ *and* $\mathscr{C} \in \mathsf{Tr}(\mathrm{N_F} \to \mathrm{N_F})$ *be a single Kraus transformation with Kraus C having Jordan-Wigner rapresentative* $J(C) = U \otimes I_{\mathrm{K}_F}$*, U acting on the first L qubits. Then* $\mathscr{C}$ *is local on the first L modes.*

Proof Due to Proposition 1, the Kraus operator of $\mathscr{C}$ can be written as $C = \sum_i C_i$, where either each C_i is a product of an even number of field operators, or each C_i is a product of an odd one. The set $\{C_i\}$ can be taken to be linearly independent without loss of generality. Let us assume by contradiction that $\mathscr{C}$ is not local on the first L modes. Therefore, since a set of independent operators generating the algebra of the j-th mode is $\{\varphi_j, \varphi_j^\dagger, \varphi_j^\dagger\varphi_j, \varphi_j^\dagger\varphi_j + \varphi_j\varphi_j^\dagger\}$, there exists at least one product C_i that contains one of the factors φ_j, $\varphi_j^\dagger$, or $\varphi_j\varphi_j^\dagger$, for some mode j of the system $\mathrm{K_F}$. Let $j(i)$ be the mode with largest label in the chosen ordering of the $N = L + K$ modes, such that the corresponding factor in the product C_i is not the identity (i.e. $\varphi_j^\dagger\varphi_j + \varphi_j\varphi_j^\dagger$). Accordingly, one has that the Jordan-Wigner representative of C_i is of the form

$$J(C_i) = K \otimes O_{j(i)} \otimes \left(\bigotimes_{l=j(i)+1}^{N} I_l \right),$$

where K is an operator on the first $1, \ldots, j(i) - 1$ qubits, and $O_{j(i)}$ is one of the factors $\sigma^+_{j(i)}, \sigma^-_{j(i)}, \sigma^+_{j(i)}\sigma^-_{j(i)}$ on the j-th qubit. This contradicts the hypothesis on the form of $J(C)$. □

Appendix 3: Jordan-Wigner Independence

In this appendix we show the consistency of Definitions 3, 4, 5 and 6 given in text. In particular, we check that they are independent of the particular choice of the order of the fermionic modes, which defines the Jordan-Wigner transform. We remember that all Jordan-Wigner representations are unitarily equivalent.

Lemma 3 *Let ρ be a fermionic state. The square root and the logarithm of ρ are well defined.*

Proof Once we have fixed the ordering of the modes, the square root of a fermionic state ρ is defined via its Jordan-Wigner representative as follows

$$\rho^{\frac{1}{2}} := J^{-1}[J(\rho)^{\frac{1}{2}}]$$

If $\tilde{J}$ is the Jordan-Wigner isomorphism associated to a different ordering, then consider $X := \tilde{J}^{-1}[\tilde{J}(\rho)^{\frac{1}{2}}]$. We can now prove that $X = \rho^{\frac{1}{2}}$ and then independence of the square root from the ordering. Indeed, one has

$$\tilde{J}(X)^2 = \tilde{J}(\rho) = UJ(\rho)U^\dagger,$$

with U unitary. It follows that

$$J(\rho) = U^\dagger \tilde{J}(X)UU^\dagger\tilde{J}(X)U = J(X)^2 \implies J(X) = J(\rho)^{\frac{1}{2}}.$$

Since J is an isomorphism, by taking J^{-1} we finally get

$$X = J^{-1}[J(\rho^{\frac{1}{2}})] = \rho^{\frac{1}{2}}.$$

Analogously, the logarithm of a fermionic state is defined through its Jordan-Wigner representative

$$\log_2(\rho) := J^{-1}[\log_2(J(\rho))]$$

Again, let $\tilde{J}$ be the Jordan-Wigner isomorphism corresponding to a different ordering, and let $X = \tilde{J}^{-1}[\log_2(\tilde{J}(\rho))]$. Firstly we notice that

$$\log_2[\tilde{J}(\rho)] = \log_2[UJ(\rho)U^\dagger] = U\log_2[J(\rho)]U^\dagger$$

since U is unitary (we remind that the logarithm of a positive operator is defined via its spectral decomposition, and a unitary map preserves the spectrum). Therefore, we find

$$J(X) = \log_2[J(\rho)] \implies X = J^{-1}[\log_2(J(\rho))] = \log_2(\rho).$$

that concludes the proof. □

Based on the above lemma we have the following proposition.

Proposition 6 *Let ρ and σ be two fermionic states. The Uhlmann fidelity $F(\rho, \sigma)$ and the von Neumann entropy $S_f(\rho)$ of Definitions 4 and 6 are well defined.*

Proof These two quantities are given by a trace of two well defined operators, as proved in the previous lemma. Moreover, since a reordering of the modes corresponds to a unitarily change of basis, the trace is Jordan-Wigner independent, and so are $F(\rho, \sigma)$ and $S_f(\rho)$. □

References

1. A. Khrennikov, *Probability and Randomness* (Imperial College Press, 2016)
2. A. Khrennikov, *Beyond quantum* (CRC Press, 2014)
3. C.E. Shannon, A mathematical theory of communication. Bell Syst. Tech. J. **27**(3), 379–423 (1948)
4. B. Schumacher, Quantum coding. Phys. Rev. A **51**, 2738–2747 (1995)
5. C.A. Fuchs, J. van de Graaf, Cryptographic distinguishability measures for quantum-mechanical states. IEEE Trans. Inf. Theory **45**(4), 1216–1227 (1999)
6. G. Chiribella, G.M. D'Ariano, P. Perinotti, Probabilistic theories with purification. Phys. Rev. A **81**, 062348 (2010)
7. G. Chiribella, G.M. D'Ariano, P. Perinotti, Informational derivation of quantum theory. Phys. Rev. A **84**, 012311 (2011)

8. G.M. D'Ariano, G. Chiribella, P. Perinotti, Giacomo Mauro D'Ariano, *Quantum Theory from First Principles: An Informational Approach* (Cambridge University Press, Cambridge, 2017)
9. H. Araki, On a characterization of the state space of quantum mechanics. Commun. Math. Phys. **75**(1), 1–24 (1980)
10. B. Dakić, Č. Brukner, *Quantum Theory and Beyond: Is Entanglement Special?* (Cambridge University Press, 2011), pp. 365–392
11. L. Masanes, M.P. Müller, A derivation of quantum theory from physical requirements. New J. Phys. **13**(6), 063001 (2011)
12. H. Barnum, A. Wilce, Local tomography and the Jordan structure of quantum theory. Found. Phys. **44**(2), 192–212 (2014)
13. G.M. D'Ariano, P. Perinotti, A. Tosini, Information and disturbance in operational probabilistic theories. Quantum **4**, 363 (2020)
14. S.B. Bravyi, Yu. Alexei, Fermionic quantum computation. Ann. Phys. **298**(1), 210–226 (2002)
15. M.M. Wolf, Violation of the entropic area law for fermions. Phys. Rev. Lett. **96**(1), 010404 (2006)
16. M.-C. Bañuls, J.I. Cirac, M.M. Wolf, Entanglement in fermionic systems. Phys. Rev. A **76**(2), 022311 (2007)
17. N. Friis, A.R. Lee, D.E. Bruschi, Fermionic-mode entanglement in quantum information. Phys. Rev. A **87**(2), 022338 (2013)
18. G.M. D'Ariano, F. Manessi, P. Perinotti, A. Tosini, Fermionic computation is non-local tomographic and violates monogamy of entanglement. EPL (Europhys. Lett.) **107**(2), 20009 (2014)
19. T. Debarba, F. Iemini, G. Giedke, N. Friis, Teleporting quantum information encoded in fermionic modes. Phys. Rev. A **101**, 052326 (2020)
20. E.P. Jordan, E.P. Wigner, Über das paulische äquivalenzverbot. Z. Phys. **47**(9–10), 631–651 (1928)
21. Frank Verstraete and Juan Ignacio Cirac, Mapping local Hamiltonians of fermions to local Hamiltonians of spins. J. Stat. Mech: Theory Exp. **2005**(09), P09012–P09012 (2005)
22. C. Pineda, T. Barthel, J. Eisert, Unitary circuits for strongly correlated fermions. Phys. Rev. A **81**(5), 050303 (2010)
23. N. Schuch, F. Verstraete, J.I. Cirac, Nonlocal resources in the presence of superselection rules. Phys. Rev. Lett. **92**(8), 087904 (2004)
24. A. Kitaev, D. Mayers, J. Preskill, Superselection rules and quantum protocols. Phys. Rev. A **69**(5), 052326 (2004)
25. N. Schuch, F. Verstraete, J.I. Cirac, Quantum entanglement theory in the presence of superselection rules. Phys. Rev. A **70**(4), 042310 (2004)
26. G.M. D'Ariano, F. Manessi, P. Perinotti, A. Tosini, The Feynman problem and fermionic entanglement: fermionic theory versus qubit theory. Int. J. Mod. Phys. A **29**(17), 1430025 (2014)
27. G.M. D'Ariano, F. Manessi, P. Perinotti, A. Tosini, The Feynman problem and fermionic entanglement: fermionic theory versus qubit theory. Int. J. Mod. Phys. A **29**(17), 1430025 (2014)
28. M. Lugli, P. Perinotti, A. Tosini, Fermionic state discrimination by local operations and classical communication. Phys. Rev. Lett. **125**, 110403 (2020)
29. M. Lugli, P. Perinotti, A. Tosini, Unambiguous discrimination of fermionic states through local operations and classical communication. Phys. Rev. A **103**, 012416 (2021)
30. M.A. Nielsen, I.L. Chuang, *Quantum Computation and Quantum Information*, 10th edn. (Cambridge University Press, USA, 2011)
31. G. Kimura, K. Nuida, H. Imai, Distinguishability measures and entropies for general probabilistic theories. Rep. Math. Phys. **66**(2), 175–206 (2010)
32. H. Barnum, J. Barrett, L.O. Clark, M. Leifer, R. Spekkens, N. Stepanik, A. Wilce, R. Wilke, Entropy and information causality in general probabilistic theories. New J. Phys. **12**(3), 033024 (2010)
33. A.J. Short, S. Wehner, Entropy in general physical theories. New J. Phys. **12**(3), 033023 (2010)

Paolo Perinotti is Associate Professor at Pavia University where he teaches Quantum Information Theory, Statistical Mechanics and Quantum Mechanics. After graduating in 1999 in Physics at Pavia University with a thesis in quantum optics under the supervision of prof. G. M. D'Ariano, he did his Ph.D. at Milan University in quantum foundations, under the supervision of prof. L. Lanz. He published more than 110 papers on various subjects ranging from quantum information and quantum estimation theory to quantum networks and cellular automata, quantum foundations and foundations of quantum field theory. Along with Mauro D'Ariano and Giulio Chiribella he introduced quantum combs, higher-order quantum theory and indefinite causal structures, operational probabilistic theory, and formulated quantum theory in terms of six operational-informational axioms. With the same authors he published in 2017 the book "Quantum Theory from First Principles. An informational approach", for Cambridge University Press. His research activity is presently focused on foundations of quantum information, quantum mechanics and quantum field theory. He is a member of the Foundational Questions Institute (FQXi), and of the International Quantum Structures Association. In 2016 he was awarded the Birkhoff-von Neumann prize for research in quantum foundations. In 2018–2022 he was councillor of the International Quantum Structures Association.

Alessandro Tosini received his M.Sc. degree and the Ph.D. in physics from the University of Pavia in 2009 and 2013, respectively. Since 2020 he is Research Assistant at the University of Pavia. He published more than 25 articles in international-level refereed journals. His research aims are: theoretical and mathematical aspects of quantum computation and quantum information theory, quantum causal networks and quantum simulators, foundations of quantum theory.

Leonardo Vaglini obtained his Master's Degree at Pisa university in 2018, with a thesis in particle physics phenomenology. Then he moved to Pavia in 2019, where he is getting his Ph.D. under the supervision of Pr. Paolo Perinotti at Pavia University. At present, his research activity is mainly focused on quantum information theory and foundations of quantum theory within the context of operational probabilistic theories.

Conjugate Logic

Niklas Johansson, Felix Huber, and Jan-Åke Larsson

Abstract We propose a *conjugate logic* that can capture the behavior of quantum and quantum-like systems. The proposal is similar to the more generic concept of epistemic logic: it encodes knowledge or perhaps more correctly, predictions about outcomes of future observations on some systems. For a quantum system, these predictions are statements about future outcomes of measurements performed on specific degrees of freedom of the system. The proposed logic will include propositions and their relations, including connectives, but importantly also transformations between propositions on conjugate degrees of freedom of the systems. A key point is the addition of a transformation that allows to convert propositions about single systems into propositions about correlations between systems. We will see that subtle choices of the properties of the transformations lead to drastically different underlying mathematical models; one choice gives stabilizer quantum mechanics, while another choice gives Spekkens' toy theory. This points to a crucial basic property of quantum and quantum-like systems that can be handled within the present conjugate logic by adjusting the mentioned choice. It also enables a discussion on what behaviors are properly quantum or only quantum-like, relating to that choice and how it manifests in the system under scrutiny.

1 Introduction

Due to the current fast development of technology towards quantum information processing, there is great interest in different tools for understanding the behavior of quantum systems. To model this processing, a digital information representation

N. Johansson · J.-Å. Larsson (✉)
Department of Electrical Engineering, Linköping University, SE-581 83 Linköping, Sweden
e-mail: jan-ake.larsson@liu.se

N. Johansson
e-mail: niklas.johansson@liu.se

F. Huber
Atomic Optics Department, Jagiellonian University, 30-348 Kraków, Poland
e-mail: felix.huber@uj.edu.pl

A. Plotnitsky and E. Haven (eds.), *The Quantum-Like Revolution*,
https://doi.org/10.1007/978-3-031-12986-5_7

is often used. This calls for a suitable associated logic. Of course, there are already well-known logics to choose from (see e.g., Ref. [6], and references therein) .

Quantum logic was proposed already in Ref. [1], by extending the language of standard logic to encompass Hilbert space structure. This proposal starts from the Hilbert space description itself and describes a logic that, for example, replaces logical negation with orthogonality in the Hilbert space. More recently, there has been a drive to avoid using any Hilbert space structure as a postulate, and to look instead at other ways of building up a mathematical structure that eventually *arrives* at quantum mechanical behavior. This has grown into an entire field of scientific investigation, see e.g., [5, 10, 16, 18]. Part of the discussion is the difficulty to avoid a direct postulate of an underlying Hilbert space. Here we will make yet another attempt to build up a structure that avoids inserting Hilbert space structure by hand. We will, however, still retain propositions that relate to phase space as in [1], but not the direct relation used there where propositions *are* subsets of phase space.

Our propositions will instead concern predictions of future measurement outcomes of conjugate degrees of freedom. There is a connection to the concept of *conjugate coding* as proposed already in the 1970s by Wiesner (but only published in 1983 [26]), but there is also an important difference. Conjugate coding is about encoding data into conjugate degrees of freedom. Here we aim for the converse: *conjugate logic* is intended to describe knowledge about the system in the form of predictions of future measurement outcomes of conjugate degrees of freedom, making this a more epistemic approach.

Epistemic logic has already been studied as a subfield of epistemology concerned with logical approaches to knowledge, belief and related notions. While *any* logic with an epistemic interpretation may be called an epistemic logic, the most widespread type of epistemic logic is that of modal logic [24]. Epistemic modal logic is concerned with agents, their knowledge, and beliefs. It formally encodes for example, one agent's belief about another agent's knowledge by adding knowledge predicates on top of standard logic. Unfortunately this is not well adapted to the quantum-mechanical situation, where so-called quantum contextuality [14] prohibits assigning a consistent set of values to the complete set of propositions in the underlying standard logic.

For this reason, the logic we aim for here is intended to capture fundamental uncertainty within the logic itself: that statements not only have an unknown truth value, but simply do not possess any intrinsic truth value. In this sense it is no longer a question of incomplete knowledge about an existing truth value, but there are composite propositions that are fundamentally uncertain. The *conjugate logic* we propose here will allow but does not force quantum behavior, and will capture the notion of incomplete knowledge in the strongest possible sense: it will contain fundamentally uncertain composite propositions.

In this logic, propositions will concern predictions of quantum-like measurement outcomes on physical systems. Logic propositions then discretize the measurement outcomes into true/false, in general being sentences on the form "the measurement outcome will lie between a and b". Here, we will restrict ourselves to dichotomic measurement outcomes, without loss of generality. Such a dichotomic measurement is usually associated with a spin measurement along the Z axis in three-dimensional space, forming the first coordinate in a two-dimensional discrete phase space [27]. The second coordinate is usually associated with a spin-X measurement, and diagonal lines in the discrete phase space are associated with spin-Y measurement.

Our proposed conjugate logic will contain these conjugate degrees of freedom, and importantly also transformations between them, not only on single systems but also transformations involving correlations between systems. This makes our approach different from other three-valued logics that attempt to capture aspects of quantum mechanics, but do not include conjugate degrees of freedom as an essential part, nor transformations between them, see e.g., Ref. [21]. In addition, the inclusion of these transformations enables a logic-language basis for stabilizer quantum mechanics complete with Clifford-group transformations [4, 8, 9].

Stabilizer quantum mechanics is an important tool in quantum information processing. Our construction will be able to reproduce the behavior of this subset of quantum mechanics, but will avoid postulating an underlying Hilbert space. It will also enable a logic-language basis for Spekkens' toy theory and extensions of it [2, 12, 13, 15, 23, 25] that are useful in foundational considerations on these issues. Let us now turn to the explicit construction.

2 Conjugate Logic Propositions, Negation, Conjunction and Disjunction

We start with something that looks reasonably familiar; a three-valued logic where one truth value will denote an *indeterminate* outcome. Note that we immediately deviate from standard logic by adding several conjugate degrees of freedom intending to capture the idea that not all measurements are compatible, meaning simultaneously predictable. Standard propositional logic contains propositions, usually denoted p and q, that can have truth values in $\{$false,true$\}$ or $\{0,1\}$. In the new logic an atomic proposition is a prediction for the outcome of a measurement, that is, a statement on the form "measurement of the Z degree of freedom would give the outcome 0", denoted $\langle Z\rangle$. We choose "outcome 0" because this gives a natural correspondence to the stabilizer formalism where the notation $\langle\cdot\rangle$ is used for a *stabilizer*, a transformation that keeps a quantum state unchanged. There, $\langle Z\rangle$ is the notation for the stabilizer of the quantum state $|0\rangle$, which gives a direct connection to the proposition above.

Similarly, we add propositions to two other conjugate degrees of freedom whose corresponding propositions $\langle X \rangle$ and $\langle Y \rangle$ are statements on predictions about these measurement outcomes. Propositions of this kind are atomic much like simple propositions of ordinary propositional logic, and serve as building blocks for more complex expressions below. Statements on measurement outcomes of different systems will be denoted by numerical indexes Z_j or by position in a string of symbols.

Propositions can be true or false just as in standard propositional logic. They may also hold no truth value meaning that the measurement outcome is indeterminate, uncertain, unknown, or cannot be predicted, denoted "?". In some presentations of propositional logic one can see "T" for true (1) and "F" for false (0), so we could add "I" for indeterminate [21], in what follows we will use "?" for readability. Importantly, the conjugate propositional logic we construct here will contain statements that *cannot* all have truth values simultaneously. Although we will assume in what follows that at least one statement per system is allowed to have a truth value.

Just as in standard logic we can form connectives. In standard logic, the unary connective NOT ($\neg$) converts a proposition into its negation. It is natural to extend this notion such that the conjugate logic NOT gives truth value 0 if the negated proposition has truth value 1, and gives 1 if the negated proposition have truth value 0; finally, it gives the indeterminate "?" truth value if the negated proposition is indeterminate "?".

Now consider the standard logic conjunction AND ($\wedge$): it gives truth value 0 if one of the constituent propositions has truth value 0, and 1 if both the constituent propositions have truth value 1. It is natural to extend this notion so that the conjugate logic conjunction has exactly this behavior and that it gives indeterminate "?" otherwise, i.e., in the case when at least one constituent is indeterminate and the other is either indeterminate or 1. For example, we denote "[measurement of the Z_1 degree of freedom would give the outcome 0] AND [measurement of the X_2 degree of freedom would give the outcome 0]" by $\langle Z_1 \rangle \wedge \langle X_2 \rangle$; or sometimes $\langle ZI \rangle \wedge \langle IX \rangle$ or $\langle ZI, IX \rangle$.

Conjugate logic disjunction (OR, $\vee$) can be defined similarly by extending the standard truth table to cases with indeterminate constituent propositions. Thus, the conjugate logic disjunction gives truth value 1 if one of the constituent propositions has truth value 1, truth value 0 if both the constituent propositions have truth value 0, and gives indeterminate "?" otherwise, i.e., if one of the constituents is indeterminate and the other is either indeterminate or 0. We see that the conjugate logic conjunction and disjunction mirror each other just as they do in standard logic (see "de Morgan's laws" in Table 1).

Finally, conjugate logic exclusive disjunction (XOR, $\veebar$) is also an extension of the standard logic truth value 1 if one of the constituent propositions has truth value 1 and the other 0, and truth value 0 if both the constituent propositions have truth value 0 or 1. We simply let it give indeterminate "?" if either or both of the constituents are indeterminate.

Table 1 Conjugate logic connectives

p	q	$\neg p$	$p \wedge q$	$p \vee q$	$p \veebar q$	$p \rightarrow q$	$p \leftrightarrow q$
0	0	1	0	0	0	1	1
0	?	1	0	?	?	1	0
0	1	1	0	1	1	1	0
?	0	?	0	?	?	0	0
?	?	?	?	?	?	1	1
?	1	?	?	1	?	1	0
1	0	0	0	1	1	0	0
1	?	0	?	1	?	0	0
1	1	0	1	1	0	1	1

3 Conjugate Logic Material and Logical Conditionals, Tautology and Contradiction

We have now come to the conjugate logic *material conditional* and *material biconditional* ($\rightarrow$ and $\leftrightarrow$), where the qualifier "material" is used to distinguish the connective ($\rightarrow$) from the "formal" conditional ($\Rightarrow$) as [22] writes, or the "logical" conditional as it is now more commonly known. Perhaps jokingly, one could add that a better name would be *immaterial conditional* instead of *material conditional* when describing quantum and quantum-like systems. We will not press the issue further here, after all, *quantum systems are the dreams that stuff is made of*.

To construct material conditional and biconditional that incorporates an intrinsic uncertainty, we will take a closer look at two alternatives. One is to construct conditionals such that they follow identities from ordinary propositional logic. For example, then $p \leftrightarrow q$ would be identical to $(p \wedge q) \vee (\neg p \wedge \neg q)$. However, this does not capture the desired behavior well: the latter expression gives an indeterminate value "?" if both p and q are indeterminate, while the material biconditional should compare if the truth values are equal. Thus, in our view, a better alternative is to have the material biconditional (EQUIVALENT TO, $\leftrightarrow$) compare the two constituent propositions and have $p \leftrightarrow q$ be 1 if p and q have the same truth value, and 0 otherwise. In particular, $p \leftrightarrow q$ is 1 if both p and q are indeterminate. The logical biconditional between two composite propositions $p \Leftrightarrow q$ can now be defined as usual, namely as the situation in which $p \leftrightarrow q$ is always true. This now coincides with the statement that p and q have identical truth table entries.

The material conditional (IMPLIES, $\rightarrow$) can similarly be extended to have $p \rightarrow q$ be 1 if p is 0, if q is 1, or (in addition to the standard definition) if p and q have the same truth value, and 0 otherwise. The additional clause captures the notion of implication in the case when both p and q are indeterminate, encoding that when

p is indeterminate we cannot draw any conclusion about q so that an indeterminate truth value is acceptable by the connective. This completes the list of connectives in Table 1, and allows us to define the logical conditional between two composite propositions $p \Rightarrow q$ as the situation in which $p \rightarrow q$ is always true.

Using the notation $\langle\neg Z\rangle$ for the statement "measurement of the Z degree of freedom would give the outcome 1" we note that

$$\neg\langle Z\rangle \Leftrightarrow \langle\neg Z\rangle, \tag{1}$$

where the equivalence holds also for indeterminate truth values. The propositions $\langle I\rangle$ and $\langle\neg I\rangle$ represent tautology and contradiction, respectively, corresponding to trivial measurements that always have the outcome 0 or 1.

4 Equivalence Relations and Implication Laws

In standard logic there are a number of logical equivalence relations and implication laws. Some of these are important properties of logic expressions like reflexivity, symmetry, and transitivity, and these are retained by conjugate logic as defined here. However, some equivalences that are common tools in standard propositional logic no longer hold. The most prominent example is the "inverse law," which in standard propositional logic tells you that $(p \vee \neg p)$ is a tautology and that $(p \wedge \neg p)$ is a contradiction. These relations fail because of the indeterminate values involved, see the relevant truth tables in Table 2. It follows that seemingly innocuous simplifications are no longer available to us, for example,

$$\langle ZI, IX\rangle \vee \langle\neg ZI, IX\rangle \Leftrightarrow \Big(\langle ZI\rangle \vee \langle\neg ZI\rangle\Big) \wedge \langle IX\rangle \not\Leftrightarrow \langle IX\rangle. \tag{2}$$

It is a simple exercise to verify that for generic propositions p, q, and r, the following equivalences hold, and do not hold (see Appendix 12 for details).

Table 2 Truth table for the inverse law

p	$\neg p$	$p \vee \neg p$	$\langle I\rangle$	$p \wedge \neg p$	$\langle\neg I\rangle$
0	1	1	1	0	0
1	0	1	1	0	0
?	?	?	1	?	0

(E1)	**Double negation**	$\neg\neg p \Leftrightarrow p$
(E2)	**De Morgan's laws**	$\neg(p \wedge q) \Leftrightarrow \neg p \vee \neg q$
		$\neg(p \vee q) \Leftrightarrow \neg p \wedge \neg q$
(E3)	**Commutative laws**	$p \wedge q \Leftrightarrow q \wedge p$
		$p \vee q \Leftrightarrow q \vee p$
(E4)	**Associative laws**	$p \wedge (q \wedge r) \Leftrightarrow (p \wedge q) \wedge r$
		$p \vee (q \vee r) \Leftrightarrow (p \vee q) \vee r$
(E5)	**Distributive laws**	$p \wedge (q \vee r) \Leftrightarrow (p \wedge q) \vee (p \wedge r)$
		$p \vee (q \wedge r) \Leftrightarrow (p \vee q) \wedge (p \vee r)$
(E6)	**Idempotence**	$p \wedge p \Leftrightarrow p$
		$p \vee p \Leftrightarrow p$
(E7)	**Identity laws**	$p \wedge \langle I \rangle \Leftrightarrow p$
		$p \vee \langle \neg I \rangle \Leftrightarrow p$
(E8)	**Domination laws**	$p \wedge \langle \neg I \rangle \Leftrightarrow \langle \neg I \rangle$
		$p \vee \langle I \rangle \Leftrightarrow \langle I \rangle$
(E9)	~~Inverse laws~~	$p \wedge \neg p \nLeftrightarrow \langle \neg I \rangle$
		$p \vee \neg p \nLeftrightarrow \langle I \rangle$
(E10)	**Absorption laws**	$p \wedge (p \vee q) \Leftrightarrow p$
		$p \vee (p \wedge q) \Leftrightarrow p$
(E11)	~~Implication law~~	$p \to q \nLeftrightarrow \neg p \vee q$
(E12)	**Contrapositive law**	$p \to q \Leftrightarrow \neg p \to \neg q$
(E13)	**Equivalence law**	$p \leftrightarrow q \Leftrightarrow (p \to q) \wedge (q \to p)$

An example of how the inverse law breaks in quantum mechanics is the double slit experiment, where the proposition $p =$ "the particle passes through the right-hand slit" corresponds to a dichotomic measurement. If interference is desired from the experiment, it must be arranged so that p is indeterminate "?", in which case $p \wedge \neg p$ is indeterminate "?" so not a contradiction, and $p \vee \neg p$ is indeterminate "?" so not a tautology. We find that this describes the situation better than the popular-science "the particle passes through both of the slits (both p and $\neg p$ are true so $p \wedge \neg p$ is true) and simultaneously none of the slits (both p and $\neg p$ are false so $p \vee \neg p$ is false)," which at best is just confusing.

For the inverse laws only the trivial implications remain, while the implication law does not hold in either direction. A similar exercise for logical implications gives the following list, details in Appendix 12.

(I1)	**Modus ponens**	$(p \to q) \wedge p \Rightarrow q$
(I2)	**Law of syllogism**	$(p \to q) \wedge (q \to r) \Rightarrow (p \to r)$
(I3)	**Modus tollens**	$(p \to q) \wedge \neg q \Rightarrow \neg p$
(I4)	**Conjunctive simpl.**	$p \wedge q \Rightarrow p$
(I5)	**Disjunctive ampl.**	$p \Rightarrow p \vee q$
(I6)	~~Disjunctive syllogism~~	$(p \vee q) \wedge \neg q \nRightarrow p$
(I7)	**Proof by contradiction**	$(\neg p \to \langle \neg I \rangle) \Rightarrow p$
(I8)	**Proof by cases**	$(p \to r) \wedge (q \to r) \Rightarrow (p \vee q) \to r$

Most of the standard rules that we use when proving theorems do still hold, the exception is disjunctive syllogism, that succumbs to the same problem as the implication law: when q is indeterminate, we cannot draw a conclusion about p. Note that modus tollens still holds, encompassing a slightly stronger requirement.

5 Fundamental Uncertainty

We now arrive at a crucial point in the construction, the very reason to include the indeterminate value for the propositions used. So far, the indeterminate value could correspond to lack of knowledge about the "actual" ontic (existing) value of the property being measured. But when making statements about quantum systems, one should take into account that the standard mathematical description does not contain such ontic values, but rather, only allows calculation of probabilities after specifying which property is to be measured. The debate goes back to the founding fathers of quantum theory [3, 7]. We do not wish to take a stance on that particular issue here, but merely describe a truly epistemic logic, that encompasses the possibility that indeterminate values "?" not just denote lack of knowledge but may be *fundamentally uncertain*.

In quantum mechanics, not all measurement outcomes can simultaneously be predicted with certainty (probability 1 as EPR [7] write). Also in quantum-like systems, like Spekkens' toy theory, not all measurement outcomes can be simultaneously predicted with certainty. It is then natural to require that in the present conjugate logic that not all propositions can hold truth values simultaneously. In particular, the uncertainty principle holds in the form of a bound on the descriptional power of this conjugate propositional logic concerning atomic propositions, i.e., propositions on single degrees of freedom of single systems.

Postulate: Bound on descriptional power
No more than a single atomic proposition can be true or false for any single system.

We can immediately conclude that some composite propositions on a single system cannot have a definite truth value, for example,

$$\langle Z \rangle \veebar \langle X \rangle \Leftrightarrow ?\,. \tag{3}$$

This bound on descriptional power corresponds directly to the uncertainty relation in quantum mechanics [11] and to the knowledge balance principle of Spekkens' toy theory [23]. A set of propositions that *can* have truth values simultaneously we will call *compatible*. Since not all propositions can have simultaneous truth values, not all measurements are compatible.

The postulate has consequences on what we can say about systems in general, for example about correlations between systems. However, to arrive at a precise statement we will need transformations between propositions; these correspond to physical operations performed on the physical systems that we study.

6 Transformations of Propositions

Our propositions concern predicted outcomes of measurements performed on physical systems. As such, one of our systems can be subjected to a range of physical transformations, the simplest case is 180° rotation of the system around one of the conjugate coordinate axes. Such a rotation will conserve the proposition that concerns the axis, but perform a negation of the other two. It is therefore natural to use the conserved axis as label, in complete parallel with the stabilizer formalism:

$$\varphi_X\langle X\rangle \Leftrightarrow \langle X\rangle,\quad \varphi_X\langle Y\rangle \Leftrightarrow \langle\neg Y\rangle,\ \varphi_X\langle Z\rangle \Leftrightarrow \langle\neg Z\rangle, \tag{4}$$

$$\varphi_Y\langle X\rangle \Leftrightarrow \langle\neg X\rangle,\ \varphi_Y\langle Y\rangle \Leftrightarrow \langle Y\rangle,\quad \varphi_Y\langle Z\rangle \Leftrightarrow \langle\neg Z\rangle, \tag{5}$$

$$\varphi_Z\langle X\rangle \Leftrightarrow \langle\neg X\rangle,\ \varphi_Z\langle Y\rangle \Leftrightarrow \langle\neg Y\rangle,\ \varphi_Z\langle Z\rangle \Leftrightarrow \langle Z\rangle. \tag{6}$$

Since these are equivalence relations, the transformations preserve truth values and compatibility relations. The "Phase rotation" S comes in two variants. The quantum phase rotation corresponds to a 90° rotation around the Z axis, so that

$$\varphi_S\langle X\rangle \Leftrightarrow \langle Y\rangle,\ \varphi_S\langle Y\rangle \Leftrightarrow \langle\neg X\rangle,\ \varphi_S\langle Z\rangle \Leftrightarrow \langle Z\rangle. \tag{7a}$$

An alternative transformation is used in Spekkens' toy theory [20] where the rotation is followed by an inversion along the Z axis,

$$\varphi_S\langle X\rangle \Leftrightarrow \langle Y\rangle,\ \varphi_S\langle Y\rangle \Leftrightarrow \langle\neg X\rangle,\ \varphi_S\langle Z\rangle \Leftrightarrow \langle\neg Z\rangle. \tag{7b}$$

The effect of this difference is small, note that in both cases $\varphi_S\varphi_S = \varphi_Z$.

We now arrive at a crucial step in the construction, the "Hadamard" transformation. This transformation also comes in two variants with a seemingly small difference in the transformation itself, but this difference will instead have very important consequences for the type of model that can be used to describe the system, we will expand on this below. The quantum-mechanical Hadamard corresponds to a physical rotation around an axis 45° between X and Z, giving here a transformation that interchanges X and Z and inverts Y,

$$\varphi_H\langle X\rangle \Leftrightarrow \langle Z\rangle,\ \varphi_H\langle Y\rangle \Leftrightarrow \langle\neg Y\rangle,\ \varphi_H\langle Z\rangle \Leftrightarrow \langle X\rangle. \tag{8a}$$

The alternative Hadamard transformation [20] used in Spekkens' toy theory corresponds to an mirror operation over the plane spanned by the mentioned axis 45°

between X and Z and the Y axis, i.e., a transformation that interchanges X and Z and preserves Y,

$$\varphi_H\langle X\rangle \Leftrightarrow \langle Z\rangle,\ \varphi_H\langle Y\rangle \Leftrightarrow \langle Y\rangle,\ \varphi_H\langle Z\rangle \Leftrightarrow \langle X\rangle. \tag{8b}$$

Finally, the identity transformation φ_I leaves all propositions unchanged. For both versions of the Phase rotation and Hadamard, we arrive at a noncommutative group generated by φ_S and φ_H, since

$$\varphi_S\varphi_S = \varphi_Z,\quad \varphi_H\varphi_Z\varphi_H = \varphi_X \quad\text{and}\quad \varphi_S\varphi_X\varphi_S^{-1} = \varphi_Y. \tag{9}$$

This also generates all single-system propositions from one, say $\langle Z\rangle$.

7 Composite System Propositions and Transformations

For composite systems, there remains to handle joint measurements, which enable statements about correlation between measurement outcomes without making statements about the individual measurement outcomes. The statement "a joint measurement of the XOR between Z_1 and Z_2 would give outcome 0" can be written $\langle Z_1 \veebar Z_2\rangle$, or to conform with standard stabilizer notation $\langle ZZ\rangle$, we will use the latter notation below.

To avoid confusion please note that $\langle ZZ\rangle$ corresponds to a joint measurement of a single dichotomic value that gives the XOR, it does not correspond to individual measurement of two dichotomic values followed by calculation of the XOR between them, this would be denoted $\langle ZI\rangle \veebar \langle IZ\rangle$. These are different procedures, and have different consequences, in particular

$$\langle ZI\rangle \veebar \langle IZ\rangle \Rightarrow \langle ZZ\rangle \quad\text{but}\quad \langle ZI\rangle \veebar \langle IZ\rangle \not\Leftarrow \langle ZZ\rangle. \tag{10}$$

There are some simple equivalences since the conjunction of a statement about a single system and a statement about its correlation to another system, is equivalent to a conjunction of individual statements about the two systems, for example,

$$\langle XI, XX\rangle \Leftrightarrow \langle XI, IX\rangle,\quad \langle \neg XI, XX\rangle \Leftrightarrow \langle \neg XI, \neg IX\rangle. \tag{11}$$

This simplification applies to conjunctions. We will now add another transformation that maps a *single* proposition on the outcome of a correlation measurement to a proposition on some outcome of a single-system measurement. One natural choice is the CZ ("controlled-Z"), because its action is symmetric on both subsystems, and can be summarized as

$$\begin{aligned}
\varphi_{CZ}\langle IX\rangle \Leftrightarrow \langle ZX\rangle,\ \varphi_{CZ}\langle IY\rangle \Leftrightarrow \langle ZY\rangle,\ \varphi_{CZ}\langle IZ\rangle \Leftrightarrow \langle IZ\rangle,\\
\varphi_{CZ}\langle XI\rangle \Leftrightarrow \langle XZ\rangle,\ \varphi_{CZ}\langle YI\rangle \Leftrightarrow \langle YZ\rangle,\ \varphi_{CZ}\langle ZI\rangle \Leftrightarrow \langle ZI\rangle,\\
\varphi_{CZ}\langle ZZ\rangle \Leftrightarrow \langle ZZ\rangle,\ \varphi_{CZ}\langle XX\rangle \Leftrightarrow \langle YY\rangle,\ \varphi_{CZ}\langle XY\rangle \Leftrightarrow \langle \neg YX\rangle.
\end{aligned} \tag{12}$$

The motivation for this definition (which (12) should be viewed as) is the standard gate-based description of a CZ, that applies a φ_Z transformation on the second system if "measurement of the Z_1 degree of freedom would give the outcome 1", and φ_I if "measurement of the Z_1 degree of freedom would give the outcome 0". It is immediate that $\langle II\rangle$, $\langle IZ\rangle$, $\langle ZI\rangle$, and $\langle ZZ\rangle$ are unaffected by φ_Z on either system. To derive the effect of this map on the proposition "measurement of X_2 would give the outcome 0," $\langle IX\rangle$, we have

$$\begin{cases}
\varphi_{CZ}\langle ZI, IX\rangle \Leftrightarrow \ \varphi_{II}\langle ZI, IX\rangle \ \Leftrightarrow \langle ZI, IX\rangle,\\
\varphi_{CZ}\langle \neg ZI, IX\rangle \Leftrightarrow \varphi_{IZ}\langle \neg ZI, IX\rangle \Leftrightarrow \langle \neg ZI, \neg IX\rangle.
\end{cases} \tag{13}$$

Since we are looking for the unique statement that $\langle IX\rangle$ maps to, we use the equivalences in Eq. (11) to obtain

$$\begin{cases}
\varphi_{CZ}\langle ZI, IX\rangle \Leftrightarrow \langle ZI, ZX\rangle,\\
\varphi_{CZ}\langle \neg ZI, IX\rangle \Leftrightarrow \langle \neg ZI, ZX\rangle,
\end{cases} \tag{14}$$

so that, for each measurement outcome of Z_1 the map transforms "measurement of X_2 would give the outcome 0" into "measurement of the XOR between Z_1 and X_2 would give the outcome 0". Even though the "inverse law" is not available, which would have directly given us $\varphi_{CZ}\langle IX\rangle \Leftrightarrow \langle ZX\rangle$, this is the only remaining possibility for a well-defined map φ_{CZ}. Similar reasoning and the symmetry of CZ gives the transformation output of the remaining single-system propositions of (12).

Some further elaboration is needed for the two final entries in the table. These can be derived from the transformation property that compatible propositions are transformed into compatible propositions, using only three of the just established single-system transformation outputs. We know that $\langle IX\rangle$ and $\langle XI\rangle$ are compatible, and therefore $\langle ZX\rangle \Leftrightarrow \varphi_{CZ}\langle IX\rangle$ and $\langle XZ\rangle \Leftrightarrow \varphi_{CZ}\langle XI\rangle$ are compatible. Conversely, $\langle IX\rangle$ and $\langle IY\rangle$ are incompatible, and therefore $\langle ZX\rangle \Leftrightarrow \varphi_{CZ}\langle IX\rangle$ and $\langle ZY\rangle \Leftrightarrow \varphi_{CZ}\langle IY\rangle$ are incompatible. It is now possible to use single-system transformations to deduce whether a given pair of two-system propositions are compatible or not. For example, $\langle ZX\rangle$ is compatible with $\langle ZI\rangle$, $\langle IX\rangle$, $\langle XZ\rangle$, $\langle YY\rangle$, $\langle XY\rangle$, $\langle YZ\rangle$, and no other two-system propositions. It is of particular interest that the pair $\{\langle ZX\rangle,\langle XZ\rangle\}$ is only compatible with $\langle YY\rangle$ (and $\langle II\rangle$), which implies that the set $\{\langle IX\rangle, \langle XI\rangle, \langle XX\rangle\}$ of three pairwise compatible propositions must be transformed into the set $\{\langle ZX\rangle, \langle XZ\rangle, \langle YY\rangle\}$ or possibly $\{\langle ZX\rangle, \langle XZ\rangle, \langle \neg YY\rangle\}$. The latter choice will give inconsistencies, for details see Appendix A 12.

Both quantum mechanics and Spekkens' toy theory use the choice

$$\varphi_{CZ}\langle XX\rangle \Leftrightarrow \langle YY\rangle. \tag{15}$$

Then, the identity $\varphi_S\varphi_Z = \varphi_S\varphi_S\varphi_S = \varphi_Z\varphi_S$ fixes

$$\varphi_{CZ}\langle XY\rangle \Leftrightarrow \varphi_{CZ}\varphi_{IS}\langle XX\rangle \Leftrightarrow \varphi_{IS}\langle YY\rangle \Leftrightarrow \langle \neg YX\rangle \tag{16}$$

This finishes the construction of φ_{CZ}, and enables generating the whole Clifford group, e.g., $\varphi_{CNOT} = \varphi_{IH}\varphi_{CZ}\varphi_{IH}$. Furthermore, we can now reproduce the behavior of stabilizer quantum mechanics if the choices of phase rotation and Hadamard are made as in Eqs. (7a) and (8a); we reproduce the behavior of Spekkens' toy theory if the choices are as in Eqs. (7b) and (8b).

8 Clifford Reduction

To transform any non-"identity" proposition on a many-system correlation into a single-system proposition, it is enough to follow these steps.

(i) WLOG the second proposition has a nontrivial letter at the first position. Apply φ_S (transforms Y to $\neg X$) and φ_H (transforms X to Z) to transform the first position to X and the following positions to identity I or Z.
(ii) Then use φ_{CZ} repeatedly to reduce the last $n-1$ positions containing Z to I, to create a string with a single X at the first position.

We will call this a *Clifford reduction* of a joint measurement. As an example,

$$\begin{aligned}\varphi_{ISIIIS}\langle XYZIZY\rangle &\Leftrightarrow \langle XXZIZX\rangle\\ \varphi_{IHIIIH}\langle XXZIZX\rangle &\Leftrightarrow \langle XZZIZZ\rangle\\ \varphi_{CZ_{12}}\varphi_{CZ_{13}}\varphi_{CZ_{15}}\varphi_{CZ_{16}}\langle XZZIZZ\rangle &\Leftrightarrow \langle XIIIII\rangle\,.\end{aligned} \tag{17}$$

A simultaneous Clifford reduction of two propositions into single-system propositions can also be performed.

(i) Reduce the first proposition and perform the same transformations on the second. The first proposition now reads $\langle XI\dots I\rangle$. Either the second proposition has also been reduced to the form $\langle \cdot I\dots I\rangle$ in which case we are done, or WLOG the second proposition has a nontrivial letter at the second position.
(ii) Reduce the second proposition excluding the very first position, to put it on the form $\langle \cdot XI\dots I\rangle$.
(iii) If the first position of the second proposition reads I, we have two compatible single-system expressions $\langle XII\dots I\rangle$ and $\langle IXI\dots I\rangle$. If the first position reads X, the two propositions can be reduced to compatible single-system propositions using Eq. (11). If the first index reads Y or Z, the two propositions can be reduced to two incompatible single-system propositions, using Hadamards on the two first positions of both systems followed by CZ. Then

we are done, and the two propositions have been reduced to either compatible or incompatible single-system propositions.

We have now shown that two propositions are always simultaneously reducible to one-system propositions. The reduction is to two compatible propositions (for different systems) if and only if the two original propositions are compatible (but not equivalent), and otherwise the reduction gives two incompatible propositions (for the same system).

The previous discussion shows that if two propositions are compatible, then they can be simultaneously reduced to the form $\langle XII\ldots I\rangle$ and $\langle IXI\ldots I\rangle$. The same is true for n compatible propositions: they can be simultaneously reduced to contain one X and identities I on the remaining positions. This is proven by induction as follows. Suppose that m compatible propositions have been reduced to one-index X's on the first m systems using a simultaneous Clifford reduction. If $m < n$, consider an additional proposition P_{m+1} compatible with the previously reduced propositions, to which the the same Clifford operations have been applied as used to reduce the m first propositions. Since P_{m+1} is compatible with the previously reduced propositions, all indices $i \leq m$ only contain I or X, because otherwise P_{m+1} would be incompatible with some P_i, $i \leq m$. The indices $i \leq m$ that contain X can now be converted into I using Eq. (11). Now reduce P_{m+1} excluding the m first indices, putting it on the form $\langle I\ldots IXI\ldots I\rangle$. This gives us a set of $m+1$ compatible single-system propositions. If $m+1=n$ we are done, otherwise repeat the process.

If the string length equals n the last step is trivial, reducing P_n excluding the first $n-1$ indices. It follows that the maximal number of compatible independent propositions is equal to the string length, since any single system allows for the truth value of at most one proposition. Furthermore, any set of compatible propositions can be augmented to size equal to string length by performing simultaneous Clifford reduction, and then specifying truth values for single systems whose truth values are yet undefined.

9 Predictions

We have already seen that some conjunctions are equivalent to other conjunctions. This also enables us to make predictions. The equivalence in Eq. (11) is the basic tool, for example, given the proposition $\langle XI, IX\rangle$, the measurement of $\langle XX\rangle$ will lead to no new knowledge; it can be obtained from the conjunction $\langle XI, IX\rangle$ already. It is also true that

$$\langle XZ, ZX\rangle \Leftrightarrow \varphi_{CZ}\langle XI, IX\rangle \Rightarrow \varphi_{CZ}\langle XX\rangle \Leftrightarrow \langle YY\rangle \tag{18}$$

which is perhaps less straightforward to see by inspection. There is a direct link to the stabilizer formalism but we will not comment more on that link here.

We will call a logical consequence of a conjunction of propositions a *prediction*. In general, to obtain all predictions from a collection of propositions, one would need to perform a joint Clifford reduction, generate the set of predictions by repeated use of Eq. (11), and then invert the Clifford reduction to perform a Clifford *expansion*. This will restore the initial collection and create the full set of predictions. As a consequence, a collection of n systems allows for at most 2^n propositions on single systems or correlations to be simultaneously true. To see why, recall that at most n independent single-system propositions can be simultaneously true. Every pair of single-system or correlation propositions yields a new single-system or correlation prediction; thus one arrives at in total 2^n propositions.

10 Measurements Affect Propositions

For a single system, no more than a single atomic proposition can be true or false at the same time. For multiple systems this generalizes to the statement that only mutually compatible propositions can simultaneously be known: at most 2^n general propositions of which there are n independent ones.

We thus need to specify what happens when some degree of freedom is measured that belongs to an incompatible proposition. Here we take the approach that any measurement must yield a definite outcome, and that measurement enables prediction of subsequent measurement outcomes. Given that there is a descriptional power bound, the question is how to incorporate this new prediction. It is reasonable that any previous proposition that is compatible with the new one remains, whereas incompatible propositions cannot coexist with the new proposition and must therefore become indeterminate.

More formally, suppose that the conjunction $P = p_1 \wedge \cdots \wedge p_m$ holds. Now measure some degree of freedom of the system. Since by our assumption the outcome enables prediction of a future measurement outcome, we capture that in a proposition q. Denote by

$$P_q = \bigwedge_{p_i \text{ compatible with } q} p_i \,. \tag{19}$$

After measurement, we have the conjunction

$$P_q \wedge q \,, \tag{20}$$

and all predictions possible from it. Note that if the measurement gives a proposition q that is a prediction of P, nothing changes. This captures the behavior of both quantum mechanics and Spekkens' toy theory. We arrive at the following consequence of the postulate of bounded descriptional power:

Consequence: Action of Measurements
A measurement determines the truth value of a proposition of the system and renders all incompatible propositions indeterminate.

11 Contextuality, and Noncontextuality

The conjugate logic constructed here allows treatment of quantum and quantum-like systems within the same formalism, capturing not only fundamental uncertainty, but also transformations between different degrees of freedom. These transformations are needed for the emergence of properly quantum behavior, in particular transformations from knowledge on one system into knowledge about correlation between systems. It is the addition of the latter type of transformation that makes stabilizer quantum mechanics accessible, but it is in fact one of the single-system transformations that differentiate between properly quantum, and merely quantum-like, behavior. It also makes our three-valued logic more restrictive. To see this we need to consider a properly quantum phenomenon, here we will use that of quantum contextuality of the Peres-Mermin square (PM, [17, 19]):

$$\begin{matrix} \langle ZI\rangle & \langle IZ\rangle & \langle ZZ\rangle \\ \langle IX\rangle & \langle XI\rangle & \langle XX\rangle \\ \langle ZX\rangle & \langle XZ\rangle & \langle YY\rangle \end{matrix} \tag{21}$$

The rows and columns in this square consist of compatible propositions. In fact, we can make predictions for the last item in rows and columns from the preceding items, using Eqs. (11) and (18),

$$\begin{gathered} \langle ZI\rangle \wedge \langle IZ\rangle \Rightarrow \langle ZZ\rangle;\ \ \langle IX\rangle \wedge \langle XI\rangle \Rightarrow \langle XX\rangle; \\ \langle ZI\rangle \wedge \langle IX\rangle \Rightarrow \langle ZX\rangle;\ \ \langle IZ\rangle \wedge \langle XI\rangle \Rightarrow \langle XZ\rangle; \\ \langle ZX\rangle \wedge \langle XZ\rangle \Rightarrow \langle YY\rangle. \end{gathered} \tag{22}$$

The exception is the final column where the derivation of the prediction (see Sect. 9) involves a Hadamard. The two different Hadamards give different predictions. The quantum-mechanical Hadamard gives

$$\langle XX, ZZ\rangle \Leftrightarrow \varphi_{HI}\langle ZX, XZ\rangle \Rightarrow \varphi_{HI}\langle YY\rangle \Leftrightarrow \langle \neg YY\rangle, \tag{23a}$$

while Spekkens' toy theory Hadamard instead gives

$$\langle XX, ZZ\rangle \Leftrightarrow \varphi_{HI}\langle ZX, XZ\rangle \Rightarrow \varphi_{HI}\langle YY\rangle \Leftrightarrow \langle YY\rangle. \tag{23b}$$

The latter prediction (23b) allows simultaneous assignment of truth values (other than indeterminate "?") to all the propositions in the PM square (21). The former prediction (23a) does not.

This is known as *quantum contextuality* [14], where the word "context" here refers to which conjunction a proposition is contained in, e.g., row or column in the PM square. For a system with the quantum-mechanical Hadamard transformation, one needs to accept one of the following alternatives

(i) some proposition in the PM square (21) must be indeterminate "?",
(ii) some proposition in the PM square (21) must possess different values in different contexts.

The former alternative captures fundamental uncertainty, while the latter alternative would give a *contextual* ontic model. Spekkens' toy theory, on the other hand, is a *noncontextual* ontic model where propositions (i′) are determinate and (ii′) possess the same value in different contexts.

12 Conclusions

We have here constructed *conjugate logic*, that includes not only an extension to standard logic introducing an indeterminate "?" truth value, but also several conjugate degrees of freedom for each individual system. Crucially, it also includes transformations between propositions that concern different degrees of freedom, and transformations between single-system propositions and correlation propositions. The construction is intended to capture not only quantum uncertainty but more properly quantum properties of the described systems, for which the transformations are crucial.

The introduction of a limitation in predictive power, and the implied action of measurements, does make the logic contain quantum-like elements. But it is really the choice of Hadamard transformation and correlating transformation (CZ) that decide if the behavior is quantum or merely quantum-like. From the point of view of conjugate logic, it is these transformations that are responsible for the emergence of quantum contextuality.

In contrast to quantum logic, the approach presented here does not force contextuality into the logic through postulating Hilbert space structure. It is instead a generic framework that allows contextuality to occur, or not occur, as desired. We aim to use this simple language to describe other quantum and quantum-like phenomena, and hope that the framework will be useful to others in the same line of work.

Acknowledgements The authors would like to thank Andrei Khrennikov for arranging the long-running Växjö conference series on foundations of physics and probability. We believe that the friendly atmosphere and the open-minded discussions of the conference series to a large extent is because of Andrei's influence, we have all benefited greatly from participation.
FH acknowledges support by the Government of Spain (FIS2020-TRANQI and Severo Ochoa CEX2019-000910-S), Fundació Mir-Puig, Generalitat de Catalunya (AGAUR SGR 1381 and CERCA), the European Union under Horizon2020 (PROBIST 754510), and the Foundation for Polish Science through TEAM-NET (POIR.04.04.00-00-17C1/18-00).

Appendix A: Truth Tables for the Logic Equivalences and Logic Implications

Some equivalence laws are very simple to check, such as (E1) and (E3)–(E8), and the inverse law (E9) is contained in the main text. What remains are the following (Tables 3, 4, 5, 6, 7, 8, 9, 10, 11, 12, 13 and 14).

Table 3 (E2) de Morgan's laws: $\neg(p \wedge q) \Leftrightarrow \neg p \vee \neg q$; $\neg(p \vee q) \Leftrightarrow \neg p \wedge \neg q$

p	q	$p \wedge q$	$\neg p \vee \neg q$	$p \vee q$	$\neg p \wedge \neg q$
0	0	0	1	0	1
0	?	0	1	?	?
0	1	0	1	1	0
?	0	?	?	?	?
?	?	?	?	?	?
?	1	?	?	1	0
1	0	0	1	1	0
1	?	?	?	1	0
1	1	1	0	1	0

Table 4 (E10) absorption laws: $p \vee (p \wedge q) \Leftrightarrow p$; $p \wedge (p \vee q) \Leftrightarrow p$

p	q	$p \wedge q$	$p \vee (p \wedge q)$	$p \vee q$	$p \wedge (p \vee q)$
0	0	0	0	0	0
0	?	0	0	?	0
0	1	0	0	1	0
?	0	0	?	?	?
?	?	?	?	?	?
?	1	?	?	1	?
1	0	0	1	1	1
1	?	?	1	1	1
1	1	1	1	1	1

Table 5 (E11) ~~the implication law~~: $(p \to q) \not\Leftrightarrow (\neg p \vee q)$

p	q	$p \to q$	$\neg p \vee q$	$(p \to q) \to (\neg p \vee q)$	$(p \to q) \leftarrow (\neg p \vee q)$
0	0	1	1	1	1
0	?	1	1	1	1
0	1	1	1	1	1
?	0	0	?	1	?
?	?	1	?	?	1
?	1	1	1	1	1
1	0	0	0	1	1
1	?	0	?	1	?
1	1	1	1	1	1

Table 6 (E12) the contrapositive law: $p \to q \Leftrightarrow \neg q \to \neg p$

p	q	$p \to q$	$\neg p$	$\neg q$	$\neg q \to \neg p$
0	0	1	1	1	1
0	?	1	1	?	1
0	1	1	1	0	1
?	0	0	?	1	0
?	?	1	?	?	1
?	1	1	?	0	1
1	0	0	0	1	0
1	?	0	0	?	0
1	1	1	0	0	1

Table 7 (E13) the equivalence law: $p \leftrightarrow q \Leftrightarrow (p \to q) \wedge (p \leftarrow q)$

p	q	$p \leftrightarrow q$	$p \to q$	$p \leftarrow q$
0	0	1	1	1
0	?	0	1	0
0	1	0	1	0
?	0	0	0	1
?	?	1	1	1
?	1	0	1	0
1	0	0	0	1
1	?	0	0	1
1	1	1	1	1

Table 8 (I1) Modus ponens: $(p \to q) \wedge p \Rightarrow q$

p	q	$p \to q$	$(p \to q) \wedge p$	$[(p \to q) \wedge p] \to q$
0	0	1	0	1
0	?	1	0	1
0	1	1	0	1
?	0	0	0	1
?	?	1	?	1
?	1	1	?	1
1	0	0	0	1
1	?	0	0	1
1	1	1	1	1

Table 9 (I2) law of syllogism: $(p \to q) \wedge (q \to r) \Rightarrow p \to r$

p	q	r	$p \to q$	$q \to r$	$(p \to q) \wedge (q \to r)$	$p \to r$
0	0	0	1	1	1	1
0	0	?	1	1	1	1
0	0	1	1	1	1	1
0	?	0	1	0	0	1
0	?	?	1	1	1	1
0	?	1	1	1	1	1
0	1	0	1	0	0	1
0	1	?	1	0	0	1
0	1	1	1	1	1	1
?	0	0	0	1	0	0
?	0	?	0	1	0	1
?	0	1	0	1	0	1
?	?	0	1	0	0	0
?	?	?	1	1	1	1
?	?	1	1	1	1	1
?	1	0	1	0	0	0
?	1	?	1	0	0	1
?	1	1	1	1	1	1
1	0	0	0	1	0	0
1	0	?	0	1	0	0
1	0	1	0	1	0	0
1	?	0	0	0	0	0
1	?	?	0	1	0	0
1	?	1	0	1	0	0
1	1	0	1	0	0	1
1	1	?	1	0	0	1
1	1	1	1	1	1	1

Table 10 (I3) Modus tollens: $(p \to q) \wedge \neg q \Rightarrow \neg p$

p	q	$p \to q$	$(p \to q) \wedge \neg q$	$[(p \to q) \wedge \neg q] \to \neg p$
0	0	1	1	1
0	?	1	?	1
0	1	1	0	1
?	0	0	0	1
?	?	1	?	1
?	1	1	0	1
1	0	0	0	1
1	?	0	0	1
1	1	1	0	1

Table 11 (I4) Conjunctive simplification: $p \wedge q \Rightarrow p$; and (I5) Disjunctive strengthening: $p \Rightarrow p \vee q$

p	q	$p \wedge q$	$(p \wedge q) \to p$	$p \vee q$	$p \to (p \vee q)$
0	0	0	1	0	1
0	?	0	1	?	1
0	1	0	1	1	1
?	0	0	1	?	1
?	?	?	1	?	1
?	1	?	1	1	1
1	0	0	1	1	1
1	?	?	1	1	1
1	1	1	1	1	1

Table 12 (I6) ~~Disjunctive syllogism~~: $(p \vee q) \wedge \neg q \nRightarrow p$

p	q	$p \vee q$	$(p \vee q) \wedge \neg q$	$[(p \vee q) \wedge \neg q] \to p$
0	0	0	0	1
0	?	?	?	?
0	1	1	0	1
?	0	?	?	1
?	?	?	?	1
?	1	1	0	1
1	0	1	1	1
1	?	1	?	1
1	1	1	0	1

Table 13 (I7) proof by contradiction: $(\neg p \to \langle \neg I \rangle) \Rightarrow p$

p	$\neg p \to \langle \neg I \rangle$	$(\neg p \to \langle \neg I \rangle) \to p$
0	0	1
?	0	1
1	1	1

Table 14 (I8) proof by cases: $(p \to r) \wedge (q \to r) \Rightarrow (p \vee q) \to r$

p	q	r	$p \to r$	$q \to r$	$(p \to r) \wedge (q \to r)$	$(p \vee q) \to r$
0	0	0	1	1	1	1
0	0	?	1	1	1	1
0	0	1	1	1	1	1
0	?	0	1	0	0	0
0	?	?	1	1	1	1
0	?	1	1	1	1	1
0	1	0	1	0	0	0
0	1	?	1	0	0	0
0	1	1	1	1	1	1
?	0	0	0	1	0	0
?	0	?	1	1	1	1
?	0	1	1	1	1	1
?	?	0	0	0	0	0
?	?	?	1	1	1	1
?	?	1	1	1	1	1
?	1	0	0	0	0	0
?	1	?	1	0	0	0
?	1	1	1	1	1	1
1	0	0	0	1	0	0
1	0	?	0	1	0	0
1	0	1	1	1	1	1
1	?	0	0	0	0	0
1	?	?	0	1	0	0
1	?	1	1	1	1	1
1	1	0	0	0	0	0
1	1	?	0	0	0	0
1	1	1	1	1	1	1

Appendix B: Choice of *CZ* Transformation

This appendix contains a derivation of an inconsistency that would arise if the CZ transformation is chosen so that

$$\widetilde{\varphi}_{CZ}\langle XX\rangle \Leftrightarrow \langle \neg YY\rangle, \tag{24}$$

where the tilde is used to distinguish the transformation used in the main text from the one considered here. The inconsistency arises when making predictions from the conjunction $\langle \neg YYI, \neg IYY\rangle$. To generate predictions we first use the procedure of Sect. 9. This starts with the simultaneous Clifford reduction

$$
\begin{aligned}
\langle \neg YYI, \neg IYY \rangle &\Leftrightarrow \varphi_{SSS} \langle \neg XXI, \neg IXX \rangle \\
&\Leftrightarrow \varphi_{SSS} \varphi_{IHH} \langle \neg XZI, \neg IZZ \rangle \\
&\Leftrightarrow \varphi_{SSS} \varphi_{IHH} \widetilde{\varphi}_{CZ_{12}} \langle \neg XII, \neg IZZ \rangle \\
&\Leftrightarrow \varphi_{SSS} \varphi_{IHH} \widetilde{\varphi}_{CZ_{12}} \varphi_{IHI} \langle \neg XII, \neg IXZ \rangle \\
&\Leftrightarrow \varphi_{SSS} \varphi_{IHH} \widetilde{\varphi}_{CZ_{12}} \varphi_{IHI} \widetilde{\varphi}_{CZ_{23}} \langle \neg XII, \neg IXI \rangle .
\end{aligned}
\tag{25}
$$

We have the simple prediction

$$
\langle \neg XII, \neg IXI \rangle \Rightarrow \langle XXI \rangle . \tag{26}
$$

Clifford expansion now gives

$$
\begin{aligned}
&\varphi_{SSS}^{-1} \varphi_{IHH} \widetilde{\varphi}_{CZ_{12}} \varphi_{IHI} \widetilde{\varphi}_{CZ_{23}} \langle XXI \rangle \\
&\quad \Leftrightarrow \varphi_{SSS}^{-1} \varphi_{IHH} \widetilde{\varphi}_{CZ_{12}} \varphi_{IHI} \langle XXZ \rangle \\
&\quad \Leftrightarrow \varphi_{SSS}^{-1} \varphi_{IHH} \widetilde{\varphi}_{CZ_{12}} \langle XZZ \rangle \\
&\quad \Leftrightarrow \varphi_{SSS}^{-1} \varphi_{IHH} \langle XIZ \rangle \\
&\quad \Leftrightarrow \varphi_{SSS}^{-1} \langle XIX \rangle \\
&\quad \Leftrightarrow \langle YIY \rangle ,
\end{aligned}
\tag{27}
$$

which gives the prediction

$$
\langle \neg YYI, \neg IYY \rangle \Rightarrow \langle YIY \rangle . \tag{28}
$$

A second prediction from $\langle \neg YYI, \neg IYY \rangle$ starts from

$$
\langle XXI, IXX \rangle \Rightarrow \langle XIX \rangle , \tag{29}
$$

that can also be obtained as above. We now apply three $\widetilde{\varphi}_{CZ}$ transformations to reach

$$
\begin{aligned}
\langle \neg YYI, \neg IYY \rangle &\Leftrightarrow \widetilde{\varphi}_{CZ_{12}} \widetilde{\varphi}_{CZ_{23}} \widetilde{\varphi}_{CZ_{13}} \langle XXI, IXX \rangle \\
&\Rightarrow \widetilde{\varphi}_{CZ_{12}} \widetilde{\varphi}_{CZ_{23}} \widetilde{\varphi}_{CZ_{13}} \langle XIX \rangle \Leftrightarrow \langle \neg YIY \rangle .
\end{aligned}
\tag{30}
$$

Using $\widetilde{\varphi}_{CZ}$ results in two predictions (28) and (30), so that $\langle \neg YYI, \neg IYY \rangle \Rightarrow \langle YIY, \neg YIY \rangle$. While $\langle YIY, \neg YIY \rangle$ is never true, the Clifford reduction (25) tells us that the left-hand side $\langle \neg YYI, \neg IYY \rangle$ can be reduced to compatible propositions for two different systems, and therefore can be simultaneously true. This shows that $\widetilde{\varphi}_{CZ}$ gives an inconsistency.

Using φ_{CZ} instead of $\widetilde{\varphi}_{CZ}$ does not change the prediction in Eq. (28), but gives

$$\begin{aligned}\langle\neg YYI, \neg IYY\rangle &\Leftrightarrow \varphi_{CZ_{12}}\varphi_{CZ_{23}}\varphi_{CZ_{13}}\langle\neg XXI, \neg IXX\rangle \\ &\Rightarrow \varphi_{CZ_{12}}\varphi_{CZ_{23}}\varphi_{CZ_{13}}\langle XIX\rangle \Leftrightarrow \langle YIY\rangle.\end{aligned} \tag{31}$$

For this choice there is no inconsistency.

References

1. G. Birkhoff, J. von Neumann, The logic of quantum mechanics. Ann. Math. Second series **37**, 823–843 (1936). https://doi.org/10.2307/1968621
2. P. Blasiak, Quantum cube: a toy model of a qubit. Phys. Lett. A **377**, 847–850 (2013). https://doi.org/10.1016/j.physleta.2013.01.045
3. N. Bohr, Can quantum-mechanical description of physical reality be considered complete?. Phys. Rev. **48**, 696–702 (1935). https://doi.org/10.1103/PhysRev.48.696
4. A.R. Calderbank et al., Quantum error correction via codes over GF(4). IEEE Trans. Inf. Theory **44**, 1369–1387 (1998). https://doi.org/10.1109/18.681315
5. G. Chiribella, G.M. D'Ariano, P. Perinotti, Informational derivation of quantum theory. Phys. Rev. A **84**, 012311 (2011). https://doi.org/10.1103/PhysRevA.84.012311
6. B. Coecke, D. Moore, A. Wilce, Operational quantum logic: an overview, in *Current Research in Operational Quantum Logic: Algebras, Categories, Languages*, ed. by B. Coecke, D. Moore, A. Wilce. Fundamental Theories of Physics, vol. 111 (Kluwer, 2000). arXiv: quant-ph/0008019
7. A. Einstein, B. Podolsky, N. Rosen, Can quantum-mechanical description of physical reality be considered complete?. Phys. Rev. **47**, 777–780 (1935). https://doi.org/10.1103/PhysRev.47.777
8. D. Gottesman, The Heisenberg representation of quantum computers (1998). arXiv: quant-ph/9807006
9. D. Gottesman, Theory of fault-tolerant quantum computation. Phys. Rev. A **57**, 127–137 (1998). https://doi.org/10.1103/PhysRevA.57.127
10. L. Hardy, Quantum theory from five reasonable axioms. Version 4 (2001). arXiv: quant-ph/0101012
11. W. Heisenberg, Über den anschaulichen Inhalt der quantentheoretischen Kinematik und Mechanik. Z. Physik **43**, 172–198 (1927). https://doi.org/10.1007/BF01397280
12. C. Hindlycke, J.-Å. Larsson, Efficient contextual ontological model of n-qubit stabilizer quantum mechanics. Phys. Rev. Lett. **12**, 130401 (2022). https://doi.org/10.1103/PhysRevLett.129.130401. https://arxiv.org/abs/2202.05081
13. N. Johansson, J.-Å. Larsson, Quantum simulation logic, oracles, and the quantum advantage. Entropy **21**, 800 (2019). https://doi.org/10.3390/e21080800
14. S. Kochen, E.P. Specker, The problem of hidden variables in quantum mechanics. J. Math. Mech. **17**, 59–87 (1967). https://doi.org/10.1512/iumj.1968.17.17004
15. P. Lillystone, J. Emerson, A contextual ψ-epistemic model of the n-qubit stabilizer formalism (2019). arXiv:1904.04268 [quant-ph]
16. L. Masanes, M.P. Müller, A derivation of quantum theory from physical requirements. New J. Phys. **13**, 063001 (2011). https://doi.org/10.1088/1367-2630/13/6/063001
17. N.D. Mermin, Hidden variables and the two theorems of John Bell. Rev. Mod. Phys. **65**, 803–815 (1993). https://doi.org/10.1103/RevModPhys.65.803
18. M. Pawłowski et al., Information causality as a physical principle. Nature **461**, 1101–1104 (2009). https://doi.org/10.1038/nature08400
19. A. Peres, *Quantum Theory: Concepts and Methods*. Fundamental Theories of Physics (Kluwer, Dordrecht, 1993). 446 pp. https://doi.org/10.1007/0-306-47120-5

20. M.F. Pusey, Stabilizer notation for Spekkens' toy theory. Found. Phys. **42**, 688–708 (2012). https://doi.org/10.1007/s10701-012-9639-7
21. H. Reichenbach, *Philosophic Foundations of Quantum Mechanics* (University of California Press, Berkeley, 1944)
22. B. Russell, *The Principles of Mathematics* (Routledge, 1903). ISBN: 978-0-415-48741-2. https://people.umass.edu/klement/pom/
23. R.W. Spekkens, Evidence for the epistemic view of quantum states: a toy theory. Phys. Rev. A **75**, 032110 (2007). https://doi.org/10.1103/PhysRevA.75.032110
24. G.H. Von Wright, *An Essay on Modal Logic* (North-Holland, Amsterdam, 1951)
25. J.J. Wallman, S.D. Bartlett, Non-negative subtheories and quasiprobability representations of qubits. Phys. Rev. A **85**, 062121 (2012). https://doi.org/10.1103/PhysRevA.85.062121
26. S. Wiesner, Conjugate coding. SIGACT News **15**, 78–88 (1983). https://doi.org/10.1145/1008908.1008920
27. W.K. Wootters, Picturing qubits in phase space (2003). arXiv: quantph/0306135

Niklas Johansson received a B.Sc. degree in Mathematics from Linköping University in 2015, an M.Sc. degree in Applied Physics and Electrical Engineering in 2016, and a Ph.D. degree in Electrical Engineering with a Specialization in Information Coding in 2022. Niklas is currently Research Manager at Sectra Communications AB, and CEO and Co-Founder of Phase Space Computing AB. His research interest is mainly focused on questions regarding the difference between classical and quantum technology.

Felix Huber received an M.Sc. degree in physics from ETH Zürich, Switzerland, in 2012, and a Ph.D. degree in theoretical quantum optics from the Universität Siegen, Germany, in 2017. In 2018 he was a Postdoctoral Researcher at the Universität zu Köln, Germany, and from 2018 to 2020 he was a Research Fellow at ICFO Barcelona, Spain. Since 2021 he is an Adiunkt at the Jagiellonian University in Krakow, Poland. His research interests include the mathematics of quantum information, quantum error correction, quantum correlations and entanglement, mathematical optimization, and invariant theory. Dr. Huber is recipient of the 2019 Dissertation Prize by the section AMOP of the German Physical Society.

Jan-Åke Larsson received a M.Sc. degree in Applied Physics and Electrical Engineering from Linköping University in 1994, and a Ph.D. degree in Applied Mathematics in 2000. After a Postdoctoral period at Aarhus University 2001–2002, he was an Associate professor at the Department of Mathematics at Linköping University 2003–2009, and at the Department of Electrical Engineering 2010–2015. From 2015 onwards he is Professor and Head of Department at the Department of Electrical Engineering at Linköping University. His research interests include a range of questions in Quantum Information Theory and Quantum Foundations, and also cryptology and more general mathematical modelling.

Revisiting Quantum Mysteries

Philippe Grangier

Abstract In this chapter we argue that in quantum mechanics, and in opposition to classical physics, it is impossible to say that an isolated quantum system "owns" a physical property. Some properties of the system, its mass for example, belong to it in a sense close to that of classical physics; but most often a property must be attributed to the system *within a context.* We give simple motivations for adopting this point of view, and show that it clarifies many issues in quantum physics.

1 Our Quantum World

In many venues intended for the general public one encounters the assertions that in the "quantum world" an object can be in several places at the same time, or that two particles located at astronomical distances can influence each other instantaneously, or that particles pass through barriers, and so forth.

However, the world thus described is nothing but the one we live in. Those who are familiar with the popularization of quantum physics are used to hearing these sentences, which fascinate and challenge. They often lead to the ideas, quite frustrating for the uninitiated, that quantum mechanics is a discipline full of paradoxes, and incomprehensible otherwise than by its mathematical formalism; and that even the best physicists disagree on what it means [1, 2].

Given that quantum mechanics provides the basis for innumerable technologies (such as electronics, lasers, or medical imaging) which have changed our daily life, and is a theory that has never been disproved, this difficulty in communicating its physical content to the layman is surprising. It may even appear embarrassing, in a context where many countries are investing billions of Euros to develop technologies related to the "second quantum revolution", associated with new ways to process information, but even more difficult to explain than the first one.

P. Grangier (✉)
Laboratoire Charles Fabry, Institut d'Optique Graduate School, CNRS, Université Paris Saclay, 91127 Palaiseau, France
e-mail: philippe.grangier@institutoptique.fr

A. Plotnitsky and E. Haven (eds.), *The Quantum-Like Revolution*,
https://doi.org/10.1007/978-3-031-12986-5_8

A legitimate question is then: Would it be possible to formulate different statements, referring specifically to this mysterious "quantum world", without appearing as a series of contradictions or absurdities? This is what we propose here, also avoiding any excessive recourse to mathematics: it is indispensable, but will only be briefly mentioned in the last part of the article.

2 Empirical Evidence and Its Consequences

Our starting point will be a series of empirical observations, that is to say, observations that relate to experiments that are quite feasible and that have been carried out, some of them recently. We will start from concepts known at the beginning of the 20th century (such as light is a wave, or matter is made of atoms with a nucleus and electrons) and we will introduce new phenomena, which we will then explain.

The fundamental physical fact is that observations of microscopic objects show a discrete, or quantized, character. Here are some examples:

- The possible wavelengths of light emitted by atoms take only certain particular values, with each atom characterized by the "spectrum" of wavelengths that it can emit.
- Light can extract electrons (particles) from matter, but in order to describe this phenomenon, called the photoelectric effect, we are led to admit that light is also formed of particles (photons).
- We can perform experiments where objects classically considered as particles (electrons, atoms...) give rise to interference effects, which are classically associated with waves, and not with particles.

A rather inevitable conclusion of these observations is that microscopic objects (atoms, electrons, photons...) have behaviors that combine a discrete character (i.e. described by integers) with the possibility of interference (traditionally associated with continuous waves).

Quantum mechanics was built, on the basis of these observations, by a few brilliant physicists who invented a somewhat bizarre mathematical formalism, able to account for all these observations, and also to make countless predictions of new phenomena. This formalism, or even these formalisms, because there are several equivalent ones, have never been put in default, but the physical nature of the objects and properties they describe has remained obscure. The questioning on how to describe (or even define) the "physical reality", quite legitimate in our view, has given rise to innumerable interpretations and paradoxical statements—the sentences quoted at the beginning of this article are a very limited sample.

It is therefore clear that an "ingredient" is missing that would give a physical meaning to this very efficient formalism; this is what we propose in the following.

3 Underestimating Empirical Evidence?

The normal approach in physics since Newton has been to define objects, to attribute properties to them, and to measure these properties. One then asserts that the object "has" this property, for example that it has a position, a velocity, or a momentum. Does this "natural" approach work in quantum mechanics? Although physicists are extremely reluctant to admit it, the answer is clearly no—and this "no", correctly interpreted, provides the empirical element that is missing in our understanding of the quantum description of the physical world.

So, let's look at a simple experiment that shows that it is impossible to say that a quantum particle "owns" a property. Of course some particle's properties, its mass for example, belong to it in a sense very close to classical physics; but in general the situation is not so simple. One can for example measure the angular momentum of an atom along a certain direction **u**, or the polarization (equivalent to angular momentum) for a photon. Such a measurement gives a discrete result, as explained above, and we can therefore say for example that we obtain the value $+1$, in adequate units. Moreover, if we repeat immediately (ideally instantly) this same measurement on the same particle, we find $+1$ again: the result can therefore be predicted with certainty, as one would expect in classical physics.

But let's complicate things a bit, and measure the angular momentum in another direction, let's say **v**. We find for example $+2$, then we come back to **u**, and we measure again, on the same particle. In a classical framework we would expect to find the previous $+1$ result, but in fact we don't necessarily get $+1$, but a random result! In general this new result can take any value in a certain set defined by the system studied, for example $(-2, -1, 0, +1, +2)$ for a certain type of atom, each of these values having a certain probability of being obtained. One sometimes speaks of "measurement-induced perturbation", a phenomenon that also exists in classical physics; but the randomness described here is inescapable: it is not possible to eliminate it completely, nor to trace it back to a "true value" of the measured quantity.

Very often this simple but fundamental experiment is not considered as a fact, but is immediately buried in mathematical formalism, invoking operators that do not commute or other arguments that are not explanations, but mathematical descriptions. However, this experiment is essential since it shows that the value attributed to certain physical properties varies in an irremediably random way when the measurement process is modified. However, this is not an incomprehensible blur, since as long as the same measurement is repeated on the same particle, the measured value always remains the same: certainty and repeatability remain, but under more restrictive conditions than those observed in classical physics.

If one legitimately considers that certainty and reproducibility are minimum requirements for a realistic description of the physical world, the inescapable consequence of the above observations is that the object to which one must attribute physical properties is not "a system", but "a system on which a given measurement is made", since only in this case the result does not change. Such a "contextual" description is an essential difference from Newtonian physics, and it can be used as a basis to build up quantum physics, as we will now see.

4 Contexts, Systems and Modalities

These observations being very quantum but not really mysterious, we can formalize them (still without mathematics!) by introducing some definitions [3–8]:

- let us designate the devices, arranged to carry out a particular measurement on a given quantum system, by the word "context";
- let us refer to a particular measurement result, on a given system in a given context, by the word "modality". By definition a modality characterizes the system as well as it is physically possible, and, once obtained, it is ideally repeatable with certainty as long as the context is not changed;
- let us admit the following fundamental principle, called "contextual quantization": *whatever the context, a measurement on a given system gives one modality among N possible ones, where the value of N characterizes the system; these N modalities are mutually exclusive, i.e. only one can be realized at a time.*

It follows from the above that if there are other different modalities in other contexts, the link between the modalities of different contexts can only be probabilistic; otherwise one would have a "super context" with a number of mutually exclusive modalities greater than N, which would contradict the fundamental principle [5].

In this approach the probabilities in quantum mechanics are thus physically necessary, since they are imposed by the contextual quantization. It is crucial to note that the conjunction of the two ingredients, quantization and contextuality, is required: for example, many probabilistic phenomena are discrete but classical (heads or tails, dice game, and so forth), and others have contextual aspects (for example in a poll it is known that the order of the questions influences the answers) but they remain classical because quantization does not play a constraining role.

5 A Little Bit of Mathematics

The last step of our approach is to mathematize the principle of contextual quantization, to obtain the quantum formalism. It is crucial to emphasize that, as usual in physics, this step cannot be "deductive" (we cannot show that the formalism is the only one possible), but must be "inductive": we propose a formalism, the most general one possible, respecting the imposed physical conditions, and we verify deductively that it works by describing well the observed effects. The theoretical question is therefore: given a modality (among N) in an initial context, resulting from a previous measurement, what is the probability of obtaining another modality (among

N) in a final context associated with a new measurement? We are thus looking for an $N \times N$ matrix of probabilities between the initial context and the final context, which is generally called a stochastic matrix.

The fundamental (and actually unique) mathematical ingredient is then to associate a projector ($N \times N$ matrix) with the set of all modalities that are related to each other with certainty, either in the same context (it is then the same modality) or in different contexts (this link by "transfer of certainty" is observed empirically). This set constitutes an equivalence class of modalities that are linked "extracontextually", and it is therefore called an extravalence class.

By admitting this fundamentally quantum idea, and a few simple consistency arguments [5], there is in fact no longer any choice: powerful mathematical theorems allow us to show that the only possible theory is quantum mechanics. More precisely, Uhlhorn's theorem [9, 10] shows that unitary transformations between projectors are necessary to preserve the mutually exclusive character of events in each context; and Gleason's theorem [11, 12] shows that Born's rule is necessary to respect the general structure of a probability law (see details in the Appendix).

The probability matrix we are looking for has then a particular form: it is unistochastic, i.e. it is constituted by the squared moduli of the coefficients of a unitary matrix. It also follows that a context is associated to a family of N orthogonal projectors, and that a change of context is associated to a unitary transformation between these families, which reconstitutes the usual framework of quantum probabilities [5]. Within this framework, a usual quantum state $|\psi\rangle$ is predictively incomplete, because it is associated to an extravalence class, and not to a single modality; the specification of the context is needed to define a true probability distribution [6, 7], which is consistent with attributing a modality to a system within a context.

6 No Paradox Anymore?

We will not go further into the mathematical details here, but to conclude let us try to translate into this new language the paradoxical statements quoted at the beginning of this article, again avoiding any formalism:

- *an object can be in several places at the same time:* we often hear also that a quantum bit is in several states at once, and that this corresponds to the notion of quantum superposition. But what does this "at the same time" mean? In fact, it is a question of detecting a modality, therefore a certainty, but a certainty in another context. To say that a particle is "at the same time" in the two arms of the interferometer means in fact that if these two arms interfere constructively, we can predict with certainty where the particle will go. On the other hand, if the measurement is made inside the interferometer, then the result is random, according to the principles explained above: the measurements either inside or outside the interferometer correspond to different contexts.

- *two particles separated by an arbitrarily large distance can influence each other instantaneously:* this is the phenomenon of entanglement, and there is in fact no instantaneous influence at a distance, but again the manifestation of a modality, which is a certain and reproducible property of the pair of particles, e.g. that the total angular momentum of the pair is zero. This can be verified in the appropriate context (by making a joint measurement on both particles), but if one performs the measurement in a context where the particles are spatially separated one will obtain necessarily random results (because of the change of context, see above). Still Born's rule must holds when the context is changed, and for appropriate measurements on entangled particles, it predicts strongly correlated results. This is often related to "quantum non-locality", but in our framework it rather illustrates the predictive incompleteness of $|\psi\rangle$ without a context [7].
- *the quantum particles pass through the barriers:* this is the tunneling effect, an "obvious" effect for classical waves (it is due to so-called evanescent waves), but "impossible" for classical particles. But the classical notions of waves and particles are inadequate here, and one must consider probability amplitudes, which allow one to get a physical picture of Born's law in this case. The change of context corresponds then to the passage from a representation where the momentum of the particle is defined, to another "incompatible" one where its position is defined. A usual quantum wave-packet is somewhere between these two extremes, but in any case speaking about particles or waves with a classical behavior is a fiction, sometimes useful but most often misleading.

The above examples do not necessarily require to fully determine the modalities, i.e. all the system's properties, but the general ideas remain valid. It may be noticed that contextual quantization is difficult to admit, even for physicists, since it asserts that properties/modalities belong to a system within a context, and not to a system considered as "alone in the universe", as classical physics would assume. But the notion in itself is not paradoxical, and if we admit that such is the behavior of the nature in which we live, then the statements related to the "quantum world" finally cease to appear as a long series of contradictions or absurdities.

7 Quantum Computing in Context

This point of view also underlines that statements made "out of context" (such as those concerning the famous Schrödinger's cat, as being both dead and alive) are meaningless. Any statement concerning a quantum system, even a large one such as a quantum computer, is only meaningful in combination with a relevant context in which the modalities under consideration can manifest themselves and be observed.

This allows us to come back to the technological promises mentioned at the beginning of this article. It is true that a quantum computer has, in principle, a gigantic computing power, based on the manipulation of entangled states involving a very large number of quantum bits. But this manipulation will only be possible and meaningful if an adequate context is available, which will also be of a very high complexity—even if it is then a classical complexity.

Then is it certain that the ensemble formed by the quantum "heart" (the system) surrounded by classical instruments capable of interacting with it (the context) will be more efficient than a usual, classical supercomputer? The answer to this question remains open, and the scientific and rational conclusion is that we have to go forward and see what happens—it may not be exactly what we expect, but this is very common in experimental physics, and it could be just as interesting.

Acknowledgements The author thanks Alexia Auffèves, Roger Balian, Nayla Farouki and Franck Laloë for many interesting and stimulating discussions.

Appendix: Uhlhorn's and Gleason's Theorems

As stated above, the empirical facts that we want to describe mathematically are:

(i) in each context a measurement provides one modality among N possible ones, that are mutually exclusive. No measurement can provide more than N mutually exclusive modalities, and once obtained in a given context, a modality corresponds to a certain and repeatable result, as long as one remains in this same context.
(ii) the certainty and repeatability of a modality can be transferred between contexts; this fundamental property is called extracontextuality of modalities. All the modalities related together with certainty, either in the same or in different contexts, constitute an equivalence class that we call an extravalence class.

Our central mathematical ingredient is then to associate a rank-one projector P_i (a $N \times N$ hermitian matrix such as $P^2 = P$) to each extravalence class of modalities, in such a way that the N projectors associated with the N modalities within a given context are mutually orthogonal. In addition, we assume that the probability to get a given modality is a function of its associated projector only. The heuristic motivation for this choice is that it ensures that the events associated with modalities cannot be subdivided in more elementary events, as it would be the case with classical (partition-based) probabilities.

The choice of a specific orthogonal set of projectors associated with a context is not given a priori, but mutually exclusive modalities in the context should stay so, whatever choice is made for the projectors. It means that if two orthogonal projectors are associated with two mutually exclusive modalities, they should stay orthogonal whatever choice is made for the projectors associated with a "reference" (fiduciary) context. Then Uhlhorn's theorem [9, 10] warrants that the transformations between the sets of projectors associated with different contexts must be unitary or anti-unitary; anti-unitary operators are associated with time reversal, so we will omit them here and consider only unitary ones.

We thus get a major result: once a set of mutually orthogonal projectors associated with a fiduciary context has been chosen, the sets of projectors associated to all other contexts are obtained by unitary transformations, so we are simply "moving" in a Hilbert space. There are also various arguments for using unitary (complex) rather than orthogonal (real) matrices; in our framework the simplest argument is to require that all permutations of modalities within a context are continuously connected to the identity, which is not possible with orthogonal matrices, but is possible with unitary ones.

The next step is to consider the probability $f(P_i)$ to get a modality associated with projector P_i. By construction a context is such that $\sum_{i=1}^{i=N} P_i = I$, and $\sum_{i=1}^{i=N} f(P_i) = 1$ for any complete set $\{P_i\}$. But these are just the assumptions of Gleason's theorem [11, 12], so there is a density matrix ρ such that $f(P_i) = \text{Trace}(\rho P_i)$. If the value 1 is reached, then ρ is also a projector Q_j and $f(P_i) = \text{Trace}(Q_j P_i)$ which is the usual Born's formula.

We have thus reconstructed the basic probabilistic framework of quantum mechanics. In order to use it, one must define explicitly the relevant physical properties and associated contexts, that may go from space-time symmetries (Galileo group, Lorentz group) [13] to qubits registers. In any case, contextual quantization applies and sets the scene where the actual physics takes place.

An interesting question is whether this probabilistic formalism might be used outside physics, for its own virtues in a "quantum-like" analogy [14]. However, it can be observed that the requirements (i) and (ii) written above are strictly respected in quantum mechanics, hence the strong predictive power, whereas this is only loosely the case in other domains. Therefore whether or not a variant of this formalism might have a useful predictive (and not only descriptive) power outside physics remains an open—and quite stimulating—question.

References

1. F. Laloë, *Do We Really Understand Quantum Mechanics?* (Cambridge University Press, 2012)
2. A. Peres, *Quantum Theory: Concepts and Methods* (Kluwer Academic Publishers, 1995)
3. P. Grangier, Contextual objectivity: a realistic interpretation of quantum mechanics. Eur. J. Phys. **23**(3), 331 (2002), arXiv:quant-ph/0012122
4. A. Auffèves, P. Grangier, Contexts, systems and modalities: a new ontology for quantum mechanics. Found. Phys. **46**, 121 (2016), arXiv:1409.2120
5. A. Auffèves, P. Grangier, Deriving Born's rule from an inference to the best explanation. Found. Phys. **50**, 1781 (2020), arXiv:1910.13738
6. P. Grangier, Completing the quantum formalism in a contextually objective framework, to appear in Found. Phys. **51**(76) (2021), arXiv:2003.03121
7. P. Grangier, Contextual inferences, nonlocality, and the incompleteness of quantum mechanics. Entropy **23**(12), 1660 (2020), arXiv:2012.09736
8. N. Farouki, P. Grangier, The Einstein-Bohr debate: finding a common ground of understanding?. Found. Sci. (2021). https://doi.org/10.1007/s10699-020-09716-7, arXiv:1907.11267
9. U. Uhlhorn, Representation of symmetry transformations in quantum mechanics. Arkiv för Fysik **23**, 307–340 (1962). Theorem: let $\mathcal{H}$ be a complex Hilbert space with $dim(\mathcal{H}) \geq 3$, and let $P_1(\mathcal{H})$ denote the set of all rank-one projections on $\mathcal{H}$. Then every bijective map $\Phi : P_1(\mathcal{H}) \rightarrow P1(\mathcal{H})$, such that $pq = 0$ in $P_1(\mathcal{H})$ if and only if $\Phi(p)\Phi(q) = 0$, is induced by a unitary or anti-unitary operator on the underlying Hilbert space

10. G. Chevalier, Wigner-type theorems for projections. Int. J. Theor. Phys. **47**, 69–80 (2008)
11. A.M. Gleason, Measures on the closed subspaces of a Hilbert space. J. Math. Mech. **6**, 885 (1957). See also [5]
12. R. Cooke, M. Keanes, W. Moran, An elementary proof of Gleason's theorem. Math. Proc. Camb. Philos. Soc. **98**, 117 (1985)
13. F. Laloë, Symétries continues, *Collection Savoirs Actuels* (EDP Sciences, 2021)
14. A.Y. Khrennikov, *Ubiquitous Quantum Structure: From Psychology to Finance* (Springer, 2010); A.Y. Khrennikov, *Interpretations of Probability* (Walter de Gruyter, 2009)

Philippe Grangier is head of the Quantum Optics group at Institut d'Optique in Palaiseau, France, and Professor at Ecole Polytechnique. He started his research activity in 1980, on experimental tests of Bell's inequalities with Alain Aspect. Then he worked on single-photon interferences (Ph.D. thesis), and interferometry using squeezed light (postdoc in Bell Labs, 1987). After creating his own group, he realized a series of quantum optics experiments in the 1990s, including optical QND measurements. Then from 2000 he performed many experiments related to Quantum Information Processing: single atom qubits in microscopic optical tweezers, quantum cryptography using single photons or continuous variables (CVQKD), entanglement control in atomic and photonic quantum states. He authored more than 200 publications, and he has received many national and international awards, including an ERC Advanced Grant. He has been the coordinator of several large European projects in the domains of Quantum Optics and Quantum Information Processing, and he is now in charge of the DIM SIRTEQ (2017–2021), a large network federating all teams working on Quantum Technologies in the Ile-de-France region. He is also involved in several large EC projects on (CV)QKD testbeds. Since 2000 he has been writing articles on Quantum Foundations.

Quantum Measurement Theory for Systems with Finite Dimensional State Spaces

Masanao Ozawa

Abstract In this chapter, we present a general theory of finite quantum measurements, for which we assume that the state space of the measured system is a finite dimensional Hilbert space and that the possible outcomes of a measurement are a finite set of real numbers. We develop the theory in a deductive manner from the basic postulates for quantum mechanics and a few plausible axioms for general quantum measurements. We derive an axiomatic characterization of all the physically realizable finite quantum measurements. Mathematical tools necessary to describe measurement statistics such as POVMs and quantum instruments are not assumed at the outset, but we introduce them as natural consequences of our axioms. Our objective is to show that those mathematical tools can be naturally derived from obvious theoretical requirements.

1 Introduction

Theories in physics are closely related to mathematics. In most cases, their basic principles are given undisputed mathematical expressions. Hilbert's sixth problem explicitly asked to treat such mathematical formulations of physical theories in a rigorous axiomatic method that was successful in the investigations on the foundations of geometry [7]. One of the typical outcomes of this problem is von Neumann's axiomatization of quantum mechanics given in his book "Mathematical Foundations of Quantum Mechanics" (English translation) [14] originally published in 1932.

In contrast to the prevailing view that his work had completed the axiomatization of non-relativistic quantum mechanics without any superselection rules, von Neumann's axioms need to be completed by revising his treatments on quantum measurements. It is true that until the 1970s there were no problems in non-relativistic

M. Ozawa (✉)
Center for Mathematical Science and Artificial Intelligence, Chubu University Academy of Emerging Sciences, Chubu University, 1200 Matsumoto-cho, Kasugai 487-8501, Japan
e-mail: ozawa@isc.chubu.ac.jp; ozawa@is.nagoya-u.ac.jp

Graduate School of Informatics, Nagoya University, Chikusa-ku, Nagoya 464-8601, Japan

A. Plotnitsky and E. Haven (eds.), *The Quantum-Like Revolution*,
https://doi.org/10.1007/978-3-031-12986-5_9

quantum mechanics that could not be solved because of the deficiency of von Neumann's axioms. However, the technological development for precision measurements based on laser optics from the 1960s revealed that von Neumann's axioms were not sufficient.

In quantum mechanics, the notion of measurement plays an indispensable role. Nevertheless, "theory of measurement" was left incomplete in von Neumann's axiomatization. Quantum mechanics as axiomatized by von Neumann was only justified in experiments in which the object is prepared in the known state, undergoes time evolution with the known Hamiltonian, and then is subject to a measurement of one of its observables. Von Neumann's quantum postulates successfully predict the probability distribution of the outcome of a measurement. However, they are not sufficient to predict the joint probability distribution of the outcomes of successive measurements in time.

Experimental technology was not precise enough to carry out successive measurements for comparing with the theory for long. However, the emerging laser technology in the 1960s enabled us to control quantum states and measuring interactions precisely enough to make repeated or successive measurements.

For the statistics of successive measurements, we need the notion of state changes caused by measurements, often called quantum state reductions, which has been considered one of the most difficult notions in quantum mechanics. In order to deal with quantum systems to be measured sequentially, we need a general notion of quantum state reductions and mathematical methods to calculate them. For this purpose, we have to mathematically characterize all of the possible state changes caused by the most general type of measurements.

Von Neumann [14] introduced the repeatability hypothesis to determine the quantum state reduction, which implies that the state of the measured system changes to the eigenstate of the measured observable corresponding to the outcome of the measurements. This principle uniquely determines the state change for measurements of non-degenerate discrete observables, in which case all the eigenspaces are one dimensional. For degenerate observables the eigenstate is not uniquely determined. Subsequently, Lüders [10] proposed the projection postulate, called the von Neumann-Lüders projection postulate, to determine the unique eigenstate.

A mathematical problem to extend this notion to measurements of observables with continuous spectra (continuous variables) was considered. Nakamura and Umegaki [13] pointed out that the unique state change described by von Neumann's repeatability hypothesis for measurements of non-degenerate observables is an instance of Umegaki's non-commutative extension of the notion of conditional expectations [43], which is a projection map of the full operator algebra onto the subalgebra generated by the measured observable. They conjectured that von Neumann's repeatability hypothesis can be extended to continuous observables by Umegaki's notion of non-commutative conditional expectations. However, Arveson [1] proved that the corresponding conditional expectation does not exist if the measured observable has a continuous spectral.

Based on the above results, Davies and Lewis [3] proposed to abandon the repeatability hypothesis as the first principle, and introduced a general mathematical tool to describe a general measurement, called an instrument, which generalizes the notion of Umegaki's "conditional expectations" [43], the notion of "operations" described for discrete observables by Schwinger [38, 40, 41] and introduced in the context of algebraic quantum field theory by Haag and Kastler [4], the notion of "effects" introduced by Ludwig [11, 12], and the notion of "POVMs" introduced by Helstrom [6]. The Davies-Lewis (DL) instruments were considered as a general mathematical framework that may describe all the possible quantum measurements. However, the consistency problem remained. It was not clear whether the class of DL instruments is too general or not general enough to describe quantum measurements, or whether every DL instrument is consistent with other postulates for quantum mechanics, or whether every DL instrument has a quantum mechanical model for a measuring process.

In 1986, Yuen [45] explicitly proposed the problem to find a mathematical characterization of all the physically realizable quantum measurements, and he conjectured that the DL instruments are too general. Shortly before this proposal, the problem had been solved by the present author [18] in1984, showing that all the physically realizable quantum measurements are faithfully characterized by completely positive instruments, the notion which modifies the notion of DL instruments by requiring complete positivity. This completed von Neumann's axiomatization of quantum mechanics by adding the most general measurement axiom, or the most general description of quantum state reductions, consistent with the other von Neumann axioms for quantum mechanics.

In this chapter, we present a general theory of finite quantum measurements, as an introduction to a general theory of quantum measurements developed as outlined above, in a form accessible without mathematics for operators in infinite dimensional Hilbert spaces and probability theory for continuous random variables. Thus, we assume that the state space of the measured system is a finite dimensional Hilbert space and that the possible outcomes of a measurement is a finite set of real numbers. We develop the theory in a deductive manner from the basic postulates for quantum mechanics and a few plausible axioms for general quantum measurements, and we derive an axiomatic characterization of all of the physically realizable finite quantum measurements. Mathematical tools necessary to describe measurement statistics, such as POVMs and quantum instruments, are not assumed at the outset, but we introduce them as natural consequences of our intuitive axioms. Our objective is to show that those mathematical tools can be naturally derived from obvious theoretical requirements.

As a chapter of a Festschrift celebrating Andrei Khrennikov and the quantum-like revolution, the author hopes that this work would help the readers to extend the scope of quantum measurement theory based on the notion of quantum instruments beyond quantum physics.

2 Quantum Mechanics

In this chapter, we consider quantum systems described by finite dimensional Hilbert spaces. Mathematically, a finite dimensional Hilbert space is defined as any finite dimensional linear space $\mathscr{H}$ over the complex number field with inner product (ξ, η) defined for all $\xi, \eta \in \mathscr{H}$, which we assume linear in η and conjugate linear in ξ following the physics convention.

Let $\mathscr{H}$ be a finite dimensional Hilbert space. A linear operator on $\mathscr{H}$ is a linear mapping defined everywhere on $\mathscr{H}$ with values in $\mathscr{H}$. Denote by $\mathscr{L}(\mathscr{H})$ the space of linear operators on $\mathscr{H}$. The adjoint of a linear operator A on $\mathscr{H}$ is a linear operator $A^\dagger$ uniquely determined by the condition $(\xi, A^\dagger\eta) = (A\xi, \eta)$ for all $\xi, \eta \in \mathscr{H}$. A linear operator A is said to be self-adjoint if $A^\dagger = A$, or equivalently $(\xi, A\xi)$ is a real number for all $\xi \in \mathscr{H}$. A linear operator A is said to be positive, in symbols $A \geq 0$, if $(\xi, A\xi) \geq 0$ for all $\xi \in \mathscr{H}$. The trace of a linear operator A is defined by $\mathrm{Tr}[A] = \sum_j (\phi_j, A\phi_j)$ for any orthonormal basis $\{\phi_j\}$.[1] A linear operator ρ is called a density operator if positive and of unit trace, i.e., $\rho \geq 0$ and $\mathrm{Tr}[\rho] = 1$.

Axioms for quantum mechanics of finite level systems without any superselection rules are given as follows.

Axiom Q1 (Quantum systems, states, and observables). *Every quantum system* **S** *is described by a finite dimensional Hilbert space* $\mathscr{H}$ *called the state space of* **S**. *States of* **S** *are represented by density operators on* $\mathscr{H}$ *and observables of* **S** *are represented by self-adjoint operators on* $\mathscr{H}$. *Every density operator on* $\mathscr{H}$ *corresponds to a state of* **S**, *and every self-adjoint operator corresponds to an observable of* **S**.

By Axiom Q1 we shall identify states with density operators, and observables with self-adjoint operators. The state of the form $\rho = |\psi\rangle\langle\psi|$ is called a pure state.[2] If **S** is in the state $\rho = |\psi\rangle\langle\psi|$, **S** is said to be in the (vector) state ψ. We denote by $\mathscr{S}(\mathscr{H})$ the space of states, or density operators, on $\mathscr{H}$ and by $\mathscr{O}(\mathscr{H})$ the space of observables, or self-adjoint operators, on $\mathscr{H}$.

Axiom Q2 (Born statistical formula). *If an observable* A *is measured in a state* ρ, *the outcome obeys the probability distribution of* A *in* ρ *defined by the Born statistical formula (BSF)*

$$\Pr\{\mathbf{x} = x \| \rho\} = \mathrm{Tr}[P^A(x)\rho] \tag{1}$$

where $x \in \mathbb{R}$, *and* $P^A(x)$ *stands for the projection onto the subspace* $\{\psi \in \mathscr{H} | A\psi = x\psi\}$.

[1] The trace is independent of the choice of the orthonormal basis $\{\phi_j\}$. To see this let $\{\xi_k\}$ be another orthonormal basis. Then, we have $\mathrm{Tr}[A] = \sum_j (\phi_j, A\phi_j) = \sum_{j,k} (\phi_j, \xi_k)(\xi_k, A\phi_j) = \sum_{j,k} (A^\dagger\xi_k, \phi_j)(\phi_j, \xi_k) = \sum_k (A^\dagger\xi_k, \xi_k) = \sum_k (\xi_k, A\xi_k)$. Thus, the trace is independent of the choice of the orthonormal basis.

[2] For any $\xi, \eta \in \mathscr{H}$, the operator $|\xi\rangle\langle\eta|$ is defined by $(|\xi\rangle\langle\eta|)\psi = (\eta, \psi)\xi$ for all $\psi \in \mathscr{H}$.

The projection $P^A(x)$ is called the spectral projection of A for the real number x. The map $P^A : x \mapsto P^A(x)$ is called the spectral measure of A [5]. From Axiom Q2, the mean value $\langle A \rangle$ and the standard deviation $\sigma(A)$ are given by

$$\langle A \rangle = \mathrm{Tr}[A\rho], \tag{2}$$

$$\sigma(A)^2 = \left\langle A^2 \right\rangle - \langle A \rangle^2 . \tag{3}$$

The standard deviations $\sigma(A)$, $\sigma(B)$ of observables A, B in a state ρ satisfy Robertson's inequality

$$\sigma(A)\sigma(B) \geq \frac{1}{2} | \langle [A, B] \rangle |. \tag{4}$$

Axiom Q3 (Time evolution). *Suppose that a system* $\mathbf{S}$ *is an isolated system with the (time-independent) Hamiltonian* H *from time* t *to* $t + \tau$. *The system* $\mathbf{S}$ *is in a state* $\rho(t)$ *at time* t *if and only if* $\mathbf{S}$ *is in the state* $\rho(t + \tau)$ *at time* $t + \tau$ *satisfying*

$$\rho(t + \tau) = e^{-i\tau H/\hbar} \rho(t) e^{i\tau H/\hbar}, \tag{5}$$

where $\hbar$ *stands for the Planck constant divided by* 2π.

Axiom Q4 (Composite systems). *The state space of the composite system* $\mathbf{S} = \mathbf{S}_1 + \mathbf{S}_2$ *of a system* $\mathbf{S}_1$ *with the state space* $\mathcal{H}$ *and a system* $\mathbf{S}_2$ *with the state space* $\mathcal{K}$ *is given by the tensor product* $\mathcal{H} \otimes \mathcal{K}$. *The observable* A *of* $\mathbf{S}_1$ *is identified with the observable* $A \otimes I$ *of* $\mathbf{S}$ *and the observable* B *of* $\mathbf{S}_2$ *is identified with* $I \otimes B$ *of* $\mathbf{S}$.

For any orthonormal bases $\{\xi_j\}$ of $\mathcal{H}$ and $\{\eta_k\}$ of $\mathcal{K}$, the family of their tensor products $\{\xi_j \otimes \eta_k\}$ forms an orthonormal basis of $\mathcal{H} \otimes \mathcal{K}$. The tensor product of $A \in \mathcal{L}(\mathcal{H})$ and $B \in \mathcal{L}(\mathcal{K})$ is the linear operator $A \otimes B \in \mathcal{L}(\mathcal{H} \otimes \mathcal{K})$ defined by $(A \otimes B)(\xi \otimes \eta) = A\xi \otimes B\eta$ for all $\xi \in \mathcal{H}$ and $\eta \in \mathcal{K}$. Every $A \in \mathcal{L}(\mathcal{H} \otimes \mathcal{K})$ is of the form $A = \sum_{j,k} C_j \otimes D_k$ where $C_j \in \mathcal{L}(\mathcal{H})$ and $D_k \in \mathcal{L}(\mathcal{K})$. The partial traces $\mathrm{Tr}_{\mathcal{K}}[A]$ over $\mathcal{K}$ and $\mathrm{Tr}_{\mathcal{H}}[A]$ over $\mathcal{H}$ of $A \in \mathcal{L}(\mathcal{H} \otimes \mathcal{K})$ are defined by $\mathrm{Tr}_{\mathcal{K}}[A] = \sum_{j,k} C_j \otimes \mathrm{Tr}[D_k]$ and $\mathrm{Tr}_{\mathcal{H}}[A] = \sum_{j,k} \mathrm{Tr}[C_j] \otimes D_k$ if $A = \sum_{j,k} C_j \otimes D_k$. It follows from Axiom Q4 that if the system $\mathbf{S}_1 + \mathbf{S}_2$ is in a state ρ_{12}, then the system $\mathbf{S}_1$ is in the state $\rho_1 = \mathrm{Tr}_{\mathcal{K}}[\rho_{12}]$ and $\mathbf{S}_2$ is in the state $\rho_2 = \mathrm{Tr}_{\mathcal{H}}[\rho_{12}]$.

3 Statistical Properties of Measuring Apparatuses

In this section, we discuss statistical properties of measuring apparatuses. We introduce plausible axioms for statistical properties of a measuring apparatus required for every apparatus to satisfy, and then we show that statistical properties of physically realizable measuring apparatuses can be naturally described by POVMs and completely positive instruments.

3.1 Output Probability Distributions

Axiom M1 (Output probability distributions and quantum state reductions). *An apparatus* $\mathbf{A}(\mathbf{x})$ *with output variable* $\mathbf{x}$ *to measure a system* $\mathbf{S}$ *determines the probability* $\Pr\{\mathbf{x} = x \| \rho\}$ *of the outcome* $\mathbf{x} = x$ *of the measurement depending on the input state* ρ *(the state of* $\mathbf{S}$ *just before the measurement), and determines the output state* $\rho_{\{\mathbf{x}=x\}}$ *(the state of* $\mathbf{S}$ *just after the measurement) depending on the input state* ρ *and the outcome* $\mathbf{x} = x$ *of the measurement.*

The variable representing the outcome of the apparatus is called the *output variable*. Let $\mathbf{S}$ be a quantum system, to be referred to the *object*, described by a Hilbert space $\mathcal{H}$ of state vectors. Let $\mathbf{A}(\mathbf{x})$ be a *measuring apparatus* with an output variable $\mathbf{x}$ to measure the object $\mathbf{S}$. We assume that $\mathbf{x}$ takes values in the real line $\mathbb{R}$. For any real number $x \in \mathbb{R}$, we shall denote by "$\mathbf{x} = x$" the probabilistic event that the output variable $\mathbf{x}$ of apparatus $\mathbf{A}(\mathbf{x})$ takes the value x. By Axiom M1, the probability distribution of the output variable $\mathbf{x}$ is determined by the input state ρ. Denote it by $\Pr\{\mathbf{x} = x \| \rho\}$, and call it as the *output distribution* of $\mathbf{A}(\mathbf{x})$ in the input state ρ. If the state ρ is a vector state $\rho = |\psi\rangle\langle\psi|$, we also write $\Pr\{\mathbf{x} = x \| \psi\} = \Pr\{\mathbf{x} = x \| \rho\}$. For any subset $\Delta \subseteq \mathbb{R}$, we define $\Pr\{\mathbf{x} \in \Delta \| \rho\} = \sum_{x \in \Delta} \Pr\{\mathbf{x} = x \| \rho\}$. We suppose that the output distribution satisfies the following conditions.

(i) **Positivity:** $\Pr\{\mathbf{x} = x \| \rho\} \geq 0$ for every $x \in \mathbb{R}$.

(ii) **Unity**: $\sum_{x \in \mathbb{R}} \Pr\{\mathbf{x} = x \| \rho\} = 1$.

(iii) **Finiteness**: There exists a finite subset of $S \subseteq \mathbb{R}$ such that if $x \notin S$ then $\Pr\{\mathbf{x} = x \| \rho\} = 0$ for any state ρ.

The output probability distribution should satisfy the following postulate.

Axiom M2' (Mixing law of output probability). *For any apparatus* $\mathbf{A}(\mathbf{x})$, *the function* $\rho \mapsto \Pr\{\mathbf{x} = x \| \rho\}$ *is an affine function of states* ρ *for any real number* x.

This means that we have

$$\Pr\{\mathbf{x} = x \| p\rho_1 + (1-p)\rho_2\} = p \Pr\{\mathbf{x} = x \| \rho_1\} + (1-p) \Pr\{\mathbf{x} = x \| \rho_2\},$$

where ρ_1 and ρ_2 are density operators and $0 < p < 1$.

The above axiom is justified by the following interpretation of the mixture of states: The system **S** is in the state $p\rho_1 + (1-p)\rho_2$ if it is in state ρ_1 with probability p and in state ρ_2 with probability $1-p$.

3.2 Probability Operator-Valued Measures

In order to characterize output probability distributions, we introduce a mathematical definition. A mapping $\Pi : x \mapsto \Pi(x)$ of $\mathbb{R}$ into the space $\mathscr{L}(\mathscr{H})$ of linear operators on $\mathscr{H}$ is called a *probability operator-valued measure (POVM)*, also known as a *positive operator-valued measure* [15] or a *probability operator measure (POM)* [6], if the following conditions are satisfied:

(i) **Positivity**: $\Pi(x) \geq 0$ for all $x \in \mathbb{R}$.

(ii) **Unity**: $\sum_{x\in\mathbb{R}} \Pi(x) = 1$.

(iii) **Finiteness**: There exists a finite subset of $S \subseteq \mathbb{R}$ such that if $x \notin S$ then $\Pi(x) = 0$.

One of important consequences from the above postulate is the following characterization of output probability distributions [16].

Theorem 1 *Axiom M2' (the mixing law of output probability) holds if and only if for any apparatus* $\mathbf{A}(\mathbf{x})$*, there uniquely exists a POVM* Π *satisfying*

$$\Pr\{\mathbf{x} = x\|\rho\} = \mathrm{Tr}[\Pi(x)\rho] \tag{6}$$

for any real number x *and density operator* ρ.

The POVM Π defined by Eq. (6) is called the *POVM of* $\mathbf{A}(\mathbf{x})$.

3.3 The Born Statistical Formula

Let A be an observable of system **S**. According to Axiom Q2 (Born statistical formula), we say that apparatus $\mathbf{A}(\mathbf{x})$ *satisfies the Born statistical formula (BSF)* for observable A on input state ρ, if we have

$$\Pr\{\mathbf{x} = x\|\rho\} = \mathrm{Tr}[P^A(x)\rho] \tag{7}$$

for every real number x, where $P^A(x)$ is the spectral projection of A for x. From Eqs. (6) and (7), apparatus $\mathbf{A}(\mathbf{x})$ satisfies the Born statistical formula on every input state if and only if the POVM Π of $\mathbf{A}(\mathbf{x})$ is the spectral measure P^A. In this case, the apparatus $\mathbf{A}(\mathbf{x})$ is called an *A-measuring apparatus*. Naturally, we assume that *for every observable* A *of* **S** *there is at least one* A*-measuring apparatus.*

The Born statistical formula is a necessary condition for the state-dependent accuracy of measurement, or for the accurate measurements of an observable A in a single state ρ. For a necessary and sufficient condition for the state-dependent accuracy of measurement, we refer the reader to [35–37]. According to those studies of the state-dependent accuracy of measurement, an apparatus accurately measures an observable A in every state ρ if and only if the Born statistical formula holds for every state ρ. Thus, the state-dependent accuracy of measurement required for every state is consistent with the conventional state-independent accuracy of measurement.

3.4 Quantum State Reductions

According to Axiom M1, depending on the input state ρ, any apparatus $\mathbf{A}(\mathbf{x})$ determines the state $\rho_{\{\mathbf{x}=x\}}$ just after the measurement for any possible outcome $\mathbf{x} = x$ with $\Pr\{\mathbf{x} = x \| \rho\} > 0$. The operational meaning of the state $\rho_{\{\mathbf{x}=x\}}$ is given as follows. If a measurement using the apparatus $\mathbf{A}(\mathbf{x})$ on input state ρ is immediately followed by a measurement using another apparatus $\mathbf{A}(\mathbf{y})$ with output variable $\mathbf{y}$, we shall denote by $\Pr\{\mathbf{y} = y | \mathbf{x} = x \| \rho\}$ the conditional probability of the outcome $\mathbf{y} = y$ of the measurement using $\mathbf{A}(\mathbf{y})$ given the outcome $\mathbf{x} = x$ of the measurement using $\mathbf{A}(\mathbf{x})$. Then, under the condition $\mathbf{x} = x$ the state just before the measurement using $\mathbf{A}(\mathbf{y})$ is the state $\rho_{\{\mathbf{x}=x\}}$ so that we naturally have

$$\Pr\{\mathbf{y} = y | \mathbf{x} = x \| \rho\} = \Pr\{\mathbf{y} = y \| \rho_{\{\mathbf{x}=x\}}\}. \tag{8}$$

If $\Pr\{\mathbf{x} = x \| \rho\} = 0$, the state $\rho_{\{\mathbf{x}=x\}}$ is taken to be indefinite. The state $\rho_{\{\mathbf{x}=x\}}$ is called the *output state of the apparatus* $\mathbf{A}(\mathbf{x})$ *given the outcome* $\mathbf{x} = x$ *on input state* ρ.

Two apparatuses are called *statistically equivalent* if they have the same output probabilities and the same output states for any outcomes and any input states.

3.5 Joint Output Probability Distributions

If a measurement using apparatus $\mathbf{A}(\mathbf{x})$ on input state ρ is immediately followed by a measurement using apparatus $\mathbf{A}(\mathbf{y})$, then from Eq. (8) the joint probability distribution $\Pr\{\mathbf{x} = x, \mathbf{y} = y \| \rho\}$ of the output variables $\mathbf{x}$ and $\mathbf{y}$ is given by

$$\Pr\{\mathbf{x} = x, \mathbf{y} = y \| \rho\} = \Pr\{\mathbf{y} = y \| \rho_{\{\mathbf{x}=x\}}\} \Pr\{\mathbf{x} = x \| \rho\}. \tag{9}$$

Thus, the joint probability distribution of outputs of successive measurements depends only on the input state of the first measurement. We shall call the above joint probability distribution the *joint output probability distribution of* $\mathbf{A}(\mathbf{x})$ *followed by* $\mathbf{A}(\mathbf{y})$. The joint output probability distributions should satisfy the following axiom:

Axiom M2 (Mixing law of joint output probability). *For any apparatuses* $\mathbf{A}(\mathbf{x})$ *and* $\mathbf{A}(\mathbf{y})$, *the function* $\rho \mapsto \Pr\{\mathbf{x} = x, \mathbf{y} = y\|\rho\}$ *is an affine function of density operators* ρ *for any real numbers* x, y.

Since the joint probability $\Pr\{\mathbf{x} = x, \mathbf{y} = y\|\rho\}$ depends on the initial input state ρ, the above axiom is also justified by the following interpretation of the mixture of states: The system **S** is in the state $p\rho_1 + (1-p)\rho_2$ if it is in state ρ_1 with probability p and in state ρ_2 with probability $1-p$.

By summing up y in Eq. (9) and using the unity relation $\sum_{y\in\mathbb{R}} \Pr\{\mathbf{y} = y\|\rho_{\{\mathbf{x}=x\}}\} = 1$, we have

$$\sum_{y\in\mathbb{R}} \Pr\{\mathbf{x} = x, \mathbf{y} = y\|\rho\} = \Pr\{\mathbf{x} = x\|\rho\} \tag{10}$$

for any x and ρ. Thus, we conclude that *Axiom M2 (the mixing law of joint output probability) implies Axiom M2' (the mixing law of output probability).*

3.6 Instruments

Davies and Lewis [2, 3] introduced the following mathematical notion for unified description of statistical properties of measurements. A mapping $\mathscr{I} : x \mapsto \mathscr{I}(x)$ of $\mathbb{R}$ into the space $\mathscr{L}(\mathscr{L}(\mathscr{H}))$ of linear transformations on $\mathscr{L}(\mathscr{H})$ is called a *Davies-Lewis (DL) instrument*, if the following conditions are satisfied.

(i) **Positivity:** $\mathscr{I}(x)$ maps positive operators in $\mathscr{L}(\mathscr{H})$ to positive operators in $\mathscr{L}(\mathscr{H})$ for any$x \in \mathbb{R}$.

(ii) **Unity:** $\sum_{x\in\mathbb{R}} \mathscr{I}(x)$ is trace-preserving.

(iii) **Finiteness:** There exists a finite subset $S \subset \mathbb{R}$ such that $\mathscr{I}(x) = 0$ for all $x \in S$.

For any apparatus $\mathbf{A}(\mathbf{x})$, we define the mapping $\mathscr{I}(x) : \rho \mapsto \mathscr{I}(x)\rho$ by

$$\mathscr{I}(x)\rho = \Pr\{\mathbf{x} = x\|\rho\}\rho_{\{\mathbf{x}=x\}}, \tag{11}$$

where $\rho \in \mathscr{S}(\mathscr{H})$ and $x \in \mathbb{R}$. The mapping $\mathscr{I}(x)$ transforms any density operator ρ to a positive operator with the trace equal to $\Pr\{\mathbf{x} = x\|\rho\}$. It follows from Axiom M2 (the mixing law of joint output probability) that $\mathscr{I}(x)$ is an affine mapping and can be extended to a linear transformation on the space $\mathscr{L}(\mathscr{H})$ of linear operators on $\mathscr{H}$ [29, 32]. Then it is easy to see that the mapping $\mathscr{I}(x) : \rho \mapsto \mathscr{I}(x)\rho$ satisfies the Davies and Lewis definition of instruments. Conversely, if any apparatus $\mathbf{A}(\mathbf{x})$ has a Davies-Lewis instrument $\mathscr{I}$ satisfying Eq. (11), then Axiom M2 (the mixing law of joint output probability) holds. Thus, we have [32]

Theorem 2 *Axiom M2 (the mixing law of joint output probability) holds if and only if for any apparatus* $\mathbf{A}(\mathbf{x})$ *there uniquely exists a DL instrument* $\mathscr{I}$ *satisfying Eq.* (11) *for any real number* x *and density operator* ρ.

The mapping $\mathscr{I}(x)$ defined by Eq. (11) for the apparatus $\mathbf{A}(\mathbf{x})$ is called the *operation of* $\mathbf{A}(\mathbf{x})$ *given the outcome* $\mathbf{x} = x$. The mapping $\mathscr{I}$ is called the *instrument of* $\mathbf{A}(\mathbf{x})$. Then, the output probability and the output state can be expressed by

$$\Pr\{\mathbf{x} = x \| \rho\} = \mathrm{Tr}[\mathscr{I}(x)\rho], \tag{12}$$

$$\rho_{\{\mathbf{x}=x\}} = \frac{\mathscr{I}(x)\rho}{\mathrm{Tr}[\mathscr{I}(x)\rho]} \tag{13}$$

where the second relation assumes $\Pr\{\mathbf{x} = x \| \rho\} > 0$. Thus, if $\mathscr{I}_{\mathbf{x}}$ and $\mathscr{I}_{\mathbf{y}}$ are the instruments of $\mathbf{A}(\mathbf{x})$ and $\mathbf{A}(\mathbf{y})$, respectively, then the joint output probability distribution can be expressed by

$$\Pr\{\mathbf{x} = x, \mathbf{y} = y \| \rho\} = \mathrm{Tr}[\mathscr{I}_{\mathbf{y}}(y)\mathscr{I}_{\mathbf{x}}(x)\rho] \tag{14}$$

for any state ρ and any real numbers x, y.

Both the output probability distribution and the output states are determined by the instrument. Thus, two apparatuses are statistically equivalent if and only if they have the same instrument.

For any linear transformation T on the space $\mathscr{L}(\mathscr{H})$ of linear operators on $\mathscr{H}$, the *dual* of T is defined to be the linear transformation T^* on $\mathscr{L}(\mathscr{H})$ satisfying

$$\mathrm{Tr}[A(T\rho)] = \mathrm{Tr}[(T^*A)\rho] \tag{15}$$

for any $A, \rho \in \mathscr{L}(\mathscr{H})$. The dual of the operation $\mathscr{I}(x)$ is called the *dual operation* $\mathscr{I}(x)^*$ *given* $\mathbf{x} = x$; by Eq. (15) it is defined by the relation

$$\mathrm{Tr}[A\mathscr{I}(x)\rho)] = \mathrm{Tr}\{[\mathscr{I}(x)^*A]\rho\} \tag{16}$$

for any $A, \rho \in \mathscr{L}(\mathscr{H})$.

The operator $\mathscr{I}(x)^*I$ obtained by applying the dual operation $\mathscr{I}(x)^*$ to the identity operator I is called the *effect of the operation* $\mathscr{I}(x)$. By Eqs. (12) and (15) we have

$$\Pr\{\mathbf{x} = x \| \rho\} = \mathrm{Tr}[(\mathscr{I}(x)^*I)\rho]. \tag{17}$$

Since ρ is arbitrary, comparing with Eq. (6), we have

$$\Pi(x) = \mathscr{I}(x)^*I \tag{18}$$

for any real number x. Thus, the POVM of $\mathbf{A}(\mathbf{x})$ is determined by the effects of the instrument $\mathscr{I}$.

Let $\mathscr{I}_{\mathbf{x}}$ and $\mathscr{I}_{\mathbf{y}}$ be the instruments of $\mathbf{A}(\mathbf{x})$ and $\mathbf{A}(\mathbf{y})$, respectively, and let $\Pi_{\mathbf{y}}$ be the POVM of $\mathbf{A}(\mathbf{y})$. Then, we have

$$
\begin{aligned}
\mathrm{Tr}[\mathscr{I}_{\mathbf{y}}(y)\mathscr{I}_{\mathbf{x}}(x)\rho] &= \mathrm{Tr}\{[\mathscr{I}_{\mathbf{y}}(y)^* I][\mathscr{I}_{\mathbf{x}}(x)\rho]\} \\
&= \mathrm{Tr}\{[\Pi_{\mathbf{y}}(y)[\mathscr{I}_{\mathbf{x}}(x)\rho]\} \\
&= \mathrm{Tr}\{[\mathscr{I}(x)^* \Pi_{\mathbf{y}}(y)]\rho\} \qquad (19)
\end{aligned}
$$

Thus, the joint output probability distribution can be expressed by

$$
\Pr\{\mathbf{x} = x, \mathbf{y} = y \| \rho\} = \mathrm{Tr}\{[\mathscr{I}(x)^* \Pi_{\mathbf{y}}(y)]\rho\} \qquad (20)
$$

for any $x, y \in \mathbb{R}$.

3.7 Selective Quantum State Reduction

For any subset Δ of $\mathbb{R}$, the outcome event "$\mathbf{x} \in \Delta$" means that the outcome of the measurement is an element of Δ. The probability of outcome event $\mathbf{x} \in \Delta$ is given by

$$
\Pr\{\mathbf{x} \in \Delta \| \rho\} = \sum_{x \in \Delta} \Pr\{\mathbf{x} = x \| \rho\}. \qquad (21)
$$

If the input state is ρ, the state just after the measurement given the outcome event $\mathbf{x} \in \Delta$ is denoted by $\rho_{\{\mathbf{x} \in \Delta\}}$. This state is determined as follows. Let A be an observable of the object $\mathbf{S}$. Suppose that the observer measures the object $\mathbf{S}$ in the state $\rho_{\{\mathbf{x} \in \Delta\}}$ using another apparatus $\mathbf{A}(\mathbf{y})$ with the POVM $\Pi_{\mathbf{y}} = P^A$. Then we have

$$
\begin{aligned}
\Pr\{\mathbf{x} \in \Delta, \mathbf{y} = y \| \rho\} &= \Pr\{\mathbf{y} = y \| \rho_{\{\mathbf{x} \in \Delta\}}\} \Pr\{\mathbf{x} \in \Delta \| \rho\} \\
&= \mathrm{Tr}[P^A(y) \Pr\{\mathbf{x} \in \Delta \| \rho\} \rho_{\{\mathbf{x} \in \Delta\}}] \qquad (22)
\end{aligned}
$$

On the other hand, by Eq. (9) we have

$$
\begin{aligned}
\Pr\{\mathbf{x} \in \Delta, \mathbf{y} = y \| \rho\} &= \sum_{x \in \Delta} \Pr\{\mathbf{x} = x, \mathbf{y} = y \| \rho\} \\
&= \sum_{x \in \Delta} \Pr\{\mathbf{y} = y \| \rho_{\{\mathbf{x} = x\}}\} \Pr\{\mathbf{x} = x \| \rho\} \\
&= \mathrm{Tr}[P^A(y) \sum_{x \in \Delta} \Pr\{\mathbf{x} = x \| \rho\} \rho_{\{\mathbf{x} = x\}}]. \qquad (23)
\end{aligned}
$$

Since A is an arbitrary observable, by comparing Eqs. (22) and (23), we have

$$
\Pr\{\mathbf{x} \in \Delta \| \rho\} \rho_{\{\mathbf{x} \in \Delta\}} = \sum_{x \in \Delta} \Pr\{\mathbf{x} = x \| \rho\} \rho_{\{\mathbf{x} = x\}}. \qquad (24)
$$

For any subset Δ of $\mathbb{R}$, we write $\mathscr{I}(\Delta) = \sum_{x\in\Delta} \mathscr{I}(x)$ and $\Pi(\Delta) = \sum_{x\in\Delta} \Pi(x)$. Let $\mathscr{I}$ be the instrument of an apparatus $\mathbf{A}(\mathbf{x})$. For any state ρ we have

$$\mathscr{I}(\Delta)\rho = \Pr\{\mathbf{x} \in \Delta \| \rho\}\rho_{\{\mathbf{x}\in\Delta\}}. \tag{25}$$

$\mathscr{I}(\Delta)$ is called *the operation given the outcome event* $\mathbf{x} \in \Delta$ of the apparatus $\mathbf{A}(\mathbf{x})$.

The state change from the state ρ to the state $\rho_{\{\mathbf{x}=x\}}$ is called an *(individual) quantum state reduction*. The state change from the state ρ to the state $\rho_{\{\mathbf{x}\in\Delta\}}$ is called a *selective quantum state reduction*. On the other hand, the state change $\rho \mapsto \rho_{\{\mathbf{x}\in\mathbb{R}\}}$ is called a *non-selective quantum state reduction*. For the instrument $\mathscr{I}$ of the apparatus $\mathbf{A}(\mathbf{x})$, the operation $T = \mathscr{I}(\mathbb{R})$ is called the *non-selective operation* of $\mathbf{A}(\mathbf{x})$, and $T^* = \mathscr{I}(\mathbb{R})^*$ is called the *non-selective dual operation* of $\mathbf{A}(\mathbf{x})$. In general a linear transformation T on $\mathscr{L}(\mathscr{H})$ is called a *positive map* if $T\rho \geq 0$ for all $\rho \geq 0$. For any DL instrument $\mathscr{I}$, $\mathscr{I}(\Delta)$ is a positive map. The non-selective operation T is trace-preserving, i.e.,

$$\mathrm{Tr}[T\rho] = \mathrm{Tr}[\rho], \tag{26}$$

for any $\rho \in \mathscr{L}(\mathscr{H})$, while the non-selective dual operation T^* is unit-preserving,

$$T^* I = I. \tag{27}$$

3.8 Repeatability Hypothesis

In the early days of quantum mechanics, only a restricted class of measurements was seriously studied. The following axiom was broadly accepted in the 1930s.

(M) Measurement axiom. If an observable A is measured in a system $\mathbf{S}$ to obtain the outcome a, then the system $\mathbf{S}$ is left in an eigenstate of A for the eigenvalue a.

Von Neumann [14] showed that this assumption is equivalent to the following assumption called the *repeatability hypothesis* [14, p. 335], posed with a clear operational condition generalizing a feature of the Compton-Simons experiment [14, pp. 212–214].

(R) Repeatability hypothesis. *If an observable A is measured twice in succession in a system* $\mathbf{S}$, *then we get the same value each time.*

It can be seen from the following definition of measurement due to Schrödinger given in his famous "cat paradox" paper [39] that von Neumann's repeatability hypothesis was broadly accepted in the 1930s.

> The systematically arranged interaction of two systems (measured object and measuring instrument) is called a measurement on the first system, if a directly-sensible variable feature of the second (pointer position) is always reproduced within certain error limits when the process is immediately repeated (on the same object, which in the meantime must not be exposed to any additional influences) [39].

The repeatability hypothesis uniquely determines the state after the measurement if the measured observable A is non-degenerate, i.e., every eigensubspace is one-dimensional. Let $A = \sum_n a_n |\phi_n\rangle\langle\phi_n|$ be a non-degenerate observable, where $|\phi_n\rangle\langle\phi_n|$ stands for the projection onto the subspace spanned by the eigenvector ϕ_n for the eigenvalue a_n. Then the measuring apparatus satisfies the repeatability hypothesis if and only if the corresponding instrument is of the form:

$$\mathscr{I}(x)\rho = |\phi_n\rangle\langle\phi_n|\,\rho\,|\phi_n\rangle\langle\phi_n| \tag{28}$$

for any $\rho \in \mathscr{S}(\mathscr{H})$ if $x = a_n$; and $\mathscr{I}(x) = 0$ otherwise.

If the measured observable is degenerate, the repeatability hypothesis does not determine the unique eigenstate as the state after the measurement. Lüders [10] proposed the projection postulate to determine the eigenstate uniquely.

(P) The von Neumann-Lüders projection postulate. If a measurement of an observable A in a state ρ leads to the outcome $\mathbf{x} = x$, the state $\rho_{\{\mathbf{x}=x\}}$ just after the measurement is given by

$$\rho_{\{\mathbf{x}=x\}} = \frac{P^A(x)\rho P^A(x)}{\mathrm{Tr}[P^A(x)\rho]}. \tag{29}$$

Thus, the projection postulate uniquely determines the instrument for measurement of A as

$$\mathscr{I}(x)\rho = P^A(x)\rho P^A(x) \tag{30}$$

for all $x \in \mathbb{R}$ and $\rho \in \mathscr{S}(\mathscr{H})$.

It is well known that the same observable can be measured with many different ways that do not satisfy the projection postulate. Thus, the von Neumann-Lüders projection postulate should not be taken as a universal postulate for quantum mechanics but should be taken as a defining condition for a class of measurements called *projective measurements*.

For any sequence of projective measurements, we can determine the joint probability distribution of the outcomes of measurements [44].

Theorem 3 (Wigner formula) *Let $A_1, \ldots, A_n$ be observables of a system* **S** *in a state ρ at time 0. If one carries out projective measurements of observables $A_1, \ldots, A_n$ at times $(0 <)t_1 < \cdots < t_n$ and otherwise leaves the system* **S** *isolated with the Hamiltonian H, then the joint probability distribution of the outcomes $\mathbf{x}_1, \ldots, \mathbf{x}_n$ of those measurements is given by*

$$\Pr\{\mathbf{x}_1 = x_1, \ldots, \mathbf{x}_n = x_n \| \rho\} = \mathrm{Tr}[P^{A_n}(x_n) \cdots U(t_2 - t_1) P^{A_1}(x_1) U(t_1) \rho \times U(t_1)^\dagger P^{A_1}(x_1) U(t_2 - t_1)^\dagger \cdots P^{A_n}(x_n)], \tag{31}$$

where $U(t) = e^{-iHt/\hbar}$.

3.9 Abandoning the Repeatability Hypothesis

The repeatability hypothesis applies only to a restricted class of measurements and does not generally characterize the state changes caused by quantum measurements. In fact, there exist commonly used measurements of discrete observables, such as photon counting, that do not satisfy the repeatability hypothesis [8]. Moreover, it has been shown that the repeatability hypothesis cannot be generalized to continuous observables in the standard formulation of quantum mechanics [18, 20, 23, 42]. In 1970, Davies and Lewis [3] proposed abandoning the repeatability hypothesis and introduced a new mathematical framework to treat all the physically realizable quantum measurements.

> One of the crucial notions is that of repeatability which we show is implicitly assumed in most of the axiomatic treatments of quantum mechanics, but whose abandonment leads to a much more flexible approach to measurement theory [3, p. 239].

The proposal of Davies and Lewis [3] can be stated as follows.

(DL) The Davies-Lewis thesis. *For every measuring apparatus* $\mathbf{A}(\mathbf{x})$ *with output variable* $\mathbf{x}$ *there exists a unique DL instrument* $\mathscr{I}$ *satisfying*

$$\Pr\{\mathbf{x} = x \| \rho\} = \mathrm{Tr}[\mathscr{I}(x)\rho], \tag{32}$$

$$\rho \to \rho_{\{\mathbf{x}=x\}} = \frac{\mathscr{I}(x)\rho}{\mathrm{Tr}[\mathscr{I}(x)\rho]}. \tag{33}$$

We have previously shown that under Axiom M1, the Davies-Lewis thesis is equivalent to Axiom M2 (the mixing law of joint output probability).

3.10 Complete Positivity

Is every Davies-Lewis instrument physically relevant? We shall show that this is not the case. In physics various phenomena can be described by mathematical models, and even in a single physical theory a single phenomenon can be modeled by various different mathematical models. Nevertheless, we should have consistency relations among all the models describing a single physical phenomenon such as invariance under the change of coordinate systems. In quantum measurement theory, a single measuring apparatus can have different models even with a fixed coordinate system, according to the arbitrariness of the spacial boundary of the measured object. As described by Axiom Q4, the observable A in a system $\mathbf{S}$ can have a different mathematical representative $A \otimes I$ in a larger system $\mathbf{S}+\mathbf{S}'$. Thus, an apparatus measuring the observable A is accompanied by another model describing it as an apparatus measuring the observable $A \otimes I$. Namely, any model of an apparatus measuring the system $\mathbf{S}$ is always accompanied by the model for an apparatus measuring the system $\mathbf{S}+\mathbf{S}'$. It is interesting that this rather obvious fact leads to an important common property of measuring apparatuses.

Axiom M3 (Extendability). *For any apparatus $\mathbf{A}(\mathbf{x})$ measuring a system $\mathbf{S}$ and any quantum system $\mathbf{S}'$ not interacting with $\mathbf{A}(\mathbf{x})$ nor $\mathbf{S}$, there exists an apparatus $\mathbf{A}(\mathbf{x}')$ measuring system $\mathbf{S}+\mathbf{S}'$ with the following statistical properties:*

$$\Pr\{\mathbf{x}' = x \| \rho \otimes \rho'\} = \Pr\{\mathbf{x} = x \| \rho\}, \tag{34}$$

$$(\rho \otimes \rho')_{\{\mathbf{x}'=x\}} = \rho_{\{\mathbf{x}=x\}} \otimes \rho' \tag{35}$$

for any $x, x' \in \mathbb{R}$, any state ρ of $\mathbf{S}$, and any state ρ' of $\mathbf{S}'$.

The above postulate is justified as follows. Suppose that apparatus $\mathbf{A}(\mathbf{x})$ measures object $\mathbf{S}$. Let $\mathbf{S}'$ be any system, described by a Hilbert space $\mathscr{H}'$, remote from $\mathbf{S}$ and $\mathbf{A}(\mathbf{x})$. Naturally, the apparatus $\mathbf{A}(\mathbf{x})$ makes no measurement on $\mathbf{S}'$ but then the same apparatus can be formally described as an apparatus $\mathbf{A}(\mathbf{x}')$ measuring the system $\mathbf{S}+\mathbf{S}'$ with the following statistical properties:

$$\Pr\{\mathbf{x}' = x \| \rho \otimes \rho'\} = \Pr\{\mathbf{x} = x \| \rho\}, \tag{36}$$

$$(\rho \otimes \rho')_{\{\mathbf{x}'=x\}} = \rho_{\{\mathbf{x}=x\}} \otimes \rho' \tag{37}$$

for any real number x, any state ρ of $\mathbf{S}$ and any state ρ' of $\mathbf{S}'$. Let $\mathscr{I}'$ be the instrument of the apparatus $\mathbf{A}(\mathbf{x}')$. Then, we have

$$\begin{aligned}\mathscr{I}'(x)(\rho \otimes \rho') &= \Pr\{\mathbf{x}' = x \| \rho \otimes \rho'\}(\rho \otimes \rho')_{\{\mathbf{x}'=x\}}\\ &= \Pr\{\mathbf{x} = x \| \rho\}\rho_{\{\mathbf{x}=x\}} \otimes \rho'\\ &= [\mathscr{I}(x)\rho] \otimes \rho'.\end{aligned} \tag{38}$$

It follows that the operation $\mathscr{I}'(x)$ of the extended apparatus $\mathbf{A}(\mathbf{x}')$ given $\mathbf{x}' = x$ is represented by $\mathscr{I}'(x) = \mathscr{I}(x) \otimes \mathrm{id}$, where id stands for the identity map on $\mathscr{L}(\mathscr{H}')$.

Let $\mathscr{H}'$ be a finite dimensional Hilbert space. Any linear transformation T on $\mathscr{L}(\mathscr{H})$ can be extended naturally to the linear transformation $T \otimes \mathrm{id}_{\mathscr{H}'}$ on $\mathscr{L}(\mathscr{H} \otimes \mathscr{H}') = \mathscr{L}(\mathscr{H}) \otimes \mathscr{L}(\mathscr{H}')$ by

$$(T \otimes \mathrm{id})(\sum_j \rho_j \otimes \rho'_j) = \sum_j T(\rho_j) \otimes \rho'_j \tag{39}$$

for any $\rho_j \in \mathcal{L}(\mathcal{H})$ and $\rho'_j \in \mathcal{L}(\mathcal{H}')$. Then, T is called *completely positive (CP)*, if $T \otimes \mathrm{id}$ maps positive operators in $\mathcal{L}(\mathcal{H} \otimes \mathcal{H}')$ to positive operators in $\mathcal{L}(\mathcal{H} \otimes \mathcal{H}')$ for any $\mathcal{H}'$. A DL instrument $\mathcal{I}$ is called a *completely positive (CP) instrument*, if the operation $\mathcal{I}(x)$ is CP for every $x \in \mathbb{R}$.

Then, from the positivity of the operation $\mathcal{I}'(x)$, the complete positivity of the original operation $\mathcal{I}(x)$ follows. Thus, Axiom M3 (Extendability) is equivalent to the following Axiom M3' (Complete positivity).

Axiom M3' (Complete positivity). *The instrument of every apparatus should be a CP instrument.*

We have posed two plausible requirements for the measurement statistics to be satisfied by any apparatus, the mixing law of joint output probability distributions $\Pr\{\mathbf{x} = x, \mathbf{y} = y \| \rho\}$ and the extendability of measurement statistics, as a set of necessary conditions for every apparatus to satisfy. Under these conditions, we have shown that every apparatus corresponds uniquely to a CP instrument that determines the output probability distributions and the quantum state reduction caused by the apparatus. Thus, the problem of determining physically possible output probability distributions and quantum state reductions is reduced to the problem as to which CP instrument corresponds to a physically realizable apparatus. This problem will be discussed in the next section and it will be shown that every CP instruments corresponds to a physically realizable apparatus.

Now we note that there indeed exists a DL instrument that is not a CP instrument. The transpose operation of matrices in a fix basis is a typical example of a positive linear map which is not CP [15]. Let T be a transpose operation on $\mathcal{L}(\mathcal{H})$ for $\mathcal{H}$, and let $\mu(x)$ be any probability distribution supported in a finite subset of $\mathbb{R}$, i.e., there exists a finite subset $S \subseteq \mathbb{R}$ such that $\mu(x) = 0$ if $x \notin S$. Then, the relation

$$\mathcal{I}(x)\rho = \mu(x)T(\rho) \tag{40}$$

for any $x \in \mathbb{R}$ and any operator ρ defines a DL instrument. However, since T is not CP, the operation $\mathcal{I}(x)$ is not CP, so the $\mathcal{I}$ is not a CP instrument. The extendability postulate implies that there is no physically realizable apparatus corresponding to the above instrument.

4 Measuring Processes

In this section, we discuss measuring processes. We introduce indirect measurement models, a class of universal models for measuring processes, carried out by physically realizable measuring apparatuses. We analyze them according to quantum mechanics, and show that every indirect measurement models uniquely determines the instrument of the apparatus that is a completely positive instrument, and conversely that every completely positive instrument is described by an indirect measurement model of an apparatus. Since any apparatus described by an indirect measurement model is considered physically realizable, in principle, we conclude that an apparatus is physically realizable if and only if its statistical properties are described by a CP instrument.

4.1 Indirect Measurement Models

Let $\mathbf{A}(\mathbf{x})$ be a measuring apparatus with the macroscopic output variable $\mathbf{x}$ to measure the object $\mathbf{S}$. The measuring interaction turns on at time t, the *time of measurement*, and turns off at time $t + \Delta t$ between object $\mathbf{S}$ and apparatus $\mathbf{A}(\mathbf{x})$. We assume that *the object and the apparatus do not interact each other before t nor after t* $+ \Delta t$ *and that the composite system* $\mathbf{S} + \mathbf{A}(\mathbf{x})$ *is isolated in the time interval* $(t, t + \Delta t)$. The *probe* $\mathbf{P}$ is defined to be the minimal part of apparatus $\mathbf{A}(\mathbf{x})$ such that the composite system $\mathbf{S} + \mathbf{P}$ is isolated in the time interval $(t, t + \Delta t)$. By minimality, we naturally assume that probe $\mathbf{P}$ is a quantum system represented by a Hilbert space $\mathcal{K}$. Denote by U the unitary operator on $\mathcal{H} \otimes \mathcal{K}$ representing the time evolution of $\mathbf{S} + \mathbf{P}$ for the time interval $(t, t + \Delta t)$.

At the time of measurement the object is supposed to be in an arbitrary input state ρ and the probe is supposed to be prepared in a fixed state σ. Thus, the composite system $\mathbf{S} + \mathbf{P}$ is in the state $\rho \otimes \sigma$ at time t and in the state $U(\rho \otimes \sigma)U^\dagger$ at time $t + \Delta t$. Just after the measuring interaction, the object is separated from the apparatus, and the probe is subjected to a local interaction with the subsequent stages of the apparatus. The last process is assumed to measure an observable M, called the *meter observable*, of the probe, without further interacting with the object $\mathbf{S}$, and the output is represented by the value of the output variable $\mathbf{x}$.

Any physically realizable apparatus $\mathbf{A}(\mathbf{x})$ can be modeled as above by a quadruple $(\mathcal{K}, \sigma, U, M)$, called an *indirect measurement model*, of a Hilbert space $\mathcal{K}$, a density operator σ in $\mathcal{K}$, a unitary operator U on $\mathcal{K} \otimes \mathcal{H}$, and a self-adjoint operator M on $\mathcal{K}$, where $\mathcal{K}$ represents the state space of the probe, σ the preparation of the probe, U the interaction between the object and the probe, and M the meter observable to be detected. An indirect measurement model $(\mathcal{K}, \sigma, U, M)$ is called *pure*, if σ is a pure state; we shall write $(\mathcal{K}, \sigma, U, M) = (\mathcal{K}, \xi, U, M)$, if $\sigma = |\xi\rangle\langle\xi|$.

4.2 Output Probability Distributions

Let $\mathbf{A}(\mathbf{x})$ be an apparatus with indirect measurement model $(\mathcal{K}, \sigma, U, M)$. Since the outcome of this measurement is obtained by the measurement of the meter observable M at time $t + \Delta t$, by the BSF (1) for observable M on input state $U(\rho \otimes \sigma)U^\dagger$ the output probability distribution of $\mathbf{A}(\mathbf{x})$ is determined by

$$\Pr\{\mathbf{x} = x \| \rho\} = \mathrm{Tr}\{[I \otimes P^M(x)]U(\rho \otimes \sigma)U^\dagger\}. \tag{41}$$

By linearity of operators and the trace, it is easy to check that the output probability distribution of $\mathbf{A}(\mathbf{x})$ satisfies the mixing law of output probability. Thus, by Theorem 1 there exists the POVM Π of $\mathbf{A}(\mathbf{x})$. To determine Π, using the partial trace operation $\mathrm{Tr}_{\mathscr{K}}$ over $\mathscr{K}$ we rewrite Eq. (41) as

$$\Pr\{\mathbf{x} = x \| \rho\} = \mathrm{Tr}[\mathrm{Tr}_{\mathscr{K}}\{U^\dagger[I \otimes P^M(x)]U(I \otimes \sigma)\}\rho]. \tag{42}$$

Since ρ is arbitrary, comparing Eqs. (6) and (42), POVM of $\mathbf{A}(\mathbf{x})$ is determined as

$$\Pi(x) = \mathrm{Tr}_{\mathscr{K}}\{U^\dagger[I \otimes P^M(x)]U(I \otimes \sigma)\} \tag{43}$$

for any $x \in \mathbb{R}$.

4.3 Quantum State Reductions

Since the composite system $\mathbf{S} + \mathbf{P}$ is in the state $U(\rho \otimes \sigma)U^\dagger$ at time $t + \Delta t$, from Axiom Q4 it is standard that the object state at the time $t + \Delta t$ is obtained by tracing out the probe part of that state. Thus, the nonselective state change is determined by

$$\rho \mapsto \rho' = \mathrm{Tr}_{\mathscr{K}}[U(\rho \otimes \sigma)U^\dagger]. \tag{44}$$

In order to determine the quantum state reduction caused by apparatus $\mathbf{A}(\mathbf{x})$, suppose that at time $t + \Delta t$ the observer were to locally measure an arbitrary observable B of the same object $\mathbf{S}$. Let $\mathbf{A}(\mathbf{y})$ be a B-measuring apparatus with output variable $\mathbf{y}$. Since both the M measurement on $\mathbf{P}$ and the B measurement on $\mathbf{S}$ at time $t + \Delta t$ are local, the joint probability distribution of their outcomes satisfies the joint probability formula for the simultaneous measurement of $I \otimes M$ and $B \otimes I$ in the state $U(\rho \otimes \sigma)U^\dagger$ [33].

It follows that the joint output probability distribution of $\mathbf{A}(\mathbf{x})$ and $\mathbf{A}(\mathbf{y})$ is given by

$$\Pr\{\mathbf{x} = x, \mathbf{y} = y \| \rho\} = \mathrm{Tr}\{[P^B(y) \otimes P^M(x)]U(\rho \otimes \sigma)U^\dagger\}. \tag{45}$$

Thus, using the partial trace $\mathrm{Tr}_{\mathscr{K}}$ we have

$$\Pr\{\mathbf{x} = x, \mathbf{y} = y \| \rho\} = \mathrm{Tr}[P^B(y)\mathrm{Tr}_{\mathscr{K}}\{[I \otimes P^M(x)]U(\rho \otimes \sigma)U^\dagger\}]. \tag{46}$$

On the other hand, from Eq. (9) the same joint output probability distribution can be represented by

$$\begin{aligned}\Pr\{\mathbf{x} = x, \mathbf{y} = y \| \rho\} &= \mathrm{Tr}[P^B(y)\rho_{\{\mathbf{x}=x\}}]\Pr\{\mathbf{x} = x \| \rho\} \\ &= \mathrm{Tr}[P^B(y)\Pr\{\mathbf{x} = x \| \rho\}\rho_{\{\mathbf{x}=x\}}].\end{aligned} \tag{47}$$

Since B is chosen arbitrarily, comparing Eqs. (46) and (47), we have

$$\Pr\{\mathbf{x} = x \| \rho\}\rho_{\{\mathbf{x}=x\}} = \mathrm{Tr}_{\mathscr{K}}\{[I \otimes P^M(x)]U(\rho \otimes \sigma)U^\dagger\}. \tag{48}$$

From Eq. (48), the state $\rho_{\{\mathbf{x}=x\}}$ is uniquely determined as

$$\rho_{\{\mathbf{x}=x\}} = \frac{\mathrm{Tr}_{\mathscr{K}}\{[I \otimes P^M(x)]U(\rho \otimes \sigma)U^\dagger\}}{\mathrm{Tr}\{[I \otimes P^M(x)]U(\rho \otimes \sigma)U^\dagger\}}. \tag{49}$$

By Eq. (48) the instrument $\mathscr{I}$ of the apparatus $\mathbf{A}(\mathbf{x})$ is determined by

$$\mathscr{I}(x)\rho = \mathrm{Tr}_{\mathscr{K}}\{[I \otimes P^M(x)]U(\rho \otimes \sigma)U^\dagger\} \tag{50}$$

for any $x \in \mathbb{R}$ and any state ρ. From the above relation, it is easy to see that $\mathscr{I}(x)$ satisfies the complete positivity; as an alternative characterization, it is well-known that a linear transformation T on $\mathscr{L}(\mathscr{H})$ is completely positive if and only if

$$\sum_{ij}(\xi_i, T(\rho_i^\dagger\rho_j)\xi_j) \geq 0 \tag{51}$$

for any finite sequences $\xi_1, \ldots, \xi_n \in \mathscr{H}$ and $\rho_1, \ldots, \rho_n \in \mathscr{L}(\mathscr{H})$ [18]. In fact, we have

$$\begin{aligned}\sum_{ij}(\xi_i, \mathscr{I}(x)(\rho_i^\dagger\rho_j)\xi_j) &= \sum_{ij}\mathrm{Tr}[\mathscr{I}(x)(\rho_i^\dagger\rho_j)\,|\xi_j\rangle\langle\xi_i|] \\ &= \sum_{ij}\mathrm{Tr}\{[|\xi_j\rangle\langle\xi_i| \otimes P^M(x)]U(\rho_i^\dagger\rho_j \otimes \sigma)U^\dagger\} \\ &= \mathrm{Tr}[X^\dagger X] \geq 0,\end{aligned}$$

where

$$X = \sum_j U(\rho_j \otimes \sqrt{\sigma})U^\dagger\,|\xi_j\rangle\langle\phi| \otimes P^M(x)$$

for an arbitrary unit vector ϕ.

Thus, we conclude that *the instrument of any apparatus with indirect measurement model* $(\mathscr{K}, \sigma, U, M)$ *is a CP instrument.*

The converse of this assertion was proven in [18], and hence we conclude this section by the following theorem.

Theorem 4 (Realization theorem) *The instrument of any apparatus with an indirect measurement model is a CP instrument, and conversely every CP instrument is obtained in this way with a pure indirect measurement model.*

According to the above theorem, we conclude that an apparatus satisfying Axiom M1 is physically realizable if and only if it satisfied Axiom M2 (the mixing law of joint output probability) and Axiom M3 (the extendability).

5 General Measurement Axiom

For quantum systems with finite dimensional state spaces, we can now complete von Neumann's axiomatization of quantum mechanics by augmenting it by the general measurement axiom that describes all the physically realizable measurements consistent with the other axioms of quantum mechanics.

Axiom Q5 (General measurement axiom). *Every physically realizable apparatus* $\mathbf{A}(\mathbf{x})$ *for the system* $\mathbf{S}$ *with the state space* $\mathscr{H}$ *uniquely corresponds to a completely positive instrument* $\mathscr{I}$ *for* $\mathscr{H}$ *such that the statistical properties of* $\mathbf{A}(\mathbf{x})$ *are determined by*

$$\mathscr{I}(x)\rho = \Pr\{\mathbf{x} = x \| \rho\}\rho_{\{\mathbf{x}=x\}}, \tag{52}$$

or equivalently

$$\Pr\{\mathbf{x} = x \| \rho\} = \mathrm{Tr}[\mathscr{I}(x)\rho], \tag{53}$$

$$\rho_{\{\mathbf{x}=x\}} = \frac{\mathscr{I}(x)\rho}{\mathrm{Tr}[\mathscr{I}(x)\rho]}, \quad \text{if } \mathrm{Tr}[\mathscr{I}(x)\rho] > 0 \tag{54}$$

for all $x \in \mathbb{R}$ *and* $\rho \in \mathscr{S}(\mathscr{H})$. *Conversely, every completely positive instrument* $\mathscr{I}$ *for* $\mathscr{H}$ *has at least one physically realizable apparatus* $\mathbf{A}(\mathbf{x})$ *with the above statistical properties.*

According to the Kraus theorem [9], for every completely positive map T there exists a family of operators $\{M_j\}$ on $\mathscr{H}$ satisfying

$$T\rho = \sum_j M_j \rho M_j^\dagger \tag{55}$$

for every $\rho \in \mathscr{L}(\mathscr{H})$, and

$$T^* I = \sum_j M_j^\dagger M_j. \tag{56}$$

It can be easily seen that every family of operators $\{M_j\}$ defines a completely positive map T satisfying Eqs. (55), (56). It follows that for every completely positive instrument $\mathscr{I}$ for $\mathscr{H}$ there exits a family of operators $\{M_{xj}\}$ on $\mathscr{H}$, called the *measurement operators* for $\mathscr{I}$, satisfying

$$\mathscr{I}(x)\rho = \sum_j M_{xj}\rho M_{xj}^{\dagger} \tag{57}$$

for every $\rho \in \mathscr{L}(\mathscr{H})$, and

$$\Pi(x) = \sum_j M_{xj}^{\dagger} M_{xj}, \tag{58}$$

where Π is the POVM associated with the instrument $\mathscr{I}$ given by Eq. (18). i.e., $\Pi(x) = \mathscr{I}(x)^* I$ for all $x \in \mathbb{R}$, and conversely that every family of operators $\{M_{xj}\}$ satisfying

$$\sum_{xj} M_{xj}^{\dagger} M_{xj} = I \tag{59}$$

defines a completely positive instrument satisfying Eqs. (57), (58). We call any family of operators $\{M_{xj}\}$ a family of *measurement operators* if Eq. (59) is satisfied. Using measurement operators the General measurement axiom can be stated as follows.

Axiom Q5' (General measurement axiom). *Every physically realizable apparatus* $\mathbf{A}(\mathbf{x})$ *for the system* $\mathbf{S}$ *with the state space* $\mathscr{H}$ *has a family of measurement operators* $\{M_{xj}\}$ *such that the statistical properties of* $\mathbf{A}(\mathbf{x})$ *are determined by*

$$\Pr\{\mathbf{x} = x\|\rho\} = \sum_j \mathrm{Tr}[M_{xj}^{\dagger} M_{xj}\rho], \tag{60}$$

$$\rho_{\{\mathbf{x}=x\}} = \frac{\sum_j M_{xj}\rho M_{xj}^{\dagger}}{\sum_j \mathrm{Tr}[M_{xj}^{\dagger} M_{xj}\rho]}, \quad \text{if } \Pr\{\mathbf{x} = x\|\rho\} > 0 \tag{61}$$

for all $x \in \mathbb{R}$ *and* $\rho \in \mathscr{S}(\mathscr{H})$. *Conversely, every family of measurement operators* $\{M_{xj}\}$ *has at least one physically realizable apparatus* $\mathbf{A}(\mathbf{x})$ *with the above statistical properties.*

By the General measurement axiom, we can generalize the Wigner formula to arbitrary sequence of measurements, and determine the joint probability distribution of the outcomes of any sequence of measurements using physically realizable apparatuses.

Theorem 5 (Generalized Wigner formula) *Let* $\mathscr{I}_1, \ldots, \mathscr{I}_n$ *be completely positive instruments for the system with the state space* $\mathscr{H}$ *in a state* ρ *at time 0. If one carries out measurements described by* $\mathscr{I}_1, \ldots, \mathscr{I}_n$ *at times* $(0 <)t_1 < \cdots < t_n$ *and otherwise leaves the system* $\mathbf{S}$ *isolated with the Hamiltonian* H, *then the joint probability distribution of the outcomes* $\mathbf{x}_1, \ldots, \mathbf{x}_n$ *of those measurements is given by*

$$\Pr\{\mathbf{x}_1 = x_1, \mathbf{x}_2 = x_2, \ldots, \mathbf{x}_n = x_n \| \rho\}$$
$$= \mathrm{Tr}[\mathscr{I}_n(x_n)\alpha(t_n - t_{n-1}) \cdots \mathscr{I}_2(x_2)\alpha(t_2 - t_1)\mathscr{I}_1(x_1)\alpha(t_1)\rho], \quad (62)$$

for any $x_1, x_2, \ldots, x_n \in \mathbb{R}$ *and* $\rho \in \mathscr{S}(\mathscr{H})$*, where* α *is defined by* $\alpha(t)\rho = e^{-iHt/\hbar}\rho e^{iHt/\hbar}$ *for all* $t \in \mathbb{R}$ *and* $\rho \in \mathscr{S}(\mathscr{H})$.

Note that if the completely positive instruments $\mathscr{I}_1, \ldots, \mathscr{I}_n$ in the Generalized Wigner formula have the families of measurement operators $M^{(1)}_{x_1 j^{(1)}}, \ldots, M^{(n)}_{x_n j^{(n)}}$ the Generalized Wigner formula Eq. (62) can be rewritten as

$$\Pr\{\mathbf{x}_1 = x_1, \mathbf{x}_2 = x_2, \ldots, \mathbf{x}_n = x_n \| \rho\}$$
$$= \sum_{j^{(1)},\ldots,j^{(n)}} \mathrm{Tr}[M^{(n)}_{x_n j^{(n)}} e^{-iH(t_n - t_{n-1})/\hbar} \cdots M^{(2)}_{x_2 j^{(2)}} e^{-iH(t_2-t_1)/\hbar} M^{(1)}_{x_1 j^{(1)}} e^{-iHt_1/\hbar} \rho$$
$$e^{iHt_1/\hbar}(M^{(1)}_{x_1 j^{(1)}})^\dagger e^{iH(t_2-t_1)/\hbar}(M^{(2)}_{x_2 j^{(2)}})^\dagger \cdots e^{iH(t_n-t_{n-1})/\hbar}(M^{(n)}_{x_n j^{(n)}})^\dagger]. \quad (63)$$

Foundations of quantum measurement theory based on the notion of completely positive instruments and indirect measurement models have been developed in [17–37].

References

1. W. Arveson, Analyticity in operator algebras. Am. J. Math. **89**, 578–642 (1967)
2. E.B. Davies, *Quantum Theory of Open Systems* (Academic, London, 1976)
3. E.B. Davies, J.T. Lewis, An operational approach to quantum probability. Commun. Math. Phys. **17**, 239–260 (1970)
4. R. Haag, D. Kastler, An algebraic approach to quantum field theory. J. Math. Phys. **5**, 848–861 (1964)
5. P.R. Halmos, *Introduction to Hilbert Space and the Theory of Spectral Multiplicity* (Chelsea, New York, 1951)
6. C.W. Helstrom, *Quantum Detection and Estimation Theory* (Academic, New York, 1976)
7. D. Hilbert, *Grundlagen der Geometrie* (B.G. Teubner, Leipzig, 1903)
8. N. Imoto, M. Ueda, T. Ogawa, Microscopic theory of the continuous measurement of photon number. Phys. Rev. A **41**, 4127–4130 (1990)
9. K. Kraus, General state changes in quantum theory. Ann. Phys. (N.Y.) **64**, 311–335 (1971)
10. G. Lüders, Über die Zustandsänderung durch den Meßprozeß. Ann. Phys. (Leipzig) **443**, 322–328 (1950)
11. G. Ludwig, Attempt of an axiomatic foundation of quantum mechanics and more general theories II. Commun. Math. Phys. **4**, 331–348 (1967)
12. G. Ludwig, Attempt of an axiomatic foundation of quantum mechanics and more general theories III. Commun. Math. Phys. **9**, 1–12 (1968)
13. M. Nakamura, H. Umegaki, On von Neumann's theory of measurements in quantum statistics. Math. Jpn. **7**, 151–157 (1962)
14. J. von Neumann, *Mathematical Foundations of Quantum Mechanics* (Princeton UP, Princeton, 1955). [Originally published: *Mathematische Grundlagen der Quantenmechanik* (Springer, Berlin, 1932)]
15. M.A. Nielsen, I.L. Chuang, *Quantum Computation and Quantum Information* (Cambridge UP, Cambridge, 2000)

16. M. Ozawa, Optimal measurements for general quantum systems. Rep. Math. Phys. **18**, 11–28 (1980)
17. M. Ozawa, Conditional expectation and repeated measurements of continuous quantum observables, in *Probability Theory and Mathematical Statistics*, ed. by K. Itô, J.V. Prohorov, Lecture Notes in Mathematics, vol. 1021 (Springer, Berlin, 1983), pp. 518–525
18. M. Ozawa, Quantum measuring processes of continuous observables. J. Math. Phys. **25**, 79–87 (1984)
19. M. Ozawa, Concepts of conditional expectations in quantum theory. J. Math. Phys. **26**, 1948–1955 (1985)
20. M. Ozawa, Conditional probability and a posteriori states in quantum mechanics. Publ. Res. Inst. Math. Sci., Kyoto Univ. **21**, 279–295 (1985)
21. M. Ozawa, On information gain by quantum measurements of continuous observables. J. Math. Phys. **27**, 759–763 (1986)
22. M. Ozawa, Measurement breaking the standard quantum limit for free-mass position. Phys. Rev. Lett. **60**, 385–388 (1988)
23. M. Ozawa, Measuring processes and repeatability hypothesis, in *Probability Theory and Mathematical Statistics: Kyoto, Japan*, ed. by S. Watanabe, Y.V. Prohorov, July 8–14 (1986). Lecture Notes in Mathematics, vol. 1299 (Springer, Berlin, 1988), pp. 412–421. arXiv:quant-ph/1809.00400
24. M. Ozawa, Realization of measurement and the standard quantum limit, in *Squeezed and Nonclassical Light*, ed. by P. Tombesi, E.R. Pike (Plenum, New York, 1989), pp. 263–286. arXiv:quant-ph/1505.01083
25. M. Ozawa, Quantum mechanical models of position measurements. Phys. Rev. A **41**, 1735–1737 (1990)
26. M. Ozawa, Quantum limits of measurements and uncertainty principle, in *Quantum Aspects of Optical Communications*. ed. by C. Bendjaballah, O. Hirota, S. Reynaud (Springer, Berlin, 1991), pp.1–17
27. M. Ozawa, Canonical approximate quantum measurements. J. Math. Phys. **34**, 5596–5624 (1993)
28. M. Ozawa, Mathematical characterizations of measurement statistics, in *Quantum Communications and Measurement*. ed. by V.P. Belavkin, O. Hirota, R.L. Hudson (Plenum, New York, 1995), pp.109–117
29. M. Ozawa, An operational approach to quantum state reduction. Ann. Phys. (N.Y.) **259**(1), 121–137 (1997)
30. M. Ozawa, Quantum state reduction and the quantum Bayes principle, in *Quantum Communication, Computing, and Measurement*, ed. by O. Hirota, A.S. Holevo, C.M. Caves (Plenum, New York, 1997), pp. 233–241. arXiv:quant-ph/9705030
31. M. Ozawa, Quantum state reduction: an operational approach. Fortschr. Phys. **46**, 615–625 (1998)
32. M. Ozawa, Measurements of nondegenerate discrete observables. Phys. Rev. A **62**, 062101 (2000)
33. M. Ozawa, Operations, disturbance, and simultaneous measurability. Phys. Rev. A **63**, 032109 (2001)
34. M. Ozawa, Uncertainty relations for noise and disturbance in generalized quantum measurements. Ann. Phys. (N.Y.) **311**, 350–416 (2004)
35. M. Ozawa, Perfect correlations between noncommuting observables. Phys. Lett. A **335**, 11–19 (2005)
36. M. Ozawa, Quantum perfect correlations. Ann. Phys. (N.Y.) **321**, 744–769 (2006)
37. M. Ozawa, Soundness and completeness of quantum root-mean-square errors. npj Quantum Inf. **5**, 1 (2019)
38. R. Schatten, *Norm Ideals of Completely Continuous Operators* (Springer, New York, 1960)
39. E. Schrödinger, Die gegenwärtige Situation in der Quantenmechanik. Naturwissenshaften **23**, 807–812, 823–828, 844–849 (1935). [English translation by J.D. Trimmer, Proc. Am. Philos. Soc. **124**, 323–338 (1980)]

40. J. Schwinger, The algebra of microscopic measurement. Proc. Natl. Acad. Sci. U.S. **45**, 1542–1554 (1959)
41. J. Schwinger, The geometry of quantum states. Proc. Natl. Acad. Sci. U.S. **46**, 257–265 (1960)
42. M.D. Srinivas, Collapse postulate for observables with continuous spectra. Commun. Math. Phys. **71**, 131–158 (1980)
43. H. Umegaki, Conditional expectation in an operator algebra. Tôhoku Math. J. **2**(6), 177–181 (1954)
44. E.P. Wigner, The problem of measurement. Am. J. Phys. **31**, 6–15 (1963). [in *Quantum Theory and Measurement*, ed. by J.A. Wheeler, W.H. Zurek (Princeton UP, Princeton, New Jersey, 1983), pp. 324–341]
45. H.P. Yuen, Characterization and realization of general quantum measurements, in *Proceedings of the 2nd International Symposium Foundations of Quantum Mechanics*, ed. by M. Namiki et al. (The Physical Society of Japan, Tokyo, 1987), pp. 360–363

Masanao Ozawa received his doctoral degree in science from Tokyo Institute of Technology in 1979, where he studied mathematical logic and mathematical physics. In his study in mathematical logic, he developed Boolean-valued analysis, by which in 1983 he settled Kaplansky's conjecture in operator algebras using Cohen's forcing method for cardinal collapsing in set theory. His study in mathematical physics established quantum measurement theory based on the notion of quantum instruments. In 1988, he settled a controversy on gravitational wave detection, by establishing the supremacy of interferometer-type detectors over resonator-type ones. For this, he constructed a measurement model that breaks the alleged sensitivity limit for interferometer-type detectors derived from Heisenberg's uncertainty principle. In 2003, this led him to reformulate Heisenberg's uncertainty principle with a universally valid error-disturbance relation, which was experimentally confirmed in 2012. He has been serving as program committee member in international conferences and editorial board member for international journals in these research areas. He has held visiting positions at Harvard University, Northwestern University, and University of Pavia. He is currently a Professor of Mathematical Science and Artificial Intelligence at Chubu University, and a Professor Emeritus at Nagoya University. He was awarded Mathematical Society of Japan Prize in 2008, International Quantum Communication Award in 2010, and Medal with Purple Ribbon by the Cabinet Office of Japan in 2015.

Probability Distributions as Alternatives of the Density Operator and Wave Function for Describing Quantum States

Margarita A. Man'ko and Vladimir I. Man'ko

Abstract Our goal is to discuss Born's rule and the probability representation of quantum mechanics. We review the construction of tomographic-probability distributions for quantum systems with continuous variables like harmonic oscillators and photons, and present the method of quantizer–dequantizer operators to map the density operators and physical observables onto functions, including probability distributions. For systems with discrete variables, like spin systems, the probability distributions identified with qubit states are studied, in view of Born's rule. The method of quantizer–dequantizer operators is reviewed. The superposition principle of quantum states is discussed in the formalism of nonlinear addition rule of states' probability distributions. Examples of qubits and Schrödinger even and odd cat states are presented. The coherent and Fock states, as well as even and odd coherent states of free particles and their evolution are studied; the tomograms of these states are given in explicit form.

1 Introduction

Canonical formulations of quantum mechanics are based on the concept of the state of a physical system different from that in classical physics. For example, the concept of state of a classical particle is associated with the notion of the particle position q and momentum p at a given time moment t. In classical statistical mechanics, taking into account fluctuations of the particle position and momentum due to the interaction of the particle with the environment, the concept of state is associated with the probability density $f(q, p, t)$. Thus, the basic notion of probability distributions plays an important role in describing classical system states. For quantum systems like the

M. A. Man'ko (✉) · V. I. Man'ko
Lebedev Physical Institute, Leninskii Prospect 53, Moscow 119991, Russia
e-mail: mankoma@lebedev.ru

V. I. Man'ko
e-mail: manko.vi@phystech.edu

V. I. Man'ko
Moscow Institute of Physics and Technology, Dolgoprudnyi, Moscow 141700, Russia

A. Plotnitsky and E. Haven (eds.), *The Quantum-Like Revolution*,
https://doi.org/10.1007/978-3-031-12986-5_10

hydrogen atom and oscillator, beginning with the introduction of quantum mechanics in the last century, the system states are associated with complex wave functions or vectors in a Hilbert space. This notion is completely different from the notion used in classical mechanics; in view of this fact, many attempts to find the formulation of quantum mechanics with the concept of states similar to classical one were undertaken.

For example, Wigner introduced the Wigner function [1] $W(q, p, t)$, which is a real function depending on the phase-space variables q and p and time t. The Wigner function has some similarity with the probability density $f(q, p, t)$ describing the classical-particle states, but the Wigner function can take negative values, and it is not the conventional probability density used in the probability theory. Nonnegative probability density of quantum system states does not exist, in view of the uncertainty relations of the position and momentum corresponding to the impossibility to simultaneously measure both the position q and momentum p (Heisenberg [2], Robertson [3], and Schrödinger [4]). The other kinds of analogous functions depending on the phase-space coordinates q and p introduced by Glauber [5] and Sudarshan [6], as well as by Husimi [7] and Kano [8], are also the distributions called quasidistributions of such physical observables as the position and momentum, which are not simultaneously measurable. The conventional concept of probability density of two random variables q and p means that, in principle, both variables can be simultaneously measured with an arbitrary accuracy, but the quantum nature of the particles does not permit this simultaneous measurement with an arbitrary accuracy available in classical mechanics.

For random classical fields like the electromagnetic fields, the notion of field states can be associated with functions of different forms; for example, there exist Gaussian forms of the fields. For several field modes, say, two modes, the Gaussian field is described by the dispersion matrix. For quantum states, Landau [9] and von Neumann [10] have shown that, for the states of quantum particle in environment, the notion of state is associated with the density matrix ρ or the density operator $\hat{\rho}$ acting in the Hilbert space, where the Dirac state vectors $|\psi\rangle$ [11] describe the pure states corresponding to the wave functions $\psi(q, t)$ introduced by Schrödinger and obeying the Schrödinger equation. This density matrix is Hermitian, $\rho = \rho^{\dagger}$, with $\mathrm{Tr}\,\rho = 1$ and nonnegative eigenvalues.

The similarity of the quantum density matrix properties to the dispersion matrix determining the properties of classical fields in Gaussian states was discussed by Khrennikov and pre-quantum classical statistical model was studied to clarify the correspondence between classical and quantum mechanics [12, 13]. This similarity provides the possibility to find the relation of classical and quantum properties. Another similarity between quantum and classical properties of physical systems is the interference phenomenon of the electromagnetic fields, as well as other possible fields. This phenomenon also exists in quantum mechanics, where the superposition principle of systems' wave functions is an experimentally proven fact.

The Schrödinger equation for the wave function is a linear equation. A linear superposition of two wave functions is the solution to the Schrödinger equation, which corresponds to the state of the quantum system considered. The density matrix of the system states [9, 10] satisfies two equations, the linear von Neumann equation and the Gorini–Kossakowski–Sudarshan–Lindblad (GKSL) equation [14–16]. The linear mixture of the density matrices, being the sum of the density matrices, is the solution to these equations. The problem of addition of density matrices, which corresponds to the superposition principle of wave functions was discussed in [17]. This problem is related to the problem of addition of probability distributions discussed in [18–20] in connection with the interference problem in quantum mechanics. This nonlinear addition rule for probability distributions was considered in [21–23] in connection with the probability representation of quantum mechanics, where quantum states are identified with fair probability distributions completely determining quantum states either by the wave functions or by the density matrices.

The aim of this paper is to give a review of the probability representation of quantum mechanics discussed in [24, 25]. The construction of the probability representation is based on finding an invertible map of the density operator acting in a Hilbert space on the probability distributions. The map is obtained by applying the formalism of quantizer–dequantizer operators acting in a Hilbert space [26–28]. The mathematical structures of quantum physics can be applied to discuss different phenomena in biology [29] and mathematical finance [30]. In view of this fact, the probability-representation formulation of quantum states can be also applied to other branches of science, where the methods of quantum mechanics are applied, for example, as considered in [31].

This paper is organized as follows.

In Sect. 2, we discuss the probability representation of quantum states providing examples of free particle and harmonic oscillator motions. In addition, we review the integrals of motion and quantum evolution of the wave function in the momentum representation and consider the wave functions of free particle and oscillator. In Sect. 3, we obtain the even and odd coherent states (Schrödinger cat states) of free particles and their wave functions. In Sect. 4, the idea of probability representation and state tomography and the method of quantizer–dequantizer operators are discussed. The example of symplectic tomography is introduced in Sect. 5. We present the probability representation of systems with discrete variables like qubits and spin-1/2 systems in Sect. 6. The superposition principle of two states of qubit and nonlinear addition rule for probability distributions describing the states are discussed in Sect. 7. Finally our conclusions are presented in Sect. 8.

2 Probability Representation of Quantum States on Examples of Free Particle Motion and Harmonic Oscillator

2.1 *Free–Like Quantum Evolution of Wave Function in the Momentum Representation*

To discuss quantum systems in the probability representation, we recall the standard description of states by wave functions of a simple system like a free particle and a harmonic oscillator.

If a quantum system has a Hermitian Hamiltonian depending only on the momentum operator $\hat{H}(\hat{p})$, the solution to the Schrödinger equation for the wave function $\psi(p,t)$ is given in the momentum representation as follows:

$$\psi(p,t) = e^{-it\hat{H}(\hat{p})}\psi_0(p,0). \tag{1}$$

Here, an arbitrary function $\psi_0(p,0)$ is the initial value of the wave function. which can be normalized $\int |\psi_0(p,0)|^2\,dp = 1$. The normalized function $\psi(p,t)$ for $\hat{H}^\dagger(\hat{p}) = \hat{H}(\hat{p})$ differs on $\psi_0(p,0)$ only by the phase factor

$$\psi(p,t) = e^{-itH(p)}\psi(p,0), \tag{2}$$

and the probability distribution $|\psi(p,t)|^2$ at time moment t is the same as at $t=0$,

$$|\psi(p,t)|^2 = |\psi(p,0)|^2. \tag{3}$$

An example of such a Hamiltonian is the Hamiltonian of a free-particle motion. The solution to the Schrödinger equation in the form of Green function $G(p,p',t) = e^{-itH(p)}\delta(p-p')$ coincides with the initial value of the wave function up to the phase factor $e^{-itp^2/2}$. In the position representation, the wave function $\psi(x,t)$ is given by the Fourier transform as follows:

$$\psi(x,t) = \frac{1}{\sqrt{2\pi}}\int \psi(p,0)\,e^{ipx-itH(p)}\,dp. \tag{4}$$

The probability distribution of the momentum $|\psi(p,t)|^2$ does not depend on time but the probability distribution of the position $|\psi(x,t)|^2$ does depend on time.

2.2 Integrals of Motion for Free Particle and Oscillator and Wave Functions of Coherent and Fock States

We start our discussion of the probability representation of quantum states with the example of an oscillator with time-dependent frequency $\omega(t)$. This problem was considered in [32], by using the method of operators–integrals of motion depending on time in the Schrödinger representation. For simplicity, we assume $\hbar = m = \omega(0) = 1$. In the Schrödinger representation, the Hamiltonian of the oscillator is

$$\hat{H}(t) = \frac{\hat{p}^2}{2} + \omega^2(t)\frac{\hat{q}^2}{2}; \tag{5}$$

this problem has time-dependent integrals of motion of the form $\hat{a}(t)$ and $\hat{a}^\dagger(t)$,

$$\hat{a}(t) = \frac{i}{\sqrt{2}}\left(\mathscr{E}(t)\hat{p} - \dot{\mathscr{E}}(t)\hat{q}\right), \qquad \hat{a}^\dagger(t) = -\frac{i}{\sqrt{2}}\left(\dot{\mathscr{E}}^*(t)\hat{p} - \dot{\mathscr{E}}^*(t)\hat{q}\right). \tag{6}$$

Here, the complex function $\mathscr{E}(t)$ satisfies the equation

$$\ddot{\mathscr{E}}(t) + \omega^2(t)\mathscr{E} = 0, \tag{7}$$

which is the equation of motion of classical parametric oscillator with the known integral of motion, which is the Wronskian $W(t)$,

$$W(t) = \mathscr{E}^*(t)\dot{\mathscr{E}}(t) - \dot{\mathscr{E}}^*(t)\mathscr{E}(t) = W(0). \tag{8}$$

If the conditions $\mathscr{E}(0) = 1$ and $\dot{\mathscr{E}}(0) = i$ are valid, one has $W(0) = 2i$ and the commutation relation of operators $\hat{a}(t)$ and $\hat{a}^\dagger(t)$ has the form

$$\left[\hat{a}(t), \hat{a}^\dagger(t)\right] = 1. \tag{9}$$

The integrals of motion (6) satisfy the evolution equation

$$\frac{d}{dt}\hat{a}(t) = \frac{\partial \hat{a}(t)}{\partial t} + i\left[\hat{H}(t), \hat{a}(t)\right] = 0. \tag{10}$$

In the limit cases, i.e., $\omega(t) = 1$ for $t < 0$ and $\omega(t) = 0$ for $t \geq 0$, the function $\mathscr{E}(t) = e^{it}$ for $t < 0$ and $\mathscr{E}(t) = 1 + i\,t$ for $t > 0$. In two other limit cases, namely, $\omega(t) = 1$ for all $-\infty < t < \infty$, the function $\mathscr{E}(t) = e^{it}$, and if $\omega(t) = 0$, the function $\mathscr{E}(t) = 1 + i\,t$ for all $-\infty < t < \infty$. This means that, in the Schrödinger representation, for the stationary harmonic oscillator with $\omega(t) = 1$, one has two integrals of motion, linear in the position and momentum integrals of motion, depending on time; they are

$$\hat{a}(t) = e^{it}\,\frac{\hat{q} + i\,\hat{p}}{\sqrt{2}}, \qquad \hat{a}^\dagger(t) = e^{-it}\,\frac{\hat{q} - i\,\hat{p}}{\sqrt{2}}. \tag{11}$$

For free particle considered as the harmonic oscillator with zero frequency, $\omega = 0$, one has two integrals of motion,

$$\hat{a}(t) = \frac{1}{\sqrt{2}}\left[\hat{q} + \hat{p}(i - t)\right], \qquad \hat{a}^\dagger(t) = \frac{1}{\sqrt{2}}\left[\hat{q} - \hat{p}(i + t)\right]. \tag{12}$$

One can obtain the matrix representation of the integrals of motion in the Fock basis $\mid n\rangle$, where one has $\hat{a}(t=0) = \hat{a}$ and $\hat{a}^\dagger(t=0) = \hat{a}^\dagger$, along with

$$\hat{a}^\dagger\hat{a} \mid n\rangle = n \mid n\rangle,\;\; \hat{a} \mid n\rangle = \sqrt{n} \mid n-1\rangle,\;\; \hat{a}^\dagger \mid n\rangle = \sqrt{n+1} \mid n+1\rangle;\;\; n = 0, 1, 2, \ldots \tag{13}$$

For free particle, one has an explicit form of the matrix of operator $\hat{a}(t)$; it is

$$\hat{a}(t) \mid n\rangle = (1 + i\,t/2)\,\sqrt{n} \mid n-1\rangle - (i\,t/2)\,\sqrt{n+1} \mid n+1\rangle. \tag{14}$$

In the momentum representation, the integral of motion $\hat{a}(t)$ becomes

$$\hat{a}(t) = \frac{i}{\sqrt{2}}\left[\left(p + \frac{\partial}{\partial p}\right) + i\,tp\right]. \tag{15}$$

An analog of the ground state for the free-particle motion, satisfying the Schrödinger equation with the Hamiltonian $\hat{H} = \hat{p}^2/2$ and constraint $\hat{a}(t)\psi_0(p,t) = 0$, i.e.,

$$\frac{i}{\sqrt{2}}\left[\left(p + \frac{\partial}{\partial p}\right) + i\,tp\right]\psi_0(p,t) = 0, \tag{16}$$

is given by the Gaussian wave function, which is the solution of (16)

$$\psi_0(p,t) = \frac{1}{\sqrt{2\pi}}\int e^{-ipx}\psi_0(x,t)\,dx = \pi^{-1/4}e^{-p^2(1+it)/2}. \tag{17}$$

For free particle at time moment $t = 0$, the wave function of the coherent state in the momentum representation $\psi_\alpha(x,0)$, in view of the form of annihilation operator (15), for $t = 0$ is given by the expression for $\psi_\alpha(x,0)$, where the position x is replaced by the momentum p and the complex number α, by the complex number $i\alpha$. Thus, we obtain the wave function of the coherent state in the momentum representation $\psi_\alpha(p,0)$ as follows:

$$\psi_\alpha(p,0) = \pi^{-1/4}e^{-|\alpha|^2/2}\exp\left(-\frac{p^2}{2} + i\sqrt{2}\alpha p + \frac{\alpha^2}{2}\right); \tag{18}$$

analogously, for $\psi_\alpha(p,t)$, we obtain the expression

$$\psi_\alpha(p,t) = e^{-itp^2/2}\psi_\alpha(p,0) = \pi^{-1/4}e^{-|\alpha|^2/2}\exp\left(-\frac{p^2(1+it)}{2} + i\sqrt{2}\alpha p + \frac{\alpha^2}{2}\right). \tag{19}$$

In the momentum representation, the free-particle Fock states $\mid n, t\rangle$, being eigenfunctions of the integral of motion

$$\hat{n}(t) = \hat{a}^\dagger(t)\hat{a}(t) = \frac{1}{2}\left(\hat{q} - \hat{p}t - ip\right)\left(\hat{q} - \hat{p}t + i\hat{p}\right), \tag{20}$$

in view of the evolution operator $\hat{U}(t) = \exp\left(-it\hat{p}^2/2\right)$, are described by the functions $\psi_n(p,t)$ expressed in terms of Hermite polynomials and Gaussian factor as follows:

$$\psi_n(p,t) = \pi^{-1/4}e^{-p^2(1+it)/2}\frac{1}{\sqrt{2^n n!}}H_n(p). \tag{21}$$

Here, $H_n(p)$ is the Hermite polynomial.

The coherent state of the quantum-free-particle-motion state $\psi_\alpha(x,t)$ has the Gaussian form

$$\psi_\alpha(x,t) = \psi_0(x,t)\,e^{-|\alpha|^2/2}\exp\left[\frac{\sqrt{2}\,\alpha x}{1+it} + \frac{\alpha^2}{2}\frac{it-1}{it+1}\right], \tag{22}$$

where $\psi_0(x,t) = \dfrac{\pi^{-1/4}}{\sqrt{1+it}}\exp\left(-\dfrac{x^2}{2(1+it)}\right)$. Function (22) is the eigenfunction of the integral of motion (12), i.e., $\hat{a}(t)\psi_\alpha(x,t) = \alpha\psi_\alpha(x,t)$. Thus, we obtained an explicit expression for the coherent-state wave function of free particle.

3 Even and Odd Coherent States of a Free Particle

The even and odd coherent states of the (photon) oscillator (even and odd Schrödinger cat state) [33], experimentally realized [34], were introduced as symmetric and antisymmetric superpositions of two coherent states,

$$\mid\alpha\rangle_\pm = N_\pm(\alpha)\left(\mid\alpha\rangle \pm \mid -\alpha\rangle\right), \tag{23}$$

where $N_\pm(\alpha) = \left\{\frac{1}{2}\left[1 \pm \exp\left(-2|\alpha|^2\right)\right]\right\}^{-1/2}$. The normalized wave functions of the oscillator states in the momentum representation is

$$\psi_{\alpha_\pm}(p) = \frac{N_\pm(\alpha)}{\pi^{1/4}}\exp\left[\frac{\alpha^2 - |\alpha|^2 - p^2}{2}\right]\left[\exp\left(-i\sqrt{2}\alpha p\right) \pm \exp\left(i\sqrt{2}\alpha p\right)\right]. \tag{24}$$

For free particle, the wave functions of the even and odd Schrödinger cat state are

$$\psi_{\alpha\pm}(p,t) = \frac{N_\pm(\alpha)}{\pi^{1/4}} e^{[(\alpha^2-|\alpha|^2)/2]-p^2(1+it)/2} \left[\exp\left(-i\sqrt{2}\alpha p\right) \pm \exp\left(i\sqrt{2}\alpha p\right)\right]. \tag{25}$$

For the potential with $\omega^2 = 0$, the Schrödinger cat states of the harmonic oscillator become the states described in the momentum representation by the wave functions (25). In the position representation, the wave functions $\psi_{\alpha\pm}(x,t)$ are given by the Fourier transform of (25); they are

$$\psi_{\alpha\pm}(x,t) = \frac{1}{\sqrt{2\pi}} \int e^{ipx} \psi_{\alpha\pm}(p,t)\, dp. \tag{26}$$

The wave functions of even and odd coherent states (26) of free particle in the position representation are given by the relation

$$\psi_{\alpha\pm}(x,t) = \frac{e^{-|\alpha|^2/2}}{2\pi^{1/4}\sqrt{1+it}} \left[\frac{1 \pm \exp\left(-2|\alpha|^2\right)}{2}\right]^{-1/2}$$
$$\times \exp\left(\frac{\alpha^2}{2}\frac{it-1}{it+1} - \frac{x^2}{2(1+it)}\right) \left[\exp\left(\frac{\sqrt{2}\alpha x}{1+it}\right) \pm \exp\left(-\frac{\sqrt{2}\alpha x}{1+it}\right)\right]. \tag{27}$$

These wave functions are normalized eigenfunctions of the operator $\hat{a}^2(t)$ given by (12), i.e., $\hat{a}^2(t)\psi_{\alpha\pm}(x,t) = \alpha^2\psi_{\alpha\pm}(x,t)$, where α is an arbitrary complex number.

4 Probability Representation and State Tomography

In this section, we review the method of quantizer $\hat{D}(\mathbf{x})$ and deqiuantizer $\hat{U}(\mathbf{x})$ operators acting in a Hilbert space H, where $\mathbf{x} = (x_1, \ldots, x_N)$ are either continuous or discrete variables. These operators were introduced in [26] to provide invertible maps of operators $\hat{A}$ acting in the Hilbert space H onto functions $f_A(\mathbf{x})$, in view of the following relations:

$$f_A(\mathbf{x}) = \mathrm{Tr}\left(\hat{A}\hat{U}(\mathbf{x})\right), \qquad \hat{A} = \int f_A(\mathbf{x})\hat{D}(\mathbf{x})\, d\mathbf{x}. \tag{28}$$

In the case of discrete parameters x_j, the integral is replaced by the corresponding sum. The main aim of this construction is to find such a pair of quantizer and dequantizer operators that, for the density operators $\hat{A} \to \hat{\rho}$, the functions $f_\rho(\mathbf{x})$ are standard probability distributions, e.g., conditional probability distributions of random variables. To provide such a construction, it is sufficient to find dequantizer operators with the properties $\hat{U}^\dagger(\mathbf{x}) = \hat{U}(\mathbf{x})$, $\mathrm{Tr}\,\hat{U}(\mathbf{x}) = 1$, and $\hat{U}(\mathbf{x}) \geq 0$, i.e., these operators have nonnegative eigenvalues. In view of Born's rule, the nonnegative trace of such operators $\hat{U}(\mathbf{x})$ with the density operators $\hat{\rho}$ has the properties of probabilities.

This approach is the main aim of our work, and it is obvious that there exists an infinite number of ways to describe the quantum state by probability distributions. To demonstrate this fact, we recall that any matrix or operator can be considered as a vector, and the trace of the product of two operators can be considered as a scalar product of the vectors. We illustrate this statement by the example of a 2×2-matrix M of an operator $\hat{M}$ with four matrix elements M_{11}, M_{12}, M_{21}, and M_{22}. We can organize these matrix elements in the form of a column vector $\mid M\rangle$ with the same four vector components. In this case, the formulas for $\mathrm{Tr}\left(\hat{A}^{\dagger}\hat{B}\right)$ can be presented as a scalar product of two vectors $\langle A \mid B\rangle$. One can say that the formula for mapping the operator $\hat{A}$ onto functions $f_A(\mathbf{x})$ and the inverse formula of reconstructing the operator $\hat{A}$, if one knows the functions $f_A(\mathbf{x})$ in this picture, mean that the vectors $\mid U(\mathbf{x})\rangle$ form the basis in the set of all the vectors in the linear space, where the vectors $\mid A\rangle$ constructed of the matrix elements of the operator matrix live.

If the vectors $\mid U(\mathbf{x})\rangle$ are the probability vectors, with components being the probabilities, the scalar product $\langle \rho \mid U(\mathbf{x})\rangle$ provides the probabilities.

The reconstruction formula provides the description of the vector $\mid \rho\rangle$ if all its components are known. All transformations of basis vector give new basis vectors. There exists an infinite number of possibilities to transform basis vectors with components having the meaning of matrix elements of the density matrices onto other basis vectors. According to Born's rule, one can describe the same density operator $\hat{\rho}$ by different probabilities applying different basis vectors corresponding to transformed given basis vectors described by the density-operator-like dequantizer operators $\hat{U}(\mathbf{x})$. Thus, the problem of obtaining the representations of quantum states by probabilities is reduced to the problem of finding all complete sets of basis vectors with components, which can be interpreted as matrix elements of Hermitian nonnegative matrices with trace equal to one. There exists an infinite number of such possibilities. For example, if one finds the dequantizer $\hat{U}(\mathbf{x})$, the unitary transform $\hat{u}$ of the dequantizers $\hat{U}(\mathbf{x}) \to \hat{u}\hat{U}(\mathbf{x})\hat{u}^{-1}$ gives the other appropriate basis. Also, all the transforms, where, instead of the unitary map of dequantizers, one uses any Kraus-type transform of the dequantizers, provide another set of basis vectors.

In the next section, we consider the symplectic-tomography scheme [24] of the probability representation of the oscillator-like state of a free particle discussed in the previous section.

5 Example of Symplectic Tomography

We construct the probability representation of systems with continuous variables like oscillators or free particles by adopting the idea [35] to use dequantizers $\hat{U}(\mathbf{x})$, with $\mathbf{x} = (X, \mu, \nu)$, in the form

$$\hat{U}(X, \mu, \nu) = \delta\left(X\hat{\mathbf{1}} - \mu\hat{q} - \nu\hat{p}\right), \tag{29}$$

where $\hat{q}$ is the position operator and $\hat{p}$ is the momentum operator. (All dimensional constants like Planck's constant are assumed to be equal to unity.) The quantizers $\hat{D}(X, \mu, \nu)$ are chosen as follows:

$$\hat{D}(X, \mu, \nu) = \frac{1}{2\pi} \exp\left[i \left(X\hat{\mathbf{1}} - \mu\hat{q} - \nu\hat{p}\right)\right]. \tag{30}$$

One can check that, for an arbitrary density operator $\hat{\rho}$ of the system states under discussion, i.e., oscillators and free particles, $w(X \mid \mu, \nu)=\mathrm{Tr}\left[\hat{\rho}\delta\left(X\mathbf{1} - \mu\hat{q} - \nu\hat{p}\right)\right]$; $-\infty < X, \mu, \nu < \infty$ is the probability density of the position X called symplectic tomogram of the state with density operator $\hat{\rho}$. For the pure state $\mid \psi\rangle\langle\psi \mid$, one has the following explicit relation determined by fractional Fourier transform of the wave function [36]

$$w_\psi(X \mid \mu, \nu) = \frac{1}{2\pi|\nu|} \left| \int \psi(y) \exp\left(i\frac{\mu y^2}{2\nu} - i\frac{Xy}{\nu}\right) dy \right|^2. \tag{31}$$

In the position representation, symplectic tomogram can be expressed through the density matrix $\rho(x, x') = \langle x \mid \rho \mid x' \rangle$, it reads

$$w_\rho(X \mid \mu, \nu) = \frac{1}{2\pi|\nu|} \int \rho(y, y') \exp\left(i\frac{\mu y^2}{2\nu} - i\frac{Xy}{\nu} - i\frac{\mu y'^2}{2\nu} + i\frac{Xy'}{\nu}\right) dy\, dy'; \tag{32}$$

the inverse formula is also available; it is

$$\hat{\rho} = \frac{1}{2\pi} \int w_\rho(X \mid \mu, \nu) \exp\left[i\left(X\hat{\mathbf{1}} - \mu\hat{q} - \nu\hat{p}\right)\right] dX\, d\mu\, d\nu. \tag{33}$$

It is easy to check that, if one has two density operators $\hat{\rho}_1$ and $\hat{\rho}_2$, the fidelity or the probability given by Born's rule $\mathscr{P}_1^{(2)} = \mathrm{Tr}\left(\hat{\rho}_1\hat{\rho}_2\right)$ in terms of tomograms of the states with density operators $\hat{\rho}_1$ and $\hat{\rho}_2$ has the form

$$\mathscr{P}_1^{(2)} = \frac{1}{2\pi} \int w_{\rho_1}(X \mid \mu, \nu) w_{\rho_2}(Y \mid -\mu, -\nu)\, e^{i(X+Y)}\, dX\, dY\, d\mu\, d\nu. \tag{34}$$

The purity parameter with the state with density operator $\hat{\rho}$ is

$$\mathrm{Tr}\,\hat{\rho}^2 = \frac{1}{2\pi} \int w_\rho(X \mid \mu, \nu) w_\rho(Y \mid -\mu, -\nu)\, e^{i(X+Y)}\, dX\, dY\, d\mu\, d\nu, \tag{35}$$

and the Wigner function $W_\rho(q, p)$ is

$$W_\rho(q, p) = \frac{1}{2\pi} \int w_\rho(X \mid \mu, \nu) \exp\left[i\left(X - \mu q - \nu p\right)\right] dX\, d\mu\, d\nu. \tag{36}$$

In view of the Radon transform, one can introduce the tomographic-probability distribution $w_{\mathrm{cl}}(X \mid \mu, \nu)$ for the classical state identified with the probability density $f(q, p) \geq 0$, where $\int f(q, p)\, dq\, dp = 1$; it is

$$w_{\mathrm{cl}}(X \mid \mu, \nu) = \int f(q, p)\, \delta\,(X - \mu q - \nu p)\, dq\, dp, \tag{37}$$

and the inverse transform is

$$f(q, p) = \frac{1}{4\pi^2} \int w_{\mathrm{cl}}(X \mid \mu, \nu)\, \exp[i\,(X - \mu q - \nu p)]\, dX\, d\mu\, d\nu. \tag{38}$$

The difference in the properties of classical and quantum probability densities is related to the Heisenberg quantum inequality $(\delta q)^2\,(\delta p)^2 \geq 1/4$, which for the classical state can be violated. On the other hand, the classical density, a quantum analog of the Wigner function, must be nonnegative. The Wigner function given by (36) can take negative values.

Now we present symplectic tomogram of the free-motion coherent state with the wave function (33); it is the normal probability distribution

$$\begin{aligned} w_\alpha(X \mid \mu, \nu, t) &= \frac{1}{\sqrt{\pi(\mu^2 + (\nu + \mu t)^2)}} \exp\left[-\frac{(X - \bar{X})^2}{\mu^2 + (\nu + \mu t)^2}\right], \\ \bar{X}_\alpha &= \sqrt{2}\,[\mu \mathrm{Re}\,\alpha + (\nu + \mu t)\,\mathrm{Im}\,\alpha]\,. \end{aligned} \tag{39}$$

The symplectic tomogram of the even and odd Schrödinger cat states is determined by Gaussian integrals; they are

$$w_{\alpha_\pm}(X \mid \mu, \nu, t) = \frac{1}{2\pi\,|\nu|} \left| \int \psi_{\alpha_\pm}(y, t) \exp\left(\frac{i\mu y^2}{2\nu} - i\frac{Xy}{\nu}\right) dy \right|^2, \tag{40}$$

where $\psi_{\alpha_\pm}(y, t)$ is given by (26). Thus, one can describe nonclassical even and odd coherent states of free particle (oscillator with zero frequency, $\omega = 0$) by the time-dependent probability distributions.

Also symplectic tomogram of the free-motion Fock state reads

$$\begin{aligned} w_n(X \mid \mu, \nu, t) = {} & \frac{1}{\sqrt{\pi(\mu^2 + (\nu + \mu t)^2)}} \exp\left[-\frac{X^2}{\mu^2 + (\nu + \mu t)^2}\right] \\ & \times \frac{1}{2^n n!} H_n^2 \left(\frac{X}{\sqrt{\mu^2 + (\nu + \mu t)^2}}\right). \end{aligned} \tag{41}$$

In the case of N free particles, their symplectic tomogram determined by the dequantizers $\prod_{j=1}^{N} \delta\left(X_j - \mu_j \hat{q}_j - \nu_j \hat{p}_j\right)$ provides the generalization of (39) and (41) to yield symplectic tomogram of the multimode coherent state

$$w_{\alpha}\left(\mathbf{X}\mid\mu,\nu,t\right)=\prod_{j=1}^{N}w_{\alpha_j}\left(X_j\mid\mu_j,\nu_j,t\right),\tag{42}$$

where $\alpha=(\alpha_1,\alpha_2,\ldots,\alpha_N)$, $\mathbf{X}=(X_1,X_2,\ldots,X_N)$, $\mu=(\mu_1,\mu_2,\ldots,\mu_N)$, and $\nu=(\nu_1,\nu_2,\ldots,\nu_N)$, while $w_{\alpha_j}\left(X_j\mid\mu_j,\nu_j,t\right)$ is given by (39).

Analogously, symplectic tomogram of the multimode Fock state of a free particle with N degrees of freedom is

$$w_{\mathbf{n}}\left(\mathbf{X}\mid\mu,\nu,t\right)=\prod_{j=1}^{N}w_{n_j}\left(X_j\mid\mu_j,\nu_j,t\right).\tag{43}$$

6 Probability Representation of Qubit States

For qubits (spin-1/2 particle, two-level atom), the states are described by the density matrix

$$\rho=\begin{pmatrix}\rho_{11}&\rho_{12}\\\rho_{21}&\rho_{22}\end{pmatrix}.\tag{44}$$

The spin projections onto three perpendicular directions x, y, and z are associated with Pauli matrices $\frac{1}{2}\sigma_x$, $\frac{1}{2}\sigma_y$, and $\frac{1}{2}\sigma_z$, where

$$\sigma_x=\begin{pmatrix}0&1\\1&0\end{pmatrix},\qquad\sigma_y=\begin{pmatrix}0&-i\\i&0\end{pmatrix},\qquad\sigma_z=\begin{pmatrix}1&0\\0&-1\end{pmatrix}.\tag{45}$$

The states with spin projections $m=+1/2$ onto directions x, y, and z are described by the density matrices of pure states

$$\rho_1=\begin{pmatrix}1/2&1/2\\1/2&1/2\end{pmatrix},\qquad\rho_2=\begin{pmatrix}1/2&-i/2\\i/2&1/2\end{pmatrix},\qquad\rho_3=\begin{pmatrix}1&0\\0&0\end{pmatrix}.\tag{46}$$

According to Born's rule, values p_j of $\mathrm{Tr}\left(\rho\rho_j\right)$; $j=1,2,3$ are the probabilities to have spin projections equal to $+1/2$ in the state with the density matrix ρ. This means that the density matrix (44) can be completely described by the probabilities p_1, p_2, and p_3 [37], i.e.,

$$\rho=\begin{pmatrix}p_3&(p_1-1/2)-i(p_2-1/2)\\(p_1-1/2)+i(p_2-1/2)&1-p_3\end{pmatrix},\tag{47}$$

or by three dichotomic probability distributions $(p_1,1-p_1)$, $(p_2,1-p_2)$, and $(p_3,1-p_3)$ satisfying the condition

$$(p_1 - 1/2)^2 + (p_2 - 1/2)^2 + (p_3 - 1/2)^2 \leq 1/4. \tag{48}$$

This condition means the nonnegativity of the eigenvalues of any state density matrix (Silvester criterion).

By using the example of a qubit, one can demonstrate that the Schrödinger equation for qubit energy levels and the von Neumann evolution equation for the density matrix can be written in the form of equations for probability distributions.

In fact, the energy quantization equation for a qubit with the Hamiltonian matrix $H = \begin{pmatrix} H_{11} & H_{12} \\ H_{21} & H_{22} \end{pmatrix}$ of the form

$$H \mid \psi\rangle = E \mid \psi\rangle, \tag{49}$$

rewritten in the form of equation for the matrix $\rho = \mid \psi\rangle\langle\psi \mid$, is

$$\begin{pmatrix} H_{11} & H_{12} \\ H_{21} & H_{22} \end{pmatrix} \begin{pmatrix} \rho_{11} & \rho_{12} \\ \rho_{21} & \rho_{22} \end{pmatrix} = E \begin{pmatrix} \rho_{11} & \rho_{12} \\ \rho_{21} & \rho_{22} \end{pmatrix}. \tag{50}$$

Here, the matrix ρ is expressed in terms of probabilities.

We arrive at a new form of the energy quantization equation for the probability vector; it reads

$$\begin{pmatrix} [H_{11} - E] & 0 & H_{12} & 0 \\ 0 & [H_{11} - E] & 0 & H_{12} \\ H_{21} & 0 & [H_{22} - E] & 0 \\ 0 & H_{21} & 0 & [H_{22} - E] \end{pmatrix} \begin{pmatrix} p_3 \\ (p_1 - 1/2) - i(p_2 - 1/2) \\ (p_1 - 1/2) + i(p_2 - 1/2) \\ 1 - p_3 \end{pmatrix} = 0. \tag{51}$$

Thus, the Schrödinger energy quantization equation can be presented as the equation for probability distributions determining the qubit quantum state. An analogous possibility also exists for the von Neumann evolution equation

$$\frac{\partial \rho}{\partial t} + i\,[H, \rho] = 0. \tag{52}$$

In view of the probability parameterization of the density matrix ρ (47), we can rewrite the quantum evolution equation in the form of a classical-like kinetic equation for probability distributions; it is

$$\begin{pmatrix} \frac{\partial p_3}{\partial t} & \frac{\partial p_1}{\partial t} - i\frac{\partial p_2}{\partial t} \\ \frac{\partial p_1}{\partial t} + i\frac{\partial p_2}{\partial t} & -\frac{\partial p_3}{\partial t} \end{pmatrix}$$
$$+i\left[\begin{pmatrix} H_{11} & H_{12} \\ H_{21} & H_{22} \end{pmatrix}, \begin{pmatrix} p_3 & (p_1 - 1/2) - i(p_2 - 1/2) \\ (p_1 - 1/2) + i(p_2 - 1/2) & 1 - p_3 \end{pmatrix}\right] = 0. \tag{53}$$

Analogous results can be formulated for arbitrary qudit states.

The above matrix differential equation means that the complex variable $z(t) = \big(p_1(t) - 1/2\big) - i\,\big(p_2(t) - 1/2\big)$ and the variable $p_3(t)$, in the case of a Hamiltonian $H_{11} = 1$ and $H_{12} = H_{22} = H_{21} = 0$, satisfy the equations $\dot{p}_3 = 0$ and $\dot{z}(t) = i\,z(t)$. Thus, we have the evolution of the state in the form of evolution equations of the probabilities,

$$p_1(t) = \frac{1}{2} + (\cos\, t)\big(p_1(0) - 1/2\big) + (\sin\, t)\big(p_2(0) - 1/2\big), \tag{54}$$

$$p_2(t) = \frac{1}{2} + (\cos\, t)\big(p_2(0) - 1/2\big) + (\sin\, t)\big(p_1(0) - 1/2\big). \tag{55}$$

The linear evolution equation for the density matrix of open system (GKSL equation [14–16])

$$\dot{\rho} = -i[H, \rho] + \frac{1}{2}\sum_k \Big([V_k\rho, V_k^\dagger] + [V_k, \rho V_k^\dagger]\Big) \tag{56}$$

can be rewritten as a kinetic equation for probabilities organized as N^2-vector $\mid p\rangle$ with components ρ_{jk}, such that $p_1 = \rho_{11}, \ldots, p_N = \rho_{1N}, p_{N+1} = \rho_{21}, \ldots, p_{N^2} = \rho_{NN}$; see, for example, [38]. In the linear equation (56) for the density matrix ρ of any dimension, the matrix H is the Hamiltonian matrix and matrices V_k are arbitrary matrices. The sum over index k is either finite or infinite one. As a result, we arrive at

$$\frac{d}{dt}|p\rangle = \mathscr{M}|p\rangle, \tag{57}$$

where matrix $\mathscr{M}$ is

$$\mathscr{M} = -i\left(H \otimes \mathbf{1} - \mathbf{1} \otimes H^\dagger\right) + \sum_k \left(V_k \otimes V_k^* - (1/2)V_k^\dagger V_k \otimes \mathbf{1} - (1/2)\mathbf{1} \otimes (V_k^{\mathrm{tr}} V_k^*)\right). \tag{58}$$

In the case of qubit, the probability vector $\mid p\rangle$ is a four-vector of the form

$$\mid p\rangle = \begin{pmatrix} p_3 \\ (p_1 - 1/2) - i\,(p_2 - 1/2) \\ (p_1 - 1/2) + i\,(p_2 - 1/2) \\ 1 - p_3 \end{pmatrix}. \tag{59}$$

Thus, we described the evolution of qubit state by linear kinetic equations for probabilities $p_1(t)$, $p_2(t)$, and $p_3(t)$ with the solutions given by (54) for a particular Hamiltonian and zero matrices V_k.

7 Superposition Principle of Qubit States in the Probability Representation

Now we present the addition rule of probabilities determining quantum states in the case of superposition of two pure states $\mid \psi_1\rangle$ and $\mid \psi_2\rangle$ and assume that the states are normalized and orthogonal, i.e., $\langle\psi_1 \mid \psi_2\rangle = 0$. These states have density operators (projectors) $\hat{\rho}_1 =\mid \psi_1\rangle\langle\psi_1 \mid$ and $\hat{\rho}_2 =\mid \psi_1\rangle\langle\psi_2 \mid$.

The superposition state $\mid \psi\rangle = C_1 \mid \psi_1\rangle + C_2 \mid \psi_2\rangle$, where $|C_1|^2 + |C_2|^2 = 1$, $|C_1|^2 = \Pi_1$, and $|C_2|^2 = \Pi_2$, has the normalized density operator (projector) $\hat{\rho}^2 = \hat{\rho} =\mid \psi\rangle\langle\psi \mid$ of the form [39]

$$\hat{\rho} = \Pi_1\hat{\rho}_1 + \Pi_2\hat{\rho}_2 + \sqrt{\Pi_1\Pi_2}\,\frac{\hat{\rho}_1\hat{\rho}_0\hat{\rho}_2 + \hat{\rho}_2\hat{\rho}_0\hat{\rho}_1}{\sqrt{\mathrm{Tr}\left(\hat{\rho}_1\hat{\rho}_0\hat{\rho}_2\hat{\rho}_0\right)}}, \tag{60}$$

where $0 \le \Pi_1 \le 1$, $0 \le \Pi_2 \le 1$, and $\hat{\rho}_0^2 = \hat{\rho}_0$ is the projector corresponding to parameters C_1 and C_2 of the superposition state $\mid \psi\rangle$. Formula (60) can be rewritten with all density matrices of operators $\hat{\rho}_0$, $\hat{\rho}_1$, and $\hat{\rho}_2$ given in the probability representation, i.e., ρ_j; $j = 0, 1, 2$ are expressed by (47) in terms of dichotomic probabilities $p_{1,2,3}^{(k)}$, and the matrix ρ is expressed in terms of probabilities $p_{1,2,3}$. As a result, we arrive at a relation of probabilities $p_{1,2,3}$ in terms of probabilities $p_{0,1,2}^{(k)}$ and the probability parameters Π_1 and Π_2 corresponding to the numbers C_1 and C_2; the matrix relation reads

$$\begin{aligned}\begin{pmatrix} p_3 & p^* \\ p & 1-p_3 \end{pmatrix} &= \Pi_1 \begin{pmatrix} p_3^{(1)} & p^{*(1)} \\ p^{(1)} & 1-p_3^{(1)} \end{pmatrix} + \Pi_2 \begin{pmatrix} p_3^{(2)} & p^{*(2)} \\ p^{(2)} & 1-p_3^{(2)} \end{pmatrix} \\ &+K \begin{pmatrix} p_3^{(1)} & p^{*(1)} \\ p^{(1)} & 1-p_3^{(1)} \end{pmatrix}\begin{pmatrix} p_3^{(0)} & p^{*(0)} \\ p^{(0)} & 1-p_3^{(0)} \end{pmatrix}\begin{pmatrix} p_3^{(2)} & p^{*(2)} \\ p^{(2)} & 1-p_3^{(2)} \end{pmatrix} \\ &+K \begin{pmatrix} p_3^{(2)} & p^{*(2)} \\ p^{(2)} & 1-p_3^{(2)} \end{pmatrix}\begin{pmatrix} p_3^{(0)} & p^{*(0)} \\ p^{(0)} & 1-p_3^{(0)} \end{pmatrix}\begin{pmatrix} p_3^{(1)} & p^{*(1)} \\ p^{(1)} & 1-p_3^{(1)} \end{pmatrix},\end{aligned} \tag{61}$$

where

$$\begin{aligned}K = \sqrt{\Pi_1\Pi_2}\Bigg[\mathrm{Tr}&\begin{pmatrix} p_3^{(1)} & p^{*(1)} \\ p^{(1)} & 1-p_3^{(1)} \end{pmatrix}\begin{pmatrix} p_3^{(0)} & p^{*(0)} \\ p^{(0)} & 1-p_3^{(0)} \end{pmatrix} \\ &\times \begin{pmatrix} p_3^{(2)} & p^{*(2)} \\ p^{(2)} & 1-p_3^{(2)} \end{pmatrix}\begin{pmatrix} p_3^{(0)} & p^{*(0)} \\ p^{(0)} & 1-p_3^{(0)} \end{pmatrix}\Bigg]^{-1/2}.\end{aligned} \tag{62}$$

In (61) and (62), $p = (p_1 - 1/2) + i(p_2 - 1/2)$, $p^{(k)} = (p_1^{(k)} - 1/2) + i(p_2^{(k)} - 1/2)$; $k = 0, 1, 2$.

Relations (61) and (62) provide the probabilities determining the superposed qubit state $\left(p_{1,2,3}\right)$ as functions of probabilities $p_{0,1,2}^{(k)}$ and the probability parameters

Π_1 and Π_2. One can check that, in the case of projectors $\hat{\rho}_k^2 = \hat{\rho}_k$; $k = 0, 1, 2$, relations (61) and (62) provide the condition $\hat{\rho}^2 = \hat{\rho}$. An analogous result can be obtained for the qudit-state superpositions.

8 Conclusions

To conclude, we formulate the main results presented here.

We reviewed the probability representation of conventional quantum mechanics. For systems with continuous variables (free particle, oscillator), the tomographic-probability distributions determining quantum states were constructed for pure states with wave functions, as well as for mixed states with density matrices. Also analogous probabilities determining the qudit-state density operators were obtained. The construction of the probability representation is based on the method of quantizer–dequantizer operators. The dequantizer operators, which have the properties of density operators, map the system-state density operators onto probabilities (fidelities), in view of Born's rule. The dequantizer operators are used to provide the inverse map of the obtained probabilities onto the density operators of systems' quantum states.

The quantum energy-level equations (Schrödinger equations) are shown to have the form of equations for probabilities determining stationary quantum states for a given Hamiltonian. The quantum evolution equations are written in the form of kinetic equations for the probabilities constructed.

The superposition principle of quantum-state wave functions is expressed in terms of nonlinear addition rule of the probabilities determining the pure states of quantum systems. Examples of the free-particle coherent states and qubit (spin-1/2 system) are discussed.

The problem of new entropic-information characteristics of quantum systems associated with the probabilities considered here will be presented in the future publication. The formalism suggested can be applied in studies in other sciences where the techniques of quantum mechanics are used, as in [40, 41]. The problem of nonlinear coherent states (f-oscillators) [42] and the problem of photon distributions corresponding to nonlinearity of electromagnetic vibrations [43] can be successfully studied using the probability representation of quantum mechanics demonstrated in this paper.

Acknowledgements We are grateful to the Editors for the invitation to contribute to a Festchrift honoring Andrei Khrennikov.

References

1. E. Wigner, On the quantum correction for thermodynamic equilibrium. Phys. Rev. **40**, 749–759 (1932)
2. W. Heisenberg, Uber den anschaulichen Inhalt der quantentheoretishen kinematik und mechanik. Z. Phys. **43**, 172–198 (1927)
3. H.P. Robertson, The uncertainty principle. Phys. Rev. **34**, 163–164 (1929)
4. E. Schrödinger, Zum Heisenbergschen Unschärfeprinzip, Sitz. Preus. Acad. Wiss. Phys.-Math. Klasse. **19**, 296-303 (1930)
5. R.J. Glauber, Coherent and incoherent states of the radiation field. Phys. Rev. **131**, 2766–2788 (1963)
6. E.C.G. Sudarshan, Equivalence of semiclassical and quantum-mechanical descriptions of statistical light beams. Phys. Rev. Lett. **1963**(10), 277–279 (1963)
7. K. Husimi, Some formal properties of the density matrix. Proc. Phys. Math. Soc. Jpn. **22**, 264–314 (1940)
8. Y. Kano, Theory of the electromagnetic field in the statistical theory of the electromagnetic field. J. Math. Phys. **6**, 1913–1915 (1965)
9. L. Landau, Das dämpfungsproblem in der wellenmechanik. Z. Phys. **45**, 430–441 (1927)
10. J. von Neumann, Wahrscheinlichkeitstheoretischer Aufbau der Quantenmechanik, *Nachrichten von der Gesellschaft der Wissenschaften zu Gottingen. Math.-Phys. Klasse*, Z. 245-272 (1927)
11. P.A.M. Dirac, *The Principles of Quantum Mechanics* (Clarendon Press, Oxford, UK, 1981)
12. A. Khrennikov, A pre-quantum classical statistical model with infinite-dimensional phase space. J. Phys. A: Math. Gen. **38**, 9051–9073 (2005)
13. A. Khrennikov, Towards a field model of prequantum reality. Found. Phys. **42**, 725–741 (2012)
14. A. Kossakowski, On quantum statistical mechanics of non-hamiltonian systems. Rep. Math. Phys. **3**, 247–274 (1972)
15. V. Gorini, A. Kossakowski, E.C.G. Sudarshan, Completely positive semigroups of N-level systems. J. Math. Phys. **17**, 821–825 (1976)
16. G. Lindblad, On the generators of quantum dynamical semigroups. Commun. Math. Phys. **48**, 119–130 (1976)
17. V.I. Man'ko, G. Marmo, E.C.G. Sudarshan, F. Zaccaria, On the relation between Schrödinger and von Neumann equations. J. Russ. Laser Res. **20**, 421–437 (1999)
18. A. Khrennikov, Hyperbolic quantum mechanics, arXiv:quant-ph/0101002v1, Accessed on 31 Dec 2000
19. A. Khrennikov, Born's formula from statistical mechanics of classical fields and theory of hitting times. Phys. A. **393**, 207–221 (2014)
20. A. Khrennikov, Roots of quantum computational supremacy: Superposition? Entanglement or complementarity? Preprint from Preprints.org (02 Dec 2019) https://doi.org/10.20944/preprints201912.0006.v1PPR: PPR104145
21. V.I. Man'ko, G. Marmo, E.C.G. Sudarshan, F. Zaccaria, Interference and entanglement: An intrinsic approach. J. Phys. A: Math. Gen. **35**, 7137–7158 (2002)
22. V.I. Man'ko, G. Marmo, E.C.G. Sudarshan, F. Zaccaria, Inner composition law of pure-spin states. AIP Conf. Proc. **545**, 92–97 (2000)
23. M.A. Man'ko, V.I. Man'ko, Observables, interference phenomenon and Born's rule in the probability representation of quantum mechanics. Int. J. Quantum Inf. **18**, 1941021 (2020)
24. S. Mancini, V.I. Man'ko, P. Tombesi, Symplectic tomography as classical approach to quantum systems. Phys. Lett. A. **213**, 1–6 (1996)
25. V.I. Man'ko, O.V. Man'ko, V.N. Chernega, Conventional quantum statistics with a probability distribution describing quantum system states. Phys. Part. Nuclei. **51**, 772–780 (2020)
26. O.V. Man'ko, V.I. Man'ko, G. Marmo, Alternative commutation relations, star products and tomography. J. Phys. A: Math. Gen. **35**, 699–719 (2002)
27. F.M. Ciaglia, F. Di Cosmo, A. Ibort, G. Marmo, Dynamical aspects in the quantizer-dequantizer formalism. Ann. Phys. **385**, 769–781 (2017)

28. P. Adam, V.A. Andreev, M.A. Man'ko, V.I. Man'ko, M. Mechler, SU(2) symmetry of qubit states and Heisenberg-Weyl symmetry of systems with continuous variables in the probability representation of quantum mechanics. Symmetry **12**, 1099 (2020)
29. M. Asano, A. Khrennikov, M. Ohya, Y. Tanaka, I. Yamato, *Quantum Adaptivity in Biology: From Genetics to Cognition* (Springer, Heidelberg, Berlin, New York, 2015)
30. I. Basieva, V.H. Cervantes, E.N. Dzhafarov, A. Khrennikov, True contextuality beats direct influences in human decision making, arXiv:1807.05684 [q-bio.NC]
31. A. Khrennikov, *Ubiquitous Quantum Structure: From Psychology to Finances* (Springer, Berlin, Heidelberg, New York, 2010)
32. V.V. Dodonov, V.I. Man'ko, *Invariants and Evolution of Nonstationary Quantum Systems, Proceedings of Lebedev Physical Institute*, Vol. 183 (Nova Science, Commack, New York, 1989)
33. V.V. Dodonov, I.A. Malkin, V.I. Man'ko, Even and odd coherent states and excitations of a singular oscillator. Phys. **72**, 597–615 (1974)
34. M. Brune, E. Hagley, J. Dreyer, V. Maitre, A. Maali, C. Wunderlich, J.A. Raimond, S. Haroche, Observing the progressive decoherence of the meter in a quantum measurement. Phys. Rev. Lett. **77**, 4887–90 (1996)
35. S. Mancini, V.I. Man'ko, P. Tombesi, Classical-like description of quantum dynamics by means of symplectic tomography. Found. Phys. **27**, 801–824 (1997)
36. V.I. Man'ko, R.V. Mendes, Non-commutative time-frequency tomography. Phys. Lett. A. **263**, 53–61 (1999)
37. M.O. Terra-Cunha, V.I. Man'ko, M.O. Scully, Quasiprobability and probability distributions for spin-1/2 states. Found. Phys. Lett. **14**, 103–117 (2001)
38. O.V. Man'ko, V.I. Man'ko, Probability representation of quantum states. Entropy **23**, 549 (2021)
39. E.C.G. Sudarshan, Search for purity and entanglement. J. Russ. Laser Res. **24**, 195–2003 (2003)
40. A. Khrennikov, A. Alodjants, classical (local and contextual) probability model for Bohm-Bell type experiments: no-signaling as independence of random variables. Entropy **21**, 157 (2019)
41. A. Khrennikov, Formalization of Bohr's contextuality within the theory of open quantum systems. J. Russ. Laser Res. **42**(3), (2021)
42. V.I. Man'ko, G. Marmo, E.C.G. Sudarshan, F. Zaccaria, Entanglement structure of the adjoint representation of the unitary group and tomography of quantum states. J. Russ. Laser Res. **24**, 507–543 (2003)
43. R. Lopez-Pena, V.I. Man'ko, G. Marmo, E.C.G. Sudarshan, F. Zaccaria, Photon distribution in nonlinear coherent states. J. Russ. Laser Res. **21**, 305–316 (2000)

Margarita A. Man'ko was born in Moscow on December 4, 1939. In 1957, she graduated with Gold Medal from high school in Moscow and became a student of the Faculty of Physics of Lomonosov Moscow State University. She studied under the Chair of Physics and Theory of Oscillations. From 1961 she worked on her Diploma Project "Maser with Sequential Resonator" in the group of Prof. Nikolay G. Basov (1964 Nobel Prize Winner) in the Laboratory of Oscillations headed by Prof. Alexander M. Prokhorov (1964 Nobel Prize Winner) of the Lebedev Physical Institute in Moscow. Since 1963 her scientific career is connected with the Lebedev Physical Institute, where she defended Ph.D. Thesis "Spectral Characteristics of Semiconductor Lasers and Methods of Mode Selection" in 1970. She has permanent position at the Quantum Radiophysics Department as Senior Scientific Researcher. She has more than 300 scientific articles in the field of quantum physics, quantum mechanics, and quantum optics published in International and Russian journals.

In addition to experimental work, she dedicated a significant part of her time to the publication activity. For more than 20 years she was Deputy Editor-in-Chief of the Proceedings of the Lebedev Institute published by Nauka in Moscow and Nova Science in New York. Since 1980, Margarita A. Man'ko is Scientific and Managing Editor of Journal of Soviet (now Russian) Laser

Research published by Springer Nature in New York. In 2001–2020, Margarita A. Man'ko was Coordinator of IoPP Division at the Lebedev Institute and Associate Editor and Editorial Board Member of Physica Scripta of the Royal Swedish Academy of Sciences published by IoP Publishing. Now she is Editorial Board Member of Quantum Reports, Symmetry, and Science of MDPI, Basel.

Margarita A. Man'ko participated in many international conferences and visited a great number of research groups in Europe, Asia, and both Americas, establishing fruitful collaborations every time. She was among the organizers of many important scientific events, such as International Conferences on Squeezed States and Uncertainty Relations and Central European Workshops on Quantum Optics.

Margarita A. Man'ko was awarded International Diplomas for the Study of Single-Mode Diode Lasers within the framework of collaboration between the Russian Academy of Sciences and the German Academy of Sciences and between the Russian Academy of Sciences and the National Center of Science and Technology of Vietnam. In 2011, she got the Certificate of Gratitude from IoP Publishing for 10 years of successful cooperation. In 2014, Margarita A.Man'ko was awarded the Leonard Euler Prize to honor her scientific lifework and to honor her contributions to physics and mathematics and she was also awarded the Honor Diploma of the Lebedev Physical Institute of the Russian Academy of Sciences for Substantial Contribution to the Development of the Russian Science.

More detailed description of scientific activity of Margarita A. Man'ko is available on https://iopscience.iop.org/article/10.1088/0031-8949/90/7/070301.

Vladimir I. Man'ko was born on April 2, 1940 in Moscow. In 1957, he graduated from high school in Moscow with the Silver Medal and became the student of the Faculty of Physics of M.V. Lomonosov Moscow State University. He studied under the Chair of Statistical Physics headed by Prof. N. N. Bogolyubov. In 1961, he passed the Landau Theoretical Minimum Examination and was accepted as Magister student to the Landau Theoretical Group under supervision of Prof. I. E. Dzyaloshinsky. In 1963, V.I. Man'ko defended the Magister Thesis "Hidden Anti-ferromagnetism in Cubic Crystals" and became a Ph.D. student at the P.N. Lebedev Physical Institute in Moscow. In 1966, V.I. Man'ko defended the Ph.D. Thesis "Some Aspects of Applying Group-Theoretical Methods in Physics" in the Lebedev Physical Institute. Starting in 1966, he obtained a permanent research position at the Lebedev Physical Institute as the Chief Researcher. In 1972, V.I. Man'ko defended the Habilitation Thesis "Dynamical Symmetries and Coherent States of Quantum Systems." V.I. Man'ko has also position as the Chair of Theoretical Physics of Moscow Institute of Physics and Technology (State University), being Assistant Professor from 1968–1980, Associate Professor from 1980–1989 and from 1989, he became Full Professor.

V.I. Man'ko with I A Malkin wrote the book "Dynamical Symmetry and Coherent States of Quantum Systems" Nauka Publisher, Moscow 1979. With V.V. Dodonov he published the book "Invariants and the Evolution of Nonstationary Quantum Systems" Nova Sience. New York.

In addition, V.I. Man'ko is the author of more than 500 scientific articles in Russian and International scientific journals. He is the co-organizer of more than 17 international conferences. Since 1976, he was the member of the Standing Committee of the International Colloquium "Group Theoretical Methods in Physics." For several years he has been a member of the Institute of Physics, Bristol (since 2002) and Vice-Editor-in-Chief of the Journal of Optics B: Quantum and Semiclassical Optics; Editor of the International Journal on Group Theory in Physics (1989–1997). Now he is the Editorial Board Member of the Journal of Russian Laser Research published by Springer in New York and the Editorial Board Member of Entropy, MDPI, Basel.

V.I. Man'ko also gave invited talks and did joint research with Universities and scientific institutions in Germany, Italy, USA, France, Poland, Mexico, Venezuela, Portugal, Spain, India, Turkey, Hungary, Serbia, Netherlands, Austria, Sweden, Vietnam, Bulgaria, South Korea, Brasilia, Great-Britain, China. He spent also time notably at Harvard University (1980), Princeton University; USA; Universities of Rome, Milano, Salerno, Pavia, Camerino, Naples and ICTP (Trieste), Italy; Universities of Ulm, Essen, Bielefeld, Berlin, Tubingen, Dortmund, Clausthal, Munich, Konstanz, Mannheim and Garching, Germany; University of Lisbon and Funchal in Portugal; Budapest Wigner Center, Hungary. In 1992–1993, V.I. Man'ko had a research position at INFN, Sez. di Napoli (Italy).

V.I. Man'ko participated in the supervision of the following Ph.D. and habilitation Ph.D. Theses: D. Trifonov, E. Doktorov, V. Dodonov, E. Kurmyshev, S. Krivoshlykov, S. Chumakov, A. L'vov, E. Akhundova, A. Zhuk, A. Mamedov, D. Ossipov, A. Klimov, S. Filippov, M. R. Bazrafkan, L. Markovich, E. Zhebrak, Ya. Korennoy, O. Pilyavets, I Dudinets. Habilitation: N. Gromov, G. Amosov.

More detailed description of scientific activity of Vladimir I. Man'ko is available on https://iopscience.iop.org/article/10.1088/0031-8949/90/7/070301.

Alternative Quantum Mechanics?

Dagomir Kaszlikowski and Paweł Kurzyński

Abstract Probability amplitudes and Born rule are the most frequently used tools to interpret quantum mechanical phenomena. However, quantum states and operations are equivalent to negative probability distributions and stochastic processes via a mathematical mapping called 'frames'. For various reasons, historical forces pushed this alternative out of the mainstream with a few notable exceptions such as Wigner or Hussimi functions. This is most unfortunate because negative probabilities, exploited as a computational tool, can offer insights into quantum theory where the orthodox, Hilbert space approach may be cumbersome. Motivated by this, we study a binary system with a negative probability distribution called *negative bit* (nebit). We embed nebits in a classical random walk to simulate certain non-classical aspects of discrete quantum dynamics and show possibilities that transcend quantum mechanics. The paper ends with open problems and prospects of the nebit approach.

1 Introduction

We all know the story. It has been passed from generation to generation of physicists through books, journal articles' intros, conversations with our mentors and late night musings about the meaning of it all. It goes more or less like this. The ultraviolet catastrophe forced Planck to introduce quanta, Einstein got Nobel prize for photoelectric effect, pushing Planck's ideas further until another bunch of young and eager to prove themselves fellows came up with Schröedinger and Heisenberg equations whose interpretation needed Born's rule to make any sense. The times were hot. New experiments needed quick explanations and acolytes of quanta, juggling probability

D. Kaszlikowski
Centre for Quantum Technologies and Department of Physics, National University of Singapore, 3 Science Drive 2, Singapore 117543, Singapore
e-mail: phykd@nus.edu.sg

P. Kurzyński (✉)
Faculty of Physics, Adam Mickiewicz University, Uniwersytetu Poznańskiego 2, 61-614 Poznań, Poland
e-mail: pawel.kurzynski@amu.edu.pl

A. Plotnitsky and E. Haven (eds.), *The Quantum-Like Revolution*,
https://doi.org/10.1007/978-3-031-12986-5_11

amplitudes, pulled out answers, like rabbits from a magician's hat. Whatever was thrown at them, they had an answer and they could use their brand new toy to predict a plethora of undiscovered phenomena, later verified by increasingly sophisticated experiments. However some founders noticed peculiarities within the theory. Einstein pained for old good classical sanity where "the moon is there if no one looks at it" and God dislikes dice. Schröedinger found a potent peculiarity that he called 'Verschrankung', transfigured to 'entanglement' and later dramatised by Einstein, Podolsky and Rosen in the notorious EPR paper that gave birth to many scientific careers and quantum computing bonanza. Oh, and we cannot of course forget about rebellious Bell whose little inequality and, for some of us, courage to call von Neumann's calculations silly, are one of the most interesting contributions to modern science.

This folklore is gross oversimplification of reality, but it gets some things right. The theory of quanta evolved rapidly, important answers were needed at brief notice, and there was little incentive to think about alternatives. Many physicists acknowledged weirdness of quanta, but its predictive powers made them 'shut up and calculate' [1]. Ultimately, only a few geezers refused blind calculations (notorious Einstein for instance) but they were swept aside by a tsunami of wave function pragmatists. But no matter how powerful the flood is, it can never drown curiosity. Defiant hippies like Feynman, Wiesner, Bell, Bennett, Deutsch and many more, came back surfing on the rising wave of quantum information they generated, bringing new paradigms and solution but also many important questions. Perhaps some of those questions can be answered only if one embraces alternatives.

So, let's imagine an alternative history, something, ironically, quite natural for the many worlds crowd. In this parallel universe, Schröedinger never discovers his wave equation, but he comes up with an idea of a stochastic theory with negative probabilities. It's already known that quantum theory is equivalent to such a theory via a simple mapping from Hilbert space to negative probabilities, called frames [2–5], so we are on the solid ground here. We do not have time to explain frames but the reader can think of Wigner function as a blueprint. In a nutshell, it is a theory of particles hopping around with negative transition probabilities. But how to interpret negative probabilities operationally? We think it's not a problem if they are positive. But does one interpret $-\frac{1}{2}$ probability of Donald Trump not running in the next presidential elections? Nobody seems to know, but there is a simple pragmatic solution. The theory is such that negative probabilities never appear at the input and output of experiments, coming into existence only in the intermediate stages, inaccessible to us [6]. Having removed this hurdle, we would use this theory to analyse natural phenomena, and in time negative probabilities, exactly like amplitudes of probability, would become natural to us. We would develop intuitions and mental shortcuts like we did with probability amplitudes and Born rule. Stochastic processes would replace unitary transformations and no one would get lost in Hilbert space again. And of course, hidden negative probabilities would not be any stranger or counterintuitive than the wave-particle duality or complementarity, or any other counterintuitive aspects of quantum theory. This alternative history is not as implausible as devoted Hilbert space followers may think. In fact, some researchers followed this line of thought and got interesting results (the ones that influenced us include [6–15]). One of them is Andrei Khrennikov [9], to whom this Festschrift is dedicated. Their work and ideas they developed strongly suggest that negative probabilities offer valuable insights into quantum theory, if not more. We hope that this paper will contribute something new to this growing body of research.

2 Nebit as a Resource

In this paper we introduce and discuss *nebit*, an abbreviation for negative bit [16]. The concept of nebit may offer a different point of view on already well established representations of quantum theory as that of stochastic processes with negative probabilities called *frames* [2–5]. We do not know yet if nebit is something new or just a spin on frames. However, we can already show that nebit fits into generalised probability theories (GPT) and thus transcends quantum theory in its present form. A famous example of a GPT are Popescu-Rorlich boxes [17] that play a pivotal role in our understanding of quantum non-locality and quantum contextuality.

Frames is a collective name for mathematical mapping from Hilbert space to a stochastic theory with negative probabilities. In this approach quantum states become negative probability distributions and quantum evolutions become stochastic processes with negative transition probabilities. A simple argument shows that negativities are unavoidable to describe unitary evolutions. For any unitary evolution U, we have a reverse process U^{-1}. Suppose that $S(U)$ is a 'proper' stochastic process with positive transition probabilities. It will generate entropy and thus $S(U^{-1})$ must remove it. To see this, let us consider a classical bit and a bi-stochastic process given by a 2×2 matrix $S_{ij} = p(i|j)$ with $\sum_i S_{ij} = \sum_j S_{ij} = 1$. These constraints imply $S_{00} = S_{11} = p$ and $S_{01} = S_{10} = 1 - p$. Suppose that at time $t = 0$ the bit is in the state 0. After one time step generated by S, at $t = 1$, the probability of finding it in the state 0 is p and $1 - p$ in the state 1. The Shannon entropy increased from nil to $-p \log p - (1 - p) \log (1 - p)$. One can formally find the inverse of S, getting a bi-stochastic S^{-1}, with either a negative entry on the diagonal $\frac{p}{2p-1}$ or in the corners $\frac{p-1}{2p-1}$, depending if p is larger or smaller than $\frac{1}{2}$. This negative stochastic process annihilates the entropy $-p \log p - (1 - p) \log (1 - p)$ acquired by the bit at time $t = 1$.

We can simulate S as the $CNOT$ gate where the control bit has the probability distribution $p, 1 - p$. Note that $CNOT$ is a bi-stochastic matrix too. How about S^{-1}? It is easy to check that it is the same $CNOT$ but now the control bit has the negative probability distribution $\frac{p}{2p-1}, \frac{p-1}{2p-1}$. In the long-standing tradition of quantum information business, it is only natural to call it nebit, hoping to achieve immortality in the literature on the subjects. Jokes aside, this simple example hints at nebit serving as a basic unit of stochastic negativity. Subsequent sections of the paper will show you more of nebit generated 'magic'. However, before getting there, we would like to outline a more general picture we have been working on. It will also help to understand better the rest of this paper.

Consider a qubit and a SIC-POVM [18] given by the set of four operators $F_a = \frac{1}{4}(1 + \vec{n}_a \cdot \vec{\sigma})$ $(a = 1, 2, 3, 4)$, where the unit vectors $\vec{n}_a$ span a tetrahedron, i.e., $\vec{n}_a \cdot \vec{n}_b = \delta_{ab} - \frac{1}{3}(1 - \delta_{ab})$. Using frames we can now map any qubit density matrix ρ with a Bloch vector $\vec{m}$ to a positive probability distribution $p_a := Tr(\rho F_a)$. In this frame, any unitary U, represented as a rotation R of $\vec{m}$, is a bi-stochastic process with negative probabilities, given by $S(U)_{kl} = \frac{1}{4} + \frac{3}{4}(R\vec{n}_k) \cdot \vec{n}_l$. This is a nice frame indeed because all the negativity sits in one place—the process $S(U)$. A choice of a different frame could throw negativity to both, the probability distribution p_a and the process $S(U)$.

Our general idea is that it may be possible to simulate any quantum operation, unitary or not, as a stochastic process with positive probabilities and a certain number of nebits acting as a reservoir of negativity. We could then quantify how non-classical a given operation is, a problem related to rather poorly understood origins of quantum supremacy. At this moment we cannot say anything specific yet but we can already give you some, we hope, interesting results.

Before jumping into technicalities, we need to say something important to both of us. We are not interested in negative probabilities ontology, treating this concept only as an alternative model of quantum mechanics. We think we need it to better understand what quantum theory is, what makes quantum computers tick. Perhaps, in the view of this, we should not use the name negative probabilities, but, for instance, signed measures, or quasi-probabilities, or 'frameons', or some other fancy name. However, the phrase 'negative probabilities' has been around for some time (Feynman used it! [6]) so we decided to keep it in all the instances where we cannot replace it by nebits.

3 Non-classical Dynamics

Consider a non-classical single particle dynamics in discrete space and time. You can visualise it as a particle jumping along edges connecting vertices of some graph. This is a generalisation of random walk.

3.1 Preliminaries

Let us define basic concepts. A graph $G = (V, E)$ is a set of vertices $V = \{v_1, v_2, \ldots, v_n\}$ and a set of edges $E \subseteq \{(v_i, v_j) | v_i, v_j \in V\}$. If $(v_i, v_j) \in E$, then it is possible to make a one-step transition from v_j to v_i. If both (v_i, v_j) and (v_j, v_i) are in E we call it undirected graphs as opposed to directed graphs whose edges are called arcs. Any edge (v_i, v_i) that starts and ends at the same vertex is called a loop. In our case, vertices represent the particle's physical states and edges are possible transitions between them.

A random walk on a graph is described by a stochastic matrix $\mathbb{S}$ that assigns a transition probability p_{ij} to every $(v_i, v_j) \in E$ ($\mathbb{S}_{ij} = p_{ij}$). Since the sum of all outgoing probabilities needs to be normalized to one, we have $\sum_i p_{ij} = 1$, i.e., all columns of $\mathbb{S}$ sum up to one. The state of the walk after t steps is an n-dimensional probability vector

$$\mathbf{p}_t = (p_1, p_2, \dots, p_n)^T, \tag{1}$$

where p_i is the probability of the particle ending up in v_i. The walk's single step is the transformation of the probability vector

$$\mathbf{p}_{t+1} = \mathbb{S}\mathbf{p}_t. \tag{2}$$

Thus, after t steps, the initial probability distribution $\mathbf{p}_0$, evolves to

$$\mathbf{p}_t = \mathbb{S}^t \mathbf{p}_0. \tag{3}$$

3.2 *Injecting Nebits*

Let us now incorporate nebits into our generalised random walks. We assume that there is an additional, unobservable, two-level system that could be an internal property of the particle or an external coin-like system. It does not matter what it is, it only matters that it introduces a hidden structure to each vertex v_i that is now made of two sub-vertices, u_i and $\bar{u}_i$. These sub-vertices are occupied with *probabilities* q_i and $\bar{q}_i$, respectively, and the resulting observable v_i's occupation probability p_i is a sum of q_i and $\bar{q}_i$

$$p_i = q_i + \bar{q}_i. \tag{4}$$

Note that q_i and $\bar{q}_i$ can be negative. Because, p_i must be a non-negative number in order to be a valid probability, hence q_i and $\bar{q}_i$ cannot be both negative. This can be guaranteed by imposing two conditions

$$0 \leq q_i + \bar{q}_i \leq 1, \tag{5}$$

$$\sum_i (q_i + \bar{q}_i) = 1. \tag{6}$$

Now, the state at time t is given by a $2n$-dimensional vector

$$\mathbf{q}_t = (q_1, \dots, q_n, \bar{q}_1, \dots, \bar{q}_n)^T. \tag{7}$$

3.3 Negative Random Walk

We have not yet specified how the sub-vertices are connected so let us fix this by introducing a matrix $\mathbb{Q}$ that generates a single time step, i.e.,

$$\mathbf{q}_{t+1} = \mathbb{Q}\mathbf{q}_t. \tag{8}$$

The elements of $\mathbb{Q}$ correspond to transition *probabilities* $\mathbb{Q}_{ij} = p_{ij}$, where

$$i, j = 1, 2, \dots, n, \bar{1}, \bar{2}, \dots, \bar{n}.$$

They have to obey $\sum_i p_{ij} = 1$, such that the constraint (6) is satisfied with each time step. We want $\mathbb{Q}$ to be a proper (positive) stochastic matrix, pushing all negativity to the initial probability distribution. Now, the system's state after t steps is fully determined by $\mathbb{Q}$ and the initial state $\mathbf{q}_0$ as follows

$$\mathbf{q}_t = \mathbb{Q}^t\mathbf{q}_0. \tag{9}$$

In general, bits (representing the physical states of the particle) and nebit in the initial probability distribution $\mathbf{q}_0$ can be correlated but, for now, we can get away with a product state

$$\mathbf{q}_0 = \begin{pmatrix} 1+\Delta \\ -\Delta \end{pmatrix} \otimes \begin{pmatrix} p_1 \\ p_2 \\ \vdots \\ p_n \end{pmatrix}. \tag{10}$$

A bit-nebit product state is easier to understand because it is clear where negativity is located. In contrast, non-product pairs 'hide' negativity in correlations, which are much more complex to analyse.

We call this model a *negative random walk*. It is a non-trivial extension of the usual random walk since the vector $\mathbf{q}_t$ has to obey the conditions (5) and (6) for all t and the transition matrix $\mathbb{Q}$ cannot be arbitrary. Moreover, for a given matrix $\mathbb{Q}$, not all initial conditions $\mathbf{q}_0$ are allowed. This effectively limits countless non-physical models.

3.4 Example No. 1—Deterministic Dynamics

We now show how to use negative random walk to simulate an effect present in pre- and post-selected quantum walks [19]. This is an interesting phenomenon when a particle can transfer in one time step between two states that are not directly connected. Within the classical probability model, the number of steps needed for

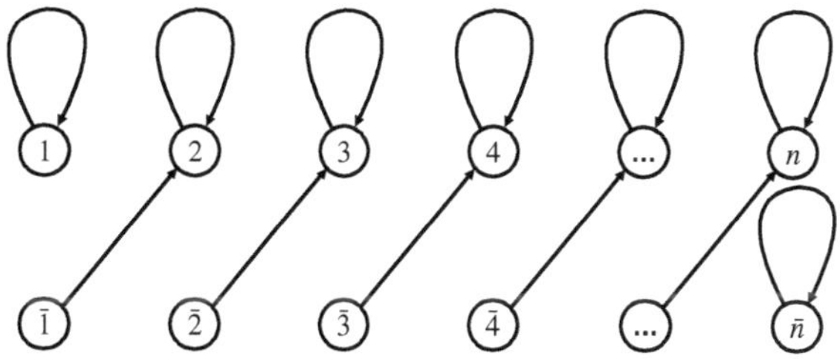

Fig. 1 Arcs representing transitions in the deterministic dynamics example

such a transfer is at least the distance between these two states, i.e., the minimal number of arcs the particle has to pass in order to reach the final vertex from the initial one.

Let us first de-clutter the previous notation and label the sub-vertices u_i by integers $i = 1, 2, \ldots, n$ and $\bar{u}_i$ by $\bar{i} = \bar{1}, \bar{2}, \ldots, \bar{n}$. One of the simplest family of deterministic transformations is given by a matrix $\mathbb{Q}$ such that for all j exactly one exit *probability* p_{ij} equals to one and the rest are zero (each column of $\mathbb{Q}$ consists of a single one and $2n - 1$ zeros). Note that this does not mean that the evolution is reversible. The number of ones in any row of $\mathbb{Q}$ can be zero or more than one, i.e., $\mathbb{Q}$ does not need to be a permutation matrix.

Specifically, we set the following transition probabilities to one: $p_{i+1,\bar{i}}$ (for $\bar{i} < \bar{n}$), p_{ii} (for all i), and $p_{\bar{n}\bar{n}}$. All the remaining probabilities are zero—see a graphical representation of this model in Fig. 1. The initial state is set to

$$\mathbf{q}_0 = (0, \underbrace{-1, \ldots, -1}_{n-1}, \underbrace{1, \ldots, 1}_{n})^T. \tag{11}$$

Note that the condition (6) is satisfied.

Recall that the observable probability of the particle being in the state i is $p_i = q_i + \bar{q}_i$. Hence at $t = 0$ the particle is in state 1 with probability $p_1 = 0 + 1 = 1$ and all the remaining states are unoccupied, i.e., $p_i = -1 + 1 = 0$. But, at $t = 1$ the system transfers to

$$\mathbf{q}_1 = (0, 0, \ldots, 0, 0, \ldots, 0, 1)^T. \tag{12}$$

This is remarkable because for this state $p_i = 0 + 0 = 0$ (for $i < n$) and $p_n = 0 + 1 = 1$, and thus the particle ends up in n with certainty, and it stays there forever. All this happens in one step from the state 1 to the state n, even though they are not connected. This is not possible in a random walk without nebits.

3.5 *Example No. 2—Unstructured Database Search*

Our next example is unstructured database search, a flagship algorithm that works faster on a quantum computer than on a classical one as shown by Grover [20].

Database entries are represented by the vertices. One vertex, which we chose to be 1 without limiting the generality, is the item we want to find. Initially, the system starts in a product state between a single nebit and the particle uniformly distributed over all the vertices (13):

$$\mathbf{q}_0 = \begin{pmatrix} 1+\Delta \\ -\Delta \end{pmatrix} \otimes \frac{1}{n} \begin{pmatrix} 1 \\ 1 \\ \vdots \\ 1 \end{pmatrix}. \tag{13}$$

The matrix $\mathbb{Q}$ looks like this

$$p_{ij} = \begin{cases} \frac{1}{n} & \text{for } i = 1, \ldots, n \text{ and } j = 2, \ldots, n \\ \frac{1}{n} & \text{for } i = \bar{1}, \ldots, \bar{n} \text{ and } j = \bar{2}, \ldots, \bar{n} \\ 1 & \text{for } i, j = 1 \\ \frac{1}{n-1} & \text{for } i = \bar{2}, \ldots, \bar{n} \text{ and } j = \bar{1} \\ 0 & \text{otherwise} \end{cases} \tag{14}$$

Note that the outgoing transition probabilities at the marked vertex are different simply because, as in many search algorithms, we use an oracle who adjusts the system's evolution if it encounters the wanted entry. For example, in the Grover's algorithm [20] the system changes its phase at the marked vertex.

It is useful to represent $\mathbb{Q}$ in the form

$$\mathbb{Q} = \mathbb{S} \oplus \bar{\mathbb{S}}, \tag{15}$$

where $\mathbb{S}$ acts on the subspace spanned by the states $i = 1, 2, \ldots, n$ and $\bar{\mathbb{S}}$ on the subspace spanned by $i = \bar{1}, \bar{2}, \ldots, \bar{n}$. The matrices $\mathbb{S}$ and $\bar{\mathbb{S}}$ are

$$\mathbb{S} = \begin{pmatrix} 1 & 1/n & 1/n & \ldots & 1/n \\ 0 & 1/n & 1/n & \ldots & 1/n \\ 0 & 1/n & 1/n & \ldots & 1/n \\ \vdots & \vdots & \vdots & \ddots & \vdots \\ 0 & 1/n & 1/n & \ldots & 1/n \end{pmatrix} \tag{16}$$

and

$$\bar{\mathbb{S}} = \begin{pmatrix} 0 & 1/n & 1/n & \ldots & 1/n \\ 1/(n-1) & 1/n & 1/n & \ldots & 1/n \\ 1/(n-1) & 1/n & 1/n & \ldots & 1/n \\ \vdots & \vdots & \vdots & \ddots & \vdots \\ 1/(n-1) & 1/n & 1/n & \ldots & 1/n \end{pmatrix}. \tag{17}$$

The system's state after a single step is

$$\mathbf{q}_1 = \begin{pmatrix} 1+\Delta \\ 0 \end{pmatrix} \otimes \mathbb{S}\frac{1}{n}\begin{pmatrix} 1 \\ 1 \\ \vdots \\ 1 \end{pmatrix} + \begin{pmatrix} 0 \\ -\Delta \end{pmatrix} \otimes \bar{\mathbb{S}}\frac{1}{n}\begin{pmatrix} 1 \\ 1 \\ \vdots \\ 1 \end{pmatrix} \tag{18}$$

$$= \begin{pmatrix} 1+\Delta \\ 0 \end{pmatrix} \otimes \frac{1}{n^2}\begin{pmatrix} 2n-1 \\ n-1 \\ \vdots \\ n-1 \end{pmatrix} + \begin{pmatrix} 0 \\ -\Delta \end{pmatrix} \otimes \frac{1}{n^2}\begin{pmatrix} n-1 \\ n+\frac{1}{n-1} \\ \vdots \\ n+\frac{1}{n-1} \end{pmatrix}, \tag{19}$$

and the resulting observable probabilities are

$$\mathbf{p}_1 = \frac{(1+\Delta)}{n^2}\begin{pmatrix} 2n-1 \\ n-1 \\ \vdots \\ n-1 \end{pmatrix} - \frac{\Delta}{n^2}\begin{pmatrix} n-1 \\ n+\frac{1}{n-1} \\ \vdots \\ n+\frac{1}{n-1} \end{pmatrix} = \frac{1}{n^2}\begin{pmatrix} (2+\Delta)n-1 \\ n-1-\Delta\frac{n}{n-1} \\ \vdots \\ n-1-\Delta\frac{n}{n-1} \end{pmatrix}. \tag{20}$$

In order for negative *probabilities* to be unobservable, the parameter Δ cannot exceed $\frac{(n-1)^2}{n}$. For $\Delta = \frac{(n-1)^2}{n}$ we get

$$\mathbf{p}_1 = (1, 0, \ldots, 0)^T. \tag{21}$$

Thus, for Δ of the order $O(n)$, the marked vertex can be found in only one step. Indeed, nebits provide a remarkable computational advantage [16]. Note, that the quantum efficiency of the Grover algorithm, i.e., finding the marked vertex in $\sqrt{n}$ steps, can be achieved if $\Delta = O(\sqrt{n})$.

4 Conclusions and Open Questions

After its inception, the theory of quanta immediately moved into the Hilbert space instead of negative probabilities domain. Was it a historical accident, or probability amplitudes and Born rule are the most 'efficient' ways to represent microscopic domain? This cannot be the reason because complexity of negative probabilities representation of quantum states is exactly the same as its Hilbert space description. What if we argue that amplitudes of probability are more intuitive? This is a fairly poor argument because an intuition about any physical systems develops through manipulation of mathematical objects representing them. Acolytes of Hilbert space start to develop intuitions about probability amplitudes and unitary operations as much as cosmologists start to 'see' space-time manifolds and event horizons after endless hours of calculations. It would be a valuable experiment if we could isolate some poor students from the outside world and teach them negative probabilities interpretation of quantum mechanics without a single mention of Hilbert space. We are quite confident they would feel at home with a probability of a photon being in the upper arm of Mach-Zehnder interferometer equal to $-1/2$. And we are certain they would finally discover an equivalent formalism in Hilbert space and argue its versatility.

We are not saying that negative probabilities should replace the Hilbert space formalism. This would be silly. However, we do think that this alternative has something to offer, and it is not necessarily more exotic than such concepts as many worlds, quantum jumps, consistent histories, etc. After all, these popular spins on quantum theory cannot explain the origin of quantum supremacy, something we will have to reckon with in the near future if we want to harness computational power of quanta.

This work shows that classical probabilistic formalism and classical stochastic processes, when enhanced with nebit (a unit of *negative probability*), begin to resemble quantum theory. It is premature to declare that this can recover entire quantum theory, but it is worth a shot.

Our studies brought up a few immediate questions we would like to investigate in the future:

1. **How 'negative' quantum theory is and why?** We demonstrated that *nebits* enable quantum behaviour in classical systems but they can also surpass it with enough negativity. Why does Nature impose such limits? How much negativity is required to fully recover quantum theory? These are problems similar to those of the Tsirelson bound, Popescu-Rorlich boxes, and quantum correlations strength.
2. **What computational advantage negative probabilities can offer?** If negativity is what makes quantum computers tick [21–23], then it is interesting to what computational problems can be solved if standard probabilistic algorithms are supplemented with nebits? Studying these problems can provide a further insight into computational complexity theory.

Acknowledgements DK is supported by National Research Foundation, Prime Ministers Office, Singapore and the Ministry of Education, Singapore under the Research Centres of Excellence programme. PK was supported by the Ministry of Science and Higher Education in Poland (science funding scheme 2016–2017 project no. 0415/IP3/2016/74).

References

1. N.D. Mermin, Phys. Today **57**, 10–11 (2004)
2. C. Ferrie, J. Emerson, J. Phys. A: Math. Theor. **41**, 352001 (2008)
3. C. Ferrie, J. Emerson, New J. Phys. **11**, 063040 (2009)
4. C. Ferrie, R. Morris, J. Emerson, Phys. Rev. A **82**, 044103 (2010)

5. C. Ferrie, Rep. Prog. Phys. **74**, 116001 (2011)
6. R.P. Feynman, in *Quantum Implications: Essays in Honour of David Bohm*, ed. by B.J. Hiley, D. Peat (Methuen, 1987), pp. 235–248
7. W. Mückenheim et al., Phys. Rep. **133**, 337 (1986)
8. G.J. Szekely, Wilmott magazine, pp. 66–68 (2005)
9. A. Khrennikov, On negative probabilities, in *Derivatives: Models on Models* (Wiley, Chichester, 2007), pp. 317–323
10. R.W. Spekkens, Phys. Rev. Lett. **101**, 020401 (2008)
11. M. Burgin, arXiv:1008.1287 (2010)
12. S. Abramsky, A. Brandenburger, New J. Phys. **13**, 113036 (2011)
13. S. Abramsky, A. Brandenburger, in *Horizons of the Mind: A Tribute to Prakash Panagaden*, ed. by F. van Breugel, E. Kashefi, C. Palamidessi, J. Rutten (Springer, 2014), pp. 59–75
14. J.A. de Barros, G. Oas, J. Math. Psychol. **74**, 34 (2016)
15. H. Zhu, Phys. Rev. Lett. **117**, 120404 (2016)
16. D. Kaszlikowski, P. Kurzynski, arXiv:1911.12820 (2019)
17. S. Popescu, D. Rohrlich, Found. Phys. **24**, 379 (1994)
18. G. Zauner, Quantendesigns - Grundzüge einer nichtkommutativen Designtheorie. Dissertation, Universität Wien, 1999
19. T. Kopyciuk, M. Lewandowski, P. Kurzynski, New J. Phys. **21**, 103054 (2019)
20. L.K. Grover, in *Proceedings of the Twenty-Eighth Annual ACM Symposium on Theory of Computing* (1996), pp. 212–219
21. M. Howard, J. Wallman, V. Veitch, J. Emerson, Nature **510**, 351 (2014)
22. V. Veitch, C. Ferrie, D. Gross, J. Emerson, New J. Phys. **14**, 113011 (2012)
23. R. Raussendorf, D.E. Browne, N. Delfosse, C. Okay, J. Bermejo-Vega, Phys. Rev. A **95**, 052334 (2017)

Dagomir Kaszlikowski is a Polish-Singaporean quantum physicist. He is interested in foundational aspects of quantum mechanics, quantum information and quantum computation theory. His recent work focuses on reconstructing quantum mechanics in phase space from basic principles without invoking Hilbert space. Dagomir is a principal investigator in the Centre for Quantum technologies in Singapore and an associate professor in the Physics department at the National University of Singapore.

Pawel Kurzynski obtained his Ph.D. in 2007 from the Department of Physics at Adam Mickiewicz University in Poznan (Poland). His Ph.D. thesis was devoted to quantum walks—a kind of a quantum lattice gas automaton. Then he shifted his scientific interests to quantum contextuality and quantum nonlocality. From 2010 to 2015 he was a postdoc at the Centre for Quantum Technologies (CQT, Singapore). Currently he is working as an associate professor at his alma mater and as a visiting senior research fellow at CQT. He is interested in foundational aspects of quantum theory and in quantum dynamics. His current research topics include: negative probabilities, quantum contextuality, quantum indistinguishability, quantum cellular automata, collective properties of composite systems, and quantum chaos.

Compositionality as We See It, Everywhere Around Us

Bob Coecke

Abstract There are different meanings of the term "compositionality" within science: what one researcher would call compositional, is not at all compositional for another researcher. The most established conception is usually attributed to Frege, and is characterised by a bottom-up flow of meanings: the meaning of the whole can be derived from the meanings of the parts, and how these parts are structured together. Inspired by work on compositionality in quantum theory, and categorical quantum mechanics in particular, we propose the notions of Schrödinger, Whitehead, and complete compositionality. Accounting for recent important developments in quantum technology and artificial intelligence, these do not have the bottom-up meaning flow as part of their definitions. Schrödinger compositionality accommodates quantum theory, and also Wittgenstein's meaning-as-context. Complete compositionality further strengthens Schrödinger compositionality in order to single out theories like ZX-calculus, that are complete with regard to the intended model. All together, our new notions aim to capture the fact that compositionality is at its best when it is 'real', 'non-trivial', and even more when it also is 'complete'. At this point we only put forward the intuitive and/or restricted formal definitions, and leave a fully comprehensive definition to future collaborative work.

1 Introduction

I was invited to write this paper for a special issue, aimed to celebrate Andrei Khrennikov's 65th birthday in recognition of his contributions to the quantum foundations community, as well as Andrei being one of the frontrunners in using quantum structures in areas other than physics—see Sect. 9 for a picture of us. In particular, we were asked to discuss why it makes sense to use certain quantum structures outside of physics, and in particular, why quantum physics provides structures that are useful outside of physics—a question initially posed by Anton Zeilinger.

B. Coecke (✉)
Compositional Intelligence team, Quantinuum, Oxford, UK
e-mail: bob.coecke@quantinuum.com

A. Plotnitsky and E. Haven (eds.), *The Quantum-Like Revolution*,
https://doi.org/10.1007/978-3-031-12986-5_12

In my work, already for some 20 years, the quantum structure that has taken the centre stage is the quantum compositional structure. By this we refer to the particular manner in which quantum systems compose, which Schrödinger singled out as the key characteristic of quantum theory. All-together, we provided a novel formalism for quantum theory entirely based on composition of systems, casting orthonormal bases [43], measurements [42], classicality [41], and (strong) complementarity [29–31], entirely in compositional terms. The formalism can now be used to give a fully compositional account of all that what one expects to get in a course on quantum foundations and quantum computing [37]. It is also now being used in quantum industry, e.g. ZX-calculus based circuit optimisation techniques such as [5, 49, 52, 79] are now internal to Cambridge Quantum's t|ket⟩ [102]. Other uses in industry include [61], and there are also efforts by major quantum computing companies to push our compositional quantum formalism as the number one educational tool for quantum computing. The key reason for that is that compositionality yields an entirely pictorial quantum formalism, usually referred to as 'quantum picturalism' [24].

There of course have been earlier attempts to provide quantum theory with a novel formalism, see e.g. [39] for an overview. While this quest was initiated by the great John von Neumann [12, 105], it is fair to say that none of those approaches have been successful—with the exception of some applications outside of physics, e.g. in NLP [108, 109]. One of the major issues of these attempts was their failure to account for quantum composition. Meanwhile, many approaches in quantum foundations now also recognise the importance of the compositional backbone [18, 19, 32, 69, 95, 98, 99]. All of this seems to strongly confirm Schrödinger's hunch that compositionality underpins quantum theory.

With respect to using quantum structures in other areas, in 2008 we used quantum compositional structure in order to solve a then open problem [22, 45]: how can we combine meanings of words in order to produce the meaning of a sentence, with meanings represented as vectors, as is standardly done in modern natural language processing (NLP) [70]. The crux was the fact that quantum compositional structure is, category-theoretically speaking, exactly the same as grammatical structure [21]. This correspondence has recently resulted in implementations of NLP on existing quantum computers, by ourselves at Cambridge Quantum [27, 28, 83, 86]. We provide more details on this in Sects. 6.1 and 7.

Now, speaking about different disciplines, the term "compositionality" has different incarnations within different communities, and even within single communities there have been contradictory conceptions. In the recently emerging and flourishing Applied Category Theory community, which considers very similar compositional structures as the ones discussed above, there does seem to be a common understanding of what compositionality means. In many cases it also manifests itself in terms of a pictorial formalism. Within this community, compositionality has become a discipline transcending vehicle, that provides a bridge for exchanging concepts, methods and results. Within one discipline it can provide a radically new perspective, putting the focus on how things interact, rather than what they are made up from. However, as far as we know, nobody within this community has ever attempted to define what the particular brands of compositionality in use are. This paper is our attempt to do so.

So in mathematical terms, compositionality is closely related to category theory. However, there are some caveats. Firstly, category theory was proposed as a meta-theory of mathematics, and the CT community at large pushed this view, while for us compositionality is all about real-world phenomena. Secondly, many subfields of category theory favour so-called cartesian categories, while compositionally speaking these are the much less interesting categories.

The multidisciplinary community mentioned above has its own journal and conference, although with different names, Applied Category Theory and Compositionality respectively. Why? This dual terminology is a result of a compromise between communities with very different cultures. We favour "Compositionality" as it clearly emphasises the key foundational concept that governs the work, while "applied [something]" somewhat goes against the spirit of being of a foundational nature. On the other hand, in defence of the term "Applied Category Theory", it refers to a clear community shift within the new generation of category-theoreticians, not just focusing on mathematics anymore. This shift has many different dimensions to it, one being the desire to solve important real-world problems.

The oldest area doing practical ACT is computer science, thinking of data-types as systems, and programs as processes [2]. There is a huge literature, and every year conferences produce a humongous amount of new papers—some more history on compositionality in computer science is in [91]. Other papers giving nice overviews of compositional structures in a variety of fields include [8, 54], and a number of recent papers showing the wide range of topics currently addressed are [7, 13, 14, 59, 60, 71, 90, 101, 103].

2 Compositionality in Linguistics and Quantum Physics

The area mostly associated with a formal definition of compositionality is formal linguistics, where its definition is usually stated as follows:

- The meaning of a whole (cf. sentence) should only depend on the meanings of its parts (cf. words) and how they are fitted together (cf. grammar).

A prominent argument for this notion, put forward by Frege [56], is that without compositionality it would be impossible for us to understand a sentence that we have never heard before. Broadly speaking, this is of course indeed the case. However, interestingly Frege himself mentions in 'Grundlagen der Arithmetik' [55, 72]:

- "Never ask for the meaning of a word in isolation, but only in the context of a sentence."

Indeed:

- In language, meanings are to a great extent induced by the context. For example, ambiguous words such as `Alice` could point at either of these two incarnations of `Alice`:

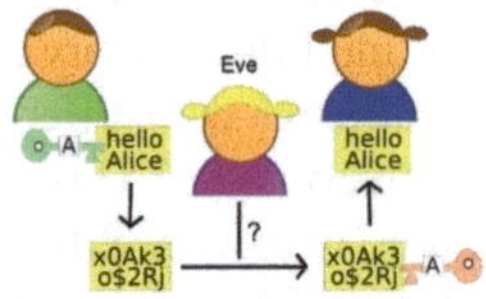

We may disambiguate `Alice` by other words in the sentence or larger text, or by non-linguistic features, like the conversation taking place between two children impersonating fairytale characters, or between security protocol experts. Even within the restricted context of gothic rock fans `Alice` could still have two meanings, two cases being:

There are some more refined formulations of linguistic compositionality that aim to account for context-dependent word ambiguities, however, the following is harder to get around:

- In language, separation between word meanings and how they grammatically fit together doesn't always make sense, as the grammar can depend on the overall meaning, which may need additional context to be disambiguated. For example, in `black metal fan` the grammar depends on the meanings:[1]

[1] We took this particular example from [38], but there are of course many more.

Using the language of [26], the respective grammatical parsings proofs are:

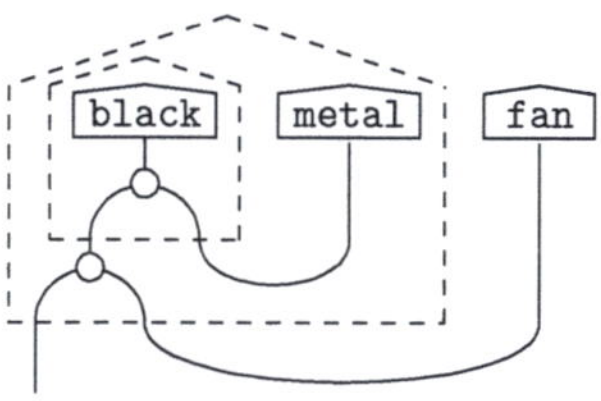

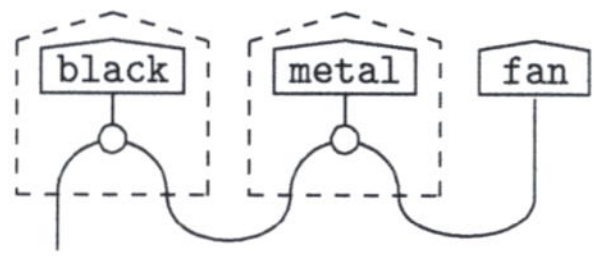

It is not as important to understand them, than just to observe that they are different, so there is no fixed grammar until the overall meaning has been unambiguously established.

In the linguistic context the fundamental problem with the above stated notion of compositionality is that it puts a special focus on bottom-up meaning-flows, the meaning of the whole being induced by the meanings of the parts. However, meaning-flows within a composite can equally well travel top-down along the same structure, that is, the meaning of the whole can contribute to meanings of parts. So what we are aiming for is some network structure that allows for such two-way meaning-flows. Maybe even better put, a network structure that allows for meaning-relationships along the different scales, cf. word verses sentence verses text.

In quantum theory the situation gets even more extreme:

- For entangled states the idea that subsystems have meanings breaks down, and so does Frege compositionality. For example, the pure Bell-state's sub-systems are in the state of total noise when conceived individually. In addition to these individual descriptions, in order to know that the overall state is the Bell-state, one needs the entire description of the Bell-state itself.

So there is no meaningful decomposition of the whole in terms of parts plus structure. Yet both language and quantum theory are perfectly compositional as far as we are concerned. How?

3 Process-Theoretic Compositionality

Let's take quantum theory as our starting point and see how we can associate a notion of compositionality to it that generalises to language. As already pointed out in the introduction, Schrödinger singled out the nature of composition of systems in quantum theory not as just a, but as <u>the</u> defining feature of the theory [96]:

> Entanglement is not one but rather the characteristic trait of quantum mechanics.

Formally, (pure-state) entanglement means:

$$|\psi_{AB}\rangle \neq |\psi_A\rangle \otimes |\psi_B\rangle \tag{1}$$

that is, the state of two systems A and B doesn't factor as a state $|\psi_A\rangle$ of A and a state $|\psi_B\rangle$ of B. We can write (1) diagrammatically as follows [23, 37]:

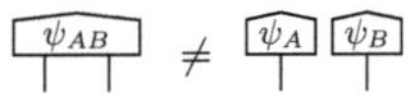

(2)

This notation of states is a special case of representing processes as boxes:

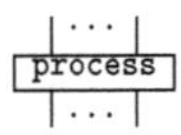

which can be composed to form *diagrams*:

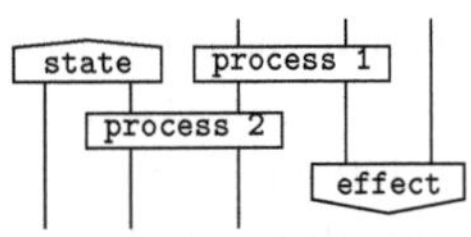

An example of a maximally entangled state/effect are the *Bell-state/effect*, which we denote as:

 $= |00\rangle + |11\rangle$  $= \langle 00| + \langle 11|$

With this state and effect our diagram becomes:

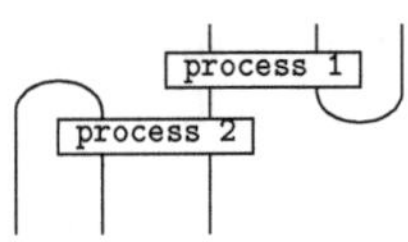

The claim we made earlier that the Bell-state's sub-systems are in the state of noise when conceived individually, diagrammatically takes the form:

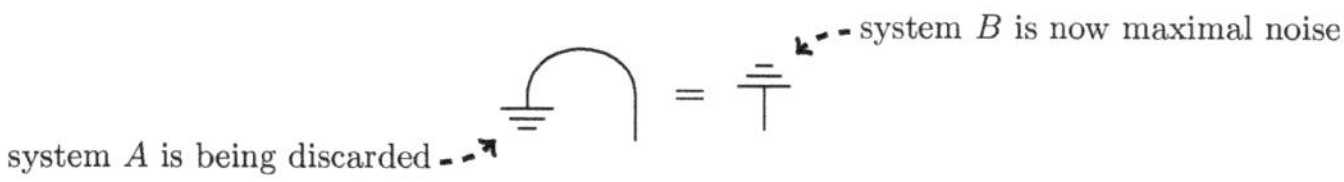

We refer to a situation where the composite of two systems cannot be meaningfully described in terms of its parts as *non-trivial composition*. Others may call it wholistic. The Bell-states, as well as any other maximally entangled states, can be shown to constitute an extremal example of non-trivial composition [33, 63]. Formally, the possible states of two systems are then isomorphically related to the processes that each system can undergo as follows:

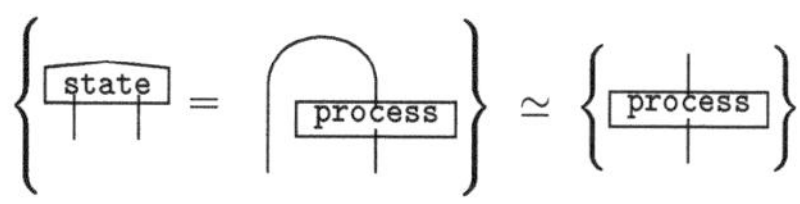

where the isomorphism is established by the following equation:

which is characteristic for extremal non-trivial composition.

What makes these kinds of states (and effects) interesting is not what they are, but what one can do with them. That is, what happens when one processes them. This for example leads to features like quantum teleportation [11, 15, 111], quantum non-locality [10, 65] and, measurement-based quantum computing [64, 93], among many more, each of these having no non-trivial classical counterparts.

This means that a general notion of compositionality should apply to general theories of processes, not just theories of states. With the advent of our compositional quantum formalism, a.k.a. categorical quantum mechanics [1, 37], a.k.a. quantum picturalism [24], we know that such a presentation leads to a much more intuitive purely diagrammatic formalism, important computational advantages, e.g. in quantum computing [50, 51, 53, 78], and bridges disciplines giving rise to new technologies such as quantum natural language processing [27, 83, 110]. The paper [34] lists recent compositionality-driven advances for the specific case of the ZX-calculus.

The idea of focusing on processes instead of states is of course not new, and even goes back to the pre-Socratic Heraclitus' "panta rhei", a.k.a. everything flows, and in more recent times was advocated by Whitehead in "Process and Reality" [107]. His stance departs from our typical Western metaphysics in which we start with a static kinematic layer with dynamics as a secondary derivative [97]. Whitehead's work however did not explicitly consider a non-trivial *parallel* composition as in (2), and therefore we will introduce two notions of compositionality, distinguishing between Whitehead and Schrödinger compositionality.

Diagrammatically, we can think of Whitehead compositionality as diagrams like this:

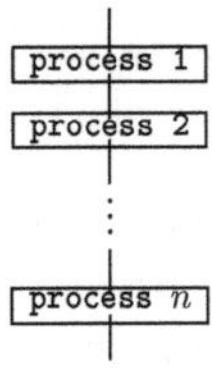

(3)

Condition (2) for two systems A and B 'separated in space', has an analogue for processes with input A and output B 'separated in time':

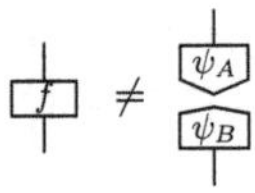

(4)

This is satisfied for any reasonable theory for processes, since otherwise we would obtain a theory only containing constant processes that produce output $|\psi_B\rangle$ for any possible input:

$$f(|\phi\rangle) = |\psi_B\rangle\langle\psi_A|\phi\rangle \simeq |\psi_B\rangle$$

Similarly, processes also enjoy non-triviality by analogy, given that what they typically do is relate varying inputs to corresponding outputs. Hence deriving them from a fixed input-output pair doesn't make much sense.

For illustrative purposes we will mainly make use of the conception of process theory as in [36, 37]. It has meanwhile been established that the process theory format provides a useful complementary presentation of quantum theory as compared the usual Hilbert space formulation. In particular, in the foundations of quantum theory a process-theoretic backbone is now a dominant feature for many developments—see e.g. [18, 19, 32, 69, 95, 98, 99].

Definition 12.1 ([36, 37]) A *process theory* is defined to consist of:

- a collection of systems (or types, or interfaces) S represented by wires,
- a collection of processes P represented by boxes, and where each process in P has a number of input wires and a number of output wires taken from S, and
- a means of composing processes, represented by wiring boxes together, and the result of doing so should again be a process in P.

While in quantum theory one can interpret these processes as taking place in space and time, this doesn't always have to be the case, for example, wirings can be physical wires:

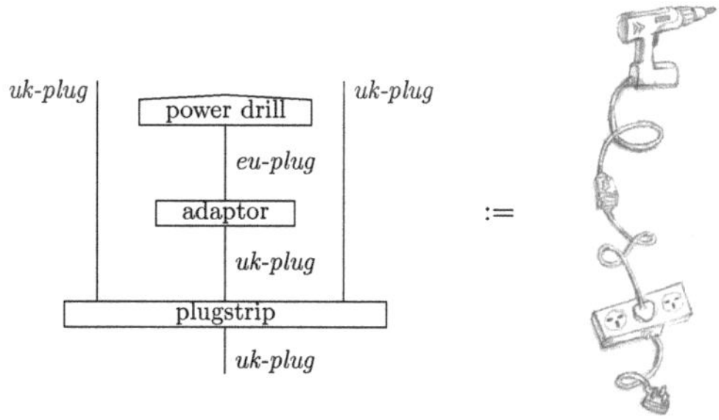

or, more abstract connections like in a flow-chart [62].

However, the definition of process theories stated above is by no means the most general notion conceivable of a compositional theory. For example, besides processes a theory could also specify transformations of processes (a.k.a. higher-order processes—see e.g. [20, 77, 89]), and it could also have transformations of interfaces etc. In the case of the latter, one could think of these diagrammatically as surfaces connecting wires, and processes transforming processes are typically denoted as 'combs' [17]:

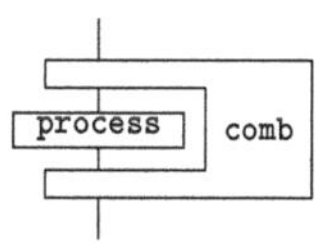

We could then also have hyper-combs transforming combs, and so on. The bottom line is that our definitions below also apply to these more general notions of compositional theories that stretch well beyond the notion of process theory stated above.

Some of the above can also be framed in the language of category theory. In the case of process theories à la Whitehead we would be talking about ordinary categories, possibly cartesian, while in the case of process theories à la Schrödinger we would be talking about proper (strict) symmetric monoidal categories [40, 84]. In the extremal case of Bell-states we are talking about compact closed categories [1, 40, 75, 76]. Going beyond process theories by considering transformations of processes and so on takes us into higher category theory [82].

4 Schrödinger Compositionality

Definition 12.2 A *Schrödinger compositional theory* is a compositional theory in the sense of the previous section, subject to two conditions:

- Composition is non-trivial, that is, the description of a whole cannot always be decomposed in any meaningful way—cf. given the description of parts, one may still require a full description of the whole in order to derive the latter.
- The ingredients of a compositional theory—e.g. the ingredients in Definition 12.1—all have clear meaningful ontological counterparts in reality.

The first condition of non-triviality is what makes the passage to a compositional description worthwhile. If it is trivial (cf. cartesian categories), then there really is not much point to elevate composition to a 1st class citizen. In fact, it is our belief that pretty much everything in the real world admits non-trivial composition when regarded in process-theoretic terms.

The second condition of the compositional structure 'being real' is essential in that it provides a model for the theory, hence establishing its consistency, and much more importantly, makes the theory relevant. It also provides us with our daily intuition as guidance for studying it, in the same way that Euclid relied on physical space to establish geometry. Finally, for the specific case of AI, compositionality then becomes a vehicle for incorporating embodiment in a compositional fashion, as for example in [13, 106].

So a slogan could be:

$$\boxed{Compositionality\, is\, at\, its\, best\, when\, it\, is\, non-trivial\, and\, real.}$$

The main purpose of the following definition is just to contrast it with Definition 12.2:

Definition 12.3 A *Whitehead compositional theory* is the same as a Schrödinger compositional theory except for the fact that composition is only one-dimensional—cf. (3).

5 Complete Compositionality

One issue with Schrödinger compositionality as we defined it is that we assumed processes to be 'black-boxes', that is, they don't necessarily interact in any obvious way. Ultimately, this boils down to what the equational structure of the process theory is: some may have a lot of equations between diagrams, other few, or none at all if they are free theories. However, there are some very special kinds of process theories around with a very rich compositional structure. They typically have a number of generating processes, which have a very easily understood interaction structure, and also generate a very rich class of processes by plugging them together. This results in a very powerful reasoning system.

One example of such a rich compositional theory is the ZX-calculus [29, 30], where, for example, a CNOT-gate arises from two generating processes [34]:

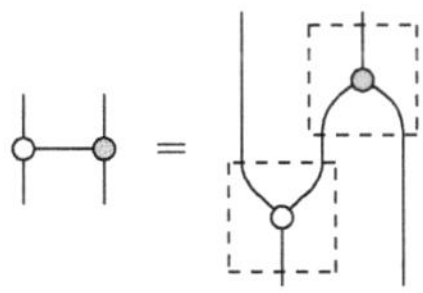

These generating processes respectively correspond to Z- and X-copying, or more generally, the respective 'spiders' they generate:

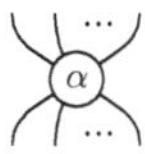

$$:= \; |0\ldots0\rangle\langle0\ldots0| + e^{i\alpha}|1\ldots1\rangle\langle1\ldots1|$$

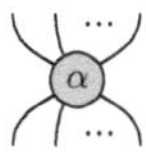

$$:= \; |+\ldots+\rangle\langle+\ldots+| + e^{i\alpha}|-\ldots-\rangle\langle-\ldots-|$$

These spiders have an interaction structure that can be written down in only 4 equations, but at the same time generate all matrices over arrays of qubits [37]. Another example of a complete compositional theory is ZW-calculus [35, 67], for which the generating processes are the GHZ-state and the W-state.

How can we qualify the richness of such a theory? We can do this in terms of how much of all the intended equations between diagrams can be derived from the interaction rules for the generating processes. In the case of the ZX-calculus and the ZW-calculus, in fact, all equations that can be derived using linear algebra can be derived from the interaction rules [68]. We say that both of these calculi are *complete*. This completeness caused quite some noise at the time, but it didn't come entirely out of the blue, some earlier work including [4, 6, 67, 73], and there also are some further elaborations [46, 104].

Definition 12.4 A *complete compositional theory* is a Schrödinger compositional theory with a complete compositional structure in the above sense.

Our slogan now becomes:

Compositionality is at its best when it is non – trivial, real and complete.

6 Examples and Non-examples

6.1 *Natural Language Processing*

Our approach to natural language from [22, 45, 66, 74] called DisCoCat, which combines meaning and linguistic structure, is Schrödinger compositional, and has equally non-trivial composition as quantum theory. Boxes in diagrams such as:

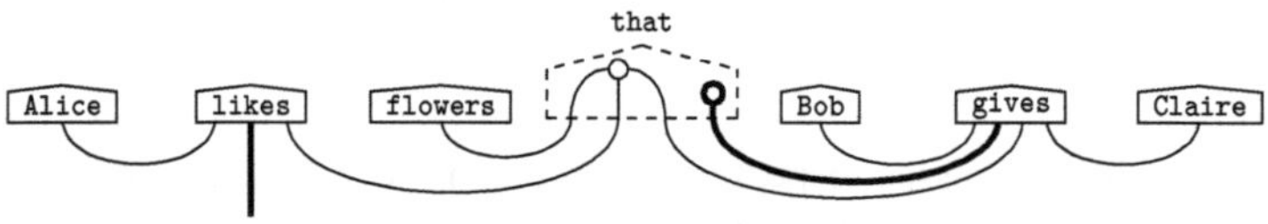

correspond to word-meanings, the wirings correspond to grammatical structure, and the resulting state is the meaning of the sentence. Hence the compositional structure is clearly 'real'.

The word-meanings themselves have composite types, and a transitive verb box such as:

evidently satisfies (2) since otherwise the subject and the object would play no role in the overall meaning of the sentence. An adjective like[2]:

cannot really be assigned any meaning until we actually know what its context is, so there clearly is no meaningful decomposition of it over parts and structure.

The wiring of `that` in terms of spiders taken from [94] can be considered as complete compositional, given that a single relative pronoun is broken down in smaller bits—formally, when interpreted in vector space, a GHZ-state and a $|+\rangle$-state. We recently proposed many more such decompositions in [48], and there is a case to be made that they are in fact 'real', as they result in new truths about grammar not previously recorded.

These language diagrams can be equivalently interpreted as quantum physics diagrams [21, 27], a connection that can only be established when conceiving both theories in the compositional manner we casted them in. This is an important feature about of shared compositional structure: it enables one to map seemingly entirely different areas onto each other, and enables an exchange of concepts and tools. In fact, in this particular case it gets even better, as this connection effectively enabled us to process natural language on a quantum computer [27, 83, 85, 86, 110]. For doing so, a language diagram, now interpreted as a quantum process, still needs to be compiled so that it fits on quantum hardware. Again this compilation is enabled by compositional structure, in this case the complete compositionality of the ZX-calculus, by 'filling' the meaning-boxes with ZX-diagrams, which can then be put in circuit-form [27]:

[2] This example was 1st pointed out to us by Fabrizio Genovese.

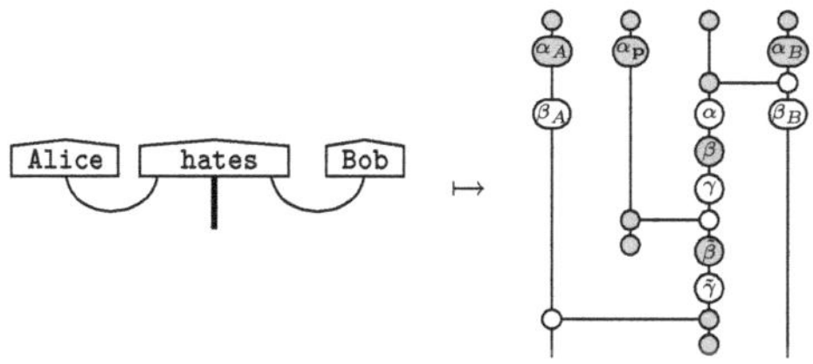

(5)

6.2 Neural Networks and Tensor Networks

In neural networks, individual neurons typically have no clear counterpart in reality, and due to how the non-linearities are inserted, have no comprehensible interaction rule. Hence standard neural networks do not satisfy the definitions of compositionality put forward here.

In the case of tensor networks in physics, nodes can have counterparts in condensed matter, but in many cases again there is no rich structure of interaction rules. Hence, tensor networks can be Schrödinger compositional, but are typically nowhere near complete compositional.

6.3 Can Maths Be Schrödinger Compositional?

Well, it kind of depends whether one conceives maths as either real or not, so it is to some extent subjective. Something that one can do is to recast mathematical structures in Schrödinger compositional terms. We for example did this with convex structures in [13], which following Gärdenfors [57, 58] can be interpreted as related to our sensual perception, and our theory then expresses how different senses compose, generating more sophisticated concepts from simple concepts. Similarly, in [106] we take 3(+1)D cartesian space, and recast it as compositional spatial relations that represent how entities within physical space are related.

7 Top-down Flow and Ambiguity

Formal linguistics is the area most associated with compositionality. But, categorical grammars such as in [3, 9, 80, 81, 87, 88] have a surprising limitation in that compositionality doesn't stretch beyond the sentence level, while, of course, when we speak/write we compose sentences in order to form larger chunks of text. DisCoCat, as described above, having been built on top of categorial grammars and is therefore also limited to relating meanings of words to those of sentences, and not to larger text.

Our attempt to fix this shortcoming was started in [26] under the name DisCoCirc and is being pushed further in [47]. The main slogan of our work is:

A sentence is a process that alters the meanings of its inhabitants.

The idea is to move from diagrams like (5) to circuits:

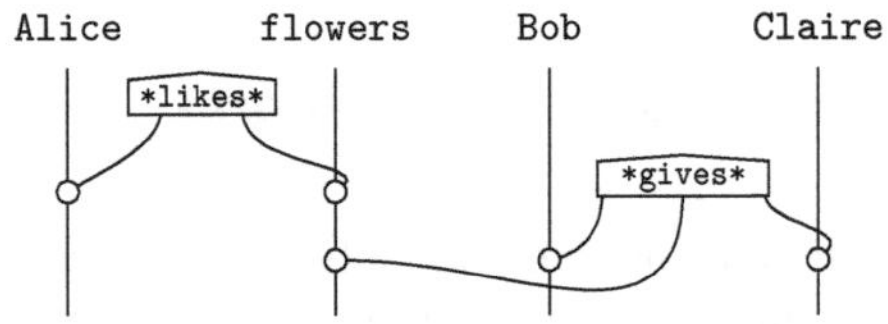

which can then be composed further:

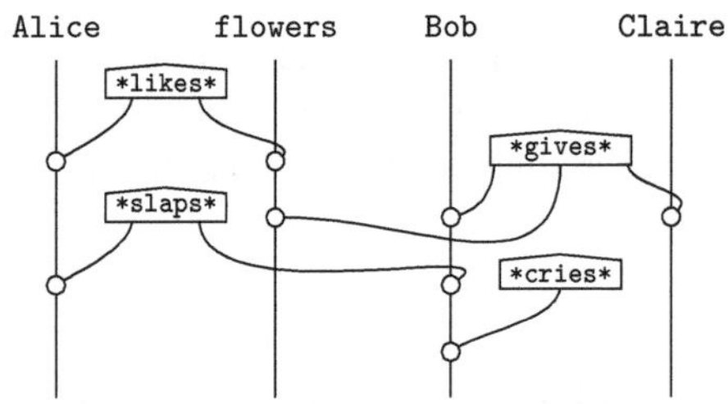

A major feature about these *language circuits* is that they can be interpreted bottom-up, as well as top-down, or a hybrid that expresses relationships between different levels. For example, we could have the meanings of a circuit as a given, and from these deduce the meanings of the words in it. This is in fact what we did in our QNLP-work [27, 83, 86], obtaining the word-meanings from a corpus of sentences we knew to be truthful. The reason for doing so in that particular case is that currently there is no better way available to load classical data on a quantum computer. So this is a case where a top-down interpretation is imposed by the technology.

As mentioned in the beginning of this paper, one of the key features of language is ambiguity, which manifests in many different ways. Evidently, when reasoning top-down, there initially will be ambiguity about the parts. In modern natural language processing meanings are represented by vectors. But wait a minute, we know how to deal with ambiguity in quantum theory: mixed states. We can now use our compositional correspondence to transfer von Neumann's quantum account on ambiguity in terms of density matrices to language [92]:

$$
\begin{aligned}
\rho_{\texttt{queen}} = & \ |\texttt{queen-royal}\rangle\langle\texttt{queen-royal}| \\
& + |\texttt{queen-band}\rangle\langle\texttt{queen-band}| \\
& + |\texttt{queen-bee}\rangle\langle\texttt{queen-bee}| \\
& + |\texttt{queen-chess}\rangle\langle\texttt{queen-chess}| \\
& + |\texttt{queen-card}\rangle\langle\texttt{queen-card}| \\
& + \ldots
\end{aligned}
$$

This can then be extended to all kinds of ambiguities, including different grammars with different meanings like in the case of `black metal fan`:

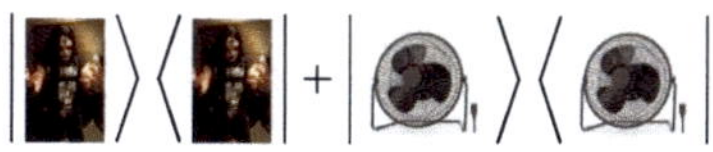

(6)

The top-down meaning-flow will then disambiguate such mixtures.

You may think that the sum in (6) takes us out of the fully compositional paradigm, but in 2005 Selinger put forward a paradigm that generates mixedness just using diagrams [100], used throughout our book [37]. It's where the 'ground' notation also entered the quantum formalism [44]:

It has even recently obtained the status of being completely compositional in [16].

8 Final Considerations

Why quantum? So why did the notion of Schrödinger compositionality emerge from quantum theory? No particular reason, really. Historically, of course, the Western metaphysical bottom-up perspective favoured reductionist interpretations. What quantum added to the picture was that it puts a case forward that simply couldn't be ignored. We also strongly believe that this is the kind of math that is very needed in humanities and social sciences, so that we finally can relieve ourselves of the mindlessness of mostly doing statistics. Ultimately, we think of these compositional structures as everywhere around us, a case that we have made at a number of previous occasions [24, 25, 28].

What's next? Admittedly, we kept things informal, adopting a restricted notion of processes, and stated the conditions on Schrödinger and complete compositionality only informally. We are currently working towards a fully-formal characterisation of several notions of compositionality.

9 Pic with Andrei

One of my first better-known papers in the area of compositionality was the transcription of a talk at Andrei Khrennikov's annual quantum foundations conference, in 2005, called Kindergarten Quantum Mechanics [23]. The next year, in 2006, Andrei got us in the local Swedish newspaper after I gave a talk "In the beginning God created ⊗", translated as "I begynnelsen skapade Gud spännmuskeln som en bild":

We thank Vincent Wang for proofreading this manuscript. At the time that this picture was taken, Vincent was still playing with teddy bears.

References

1. S. Abramsky, B. Coecke, A categorical semantics of quantum protocols, in *Proceedings of the 19th Annual IEEE Symposium on Logic in Computer Science (LICS)* (2004), pp. 415–425. arXiv:quant-ph/0402130
2. S. Abramsky, N. Tzevelekos, Introduction to categories and categorical logic, in *New Structures for Physics*. ed. by B. Coecke. Lecture Notes in Physics. (Springer, Berlin, 2011), pp.3–94
3. K. Ajdukiewicz, Die syntaktische konnexität. Studia. Philosophica **1**, 1–27 (1937)
4. M. Backens, The ZX-calculus is complete for stabilizer quantum mechanics. New J. Phys. **16**, 093021 (2014). arXiv:1307.7025
5. M. Backens, H. Miller-Bakewell, G. de Felice, J. van de Wetering, There and back again: a circuit extraction tale (2020). arXiv:2003.01664

6. M. Backens, The zx-calculus is complete for the single-qubit clifford+t group, in B. Coecke, I. Hasuo, P. Panangaden (eds.), *Proceedings of the 11th workshop on Quantum Physics and Logic, Electronic Proceedings in Theoretical Computer Science*, vol. 172 (Open Publishing Association, 2014), pp. 293–303
7. J.C. Baez, B. Fong, A compositional framework for passive linear networks (2015). arXiv:1504.05625
8. J..C.. Baez, M. Stay, Physics, topology, logic and computation: a Rosetta stone, in *New Structures for Physics*. ed. by B. Coecke. Lecture Notes in Physics. (Springer, 2011), pp.95–172
9. Y. Bar-Hillel, A quasi arithmetical notation for syntactic description. Language **29**, 47–58 (1953)
10. J.S. Bell, On the Einstein-Podolsky-Rosen paradox. Physics **1**(3), 195–200 (1964)
11. C.H. Bennett, G. Brassard, C. Crepeau, R. Jozsa, A. Peres, W.K. Wootters, Teleporting an unknown quantum state via dual classical and Einstein-Podolsky-Rosen channels. Phys. Rev. Lett. **70**(13), 1895–1899 (1993)
12. G. Birkhoff, J. von Neumann, The logic of quantum mechanics. Ann. Math. **37**, 823–843 (1936)
13. J. Bolt, B. Coecke, F. Genovese, M. Lewis, D. Marsden, R. Piedeleu, Interacting conceptual spaces I: Grammatical composition of concepts, in *Concepts and their Applications*. ed. by M. Kaipainen, A. Hautamäki, P. Gärdenfors, F. Zenker. Synthese Library, Studies in Epistemology, Logic, Methodology, and Philosophy of Science. (Springer, 2018)
14. F. Bonchi, P. Sobocinski, F. Zanasi, A categorical semantics of signal flow graphs, in *CONCUR'14: Concurrency Theory*, Lecture Notes in Computer Science, vol. 8704 (Springer, 2014), pp.435–450
15. D. Bouwmeester, J.-W. Pan, K. Mattle, M. Eibl, H. Weinfurter, A. Zeilinger, Experimental quantum teleportation. Nature **390**(6660), 575–579 (1997)
16. T. Carette, E. Jeandel, S. Perdrix, R. Vilmart, Completeness of graphical languages for mixed states quantum mechanics (2019). arXiv:1902.07143
17. G. Chiribella, G.M. D'Ariano, P. Perinotti, Quantum circuit architecture. Phys. Rev. Lett. **101**, 060401 (2008)
18. G. Chiribella, G.M. D'Ariano, P. Perinotti, Probabilistic theories with purification. Phys. Rev. A **81**(6), 062348 (2010)
19. G. Chiribella, G.M. D'Ariano, P. Perinotti, Informational derivation of quantum theory. Phys. Rev. A **84**(1), 012311 (2011)
20. G. Chiribella, G.M. D'Ariano, P. Perinotti, B. Valiron, Quantum computations without definite causal structure. Phys. Rev. A **88**(2), 022318 (2013)
21. S. Clark, B. Coecke, E. Grefenstette, S. Pulman, M. Sadrzadeh, A quantum teleportation inspired algorithm produces sentence meaning from word meaning and grammatical structure. Malays. J. Math. Sci. **8**, 15–25 (2014). arXiv:1305.0556
22. S. Clark, B. Coecke, M. Sadrzadeh, A compositional distributional model of meaning, in *Proceedings of the Second Quantum Interaction Symposium (QI-2008)* (2008), pp. 133–140
23. B. Coecke, Kindergarten quantum mechanics, in *Quantum Theory: Reconsiderations of the Foundations III*. ed. by A. Khrennikov (AIP Press, 2005), pp.81–98. arXiv:quant-ph/0510032
24. B. Coecke, Quantum picturalism. Contemp. Phys. **51**, 59–83 (2009). arXiv:0908.1787
25. B. Coecke, From quantum foundations via natural language meaning to a theory of everything, in *The Incomputable: Journeys Beyond the Turing Barrier*. ed. by S.B. Cooper, M.I. Soskova. Theory and Applications of Computability. (Springer International Publishing, 2017), pp.63–80
26. B. Coecke (2019) The mathematics of text structure. arXiv:1904.03478
27. B. Coecke, G. de Felice, K. Meichanetzidis, A. Toumi, Foundations for near-term quantum natural language processing (2020)
28. B. Coecke, G. de Felice, K. Meichanetzidis, A. Toumi, How to make qubits speak, in *Quantum Computing in the Arts and Humanities*. ed. by E.R. Miranda (Springer, 2021)

29. B. Coecke, R. Duncan, Interacting quantum observables, in *Proceedings of the 37th International Colloquium on Automata, Languages and Programming (ICALP)*, Lecture Notes in Computer Science (2008)
30. B. Coecke, R. Duncan, Interacting quantum observables: categorical algebra and diagrammatics. New J. Phys. **13**, 043016 (2011). arXiv:quant-ph/0906.4725
31. B. Coecke, R. Duncan, A. Kissinger, Q. Wang, Strong complementarity and non-locality in categorical quantum mechanics, in *Proceedings of the 27th Annual IEEE Symposium on Logic in Computer Science (LICS)* (2012). arXiv:1203.4988
32. B. Coecke, B. Edwards, R.W. Spekkens, Phase groups and the origin of non-locality for qubits. Electron. Notes Theor. Comput. Sci. **270**(2), 15–36 (2011). arXiv:1003.5005
33. B. Coecke, F. Genovese, S. Gogioso, D. Marsden, R. Piedeleu, Uniqueness of composition in quantum theory and linguistics, in Proceedings of QPL (2018). arXiv:1803.00708
34. B. Coecke, D. Horsman, A. Kissinger, Q. Wang, Kindergarden quantum mechanics graduates (... or how i learned to stop gluing lego together and love the zx-calculus). arXiv:2102.10984
35. B. Coecke, A. Kissinger, The compositional structure of multipartite quantum entanglement, in *Automata, Languages and Programming*, Lecture Notes in Computer Science (Springer, 2010), pp. 297–308. arXiv:1002.2540
36. B. Coecke, A. Kissinger, Categorical quantum mechanics I: causal quantum processes, in *Categories for the Working Philosopher*. ed. by E. Landry (Oxford University Press, 2016). arXiv:1510.05468
37. B. Coecke, A. Kissinger, *Picturing Quantum Processes. A First Course in Quantum Theory and Diagrammatic Reasoning* (Cambridge University Press, 2017)
38. B. Coecke, K. Meichanetzidis. Meaning updating of density matrices. J. Appl. Logics. **2631**(5), 745 (2020). arXiv:2001.00862
39. B. Coecke, D.J. Moore, A. Wilce, Operational quantum logic: an overview, in *Current Research in Operational Quantum Logic: Algebras, Categories and Languages*. ed. by B. Coecke, D.J. Moore, and A. Wilce, *Fundamental Theories of Physics*, vol. 111 (Springer, 2000), pp. 1–36. arXiv:quant-ph/0008019
40. B. Coecke, É.O. Paquette, Categories for the practicing physicist, in *New Structures for Physics*. ed. by B. Coecke, Lecture Notes in Physics (Springer, 2011), pp. 167–271. arXiv:0905.3010
41. B. Coecke, É. O. Paquette, D. Pavlović, Classical and quantum structuralism, in *Semantic Techniques in Quantum Computation*. ed. by S. Gay, I. Mackie (Cambridge University Press, 2010), pp. 29–69. arXiv:0904.1997
42. B. Coecke, D. Pavlovic, Quantum measurements without sums, in *Mathematics of Quantum Computing and Technology*. ed. by G. Chen, L. Kauffman, S. Lamonaco (Taylor and Francis, 2007), pp. 567–604. arXiv:quant-ph/0608035
43. B. Coecke, D. Pavlović, J. Vicary, A new description of orthogonal bases. Math. Struct. Comput Sci. to appear **23**, 555–567 (2013). arXiv:quant-ph/0810.1037
44. B. Coecke, S. Perdrix, Environment and classical channels in categorical quantum mechanics, in *Proceedings of the 19th EACSL Annual Conference on Computer Science Logic (CSL), Lecture Notes in Computer Science*, vol. 6247 (2010), pp. 230–244. Extended version: arXiv:1004.1598
45. B. Coecke, M. Sadrzadeh, S. Clark, Mathematical foundations for a compositional distributional model of meaning, in *A Festschrift for Jim Lambek*. ed. by J. van Benthem, M. Moortgat, W. Buszkowski, *Linguistic Analysis*, vol. 36 (2010), pp. 345–384. arxiv:1003.4394
46. B. Coecke, Q. Wang, ZX-rules for 2-qubit clifford+T quantum circuits, in *Reversible Computation - 10th International Conference, RC 2018, Leicester, UK, September 12-14, 2018, Proceedings*. ed. by J. Kari, I. Ulidowski, *Lecture Notes in Computer Science*, vol. 11106 (Springer, 2018), pp. 144–161
47. B. Coecke, V. Wang, Distilling grammar into circuits (or, de-humanising grammar for efficient machine use) (2021)
48. B. Coecke, V. Wang, Grammar equations. arXiv:2106.07485

49. N. de Beaudrap, X. Bian, Q. Wang, Fast and effective techniques for t-count reduction via spider nest identities, in *15th Conference on the Theory of Quantum Computation, Communication and Cryptography (TQC 2020)*. Schloss Dagstuhl-Leibniz-Zentrum für Informatik (2020). arXiv:2004.05164
50. N. de Beaudrap, D. Horsman, The zx calculus is a language for surface code lattice surgery (2017). arXiv:1704.08670
51. R. Duncan, A. Kissinger, S. Pedrix, J. van de Wetering, Graph-theoretic simplification of quantum circuits with the zx-calculus (2019). arXiv:1902.03178
52. R. Duncan, A. Kissinger, S. Perdrix, J. Van De Wetering, Graph-theoretic simplification of quantum circuits with the ZX-calculus. Quantum **4**, 279 (2020)
53. R. Duncan, S. Perdrix Rewriting measurement-based quantum computations with generalised flow, in *Proceedings of ICALP*, Lecture Notes in Computer Science (Springer, 2010), pp. 285–296
54. B. Fong, D.I. Spivak, Seven sketches in compositionality: an invitation to applied category theory (2018). arXiv:1803.05316
55. G. Frege, *Die Grundlagen der Arithmetik: eine logisch mathematische Untersuchung über den Begriff der Zahl* (W. Koebner, 1884)
56. G. Frege, Letter to Jourdain. *Philosophical and Mathematical Correspondence* (1914), pp. 78–80
57. P. Gärdenfors, *Conceptual Spaces: The Geometry of Thought* (MIT Press, 2000)
58. P. Gärdenfors, *The Geometry of Meaning: Semantics Based on Conceptual Spaces* (MIT Press, 2014)
59. N. Ghani, J. Hedges, V. Winschel, P. Zahn, Compositional game theory, in *Proceedings of the 33rd Annual ACM/IEEE Symposium on Logic in Computer Science* (ACM, 2018), pp. 472–481
60. D.R. Ghica. Geometry of synthesis: a structured approach to vlsi design, in *Proceedings of the 34th annual ACM SIGPLAN-SIGACT symposium on Principles of programming languages* (2007), pp. 363–375
61. C. Gidney, A.G. Fowler, Efficient magic state factories with a catalyzed $|CCZ\rangle$ to $2|T\rangle$ transformation. Quantum **3**, 135 (2019)
62. F.B. Gilbreth, L.M. Gilbreth, Process charts and their place in management. Mech. Eng. **70**, 38–41 (1922)
63. S. Gogioso, D. Marsden, B. Coecke, Symmetric monoidal structure with local character is a property, in *Proceedings of QPL* (2018). arXiv:1805.12088
64. D. Gottesman, I.L. Chuang, Demonstrating the viability of universal quantum computation using teleportation and single-qubit operations. Nature **402**(6760), 390–393 (1999)
65. D.M. Greenberger, M.A. Horne, A. Shimony, A. Zeilinger, Bell's theorem without inequalities. Am. J. Phys. **58**, 1131–1143 (1990)
66. E. Grefenstette, M. Sadrzadeh, Experimental support for a categorical compositional distributional model of meaning, in *The 2014 Conference on Empirical Methods on Natural Language Processing* (2011), pp. 1394–1404. arXiv:1106.4058
67. A. Hadzihasanovic, A diagrammatic axiomatisation for qubit entanglement, in *Proceedings of the 30th Annual IEEE Symposium on Logic in Computer Science (LICS)* (2015). arXiv:1501.07082
68. A. Hadzihasanovic, K.F. Ng, Q. Wang, Two complete axiomatisations of pure-state qubit quantum computing, in *Proceedings of the 33rd Annual ACM/IEEE Symposium on Logic in Computer Science* (ACM, 2018), pp. 502–511
69. L. Hardy, Foliable operational structures for general probabilistic theories, in *Deep Beauty: Understanding the Quantum World through Mathematical Innovation*. ed. by H. Halvorson (Cambridge University Press, 2011), pp. 409–442. arXiv:0912.4740
70. Z.S. Harris, Distributional structure. Word **10**(2–3), 146–162 (1954)
71. R. Hinze, D. Marsden, *Equational reasoning with lollipops, forks, cups, caps, snakes, and speedometers* (J. Log, Algebraic Methods Program, 2016)

72. T.M.V. Janssen, Frege, contextuality and compositionality. J. Logic, Lang. Inf. **10**(1), 115–136 (2001)
73. E. Jeandel, S. Perdrix, R. Vilmart, A complete axiomatisation of the zx-calculus for clifford+t quantum mechanics, in *Proceedings of the 33rd Annual ACM/IEEE Symposium on Logic in Computer Science* (2018), pp. 559–568. arXiv:1705.11151
74. D. Kartsaklis, M. Sadrzadeh, Prior disambiguation of word tensors for constructing sentence vectors, in The 2013 Conference on Empirical Methods on Natural Language Processing. (ACL, 2013), pp. 1590–1601
75. G.M. Kelly, Many-variable functorial calculus I, in *Coherence in Categories*. ed. by G.M. Kelly, M. Laplaza, G. Lewis, S. Mac Lane, *Lecture Notes in Mathematics*, vol. 281 (Springer, 1972), pp. 66–105
76. G.M. Kelly, M.L. Laplaza, Coherence for compact closed categories. J. Pure Appl. Algebra **19**, 193–213 (1980)
77. A. Kissinger, S. Uijlen, A categorical semantics for causal structure, in *2017 32nd Annual ACM/IEEE Symposium on Logic in Computer Science (LICS)* (IEEE, 2017), pp. 1–12
78. A. Kissinger, J. van de Wetering, Pyzx: large scale automated diagrammatic reasoning. arXiv:1904.04735
79. A. Kissinger, J. van de Wetering, Reducing the number of non-clifford gates in quantum circuits. Phys. Rev. A **102**, 022406 (2020)
80. J. Lambek, The mathematics of sentence structure. Am. Math. Mon. **65** (1958)
81. J. Lambek, From word to sentence. *Polimetrica* (Milan, 2008)
82. T. Leinster, *Higher Operads, Higher Categories* (Cambridge University Press, 2004)
83. R. Lorenz, A. Pearson, K. Meichanetzidis, D. Kartsalkis, B. Coecke, Qnlp in practice: running compositional models of meaning on a quantum computer (2021). arXiv:2102.12846
84. S. Mac Lane, *Categories for the Working Mathematician* (Springer, 1998)
85. K. Meichanetzidis, S. Gogioso, G. de Felice, A. Toumi, N. Chiappori, B. Coecke, Quantum natural language processing on near-term quantum computers. Accepted for QPL 2020 (2020). arXiv:2005.04147
86. K. Meichanetzidis, A. Toumi, G. de Felice, B. Coecke, Grammar-aware question-answering on quantum computers (2020). arXiv:2012.03756
87. R. Montague, Universal grammar. Theoria **36**, 373–398 (1970)
88. R. Montague, The proper treatment of quantification in ordinary English, in *Approaches to natural language* (Springer, 1973), pp. 221–242
89. O. Oreshkov, F. Costa, Č Brukner, Quantum correlations with no causal order. Nat. Commun. **3**, 1092 (2012)
90. D. Pavlovic, Monoidal computer i: basic computability by string diagrams. Inf. Comput. **226**, 94–116 (2013)
91. D. Pavlovic, Twitter replies (2021)
92. R. Piedeleu, D. Kartsaklis, B. Coecke, M. Sadrzadeh, Open system categorical quantum semantics in natural language processing, in *6th Conference on Algebra and Coalgebra in Computer Science (CALCO 2015)*. Schloss Dagstuhl-Leibniz-Zentrum fuer Informatik (2015)
93. R. Raussendorf, D.E. Browne, H.J. Briegel, Measurement-based quantum computation on cluster states. Phys. Rev. A **68**(2), 22312 (2003)
94. M. Sadrzadeh, S. Clark, B. Coecke, The Frobenius anatomy of word meanings I: subject and object relative pronouns. J. Logic Comput. **23**, 1293–1317 (2013). arXiv:1404.5278
95. D. Schmid, J.H. Selby, R.W. Spekkens, Unscrambling the omelette of causation and inference: The framework of causal-inferential theories (2020). arXiv:2009.03297
96. E. Schrödinger, Discussion of probability relations between separated systems. Math. Proc. Camb. Philos. Soc. **31**, 555–563 (1935)
97. J. Seibt, Process philosophy. *Stanford Encyclopedia of Philosophy* (2012)
98. J.H. Selby, B. Coecke, Leaks: quantum, classical, intermediate and more. Entropy **19**(4), 174 (2017)
99. J.H. Selby, C.M. Scandolo, B. Coecke, Reconstructing quantum theory from diagrammatic postulates (2018). arXiv:1802.00367

100. P. Selinger, Dagger compact closed categories and completely positive maps. Electron. Notes Theor. Comput. Sci. **170**, 139–163 (2007)
101. D. Shiebler, B. Gavranović, P. Wilson, Category theory in machine learning (2021). arXiv:2106.07032
102. S. Sivarajah, S. Dilkes, A. Cowtan, W. Simmons, A. Edgington, R. Duncan, t A retargetable compiler for NISQ devices (2020). arXiv:2003.10611
103. T. St Clere Smithe, Bayesian updates compose optically (2020). arXiv:2006.01631
104. R. Vilmart. A near-minimal axiomatisation of zx-calculus for pure qubit quantum mechanics, in *2019 34th Annual ACM/IEEE Symposium on Logic in Computer Science (LICS)* (IEEE, 2019), pp. 1–10
105. J. von Neumann, Mathematische grundlagen der quantenmechanik (Springer). Translation *Mathematical Foundations of Quantum Mechanics* (Princeton University Press, 1932), p.1955
106. V. Wang-Mascianica, B. Coecke, Talking space: inference from spatial linguistic meanings (2021). arXiv:2109.06554
107. A.N. Whitehead, *Process and Reality* (Harper & Row, 1957)
108. D. Widdows, Orthogonal negation in vector spaces for modelling word-meanings and document retrieval, in *41st Annual Meeting of the Association for Computational Linguistics*, Japan (2003)
109. D. Widdows, S. Peters, Word vectors and quantum logic: experiments with negation and disjunction. Math. Lang. **8**(141–154) (2003)
110. W. Zeng, B. Coecke, Quantum algorithms for compositional natural language processing. Electron. Proc. Theor. Comput. Sci. **221** (2016). arXiv:1608.01406
111. M. Zukowski, A. Zeilinger, M.A. Horne, A.K. Ekert, Event-ready-detectors Bell experiment via entanglement swapping. Phys. Rev. Lett. **71**, 4287–4290 (1993)

Bob Coecke is Chief Scientist at Cambridge Quantum/Quantinuum, where he also heads the Oxford-based Quantum NLP & Compositional Intelligence team, and is Emeritus Professor at Wolfson College, Oxford University. Previously he was Professor of Quantum Foundations, Logics and Structures at the Department of Computer Science at Oxford University, where he was 20 years, and co-founded, built, and led a multi-disciplinary research group of up to 50 people. He supervised 66 PhD students. He pioneered Categorical Quantum Mechanics (now in AMS's MSC2020 classification), ZX-calculus, DisCoCat natural language meaning, mathematical foundations for resource theories, Quantum Natural Language Processing, and is co-author of Picturing Quantum Processes, a book providing a fully diagrammatic treatment of quantum theory and its applications. He co-authored close to 200 research papers. He's a founding father of the QPL (Quantum Physics and Logic) and ACT (Applied Category Theory) communities, of the diamond-open-access journal Compositionality, and Cambridge University Press' Applied Category Theory book series. He was the first person to have Quantum Foundations as part of his professorial title. His work headlined in various media outlets including Forbes, New Scientist, PhysicsWorld, ComputerWeekly.

Measurement Models in Quantum Cognition

Jerome R. Busemeyer

Abstract A theoretical puzzle for quantum cognition researchers is to simultaneously account for an A-A repeat effect, question order effects, the QQ equality, and a repeat A-B-A effect. Recently, the problem was resolved in the framework of a theory of non-atomic quantum instruments by Ozawa and Khrennikov. The quantum instruments used by Ozawa-Khrennikov are not standard quantum physics. The purpose of our contribution is to present an alternative, perhaps more standard, account. The actual empirical status of all these effects remains debatable, and so we encourage further empirical research on these issues.

1 Introduction

It is a great honor to contribute to the Festschrift for Andrei Khrennikov, one of the pioneers to apply "quantum like" mathematical models outside of physics to psychology, biology, sociology, and economics. He has been a powerful force in this new direction ever since some of his earliest work on the topic [3], and his footprints are visible all across the map of this field.

One of the contributions that Andrei made to this field concerns a problem that arises when applying quantum measurement theory to human judgments. In particular, the puzzle is the following: can quantum measurement theory simultaneously account for *response repetition* and *question order* effects [4]. This is the issue we plan to address in our contribution to the Festschrift.

J. R. Busemeyer (✉)
Psychological and Brain Sciences, Indiana University, Bloomington, IN, USA
e-mail: jbusemey@iu.edu

A. Plotnitsky and E. Haven (eds.), *The Quantum-Like Revolution*,
https://doi.org/10.1007/978-3-031-12986-5_13

2 What Are Response Repetition and Question Order Effects

Survey researchers design questions to assess opinions and attitudes such as the following two examples concerning religious preferences of conservative Christians and liberals:

(A) Conservative Christians have gone too far in trying to impose their religious values on the country
(B) Liberals have gone too far in trying to keep religion out of the schools and the government.

The A-A type of response repetition effect refers to the expectation that if question A is asked repeatedly, back to back, with no other information, then a person will simply repeat their answer. For example, with binary (yes, no) responses

$$p(A = y|A = y) = 1.$$

This expectation is so strong that it is never really tested. Survey researchers have never actually empirically tested this expectation.

The question order effect refers to the effect that changing question orders has on the joint relative frequency of responses. Survey researchers are wary of question order effects and they often test them by changing the order that questions are asked to different participants: half the participants are asked question A first followed immediately by question B (A-B condition), and the other half are asked in the opposite order (B-A condition). A question order effect occurs when the joint distribution for the pair of answers in one order differs from the other order. For example, with binary responses

$$p(A = y, B = y) \neq p(B = y, A = y)$$

where $p(X = i, Y = j)$ refers to the sequential probability of observing $X = i$ and then $Y = j$. Question order effects are quite pervasive in social science research Tourangeau et al. [7].

Wang et al. [9] discovered an interesting empirical regularity that question order effects tend to obey called the QQ equality. The following identity was empirically well supported by over 80 national surveys that examined question orders in a 10 year period of time

$$\begin{aligned} QQ &= [p(A = y, B = n) + p(A = n, B = y)] \\ &\quad - [p(B = y, A = n) + p(B = n, A = y)] = 0 \end{aligned}$$

Actually this empirical regularity was discovered after it was derived from a general model of question order effects proposed earlier by Wang and Busemeyer [8], which is based on the use of projectors to represent quantum measurements.

Khrennikov et al. [4] were especially interested in the following effect called the A-B-A repetition effect. First question A is asked, followed by question B, and finally question A is repeated. This is analogous to a Stern-Gerlach type of experiment. The main question concerns whether not the participant will change answers to the repeated question about A after answering the middle question about B. The empirical status of this type of repetition effect is still under investigation. For example, with binary choices the repetition effect states that

$$p(A_2 = i | A_1 = i, B = j) = 1.$$

So far as we know, only one study actually empirically investigated this design, and this study found that people did not repeat in this situation [1]. However, comments on this work raised concerns about the way that the second question about A was asked in this study.

The quantum measurement problem. Leaving behind the empirical status of all these effects for the time being, a problem for quantum measurement occurs if the following three findings are observed: (1) a person repeats in the A-A design, (2) the person produces question order effects using A-B versus B-A designs, and (3) the person repeats answers to question A in the A-B-A design.

3 Previous Solutions to the Quantum Measurement Problem

Khrennikov et al. [4] proved that simply using projectors, or even more general positive operator valued measurements, cannot solve the above quantum measurement problem for human judgments. Positive operator valued measurements cannot satisfy the response replicability effect in the A-A design. Projectors satisfy response replicability in the A-A design, but in order to account for question order effects, the projectors must be non-commutative, which prevents them from accounting for response replicability in the A-B-A design.

Recently, the problem was resolved in the framework of a theory of non-atomic quantum instruments by Ozawa and Khrennikov [5]. However, this work left open the question whether or not this new theoretical framework could also account for the QQ equality constraint. The latter question was later answered in the affirmative by theoretical extensions developed in Ozawa and Khrennikov [6]. Thus, as Ozawa and Khrennikov put it "the black cloud on the blue sky of quantum-like modeling of cognition disappeared."

4 System Plus Environment Measurement Model

The Ozawa-Khrennikov solution was as they say "nontrivial," because the quantum instruments that they used to solve this problem are not standard quantum physics. Although we think the Ozawa-Khrennikov solution is brilliant, the purpose of our contribution is to present an alternative, perhaps more standard, account.

Our alternative follows the Ozawa-Khrennikov measurement model by using a more recent approach based a system plus environment formulation. This is formed by a tensor product space $\mathscr{H}_{S\times E} = \mathscr{H}_{\mathscr{S}} \otimes \mathscr{H}_{\mathscr{E}}$ of a system space $\mathscr{H}_{\mathscr{S}}$ and an environment space $\mathscr{H}_{\mathscr{E}}$. The system space represents the cognitive system under investigation, and the environment space represents the measurements.

With this representation, the measurement process is performed by unitary transformations see, e.g., [2]. Control U gates operate in the combined $\mathscr{H}_{S\times E}$ system plus environment space. They operate like production rules: if the system is in a particular state, then change the environment cubits to the appropriate basis states.

4.1 *Analogies Between Stern–Gerlach and A-B-A Experiments*

It is helpful to pause here to make an analogy between the application of measurements in a Stern-Gerlach physics experiment and measurements in an A-B-A psychology experiment. Imagine that we have two double major (physics and psychology) students, Angela and Bob. Angela is a very careful and conscientious student but Bob is kind of lazy.

Their physics professor asks them both to perform a Stern Gerlach experiment to test a spin particle first in the "up-down" direction, and then in the "plus-minus" direction, and then again in the "up-down" direction. Angela follows her professor's instructions and carries out the series of three tests according to instructions and reports back to her professor of obtaining a 0.50 probability of getting a spin "up" on the last "up-down" test given that the first "up-down" test produced a spin "up" result (in accordance with quantum physics of spin and Angela gets a grade of A). Bob is not only lazy but also unfamiliar with quantum theory. Bob only carries out a series of two tests, first testing in the "up-down" direction and then in the "plus-minus" direction. However, Bob sees no reason to carry out the last test in the "up-down" direction, because he already knows the result for that direction from the first test. So Bob never bothers to carry out any new quantum test in the "up-down" direction the second time and instead just uses the previously recorded answer. Consequently, he reports back to his professor of obtaining 1.0 probability of getting a spin "up" on the last "up-down" test given that the first "up-down" test produced a spin "up" result (contrary to quantum physics of spin and Bob gets a grade of F).

Then Angela and Bob "volunteer" for a psychology experiment as required by their psychology class. The psychology experimenter asks them to first judge question A, and then judge question B, and then reconsider and judge question A again. Angela follows the instructions and makes three actual judgments: she judges "yes" to the first question about A, then she judges "no" to the second question about B, and finally she reconsiders the question about A and forms a new judgment, which changes her mind to make a "no" response. Lazy Bob only makes two actual judgments: he judges "yes" to the first question about A, then judges "no" to the second question about B. However, Bob sees little point in re-evaluating and judging the last question about A again, because he already knows the answer to that question. So Bob simply reports the same "yes" answer on the second question about A that he recorded for the first question about A. Bob never bothers to re-evaluate question A the second time.

What we learn from these stories is that Angela and Bob apply different types of measurement operators for both the physics experiments and for the psychology experiments on their last test. Angela conducts a new quantum test to obtain the outcome of the last measurement in both the physics and the psychology experiments. Bob does not bother to conduct any new quantum test for the last measurement, and instead simply reports the same answer recorded on the first test. Now we will implement these two different types of measurement procedures in a system plus environment quantum measurement process.

4.2 Hilbert Space

For the A-B-A design, we need to consider three measurements, first we measure A (recording the first answer), then we measure B (recording the second answer), and finally we measure A again (recording a third answer). For simplicity, we will assume that the measurements are binary, but the basic ideas easily extend to arbitrary (but finite) measurements.

We start with a space $\mathscr{H}_{\mathscr{S}}$ representing the cognitive system spanned by a basis $\mathscr{S} = \{|S_i\rangle, i = 1, \ldots, N\}$. The environment is represented by a space $\mathscr{H}_{\mathscr{E}}$ spanned by 8 basis vectors $\{|E_j\rangle \otimes |E'_k\rangle \otimes |E''_l\rangle, j, k, l = 0, 1\}$, 0 for no, 1 for yes, representing the 3 binary environmental measurements that get recorded, where the first pair of cubits $\{|E_j\rangle, j = 0, 1\}$ represents the first response to question A, and the second pair of cubits $\{|E_k\rangle, k = 0, 1\}$ represents the response to question B, and the third pair of cubits $\{|E_l\rangle, l = 0, 1\}$ represents the second response to question A. The combined system is the $N + 8$ dimensional tensor product state $\mathscr{H}_{S\times E} = \mathscr{H}_{\mathscr{S}} \otimes \mathscr{H}_{\mathscr{E}}$.

4.3 Initial State for both Angela and Bob

Define the initial state of the cognitive system for Angela and Bob as follows

$$|\psi\rangle = \sum |S_i\rangle \langle S_i| \cdot |\psi\rangle = \sum \psi_i |S_i\rangle .$$

and the initial state of the environment system as

$$\begin{aligned} |E_+\rangle &= \frac{1}{\sqrt{2}} |E_0\rangle + \frac{1}{\sqrt{2}} |E_1\rangle , \\ |E'_+\rangle &= \frac{1}{\sqrt{2}} |E'_0\rangle + \frac{1}{\sqrt{2}} |E'_1\rangle , \\ |E''_+\rangle &= \frac{1}{\sqrt{2}} |E''_0\rangle + \frac{1}{\sqrt{2}} |E''_1\rangle . \end{aligned}$$

Then the initial state of the system plus environment is defined as

$$|\zeta\rangle = |\psi\rangle \otimes |E_+\rangle \otimes |E'_+\rangle \otimes |E''_+\rangle .$$

4.4 Projectors Operating on the Cognitive System for both Angela and Bob

We can define $\boldsymbol{T}_A$ as the projector acting on vectors in $\mathscr{H}_{\mathscr{S}}$, which projects on to the yes answer for question A, and then define $\boldsymbol{F}_A = I - \boldsymbol{T}_A$. The initial cognitive state can then be expressed as a superposition of the projection on the yes response and the no response to question A.

$$\begin{aligned} |\psi\rangle &= (\boldsymbol{T}_A + \boldsymbol{F}_A) |\psi\rangle \\ &= \boldsymbol{T}_A |\psi\rangle + \boldsymbol{F}_A |\psi\rangle . \end{aligned}$$

Likewise, we can define $\boldsymbol{T}_B$ as the projector acting on vectors in $\mathscr{H}_{\mathscr{S}}$, which projects on to the yes answer for question B, and define $\boldsymbol{F}_B = I - \boldsymbol{T}_B$. The initial cognitive state can then be expressed as a superposition of the basis states associated with the yes response and the basis state associated with the no response to question B.

$$\begin{aligned} |\psi\rangle &= (\boldsymbol{T}_B + \boldsymbol{F}_B) |\psi\rangle \\ &= \boldsymbol{T}_B |\psi\rangle + \boldsymbol{F}_B |\psi\rangle . \end{aligned}$$

4.5 Control U Gates of First Two Measurements for Angela and Bob

The control U gate CU_A for the first question A is defined to operate depending on the projections of the cognitive state for the first question A. This transformation is unitary because it linearly maps one orthonormal basis to another.

$$\begin{aligned} CU_A|\zeta\rangle &= CU_A(\boldsymbol{F}_A+\boldsymbol{T}_A)|\psi\rangle\otimes|E_+\rangle\otimes|E'_+\rangle\otimes|E''_+\rangle \\ &= CU_A(\boldsymbol{F}_A|\psi\rangle+\boldsymbol{T}_A|\psi\rangle)\otimes|E_+\rangle\otimes|E'_+\rangle\otimes|E''_+\rangle \\ &= CU_A\boldsymbol{F}_A|\psi\rangle\otimes|E_+\rangle\otimes|E'_+\rangle\otimes|E''_+\rangle \\ &+ CU_A\boldsymbol{T}_A|\psi\rangle\otimes|E_+\rangle\otimes|E'_+\rangle\otimes|E''_+\rangle \\ &= \boldsymbol{F}_A|\psi\rangle\otimes|E_0\rangle\otimes|E'_+\rangle\otimes|E''_+\rangle+\boldsymbol{T}_A|\psi\rangle\otimes|E_1\rangle\otimes|E'_+\rangle\otimes|E''_+\rangle \end{aligned}$$

Control U gate CU_B for the second question B measurement is defined to operate depending on the projections of the cognitive state for question B. This transformation is again unitary because it linearly maps one orthonormal basis to another.

$$\begin{aligned} CU_B|\zeta\rangle &= CU_B(\boldsymbol{F}_B+\boldsymbol{T}_B)|\psi\rangle\otimes|E_+\rangle\otimes|E'_+\rangle\otimes|E''_+\rangle \\ &= CU_B(\boldsymbol{F}_B|\psi\rangle+\boldsymbol{T}_B|\psi\rangle)\otimes|E_+\rangle\otimes|E'_+\rangle\otimes|E''_+\rangle \\ &= CU_B\boldsymbol{F}_B|\psi\rangle\otimes|E_+\rangle\otimes|E'_+\rangle\otimes|E''_+\rangle \\ &+ CU_B\boldsymbol{T}_B|\psi\rangle\otimes|E_+\rangle\otimes|E'_+\rangle\otimes|E''_+\rangle \\ &= \boldsymbol{F}_B|\psi\rangle\otimes|E_+\rangle\otimes|E'_0\rangle\otimes|E''_+\rangle+\boldsymbol{T}_B|\psi\rangle\otimes|E_+\rangle\otimes|E'_1\rangle\otimes|E''_+\rangle. \end{aligned}$$

Therefore, the control U gate that is applied when question A is asked first and then question B is the product

$$\begin{aligned} CU_BCU_A|\zeta\rangle &= CU_B\left(\boldsymbol{F}_A|\psi\rangle\otimes|E_0\rangle\otimes|E'_+\rangle\otimes|E''_+\rangle\right) \\ &+ CU_B\left(\boldsymbol{T}_A|\psi\rangle\otimes|E_1\rangle\otimes|E'_+\rangle\otimes|E''_+\rangle\right) \\ &= \boldsymbol{F}_B\boldsymbol{F}_A|\psi\rangle\otimes|E_0\rangle\otimes|E'_0\rangle\otimes|E''_+\rangle \\ &+ \boldsymbol{T}_B\boldsymbol{F}_A|\psi\rangle\otimes|E_0\rangle\otimes|E'_1\rangle\otimes|E''_+\rangle \\ &+ \boldsymbol{F}_B\boldsymbol{T}_A|\psi\rangle\otimes|E_1\rangle\otimes|E'_0\rangle\otimes|E''_+\rangle \\ &+ \boldsymbol{T}_B\boldsymbol{T}_A|\psi\rangle\otimes|E_1\rangle\otimes|E'_1\rangle\otimes|E''_+\rangle. \end{aligned}$$

4.6 Projectors for Measurement of First Two Questions

The projectors for measurement are defined on the system plus environment space. The measurement projectors for no versus yes to the first question are

$$\begin{aligned} A_F &= I \otimes |E_0\rangle\langle E_0| \otimes I \otimes I \\ A_T &= I \otimes |E_1\rangle\langle E_1| \otimes I \otimes I. \end{aligned}$$

Likewise the measurement projectors for no versus yes to the second question are

$$\begin{aligned} B_F &= I \otimes I \otimes |E_0'\rangle\langle E_0'| \otimes I \\ B_T &= I \otimes I \otimes |E_1'\rangle\langle E_1'| \otimes I. \end{aligned}$$

Finally, the measurement projectors for no versus yes to the third question are

$$\begin{aligned} A_F' &= I \otimes I \otimes I \otimes |E_0''\rangle\langle E_0''| \\ A_T' &= I \otimes I \otimes I \otimes |E_1''\rangle\langle E_1''|. \end{aligned}$$

4.7 Response Probabilities for First Two Questions A,B

If we measure question A and then B, then we apply CU_A and then CU_B to the initial state and compute the probability of the measurement outcomes. For example,

$$\begin{aligned} p(A = y, B = n) &= \| \boldsymbol{B}_F \boldsymbol{A}_T CU_B CU_A |\zeta\rangle \|^2 \\ &= \left\| (\boldsymbol{F}_B \boldsymbol{T}_A |\psi\rangle) \otimes |E_1\rangle \otimes |E_0'\rangle \otimes |E_+''\rangle \right\|^2 \\ &= \| \boldsymbol{F}_B \boldsymbol{T}_A |\psi\rangle \|^2 \end{aligned}$$

In the opposite order, we obtain

$$\begin{aligned} p(B = n, A = y) &= \| \boldsymbol{A}_T \boldsymbol{B}_F CU_A CU_B |\zeta\rangle \|^2 \\ &= \left\| (\boldsymbol{T}_A \boldsymbol{F}_B |\psi\rangle) \otimes |E_1\rangle \otimes |E_0'\rangle \otimes |E_+''\rangle \right\|^2 \\ &= \| \boldsymbol{T}_A \boldsymbol{F}_B |\psi\rangle \|^2. \end{aligned}$$

Thus, for the first two A, B questions, the system plus environment model produces exactly the same sequential probabilities as the original question order model Wang and Busemeyer [8] using only projectors applied to the cognitive system. Consequently, this system plus environment model also satisfies the QQ equality.

4.8 State After Answering First Two Questions

Suppose the answers to the first two questions are $A = y$, $B = n$. Then the state, conditioned on these answers equals

$$\left|\zeta_{A=1,B=0}\right\rangle = \frac{1}{\sqrt{p(A=y, B=n)}} \left(\boldsymbol{F}_B \boldsymbol{T}_A \left|\psi\right\rangle\right) \otimes \left|E_1\right\rangle \otimes \left|E'_0\right\rangle \otimes \left|E''_+\right\rangle$$
$$= \frac{1}{\sqrt{p(A=y, B=n)}} \left|\psi_{A=1,B=0}\right\rangle \otimes \left|E_1\right\rangle \otimes \left|E'_0\right\rangle \otimes \left|E''_+\right\rangle$$

The state following other pairs of answers is constructed in a similar manner, and can be written more generally as ($i = 0, 1$; $j = 0, 1$)

$$\left|\zeta_{AB}\right\rangle = \frac{1}{\sqrt{p(A=i, B=j)}} \left|\psi_{A,B}\right\rangle \otimes \left|E_i\right\rangle \otimes \left|E'_j\right\rangle \otimes \left|E''_+\right\rangle$$

4.9 Control U Gates of Last Measurement for Angela and Bob

For Angela, the Control U gate $CU_{A'}$ for the last question A, after answering the first two questions, is defined as follows:

$$\begin{aligned} CU_{A'}\left|\zeta_{AB}\right\rangle &= CU_{A'}\left(\boldsymbol{F}_A + \boldsymbol{T}_A\right)\left|\psi_{AB}\right\rangle \otimes \left|E_i\right\rangle \otimes \left|E'_j\right\rangle \otimes \left|E''_+\right\rangle \\ &= CU_{A'}\left(\boldsymbol{F}_A\left|\psi_{AB}\right\rangle + \boldsymbol{T}_A\left|\psi_{AB}\right\rangle\right) \otimes \left|E_i\right\rangle \otimes \left|E'_j\right\rangle \otimes \left|E''_+\right\rangle \\ &= CU_{A'}\boldsymbol{F}_A\left|\psi_{AB}\right\rangle \otimes \left|E_i\right\rangle \otimes \left|E'_j\right\rangle \otimes \left|E''_+\right\rangle \\ &+ CU_{A'}\boldsymbol{T}_A\left|\psi_{AB}\right\rangle \otimes \left|E_i\right\rangle \otimes \left|E'_j\right\rangle \otimes \left|E''_+\right\rangle \\ &= \boldsymbol{F}_A\left|\psi_{AB}\right\rangle \otimes \left|E_i\right\rangle \otimes \left|E'_j\right\rangle \otimes \left|E''_0\right\rangle + \boldsymbol{T}_A\left|\psi_{AB}\right\rangle \otimes \left|E_i\right\rangle \otimes \left|E'_j\right\rangle \otimes \left|E''_1\right\rangle. \end{aligned}$$

Then the probability of answering yes to the last question about A is

$$\begin{aligned} p(A = 1\ A = i, B = j) &= \left\| A'_T CU_{A'}\left|\zeta_{AB}\right\rangle \right\|^2 \\ &= \left\|\boldsymbol{T}_A\left|\psi_{AB}\right\rangle\right\|^2. \end{aligned}$$

Consequently, Angela does not necessarily repeat her response on the last question.

For Bob, the Control U gate $CU_{A''}$ for the last question about A is defined as follows:

$$\begin{aligned} CU_{A''}\left|\zeta\right\rangle &= CU_{A''}\left|\psi_{AB}\right\rangle \otimes \left|E_i\right\rangle \otimes \left|E'_j\right\rangle \otimes \left|E''_+\right\rangle \\ &= \left|\psi_{AB}\right\rangle \otimes \left|E_i\right\rangle \otimes \left|E'_j\right\rangle \otimes \left|E''_i\right\rangle. \end{aligned}$$

Then the probability of answering yes to the last question about A is

$$p(A = 1\ A = i, B = j) = \left\| A'_T C U_{A''} |\zeta_{AB}\rangle \right\|^2$$
$$= \begin{Bmatrix} 1, i = 1 \\ 0, i = 0 \end{Bmatrix}.$$

Consequently, Bob repeats his answer on the last question.

5 Conclusions

This chapter provides an alternative way to construct a quantum measurement model that can simultaneously account for a A-A repeat effect, question order effects, the QQ equality, and a repeat A-B-A effect. We don't have any empirical evidence for preferring the model proposed here over the earlier model proposal by Ozawa and Khrennikov [5] and Ozawa and Khrennikov [6]. Further empirical research is needed to distinguish these models. However, both models rely on a system plus environment representation, which we think is an important new theoretical step for measurement models in quantum cognition.

The actual empirical status of these "effects" is still debatable. Certainly, we know that question order effects occur from an abundance of evidence over years of past research Tourangeau et al., [7]. Strong evidence supporting the QQ equality was also reported in Wang et al. [9], however more research is needed to determine the possible boundaries on this empirical regularity. We do not know of any empirical test of the A-A repeat effect, but most researchers apparently believe this holds because they go through great efforts to avoid repeating questions.

We only know one empirical test of the A-B-A repeat effect, and this study actually found that people do not necessarily repeat Busemeyer and Wang, [1] their answers. Furthermore, the tendency to change answers varied strongly across different pairs of questions. However, this study used a 0–100 point rating scale rather than binary choices to make the measurement more sensitive to changes in opinion. Also comments related to this study raised the concern that the second question about A may not have been exactly the same as the first question, because the respondents were told "now that you have considered question B, what is your answer to question A." It is difficult to perform the A-B-A experiment because repeating a question strikes the participants as a strange request. They may wonder whether this repetition of the question is an implicit test of their consistency or attentiveness. For example, mechanical turk respondents are evaluated on the basis of the "quality" of their responses. Also, the statistical definition of this repetition effect needs to be stated more carefully and rigorously, because the demand that answers always exactly repeat can be falsified simply on the basis of a small amount of measurement error. Finally, we expect that there are individual differences in these effects. Clearly, some participants in a psychology experiment will behave like Angela, and some will behave like Bob. Therefore it is important for a measurement theory to account for both types of behavior.

Despite all the difficulties with empirically testing the A-B-A consistency, we think it is important to try to experimentally test this property. It is a direct analogue of the Stern–Gerlach experiment, which was so important in physics to draw attention to the quantum nature of the physical world. The A-B-A experiment might be able to draw attention to the quantum nature of the psychological world.

References

1. J.R. Busemeyer, Z. Wang, Is there a problem with quantum models of psychological measurements? PloS one **12**(11) (2017)
2. R.B. Griffiths, *Consistent Quantum Theory* (Cambridge University Press, 2003)
3. A.Y. Khrennikov, Classical and quantum mechanics on information spaces with applications to cognitive, psychological, social, and anomalous phenomena. Found. Phys. **29**, 1065–1098 (1999)
4. A.Y. Khrennikov, I. Basieva, E.N. Dzhafarov, J.R. Busemeyer, Quantum models for psychological measurements: an unsolved problem. PloS one **9**(10), e110909 (2014)
5. M. Ozawa, A. Khrennikov, Application of theory of quantum instruments to psychology: combination of question order effect with response replicability effect. Entropy **22**(1), 37 (2020)
6. M. Ozawa, A. Khrennikov, Modeling combination of question order effect, response replicability effect, and qq-equality with quantum instruments. J. Math. Psychol. **100**, 102491 (2021)
7. R. Tourangeau, L.J. Rips, K.A. Rasinski, *The Psychology of Survey Response* (Cambridge University Press, 2000)
8. Z. Wang, J.R. Busemeyer, A quantum question order model supported by empirical tests of an a priori and precise prediction. Top. Cognit. Sci. **5**(4), 689–710 (2013)
9. Z. Wang, T. Solloway, R.M. Shiffrin, J.R. Busemeyer, Context effects produced by question orders reveal quantum nature of human judgments. Proc. Natl. Acad. Sci. USA **111**(26), 9431–9436 (2014)

Jerome Busemeyer previously was a Full Professor at Purdue University before 1997, and now is Distinguished Professor in Psychological and Brain Sciences, Cognitive Science, and Statistics at Indiana University-Bloomington. His research has been funded by the National Science Foundation, and the National Institute of Mental Health, and he serve on grant review panels for these agencies. He was the Manager of the Cognition and Decision Program at the Air Force Office of Scientific Research in 2005–2007. He has published five books in decision and cognition, and over 100 journal articles across disciplines. He served as the Chief Editor of *Journal of Mathematical Psychology*, Associate Editor of *Psychological Review*, and he was the founding Chief Editor of *Decision*. He is a fellow of the Society of Experimental Psychologists and he won the prestigious Warren medal from that society in 2015. He became a fellow of the Cognitive Science Society and a fellow of the American Academy of Arts and Sciences in 2017. During his early career, he became well known for the development of a dynamic and stochastic model of human decision making called decision field theory. Later, he was one of the pioneers to develop a new approach to cognition based on principles from quantum theory. In 2012, Cambridge University Press published his book with Peter Bruza introducing this new theory applying quantum probability to model human judgment and decision-making.

Quantum Cognition: Quo Vadis?

Emmanuel M. Pothos and James M. Yearsley

Over the last 10 years there has been growing interest in the application of the quantum probability rules in the modelling of many aspects of human behaviour, including decision making, similarity, and categorization. We review the quantum tools behavioural scientists are more likely to employ and consider the special challenges in corresponding modelling work. We introduce broad categories of objectives that cognitive scientists have and consider the way quantum methods help achieve them, with three prominent examples in decision making, similarity, and constructive influences. In all cases, a key contribution from the application of quantum methods has been the introduction of new theoretical concepts and ideas, as well as analytical advances. Finally, we outline our impressions regarding priorities for this innovative research programme.

1 Introduction—the Quantum Tools Employed by Psychologists

The overwhelming majority of behavioural scientists assume a classical brain, that is, they assume the neurophysiological processes which lead to human thought are all classical [34]. Specifically, there are no cognitive models which rely on the controversial proposal that neuronal function may involve quantum processes (e.g., [24]). Given this approach, in what way could quantum theory be relevant in behavioural modelling? Together with a physical theory, quantum theory is also a probability theory, that is, a set of rules for how to assign probabilities to events. In behavioural

E. M. Pothos (✉) · J. M. Yearsley
Department of Psychology, City, University of London, London EC1R 0JD, UK
e-mail: Emmanuel.pothos.1@city.ac.uk

J. M. Yearsley
e-mail: james.yearsley@city.ac.uk

A. Plotnitsky and E. Haven (eds.), *The Quantum-Like Revolution*,
https://doi.org/10.1007/978-3-031-12986-5_14

science, a critical question concerns exactly this, that is, how naïve observers assign probabilities to events, estimate the certainty versus uncertainty of various possibilities, and in general engage in probabilistic inference. Thus, we can extract the probability rules from quantum theory (which we can call quantum probability theory, QPT) and consider their applicability in cognitive theory. This is essentially what the quantum cognition research programme is about, as instigated by many pioneering scientists a few decades ago, including Andrei Khrennikov [25, 27–29], see also [1, 7, 10, 12, 13, 43, 65]. To emphasize this key point, in QPT cognitive models, it is assumed there are no real (in the physical sense) superpositions, no real entanglement, no real supercorrelations. All these technical concepts from quantum theory are considered emergent or epiphenomenal ways to characterize behaviour. Behaviour is analogous to volume or pressure in physics; that is, behaviour is the result of scores of individual neurons interacting, but there is no sense in which behaviour has reality beyond that as an emergent property of the brain.

Behavioural scientists have employed most basic tools from QPT, but with various simplifications or assumptions. The Hilbert space is assumed to correspond to the knowledge space of the cognitive agents under study—note, the relevant cognitive agents are usually participants prior to being introduced to an experimental behavioural manipulation. Different subspaces in such a knowledge space are typically associated with the discrete answers to the questions that are part of the experimental manipulation. The mental state of these participants is typically modelled as a pure state, subject to some uninformativeness condition relevant to the questions that are part of the experimental manipulation. Of course, maximum uninformativeness relativshift from projectorse to one basis set in the Hilbert space still implies maximum informativeness relative to another basis set. Behavioural scientists have not been terribly clear to articulate the corresponding assumptions: that is, how exactly are we to understand superposition states. However, there has been a shift towards employing density states for modelling the relevant mental state of participants, and this avoids some of these interpretative problems (e.g., [56, 60]).

The computation of probabilities, given a mental state and a basis set, is usually carried out with projection operators, that is, it is assumed that projection is errorless, in that there is a faithful correspondence between outcome of measurement and projection. This is in line with the general mindset of a behavioural scientist that any prediction is going to be subject to considerable noise anyway, so that it is perhaps less crucial to separate out specific sources of error. However, in some behavioural applications, it has been shown important to shift from projectors to positive operator valued measures (POVMs), notably when one-dimensional subspaces are employed and it is pertinent to have asymmetric conditional probabilities [56] or when there are multiple questions in a manipulation so that the POVM error rate, accumulatively, may start having significant behavioural impact [62].

Behavioural scientists are especially interested in the dynamics of decisions, that is, the progression of internal states which eventually lead to a decision. Indeed, some of the most influential decision models are the drift diffusion models (e.g., [47]), which provide a framework for understanding the way internal states change as a decision outcome forms in a person's mind. Most quantum cognitive models have involved simple unitary dynamics (e.g., [42]). Unitary dynamics has obvious advantages, not least in terms of parametrization and the, relatively speaking, simplicity of the technical formulation. However, unitary dynamics have some obvious problems, notably persistent periodicity. Clearly, we do not expect human decision makers to indefinitely prevaricate between the available options (although sometimes it may seem that this is the case, especially for political decision making). Rather, there is a reasonable expectation of ambivalence followed by stabilisation. Some researchers have been adopting open systems modelling in behavioural science, though these tools have yet to become common (e.g., [4, 5, 18, 33]). For researchers continuing to adopt unitary dynamics, with the realisation of these issues, there is an increased onus to specify the cut off point in the oscillation, that is, the point at which the mental agent stops being ambivalent and makes up her mind. Inevitably, the corresponding approaches have heuristic flavour and cannot be rigorously justified within quantum theory.

There has been significant interest in entanglement and supercorrelations (e.g., [2, 11, 15]). This work has typically employed the Bell or CHSH inequalities, on pairs of binary questions for two systems. In some cases, but not all, quantum models are proposed, which are based on entangled states between the two systems. All the known research of this kind has so far concerned mental representations within the same participant, i.e., within the same cognitive system. Under such circumstances, it is of course natural to consider at least locality as being violated. It is also possible to envisage interpretations that the assumption of independence of measurement settings is violated as well. Bell tests are still useful, for example, in terms of tests of system independence, e.g., when the two systems correspond to cognitive entities for which a default assumption indeed corresponds to independence (e.g., [11]).

Several researchers in behavioural sciences have inquired as to why we have labelled this work 'quantum' cognition instead of e.g. cognitive modelling based on linear projective algebra or something such like. There are two main reasons. The first reason is that the tools that are employed in QPT cognitive modelling go beyond just tools from linear algebra. Notably, we have to invoke powerful theorems from quantum theory regarding the way to associate subspaces and projections to probabilities and how to interpret superposition states (Gleason's theorem and Kochen-Specker theorem). The second reason is that, for better or worse, the label quantum has been enabling a statement of adherence to certain methods under certain criteria for rigour and techniques. A sizeable number of papers on 'quantum' cognitive models has accumulated and it would help coherence of this research programme to continue labelling corresponding work as 'quantum-like'.

2 Levels of Explanation

A problem which is also encountered to some extent in physics and other sciences, but is particularly acutely felt in psychology, concerns levels of explanation. For example, in physics the theory of electromagnetism concerns the way electrical and magnetic forces exist in the world. It is well-known that there are more fundamental theories of electromagnetism, such as quantum electrodynamics; or the electro-weak unification which unifies electromagnetism with the weak nuclear. But, if one looks at a concept like temperature, there is a qualitative difference, in that the concept of temperature makes sense for large collections of particles, while individual particles do not 'have' temperature. Arguably, in behavioural sciences, all concepts are like temperature. The 'correct' way to describe what happens inside our heads concerns the interaction of perhaps as many as 80 billion neurons with some neurons having several thousand connections (synapses) to other neurons. Concepts such as probabilistic inference, learning, categorization or language are (mostly) meaningless at the level of individual neurons, but are good ways to describe emergent properties of large ensembles of neurons. This makes a discouraging impression regarding our chances to understand human behaviour based on what individual neurons are doing. Moreover, even if we can understand how particular neurons or combinations of neurons can encode various aspects of our environment, neurophysiology is unlikely to yield satisfactory answers regarding the reasons why human cognition appears to be biased one way or another. This is essentially the idea of levels of explanation in the study of the mind [35], see also [22]. These ideas are key in cognitive explanation and, as they are less familiar outside behavioural sciences, we consider them in detail.

Marr proposed three levels of explanation. Those presently relevant are the most abstract level, the computational level, and the algorithmic level, which is intended as intermediate between the computational level and the implementation level (this last level is about neurological/ biological processes and is not relevant here). The computational level concerns the 'what' and the 'why' questions for the system that is studied, that is "what is the goal of the computation, why is it appropriate, and what is the logic of the strategy by which it can be carried out?" (Marr [35], p. 25). Following Marr's example of studying a cash register, the what question is that the machine does arithmetic and the why is 'why arithmetic' instead of, for example, multiplication. A computational level specification of the cash register machine can then be elaborated easily, for example, given the objectives of the machine (which is to provide accurate measurement of the sum cost of a trolley of goods).

By contrast, the algorithmic level concerns a process explanation of the studied system, specifically the representations that are employed by the system and the algorithms that operate on the representations to produce an output from an input. In other words, the algorithmic level is about the 'how' question regarding the studied system. In the cash register example, an algorithmic level model could involve Arabic numerals for representation and the usual arithmetic calculus to carry out the necessary mapping from input to output information. Of course, the choices for algorithm and representation constrain each other [3, 35], and even then there may be alternative algorithmic level explanations which work.

Marr's [35] analysis has been hugely influential in psychology, but there are indications that it may be less suitable for modern cognitive models. The problem is that with an increasing sophistication of cognitive models, specification at multiple (Marr) levels is becoming more and more of a requirement. For example, consider

[64] leaky, competing, accumulator (LCA) model, which concerns the time course of a perceptual choice task. It seems clear that the main objective of the model is the algorithmic model (the specific way in which evidence accumulation at different time steps eventually leads to a decision). However, there are several overarching principles of LCA (e.g., the idea that evidence accumulation is leaky and a principle of lateral inhibition) that belong to the computational level.

These interpretational problems relate to probabilistic cognitive models and indeed Griffiths et al. [22, p. 357] provided a well-thought generalization to [35] analysis by noting that "probabilistic models of cognition pursue a top-down or 'function-first' strategy, beginning with abstract principles that allow agents to solve problems posed by the world—the functions that minds perform—and then attempting to reduce these principles to psychological and neural processes." Griffiths et al. [22] further explain the intention of a probabilistic cognitive model with a range of questions, such as in relation to the information needed, the necessary representations, and the constraints on the computation. While most of these questions belong to computational level explanations, it is also clear that some of them are connected to algorithmic level ones. Part of the problem is this: a probabilistic cognitive model (classical or quantum) embodies a default algorithmic assumption, namely the native probabilistic calculus. However, there is rarely a strong corresponding commitment. Indeed, when [48] considered the practicalities of computing Bayesian probabilities (e.g., priors), they suggested that the appropriate algorithmic level description of a Bayesian categorization model should involve various approximations regarding probabilistic computation.

Many quantum models of cognition are analogous to classical/ Bayesian models. So, their explanatory objective is best stated using [22] approach, as models focusing on the mathematical principles that characterize human causal reasoning, when participants do not adhere to classical probability constraints. This discussion should illustrate more directly how any purported quantum features of cognition, including superposition, incompatibility, entanglement etc. are epiphenomenal and would disappear should we pursue more complete explanations, e.g., based on neurophysiological and cognitive variables [61].

As a final note, it is worth pointing out that not all researchers accept that there might be a principled description of behaviour based on variables specified at the cognitive level. Without going into detail, note that [17] questioned our capacity to study in a principled way higher cognition, because of the way our beliefs are interconnected—so that, for example, one could not study the impact of e.g. one piece of information for a particular question in isolation, thus rendering the principled experimental study of cognition unworkable. Most researchers do no endorse this extreme position, even though it is generally acknowledged that behavioural science will never be associated with the same degree of precision as e.g. physics.

3 Behavioural Sciences—Other Technical Limitations

We provide a few additional comments regarding inherent technical limitations in conducting research in the behavioural sciences. These comments are relevant because they impact on our expectations for what we can reasonably expect from any framework for cognitive modelling, including quantum theory. For example, a reasonable expectation from a physics theory is accurate prediction (sometimes to astonishing degrees of accuracy). By contrast, theories in behavioural sciences are rarely able to achieve comparable levels of predictive accuracy. This is why.

Behavioural studies are constrained to small sample sizes. Most studies are conducted with human participants. For example, a decision experiment might involve presenting some information to human participants to examine the way they respond to some related questions. Such studies can be conducted in a physical laboratory, in which case at the very least per participant 10 min of research assistant time are required. Sometimes, study format allows implementation over the internet, through crowd sourcing platforms, such as mTurk or Prolific Academic. But even in these cases, per participant there would be a minimum cost of about £0.50, making it impractical to recruit sample sizes over 1000 individuals. Even when monetary limitations can be overcome, the number of individuals willing to take part in such experiments in e.g. mTurk is limited. With several researchers running related experiments, one can quickly exhaust the available participants.

In behavioural sciences, there are fundamental limitations in the sensitivity of the measuring instruments. For example, a lot of decision research relies on individual differences, that is, the systematic understanding of how the participants in a study vary along some relevant behavioural dimension (e.g., sensitivity to reward). Individual differences are typically measured using questionnaires. Despite extensive research into the validity of these questionnaires, there are inherent questions regarding both the accuracy of the purported individual differences dimensions and the extent to which overt responding to a series of related questions provides a valid corresponding measure. Analogously tests that decision making might reflect a particular set of probabilistic principles instead of another would be based on information provided to participants, with extensive potential for misinterpretation and/or additional assumptions the participants might make (e.g., see [36], for an extensive discussion of the multiple alternative explanations for a famous decision finding).

Finally, partly because cognitive theory concerns emergent properties, it is often the case that there is no unique way to approach a particular set of findings. Behavioural theory is rife with mimicries (cf. [3]) and, despite the sophistication of statistical model comparison methods, it is often very challenging to provide conclusive statements regarding model adequacy. It has to be said that some of these issues are not unique to behavioural sciences (cf. [31]), though the enormous proliferation of distinct theoretical proposals in behavioural sciences certainly provides an impression that these problems are more acute here.

4 What Are We Trying to Explain with (Quantum) Cognitive Models

Given our discussion of [35] levels of explanation, it can be hoped that by this point there is a fair idea of the objectives of quantum cognitive models (and of course any model based on formal probabilistic principles). As a guide we consider three objectives, which we shall further illustrate with concrete examples of quantum cognitive models subsequently.

In principle, we should be able to employ a quantum cognitive model to examine the rational status of a decision or a set of decisions. In many aspects of decision making, it is as important if not more important to understand not how naïve observers are likely to respond, but instead how they should respond to particular questions. For example, in the case of medical diagnosis, we are not necessarily interested in the decision that a medical professional is likely to spontaneously make, but rather in the best possible/ normative/ rational decision that the person should make, given diagnostic information (e.g., [9]). Likewise, when considering jury decision making, there is an increased emphasis on how a jury should rationally behave, instead of e.g. how they develop intuitions [40]. Note, these statements imply but do quite unpack the meaning of rationality. It would be digressive to explore this issue comprehensively here (e.g., see Wheeler et al. [66]). Briefly, rationality is typically understood in terms of consistency of a decision with principles which reflect some suitable standard of correctness of appropriate-ness. For example, a standard to justify the rational status of classical probability theory is the Dutch Book Theorem, which tells us that assigning probabilities according to classical probability theory protects us from a certain loss (i.e., a loss regardless of the way events turn out to be, [14]).

The archetypal objective of behavioural theory at the computational level is the understanding of the principles which guide cognition. Even resigning ourselves to inescapable error when trying to model human behaviour, we would like to understand the principles which best describe the way humans are likely to generate probabilistic intuitions as a result of some information. Essentially, this is the objective of predicting human behaviour. It can guide our understanding both of the way human observers are likely to react to particular pieces of information and the kind of interventions which might be more successful in nudging human behaviour in certain directions.

Finally, we should appreciate the limitations of human cognition. Clearly, as a species we have a degree of intellectual prowess; equally clearly, there is room for improvement in terms of individual and collective decision making. What are the sources of limitations in human cognition and what might be ways with which we could try to mitigate these limitations? A behavioural model at the computational level might help elucidate these key questions for the further development of our species.

5 Three Examples of Quantum Cognitive Models

In this section we present three examples of quantum cognitive models, in order with a view to illustrate the motivation for employing quantum ideas and evaluate the added-value of these quantum models, over and above existing ideas.

We start with the conjunction fallacy [59], which is one of the most famous findings in the decision literature. The conjunction fallacy is called a fallacy because, even once we have been told what is the classical, normative answer, it is hard to avoid the intuition from the 'incorrect' answer. It is this persistence in the intuition regarding the 'incorrect' answer that has made this finding so influential (cf. [19]), as well as the fact that it provides a test of psychological plausibility for one of the most fundamental constraints in Bayesian/classical probability theory (CPT).

Participants would be told of a hypothetical person, Linda, who is described very much like a feminist (F), but not a bank teller (BT). Participants are then asked to evaluate various possibilities about Linda, including Prob(BT) and Prob(F & BT). Now, by assumption the former would be very low and indeed most participants decide that just the possibility that Linda is a bank teller is highly unlikely. The interesting finding is that the conjunction is typically rated as more likely that the individual probability, even though classically it has to be the case that Prob(F & BT) $= <$ Prob(BT). These are subjective probabilities and one natural response when fiprobabilitiesrst encountering the conjunction fallacy is that with which make the set theoretic structure of the corresponding events more salient such paradoxes will be eliminated. Tentori et al. [55] asked participants to evaluate the probability that a Scandinavian person is more likely to have blue eyes vs. blond hair and blue eyes,a conjunction fallacy was still observed. Another natural response to the conjunction fallacy is that participants misunderstand the various statements about Linda. What if they understand Prob(BT & ~ F) when they read Prob(BT)? In fact, it is easy to discount such possibilities [36]. On the whole, pretty much all reasonable objections to the conjunction fallacy have been explored. It is an extremely robust finding.

The conjunction fallacy attacks the heart of the proposal that human decision making follows CPT. If it did, there is no way that such a simple judgment would resist consistency with CPT so rigidly. The original proponents of the conjunction fallacy, [59] interpreted the result as indicating that in at least some cases human decision making follows principles other than CPT, notably individual heuristics and biases. This is a conclusion with important ramifications for our understanding of human decision making. In recent years, the question of the rational status of human decision making has been mostly addressed through consistency with CPT (e.g., [38, 46]) and corresponding theorems (such as the Dutch Book theorem, [14]). Therefore, inconsistency with CPT is essentially a statement of irrationality in human decision making.

Busemeyer et al. [13] proposed a model for the conjunction fallacy based on quantum representations. The basic idea is simple. Assume that the questions about whether Linda is a bank teller and a feminist are incompatible. Such incompatibility could arise from many sources, e.g., perhaps one question provides a perspective which essentially alters the meaning of the other question (note, we are not trying to sweep this problem under the rug: it is a challenging problem,but, for now, let us assume the existence of incompatibility in this way). Then, it is straightforward to show that simple quantum representations can be set up in such a way that Prob(F & then BT) can be greater than Prob(BT). Note that, of course, for incompatible questions there is no sense of conjunction as concurrent evaluation of two questions. Instead, [13] defined a notion of sequential conjunction (as sequential projection) and associated this sequential conjunction with the participants' evaluation of the statement that Linda is a feminist and a bank teller. Given such sequential conjunctions, there is a QPT analogue of the law of total probability, which shows how a marginal can decompose to terms corresponding to the classical law of total probability plus an interference term. It is this interference term which allows consistency between QPT and the conjunction fallacy.

QPT is a probability theory with the simplest kind of interference terms, while of course CPT has no interference terms [54]. The conjunction fallacy is a situation which shows necessity for these simplest kind of interference terms. It cannot really inform the question of whether more elaborate interference terms might be needed. But bear in mind that the conjunction fallacy is a simple finding, it reduces to a percentage of endorsement for the conjunctive statement relative to the marginal. What QPT does is that it provides a formal notion of the way in which the conjunction fallacy can *make sense*. That is, it allows an alternative intuition for how probabilities could be approached in the Linda problem, in a way that the conjunction fallacy is no longer a fallacy but rather the correct response. Going further, it can be shown that QPT is consistent with the same criterion for rationality as CPT, the Dutch Book theorem, subject to certain conditions regarding the contextuality of the relevant questions [45]. Therefore, we can (almost) say that the conjunction fallacy is a rational response, under particular circumstances.

So far we have considered how the conjunction fallacy helps us address two objectives of computational level cognitive models, whether a decision is rational and the computational principles which guide a decision. We can say that the conjunction fallacy shows that there will be cases of everyday reasoning such that QPT provides a better description of behaviour, than CPT. What about limitations? The QPT account of the conjunction fallacy rests on the assumption that the relevant questions are incompatible. In behavioural sciences, incompatibility has to be understood as changes in meaning or perspective, which might arise in a sequence of questions. For example, a question about 'bugs' will have a different meaning when the previous questions concern holidays vs. spies. Therefore, the validity of QPT principles in cognition goes hand in hand with the validity of the assumption of this kind of contextuality in real world questions [45]. In most cases, this will not be the case. That is, contextuality cannot be assumed and so a conjunction fallacy will reflect irrational decision making. Thus, QPT also helps us understand the way a decision which conflicts with CPT prescription can be said to be rational vs. not rational.

The second example concerns similarity, specifically the extent to which human similarity judgments are consistent with the metric axioms, minimality, symmetry, and the triangle inequality. One might wonder why might we think that the metric axioms are relevant in the study of similarity, especially bearing in mind the intuition that judgments of similarity appear very flexible [8, 21]. However, in a very influential theoretical tradition, analysing pairwise similarity judgments can actually point towards point-based representations in metric spaces [51, 52]. Point-based representations in metric spaces have in turn being the basis for the dominant models in categorization, such as exemplar and prototype theories (e.g., [37], for an exception see [41]). Given such a representation approach to similarity, there is a clear expectation of consistency with the metric axioms. We focus on symmetry (for a discussion of triangle inequality see [63]).

A priori, symmetry is a very reasonable constraint for similarity judgments. It boils down to the expectation that the similarity between objects A and B is the same as the similarity between objects B and A. How could it be otherwise? Yet, [57, 58] reported systematic violations of symmetry, which he ascribed to differences in the amount of knowledge we have for the compared objects. For example, he reported that the similarity between (North) Korea and (Red) China (for brevity we will henceforth refer to Korea and China) was higher than that between China and Korea. His argument was that the higher knowledge his participants would have of China would render the Korea-China direction more natural than the China-Korea one, leading to higher similarity for the former. Tversky [57] presented several other examples for which extent of knowledge would be the basis for predicting similarity asymmetries.

Point-based representations in a metric space do not naturally accommodate violations of symmetry in similarity judgments. Typically, we have Sim = function (distance (A, B)). Since distance (A, B) is symmetric, then similarity has to be symmetric as well. Some researchers have suggested that such similarity models could be augmented with a directionality parameter, so that we can write Sim = function (p_{AB} * distance (A, B)), where is a p_{AB} is a directionality parameter. Clearly, p_{AB} can be different from p_{BA} so that there is potential to violate symmetry in similarity judgments. While arithmetically such an approach can work, it is not satisfactory without further insight into how these directionality parameters are determined. There have been a few other more formal approaches to accommodating violations of symmetry [6, 30] and evaluating them in detail is beyond the scope of the present work. Instead, we focus on a quantum approach to modelling similarity in a way that asymmetries can emerge quite naturally.

In quantum theory it is straightforward to associate representations of particular objects with subspaces of varying dimensionalities. Dimensionality can be naturally related to the extent of knowledge we have for different objects. For example, if we represent the China subspace with more dimensions, this can correspond to greater knowledge of features about China (cf. [57]). Then, a similarity judgment can minimally correspond to the ease of projecting from one subspace to the other, a process which can be interpreted as one of the ease of thinking about the first objects and then the second. In a simple example where all subspaces are represented with one-dimensional subspaces, then such a sequential projection would be equal to the square of the cosine between the corresponding basis vectors—analogous approaches have been adopted previously in psychology (e.g., [53]). Interestingly, when there are differences in the relative dimensionality of the subspaces, then it can be shown that, when starting from an initially neutral mental state, projecting from the low to the high dimensionality subspace leads to a higher similarity than projecting in the reverse order [44]. This quantum approach to similarity can be argued to have various advantages over existing approaches, including more coherent coverage of a number of findings in similarity judgments indicating contextuality.

The quantum approach to similarity is superficially fairly simple: it consists of adopting a representational approach involving subspaces instead of points and computing similarity as sequential projection. Yet it is this simplicity, and the naturalness with which asymmetries can emerge, that, we think, which we think makes the quantum model appealing. The quantum model allows a fairly minimal statement of the principles relevant in similarity judgments and clearly relates to existing ideas regarding similarity.

When it comes to similarity judgments, it is less obvious how to consider the question of rationality. Clearly, there is no sense in which one judgment of similarity is more rational or appropriate than another. However, we could consider whether similarity judgments are based on coherent principles and are consistent with each other. The converse possibility is that they reflect biases which would be hard to formalise and may appear as noise under any attempts to study rigorously (cf. [21]). Quantum theory offers a way to understand variation in similarity judgments in a systematic way and specifically allows us to appreciate how different contexts may lead to differences in similarity judgments.

Regarding possible limitations, it is difficult to appreciate in what way a particular way of computing similarity might be more adaptive than another. One view is that similarity ought to serve a function of 'correct' generalizations and category formation in our environment. For example, imagining our primitive ancestors, in trying out a berry that did not lead to any side effects, one would want to correctly generalize to other berries which are similar to the original one. Arguably this is the key adaptive function of similarity. From this point of view, it is harder to appreciate what would be the adaptive function of e.g. knowledge-related asymmetries in similarity judgments and other contextual effects, though it has to be noted that similarity judgments have never (to the best of our knowledge) been studied with a view to examine their adaptive value.

The final example concerns constructive influences in judgment. There is some evidence that making a decision can alter the mental states relevant to the decision. For example, suppose you are given a choice between two holidays, a holiday to Athens and a holiday to Rome. If you were to choose Athens, there is some evidence showing that your preference for the Athens option post-choice would increase (e.g., Sharot et al. [49]), possibly as a result of cognitive dissonance [16]—it would appear to be adaptive to increase preference for choices already made. Other related evidence is that a decision task based on a number of cues or factors can lead to an increase in the way the strength of cues/ factors are perceived, so as to make them more consistent with the eventual decision. For example, Glöckner et al. [20] demonstrated what they called coherence shifts, changes in subjective cue validities related to a decision, in a direction indicating greater consistency with the decision.

Such constructive influences in judgment fit extremely well with the idea of collapse in quantum theory, that is, the idea that a measurement is associated with a change in the relevant state so that the state identifies itself with one of the question outcomes. White et al. [67, 68] employed the collapse mechanism in quantum theory to make simple predictions regarding the way two stimuli are perceived, one after the other. In all cases, the second stimulus would be rated. However, the first stimulus would be rated in some cases, but not others. We can use this notation: first stimulus double rating condition (FSDR), second stimulus double rating condition (SSDR), and second stimulus single rating condition (SSSR). White et al. [67, 68] employed pairs of stimuli such that they would have fairly clear positive (P) or negative (N) valence. Pairs would conform to a PN or NP configuration. Then, in PN cases, SSDR $<$ SSSR and in NP cases SSDR $>$ SSSR, that is, in both cases the intermediate rating would result in a judgment that is more intense (more negative if the stimulus is N or more positive if it is P) for the second stimulus. They called this finding the evaluation bias. The evaluation bias could be the result of increased attention or memory for the first stimulus, when this is rated. Such explanations based on memory or attention have the disadvantage that they do not predict a unique direction for the evaluation bias. By contrast, using quantum theory, because the way the state changes as a result of a judgment is unique, so it is the case that the evaluation bias has a (fairly) unique form as well (the qualification 'fairly' arises because it is not in general possible to be entirely certain on how a stimulus will be evaluated).

The work of [67, 68] indicates that judgments or decisions are associated with a specific change in the internal state. In fact, other evidence has been accumulating that mental states change in the way predicted by QPT as a result of judgments. Yearsley and Pothos [62] have shown that increasing the number of intermediate decisions, when faced with a series of evidence, there is a slowing down of opinion change analogous to the quantum Zeno effect. Kvam et al. [32] have likewise shown that confidence ratings and decisions can interact. In all cases, the driving force of the prediction is the same assumption that the collapse of the state vector in QPT is a plausible mechanism for how the mental state changes as a result of decisions.

Quantum models for constructive influences in decision making allow us to evaluate the rational status of such influences, by analogy with the corresponding explanation in quantum physics. In quantum physics, the state of a system changes as a result of a measurement because a measurement can result in a 'disturbance' (noting that a disturbance here has to mean something different from the usual notion of one system physically impacting on another; e.g., it can mean that a decision changes the relevant mental state). Can there be decisions which likewise have an analogously disturbing influence the system? Consider the example of baby Eve from [45]. Her parents want to check whether she sleeps on her back or on her front. She usually sleeps on her back. But, when her parents check on her, this usually disturbs her, and so she is more likely to roll over. Under such circumstances, we would expect that Prob(front at 2100) is low, but Prob(back at 2000 & front at 2100) is high. The reason for this apparent conjunction fallacy is that the measurement at 2000 disturbs Eve, so she rolls over; we cannot directly compare the question of 'front at 2100' in isolation and after the event 'back at 2000', as essentially these are different questions. It seems possible that there are other questions which have analogous disturbing influences on the relevant system and for which employing QPT principles for probabilistic reasoning would be more rational than CPT ones. The corresponding limitations in cognitive process are that the mind might produce constructive influences even when they are not warranted, that is, when they are not associated with corresponding changes in the system of interest.

6 Future Directions

In this brief introduction, we hope to have illustrated the considerable capacity of quantum theory to enrich our effort to understand the mind, because of the many unique concepts and processes that are part of the theory. It might even seem that quantum theory is better suited to the study of the mind than CPT, since the latter suffers from unrealistic requirements regarding the existence of a complete joint probability distribution for the available questions. It is fair to say that the application of quantum theory in behavioural sciences is fairly well-established at this point, as a probabilistic framework well suited to cognitive modelling that is alternative to classical theory.

We conclude this introduction with two observations concerning future directions. First, the application of QPT has so far been mostly focussed on data from human experimental paradigms. Such datasets are inevitably somewhat limited and subject to all sorts of biases, including task demands [39] and the more likely engagement of 'fast' cognition [26]. An interesting exception is that of [65], who applied a quantum model to a large number of datasets. But, overall, we suggest that the real challenge of quantum cognitive models and the way in which their putative added value can be best illustrated is with applications employing decision making data from the real world. Indeed, the work of Andrei Khrennikov (e.g., [27]) points in this direction. As a recent example, we note White et al. (in press), who attempted to use a quantum model for the evaluation bias in the context of an organisational psychology questionnaire, which was part of a real organizational training workshop. But clearly the norm for tests of quantum models continues to be experimental studies, either in the lab or using surveys circulated to paid participants (e.g., through mTurk). We think there is much to be gained by a greater focus on real world data.

The second point is that QPT representations would be worth exploring in artificial intelligence (AI). There is an interesting parallel between cognitive and AI architectures, for example, both logic and probability theory have been employed for both. Quantum representations offer an intriguing alternative, which may allow more economical ways to encode a series of questions than either logic or CPT. Specifically, incompatibility means that the same dimensionality subspace can be used to squeeze in multiple basis sets, corresponding to, e.g., different sets of questions for a particular problem. Like all data reduction approaches, employing incompatibility in this way involves some error. It remains to be seen whether quantum representations can be fruitfully employed in AI, in terms of its data reduction potential, and how the corresponding approximations compare to those from other frameworks.

Acknowledgements This work was supported by the Office of Naval Research Global [grant number N62909-19-1-2000]. EMP was supported by ONRG grant N62909-19-1-2000.

References

1. D. Aerts, S. Aerts, Applications of quantum statistics in psychological studies of decision processes. Found. Sci. **1**, 85–97 (1995)
2. D. Aerts, J.A. Arguëlles, L. Beltran, S. Geriente, M.S. de Bianchi, S. Sozzo, T. Veloz, Spin and wind directions I: Identifying entanglement in nature and cognition. Found. Sci. **23**, 323–335 (2018)
3. J.R. Anderson, Arguments concerning representations for mental imagery. Psychol. Rev. **85**, 249–277 (1978)
4. M. Asano, M. Ohya, Y. Tanaka, I. Basieva, A. Khrennikov, Quantum-like model of brain's functioning: decision making from decoherence. J. Theor. Biol. **281**, 56–64 (2011). https://doi.org/10.1016/j.jtbi.2011.04.022
5. M. Asano, M. Ohya, Y. Tanaka, A. Khrennikov, I. Basieva, On application of Gorini-Kossakowski-Sudarshan-Lindblad equation in cognitive psychology. Open. Syst. Inf. Dyn. **18**, 55–69 (2011). https://doi.org/10.1142/S1230161211000042
6. G.F. Ashby, N.A. Perrin, Towards a Unified Theory of Similarity and Recognition. Psychol. Rev. **95**, 124–150 (1988)
7. H. Atmanspacher, T. Filk, A proposed test of temporal nonlocality in bistable perception. J. Math. Psychol. **54**, 314–321 (2010)
8. L.W. Barsalou, Deriving categories to achieve goals, in *The Psychology of Learning and Motivation*, vol. 27, ed. by G.H. Bower (Academic Press, New York, 1991), pp.1–64
9. G.R. Bergus, G.B. Chapman, B.T. Levy, J.W. Ely, R.A. Oppliger, Clinical diagnosis and order information. Med. Decis. Making **18**, 412–417 (1998). https://doi.org/10.1177/0272989X9801800409
10. P.D. Bruza, Z. Wang, J.R. Busemeyer, Quantum cognition: a new theoretical approach to psychology. Trends Cogn. Sci. **19**, 383–393 (2015)
11. P. Bruza, K. Kitto, D. Nelson, C. McEvoy, Is there something quantum-like about the human mental lexicon? J. Math. Psychol. **53**(5), 362–377 (2009)
12. J.R. Busemeyer, P. Bruza, *Quantum Models of Cognition and Decision Making* (Cambridge University Press, Cambridge, UK, 2011)
13. J.R. Busemeyer, E. Pothos, R. Franco, J.S. Trueblood, A quantum theoretical explanation for probability judgment 'errors.' Psychol. Rev. **118**, 193–218 (2011)
14. B. de Finetti, A. Machi, A. Smith, *Theory of Probability: A Critical Introductory Treatment* (Wiley, New York, 1993)
15. E.N. Dzhafarov, R. Zhang, J. Kujala, Is there contextuality in behavioural and social systems? Phil. Trans. R. Soc. A **374**(2058), 20150099 (2016)
16. L. Festinger, *A Theory of Cognitive Dissonance* (Stanford University Press, Stanford, 1957)
17. J.A. Fodor, *The Modularity of Mind* (The MIT Press, Cambridge, MA, 1983)
18. I.G. Fuss, D.J. Navarro, Open parallel cooperative and competitive decision processes: a potential provenance for quantum probability decision models. Top. Cogn. Sci. **5**, 818–843 (2013)
19. I. Gilboa, *Theory of Decision Under Uncertainty* (Cambridge University Press, Cambridge, UK, 2000)

20. A. Gloeckner, T. Betsch, N. Schindler, Coherence shifts in probabilistic inference tasks. J. Behav. Decis. Mak. **23**, 439–462 (2010)
21. N. Goodman, Seven strictures on similarity, in *Problems and projects*, ed. by N. Goodman (Bobbs-Merrill, Indianapolis, 1972), pp. 437–447
22. T.L. Griffiths, N. Chater, C. Kemp, A. Perfors, J.B. Tenenbaum, Probabilistic models of cognition: exploring representations and inductive biases. Trends Cogn. Sci. **14**, 357–364 (2010)
23. U. Hahn, N. Chater, L.B. Richardson, Similarity as transformation. Cognition **87**, 1–32 (2003)
24. S.R. Hammeroff, Quantum computation in brain microtubles? The Penrose—Hameroff "Orch OR" model of consiousness. Philos. Trans. R. Soc. London (A) **356**, 1869–1896 (1998)
25. E. Haven, A. Khrennikov, *Quantum Social Science* (Cambridge University Press, Cambridge, 2013)
26. D. Kahneman, *Thinking Fast and Slow* (Penguin, London, UK, 2001)
27. A. Khrennikov, *Ubiquitous Quantum Structure: from Psychology to Finances* (Springer, Berlin, Heidelberg, New York, 2010)
28. A. Khrennikov, I. Basieva, Possibility to agree on disagree from quantum information and decision making. J. Math. Psychol. 1–15 (2014)
29. A. Khrennikov, I. Basieva, E.M. Pothos, I. Yamato, Quantum probability in decision making from quantum information representation of neuronal states. Sci. Rep. **8**, 16225 (2018)
30. C.L. Krumhansl, Concerning the applicability of geometric models to similarity data: the interrelationship between similarity and spatial density. Psychol. Rev. **85**, 445–463 (1978)
31. T.S. Kuhn, *The Structure of Scientific Revolutions* (University of Chicago Press, Chicago, 1962)
32. P.D. Kvam, T.J. Pleskac, S. Yu, J.R. Busemeyer, Interference effects of choice on confidence: quantum characteristics of evidence accumulation. Proc. Natl. Acad. Sci. **112**, 10645–10650 (2015)
33. P.D. Kvam, J.R. Busemeyer, T.J. Pleskac, (2020). An open system model of decision-making and temporal oscillations in preference strength. https://doi.org/10.31234/osf.io/cb26p
34. A. Litt, C. Eliasmith, F.W. Kroon, S. Weinstein, P. Thagard, Is the brain a Quantum computer? Cogn. Sci. **30**, 593–603 (2006)
35. D. Marr, *Vision: A Computational Investigation into the Human Representation and Processing of Visual Information* (W. H. Freeman, San Francisco, 1982)
36. R. Moro, On the nature of the conjunction fallacy. Synthese **171**, 1–24 (2009)
37. R.M. Nosofsky, Relations between exemplar-similarity and likelihood models of classification. J. Math. Psychol. **34**, 393–418 (1990)
38. M. Oaksford, N. Chater, *Bayesian Rationality: The Probabilistic Approach to Human Reasoning* (Oxford University Press, Oxford, 2007)
39. M.T. Orne, On the social psychology of the psychological experiment: with particular reference to demand characteristics and their implications. Am. Psychol. **17**, 774–783 (1962)
40. N. Pennington, R. Hastie, Evidence evaluation in complex decision making. J. Pers. Soc. Psychol. **51**, 242–258 (1986)
41. E.M. Pothos, N. Chater, A simplicity principle in unsupervised human categorization. Cogn. Sci. **26**, 303–343 (2002)
42. E.M. Pothos, J.R. Busemeyer, A quantum probability explanation for violations of "rational" decision theory. Proc. R. Soc. B **276**, 2171–2178 (2009)
43. E.M. Pothos, J.R. Busemeyer, Can quantum probability provide a new direction for cognitive modeling? Behav. Brain Sci. **36**, 255–327 (2013)
44. E.M. Pothos, J.R. Busemeyer, J.S. Trueblood, A quantum geometric model of similarity. Psychol. Rev. **120**, 679–696 (2013)
45. E.M. Pothos, J.R. Busemeyer, R.M. Shiffrin, J.M. Yearsley, The rational status of quantum cognition. J. Exp. Psychol. Gen. **146**, 968–987 (2017)
46. E.M. Pothos, I. Basieva, A. Khrennikov, J.M. Yearsley, Perspectives on correctness in probabilistic inference from psychology. Spanish J. Psychol. **22**, E55 (2019)
47. R. Ratliff, P. Smith, Modeling simple decisions and applications using a diffusion model, in *The Oxford Handbook of Computational and Mathematical Psychology*. ed. by J.R. Busemeyer, Z. Wang, J.T. Townsend, A. Eidels (Oxford University Press, Oxford, 2015), pp.35–62

48. A.N. Sanborn, T.L. Griffiths, D.J. Navarro, Rational approximations to rational models: alternative algorithms for category learning. Psychol. Rev. **117**, 1144–1167 (2010)
49. Sharot et al., Defined a notion of sequential conjunction (as sequential projection) and associated this sequential conjunction with the participants' evaluation of the statement that Linda is a feminist and a bank teller (2011)
50. T. Sharot, C.M. Velasquez, R.J. Dolan, Do decisions shape preference? Evidence from blind choice. Psychol. Sci. **21**(9) 1231–1235 (2010). https://doi.org/10.1177/0956797610379235
51. R.N. Shepard, Multidimensional scaling, tree-fitting, and clustering. Science **210**, 390–398 (1980)
52. R.N. Shepard, Toward a universal law of generalization for psychological science. Science **237**, 1317–1323 (1987)
53. S.A. Sloman, Feature-based induction. Cogn. Psychol. **25**, 231–280 (1993)
54. R.D. Sorkin, Quantum mechanics as quantum measure theory. Mod. Phys. Lett. A **09**, 3119 (1994)
55. K. Tentori, N. Bonini, D. Osherson, The conjunction fallacy: a misunderstanding about conjunction? Cogn. Sci. **28**, 467–477 (2004)
56. J.S. Trueblood, J.M. Yearsley, E.M. Pothos, A quantum probability framework for human probabilistic inference. J. Exp. Psychol. Gen. **146**, 1307–1341 (2017)
57. A. Tversky, Features of similarity. Psychol. Rev. **84**, 327–352 (1977)
58. A. Tversky, I. Gati, Studies of similarity, in *Cognition and categorization.* ed. by E. Rosch, B.B. Lloyd (Lawrence Erlbaum Associates, Hillsdale, N. J., 1978), pp.79–98
59. A. Tversky, D. Kahneman, Extensional versus intuitive reasoning: The conjuctive fallacy in probability judgment. Psychol. Rev. **90**, 293–315 (1983)
60. J.M. Yearsley, J.R. Busemeyer, Quantum cognition and decision theories: a tutorial. J. Math. Psychol. **74**, 99–116 (2016)
61. J.M. Yearsley, E.M. Pothos, Challenging the classical notion of time in cognition: a quantum perspective. Proc. R. Soc. B **281**, 1471–1479 (2014)
62. J.M. Yearsley, E.M. Pothos, Zeno's paradox in decision making. Proc. R. Soc. B **283**, 20160291 (2016)
63. J.M. Yearsley, A. Barque-Duran, E. Scerrati, J.A. Hampton, E.M. Pothos, The triangle inequality constraint in similarity judgments. Prog. Biophys. Mol. Biol. **130**, 26–32 (2017)
64. M. Usher, J.L. McClelland, The time course of perceptual choice: the leaky, competing accumulator model. Psychol. Rev. **108**, 550–592 (2001)
65. Z. Wang, T. Solloway, R.M. Shiffrin, J.R. Busemeyer, Context effects produced by question orders reveal quantum nature of human judgments. Proc. Natl. Acad. Sci. **111**, 9431–9436 (2014)
66. G. Wheeler, Bounded rationality. In E. N. Zalta (Ed.), *The Stanford Encyclopedia of Philosophy* (Winter 2018 ed.) (Metaphysics Research Lab, Stanford University, 2018). https://plato.stanford.edu/archives/win2018/entries/bounded-rationality
67. L.C. White, E.M. Pothos, J.R. Busemeyer, Sometimes it does hurt to ask: the constructive role of articulating impressions. Cognition **133**(1), 48–64 (2014). https://doi.org/10.1016/j.cognition.2014.05.015
68. L.C. White, A. Barqué-Duran, E.M. Pothos, An investigation of a quantum probability model for the constructive effect of affective evaluation. Philos. Trans. A **374**, 2015014 (2015). https://doi.org/10.1098/rsta.2015.0142
69. L.C. White, E.M. Pothos, M. Jarrett, The cost of asking: how evaluations bias subsequent judgments. Decision (in press)

Emmanuel Pothos received a BSc in Physics from Imperial (1995) and DPhil in experimental psychology from Oxford (1998). At an undergraduate student at Imperial he received the Stainley Raimes Memorial Prize, for outstanding performance in first and second year mathematics.

Emmanuel has worked at several universities, including Bangor University, Edinburgh University, Crete University, Swansea University, and, since 2009, City, University of London, where he has been a professor since 2014. He has been interested in several topics in cognitive science, including learning, categorization, similarity, language, and (more recently) decision making. He has been part of the quantum cognition research community from the very early days and has contributed some of the early models, for example, concerning the disjunction effect and the conjunction fallacy. He has co-authored two major reviews of the quantum cognition research programme, for the Behavioral and Brain Sciences (2013) and the Annual Review of Psychology (2022). He continues to actively develop quantum cognitive models and explore the potential and boundaries of quantum theory for behavioural modelling. He has 120 journal articles has his work has been funded by several organisations, including the ESRC, the AFOSR, the ONRG, and the Leverhulme Trust.

James Yearsley received a BSc in Physics with Theoretical Physics (2007) and PhD in Theoretical Physics (2011) from Imperial College London. After a spell as a post-doc in the department of Applied Mathematics and Theoretical Physics at the University of Cambridge he switched to working in psychology, completing post-docs at City, University of London and Vanderbilt University, TN, before re-joining City, University of London as a lecturer in 2017. He has worked on several topics in the field of mathematical psychology, including decision making, memory, similarity, and learning processes in clinical conditions. Due to his background in physics he has a particular interest in the quantum cognition program, including working on quantum models of probabilistic inference, the first demonstration of a quantum Zeno effect in psychology, and the first experimental demonstration of the co-occurrence of conjunction fallacies and order effects in a decision making task. He has also authored two tutorial papers introducing psychologists to some of the key concepts in quantum theory relevant for applications to psychology, including the dynamics of open systems and POVMs.

Entanglement in Cognition Violating Bell Inequalities Beyond Cirel'son's Bound

Diederik Aerts, Jonito Aerts Arguëlles, Lester Beltran, Suzette Geriente, and Sandro Sozzo

Abstract We present the results of two tests where a sample of human participants were asked to make judgements about the conceptual combinations *The Animal Acts* and *The Animal eats the Food*. Both tests significantly violate the Clauser-Horne-Shimony-Holt version of Bell inequalities ('CHSH inequality'), thus exhibiting manifestly non-classical behaviour due to the meaning connection between the individual concepts that are combined. We then apply a quantum-theoretical framework which we developed for any Bell-type situation and represent empirical data in complex Hilbert space. We show that the observed violations of the CHSH inequality can be explained as a consequence of a strong form of 'quantum entanglement' between the component conceptual entities in which both the state and measurements are entangled. We finally observe that a quantum model in Hilbert space can be elaborated in these Bell-type situations even when the CHSH violation exceeds the known 'Cirel'son bound', in contrast to widespread beliefs. These findings confirm and strengthen the results we recently obtained in a variety of cognitive tests and document and image retrieval operations on the same conceptual combinations.

D. Aerts · J. A. Arguëlles · L. Beltran · S. Geriente
Center Leo Apostel for Interdisciplinary Studies, Free University of Brussels (VUB), Krijgskundestraat 33, 1160 Brussels, Belgium
e-mail: diraerts@vub.be

J. A. Arguëlles
e-mail: jonitoarguelles@gmail.com

L. Beltran
e-mail: lestercc21@gmail.com

S. Geriente
e-mail: sgeriente83@yahoo.com

S. Sozzo (✉)
Department of Humanities and Cultural Heritage (DIUM) and Centre CQSCS, University of Udine, Vicolo Florio 2/b, Udine 33100, Italy
e-mail: sandro.sozzo@uniud.it

A. Plotnitsky and E. Haven (eds.), *The Quantum-Like Revolution*,
https://doi.org/10.1007/978-3-031-12986-5_15

1 Introduction

In physics, one generally maintains that the so-called 'Bell inequalities' prove that quantum entities exhibit certain specific aspects, such as 'contextuality', 'entanglement' and 'nonseparability', which cannot be explained in terms of the mathematical structures that are typically used in classical physics. This contextuality also persists for far away quantum entities, an aspect known as 'nonlocality' after Bell's seminal paper [1, 2]. The consistent violation of Bell inequalities in physics tests performed on a variety of quantum entities, in addition to confirming the predictions of quantum theory, in particular, reveals that the statistics of repeated experiments shows connections between quantum entities that cannot be represented within a classical probabilistic model satisfying the axioms of A. Kolmogorov ('Kolmogorovian probability') (see, e.g., [3]).

The setting of a Bell-type test is relatively simple, if presented in the so-called 'Einstein-Podolsky-Rosen-Bohm form' [4, 5]. We briefly review it here, because Bell-type tests are relevant to the purposes of the present paper. One considers a composite physical entity $\mathscr{S}_{12}$, prepared in an initial state p and such that two individual entities $\mathscr{S}_1$ and $\mathscr{S}_2$ that have interacted in the past but are now far away ('space-like separation') can be recognized as parts of $\mathscr{S}_{12}$. Then, 4 coincidence experiments AB, AB', $A'B$ and $A'B'$ are performed on $\mathscr{S}_{12}$ which consist in performing measurements e_A, with outcomes A_1 and A_2, and $e_{A'}$ with outcomes A'_1 and A'_2, on entity $\mathscr{S}_1$, and measurements e_B, with outcomes B_1 and B_2, and $e_{B'}$ with outcomes B'_1 and B'_2, on entity $\mathscr{S}_2$. If the measurement outcomes can be only -1 or $+1$, then the expected values of AB, AB', $A'B$ and $A'B'$ become the correlation functions $E(A, B)$, $E(A, B')$, $E(A', B)$ and $E(A', B')$, respectively, and a specific inequality, which is the 'Clauser-Horne-Shimony-Holt version of Bell inequalities' (briefly, 'CHSH inequality')

$$-2 \leq E(A', B') + E(A', B) + E(A, B') - E(A, B) \leq 2 \tag{1}$$

can be formulated [2]. John Bell put forward the type of inequality carrying his name [1], with the aim of making experimental tests possible with respect to the strange quantum phenomenon, revealed in the Einstein-Podolsky-Rosen paradox reasoning [4], and called 'entanglement', such that, in the absence of the phenomenon of entanglement, the inequality would not be violated. The violation of the Bell inequalities can also be connected with the non-existence of a Kolmogorovian probability model for the joint entity [3] but, in the investigation we put forward here, our interest is in the original situation for which the inequalities were formulated, namely, the possible connections between individual entities recognizable within a joint entity. It is indeed the case that quantum theory makes specific predictions about Eq. (1). More specifically, one can choose the individual entities, e.g., pair of spin 1/2 quantum particles, states, e.g., entangled state (typically, the singlet spin state), and observables, e.g., pairs of spin along different directions, in such a way that Eq. (1) is violated [1, 2]. Moreover, one can show that the 'CHSH factor'

$\Delta_{CHSH} = E(A', B') + E(A', B) + E(A, B') - E(A, B)$ is bound in quantum theory by the numerical value $\Delta_{QMC} = 2\sqrt{2} \approx 2.83$, known as the 'Cirel'son bound' [6, 7]. It is important to observe, at this stage, that the existence of such a 'quantum bound' is not trivial because, from a mathematical point of view, the CHSH factor can take any value between -4 and $+4$. In addition, the derivation of the Cirel'son bound explicitly refers to measurements e_{XY}, corresponding to coincidence experiments XY, $X = A, A'$, $Y = B, B'$, on the composite entity that are represented by product self-adjoint operators whenever the states of the composite entity $\mathscr{S}_{12}$ are represented by the unit vectors of the tensor product Hilbert space of two Hilbert spaces of which their unit vectors represent the states of the possible to be recognized entities $\mathscr{S}_1$ and $\mathscr{S}_2$. This remark will play an important role in what follows.

In the last two decades, several theoretical studies, mainly inspired by quantum computation and quantum information and their burgeoning applications, have deeply analysed and extended Bell inequalities (see, e.g., [8, 9]). In addition, numerous empirical tests have followed the seminal experiments of Alain Aspect and his collaborators [10, 11]. All empirical tests that have been performed so far strongly confirm the predictions of quantum theory (see, e.g., [12–15]). The natural consequence of these results is that 'entanglement', i.e. unavoidable connection between the individual entities recognizable in a composite quantum entity, give rise to empirically observable phenomena, so much that, in particular, entanglement is nowadays considered as one of the fingerprints of quantum theory.

A growing body of theoretical and empirical research reveals that several quantum structures, such as 'contextuality', 'entanglement', 'indistinguishability', 'interference' and 'superposition', also manifest themselves in other domains than microphysics, which include, in particular, cognition (e.g., human probability and similarity judgements, decision-making and visual perception), socio-economic domains (economics and finance) and information systems (computer science and artificial intelligence) (see, e.g., [16–30] and references therein). Thus, interesting results have been obtained in the identification of quantum structures and, in particular, quantum entanglement, in the combination of natural concepts and applied computer science domains, as information retrieval and natural language processing (see, e.g., [31–47] and references therein). Along these lines of investigation, our team has produced both theoretical and empirical research on how to identify entanglement in conceptual combinations and document retrieval applications.

At an empirical level, in particular, we performed both cognitive tests on human participants [38], document retrieval tests on structured corpuses of documents [46] and image retrieval tests on the web [29]. In these tests, we considered the composite conceptual entity *The Animal Acts* as a combination of the individual conceptual entities *Animal* and *Acts*. We performed an additional document retrieval test on the web in which we considered the composite conceptual entity *The Animal eats the Food* as a combination of the individual conceptual entities *Animal* and *Food* [32]. In all these tests, we observed a systematic violation of the CHSH inequality, accompanied by a systematic violation of a property called the 'marginal law'.

In particular, the violation of the marginal law, which is believed not to occur in principle in Bell-type tests in quantum physics (a subject we discuss later), led us to study in detail a theoretical problem which is usually overlooked in physics, the 'identification problem', that is, the problem of recognising individual entities of a composite entity by performing on the latter measurements that resemble the typical coincidence experiments of Bell-type tests [41, 42]. We constructed a general theoretical framework to model any Bell-type situation in the Hilbert space formalism of quantum theory [48]. In this framework, one explicitly applies the prescription of the standard formalism of quantum theory according to which the joint entity needs to be modelled in a complex Hilbert space of which the dimension is given by the number of outcomes of the defining measurements, which is hence $\mathbb{C}^4$. Only then as a secondary step, with respect to the attempt to 'recognize' entities in the joint entity, one considers possible isomorphisms with another Hilbert space, namely, the one built as the tensor product $\mathbb{C}^2 \otimes \mathbb{C}^2$ of two-dimensional complex Hilbert spaces which need to describe the two possible entities, since the defining measurements to recognize these entities have two outcomes. As we analyzed in detail in [41, 42], there is no unique isomorphism between $\mathbb{C}^4$ and $\mathbb{C}^2 \otimes \mathbb{C}^2$, and this is the reason why, from a mathematical point of view, there are different ways to account for entanglement being present within the joint entity with respect to the entities to be recognized as individual entities. Essentially, entanglement manifests itself on the level of the probabilities of a joint measurement not being able to be written as products of probabilities of the component measurements of this joint measurement. Hence, entanglement is a property of the relation between the joint measurements and their components. Only when an extra symmetry is present connecting different joint measurements, the entanglement of these different joint measurements can be captured in a state of the joint entity. This symmetry can be detected by verifying whether the marginal law corresponding to the considered joint measurements is satisfied. In that case, the entanglement of these joint measurements can together be captured in a state of the joint entity. If this is true for all joint measurements one can prove that there is only one isomorphism connecting $\mathbb{C}^4$ with $\mathbb{C}^2 \otimes \mathbb{C}^2$ and in that case $\mathbb{C}^4$ can be substituted by $\mathbb{C}^4$ with $\mathbb{C}^2 \otimes \mathbb{C}^2$ and the joint entity can be directly modeled in $\mathbb{C}^2 \otimes \mathbb{C}^2$. With respect to 'entanglement', a property of the structure of the probabilities of a joint measurement with respect to the probabilities of the component measurements, this situation is exceptional and definitely not the general one. In [42] we work out in detail several experimental examples and show how entanglement can generally only be located in the joint measurements and only exceptionally in the state of the joint entity. It is widely accepted that, in the typically considered situations in quantum physics where entanglement is being measured and studied, this exceptional symmetry is always present and hence apriori the tensor product Hilbert space is taken to be the Hilbert space to model the joint entity, although already the first experiments of Aspect [10] showed violations of the marginal law, confirmed manifestly by later experiments [13]. For the cognition experiments that we consider in the present article where joint measurements are performed on joint concepts *The Animal Acts* and *The Animal eats the Food*, the symmetry is not fulfilled and hence no unique isomorphism exists between $\mathbb{C}^4$ and $\mathbb{C}^2 \otimes \mathbb{C}^2$. The new quantum-theoretical perspective enabled modelling of both cognitive and web tests on *The*

Animal Acts and introduced a novel mathematical ingredient in the modelling of Bell-type situations, namely, 'entangled measurements' [30, 41, 42, 47].

Entangled measurements have recently been successfully applied in performing quantum computation and information tasks (see, e.g., [9]). We show in the present paper that 'entanglement measurement also allows to identify the presence of quantum entanglement also in Bell-type tests where the violation of the Cirel'son bound occurs in addition to the violation of the CHSH inequality'. More specifically, we present the results of two cognitive tests, one on *The Animal Acts* and the other on *The Animal eats the Food* situations, that we have recently performed on a sample of 81 participants. In both tests, a significant violation of the CHSH has been observed. Furthermore, while *The Animal Acts* test has revealed a result very close to the Cirel'son bound, *The Animal eats the Food* test has revealed a significant violation of the latter bound–a result which we had already obtained in the web tests above on *The Animal Acts* [46, 47]. We successfully apply and refine the quantum-theoretical framework to *The Animal Acts* and *The Animal eats the Food* situations and get a confirmation that a 'strong form of quantum entanglement' exists between the respective individual concepts which involves both states and measurements'. Moreover, we get a further confirmation that 'such a form of entanglement captures the deeply non-classical meaning connections existing between those individual concepts'. Finally, we prove that 'quantum entanglement also exists beyond Cirel'son bound, but it is a joint effect of state-entanglement and measurement-entanglement. This final result is relevant, in our opinion, because it refutes the widespread belief that one cannot model situations exceeding Cirel'son in the Hilbert space formalism in Hilbert space (see, e.g., [6, 7, 9]). As such, this final result may also have an impact on the foundations of quantum theory.

For the sake of completeness, we summarise the content of the present paper in the following.

In Sect. 2, we briefly review the theoretical and empirical results that our research team has obtained in the last years on the identification of quantum entanglement in conceptual combinations. In Sect. 3, we report and analyse the data collected in two cognitive tests that we have recently performed on human participants about the conceptual combinations *The Animal Acts* and *The Animal eats the Food*. In Sect. 4, we explicate, refine and particularise to *The Animal Acts* and *The Animal eats the Food* situations the general quantum-theoretical framework that we have developed in [48] to model any Bell-type situation. In Sect. 5, we show that the quantum-theoretical framework allows to faithfully represent empirical data in Sect. 3. Finally, in Sect. 6, we explain how and why a violation of the CHSH inequality which also exceeds the Cirel'son bound may indicate the presence of a stronger form of quantum entanglement involving both states and measurements.

2 A Brief History of Bell–Type Tests in Conceptual Combinations

We review in this section the results that we have obtained in the last decade on the identification of entanglement in the combination of natural concepts and related applied domains, as information retrieval and natural language processing. For a detailed description of the tests and a complete analysis of the obtained results, the reader is referred to the papers cited below (see also Sect. 3).

In the first empirical test of entanglement involving concepts, we studied the combination *The Animal eats the Food*, which we regarded as a combination of the individual concepts *Animal* and *Food* [32]. We considered different items of *Animal*, namely, *Cat*, *Cow*, *Horse* and *Fish*, and different items of *Food*, namely, *Grass*, *Meat*, *Fish* and *Nuts*, and combined them to form all possible combinations, i.e. *The Cow eats the Grass*, *The Cat eats the Meat*, *The Cat eats the Fish*, *The Squirrel eats the Nuts*, and so on, which were considered as items of the combination *The Animal eats the Food*. Next, we split the 16 combinations obtained in this way into 4 groups of 4 combinations each in order to reproduce the 4 coincidence experiments described in Sect. 1. For example, coincidence experiment AB had 4 outcomes, *The Cat eats the Grass*, *The Cat eats the Meat*, *The Cow eats the Grass* and *The Cow eats the Meat*; the other experiments were constructed in an analogous way. We used the World Wide Web as a conceptual space and counted co-occurrence of words, e.g., “cow” and “grass”, “cat” and “fish”, and so on, in document retrieval operations by means of the ‘Google’ search engine. We collected relative frequencies of co-occurrences which we considered, in the large number limit, as joint probabilities and inserted them into the correlation functions in Eq. (1). We were surprised to find $\Delta_{CHSH} = 2.86$, hence a significant violation of the CHSH inequality. We proposed the presence of some type of entanglement between the concepts *Animal* and *Food* as being responsible of the observed deviation from the CHSH inequality. At that time, we also noticed, without however deepening it, the simultaneous violation of Cirel’son bound.

This empirical finding was really exciting for us, also because it was partially unexpected. Thus, we decided to deepen the investigation on entanglement in conceptual combinations and performed a cognitive test in which the direct responses of a sample of participants were collected [38]. In this case, we analysed the conceptual combination *The Animal Acts*, which we regarded as a combination of the individual concepts *Animal* and *Acts*–“acts” refers here to the action of an animal emitting a sound. We again considered different items of *Animal*, namely, *Horse*, *Bear*, *Tiger* and *Cat*, and different items of *Acts*, namely, *Growls*, *Whinnies*, *Snorts* and *Meows*, and combined them to form all possible combinations, i.e. *The Horse Growls*, *The Bear Whinnies*, *The Tiger Snorts*, *The Cat Meows*, and so on, which were considered as items of the combination *The Animal Acts*. Then, we split the 16 combinations obtained in this way into 4 groups of 4 combinations each in order to reproduce the 4 coincidence experiments in Sect. 1. For example, coincidence experiment AB had 4 outcomes, *The Horse Growls*, *The Horse Whinnies*, *The Bear Growls*, *The Bear*

Whinnies. The other experiments were constructed in a similar way. After reading an introductory text, a sample of 81 respondents had to fill in a questionnaire where, in each coincidence experiment, they had to choose which item in a list of 4 items they judged as a 'good example' of the conceptual combination *The Animal Acts*. Relative frequencies of positive responses were considered, in the large number limit, as joint probabilities and inserted into the correlation functions in Eq. (1). Also in this case, we found a significant violation the CHSH inequality, $\Delta_{CHSH} = 2.42$, which we again interpreted as the consequence of a form of entanglement in the combination *The Animal Acts*, due to the connections of meaning between the component concepts *Animal* and *Acts*.

The numerical value 2.42 obtained in the cognitive test on *The Animal Acts* resembled the values found in Bell-type tests in quantum physics (see Sect. 1). To make the analogy with physics more convincing, we then decided to construct an explicit quantum-theoretic model in Hilbert space of empirical data [41]. But, we immediately realised that we could not reproduce empirical data on *The Animal Acts* test by representing the initial state of *The Animal Acts* by an entangled state and the 4 measurements by product measurements, as one would be tempted to do in analogy with a quantum representation in physics. We discovered that the main reason of this impossibility was that, if one wants to model *The Animal Acts* situation starting from the modelling of *Animal* and *Acts*, hence using the usual tensor product representation, then the 'marginal law of probability' has to be satisfied, as it occurs in quantum physics, whereas it was systematically violated in the cognitive test. This difference between physical and cognitive realms led us to study in detail a problem that is generally overlooked in physics, the 'identification problem', that is, the problem of recognising individual entities, e.g., concepts, from measurements performed on composite entities, e.g., conceptual combinations. The investigation led us to develop a novel quantum-theoretical framework for any Bell-type situation, which allowed us to recognise systematic analogies and differences between violations of Bell inequalities in physics and cognition [48].[1] But, the quantum-theoretical framework also introduced a new fundamental element in the modelling of such situations, namely, the empirically justified use of 'entangled measurements' in addition to entangled states. Entangled measurements were indeed successfully used in the quantum representation of *The Animal Acts* situation [41].

In the meanwhile, we had also started systematically looking into possible applications of quantum structures in information retrieval and natural language processing, where representation of 'meaning entities', like concepts and conceptual combinations, play a crucial role (see, e.g., [27, 44]). In particular, we considered again *The Animal Acts* situation in a document retrieval test in which we collected data on co-occurrence of words, as we had done for *The Animal eats the Food* situation, but using known corpuses of documents instead of web search engines–corpuses of documents

[1] A closer analogy between Bell-type situations in physical and cognitive realms was identified in a cognitive test we performed on the conceptual combination *Two Different Wind Directions*, where the marginal law was not violated and we could represent data in Hilbert space using entangled states and product measurements [50, 51].

provide more reliable word counts than search engines [46]. We used 'Google Books', 'Corpus of Contemporary American English (COCA)' and 'News On Web (NOW)' as corpuses of documents and performed the coincidence experiments described in Sect. 1, finding 3.41, 3.00 and 3.33, respectively in the factor appearing in Eq. (1). These empirical findings were statistically significant and, more important, they were completely unexpected. Indeed, not only had we obtained consistent violations of the CHSH inequality across all corpuses of documents, but the numerical value of the violation systematically exceeded the Cirel'son bound. According to a widespread belief, a violation of the CHSH inequality which also violates Cirel'son bound cannot be modelled in Hilbert space formalism of quantum theory, hence does not represent a 'quantum effect' [9]. On the other side, we noticed that the derivation of this bound does not consider entangled measurements, but only product measurements [6, 7]. If one allows entangled measurements, then it is in principle possible to model in Hilbert space any violation of the CHSH inequality whose numerical value lies in the mathematically permitted interval from -4 to $+4$. Indeed, we applied in [47] the general quantum-theoretical framework and faithfully represented the web data presented in [46].

Finally, in an independent but related investigation, we performed a web test on artefacts of visual perception, such as images, using 'Google Images' as a web search engine to count images representing items of the conceptual combination *The Animal Acts* [29]. Also in this case, we obtained the value $\Delta_{CHSH} = 2.41$ for the CHSH factor in Eq. (1), which again indicated the presence of entanglement in visual perception. The corresponding quantum-theoretical modelling enabled faithful representation of empirical data and confirmed that a strong form of entanglement, involving both states and measurements, exists between the component concepts *Animal* and *Acts*, again due to their meaning connections.

We now intend to complete the investigation outlined above and consider two cognitive tests on the conceptual combinations *The Animal Acts* and *The Animal eats the Food* with the aim of 'identifying the presence of quantum entanglement beyond the limits imposed by Cirel'son bound'. We will dedicate the following sections to this purpose.

3 New Empirical Results

We report in this section the details of the two cognitive tests that we have performed on the conceptual combinations *The Animal Acts* and *The Animal eats the Food*. As we will see, their results substantially confirm the empirical patterns highlighted in Sect. 2.

Let us start by *The Animal Acts* test. As anticipated in Sect. 2, we consider the concept *The Animal Acts* as a combination of the individual concepts *Animal* and *Acts*, where "acts" refers to the sound, or noise, produced by an animal. Next, we consider two pairs of items of *Animal*, namely, (*Horse*, *Bear*) and (*Tiger*, *Cat*), and two pairs of items of *Acts*, namely, (*Growls*, *Whinnies*) and (*Snorts*, *Meows*). We are now ready to illustrate the test.

A sample of 81 individuals were presented in a 'within subjects design' a questionnaire which contained 4 coincidence experiments AB, AB', $A'B$ and $A'B'$ whose setting was similar to the typical setting of a Bell-type test sketched in Sect. 1. More specifically, participants were preliminarily asked to read an 'introductory text' where the concepts under study were introduced and a description of the tasks involved in the judgement test was provided. Then, in each coincidence experiment, participants were asked to choose which item in a list of 4 items they judged as a good example of the conceptual combination *The Animal Acts*.

In coincidence experiment AB, participants had to choose among the 4 items:
(A_1B_1) *The Horse Growls*
(A_2B_2) *The Bear Whinnies*
(A_1B_2) *The Horse Whinnies*
(A_2B_1) *The Bear Growls*

If the response was (A_1B_1) or (A_2B_2), then experiment AB was attributed outcome $+1$; if the response was (A_1B_2) or (A_2B_1), then experiment AB was attributed outcome -1.

In coincidence experiment AB', participants had to choose among the 4 items:
$(A_1B'_1)$ *The Horse Snorts*
$(A_1B'_2)$ *The Horse Meows*
$(A_2B'_1)$ *The Bear Snorts*
$(A_2B'_2)$ *The Bear Meows*

If the response was $(A_1B'_1)$ or $(A_2B'_1)$, then experiment AB' was attributed outcome $+1$; if the response was $(A_1B'_2)$ or $(A_2B'_1)$, then experiment AB' was attributed outcome -1.

In coincidence experiment $A'B$, participants had to choose among the 4 items:
(A'_1B_1) *The Tiger Growls*
(A'_1B_2) *The Tiger Whinnies*
(A'_2B_1) *The Cat Growls*
(A'_2B_2) *The Cat Whinnies*

If the response was (A'_1B_1) or (A'_2B_2), then experiment $A'B$ was attributed outcome $+1$; if the response was (A'_1B_2) or (A'_2B_1), then experiment $A'B$ was attributed outcome -1.

Finally, in coincidence experiment $A'B'$, participants had to choose among the 4 items:
$(A'_1B'_1)$ *The Tiger Snorts*
$(A'_1B'_2)$ *The Tiger Meows*
$(A'_2B'_1)$ *The Cat Snorts*
$(A'_2B'_2)$ *The Cat Meows*

If the response was $(A'_1B'_1)$ or $(A'_2B'_2)$, then experiment $A'B'$ was attributed outcome $+1$; if the response was $(A'_1B'_2)$ or $(A'_2B'_1)$, then experiment $A'B'$ was attributed outcome -1.

For each coincidence experiment AB, AB', $A'B$ and $A'B'$, we collected the relative frequencies of the obtained responses which we considered, in the large number

Table 1 The data collected in coincidence experiments on entanglement in the conceptual combination *The Animal Acts*

Experiment AB	*Horse Growls*	*Horse Whinnies*	*Bear Growls*	*Bear Whinnies*
Probability	$\mu(A_1B_1) = 0.0494$	$\mu(A_1B_2) = 0.1235$	$\mu(A_2B_1) = 0.7778$	$\mu(A_2B_2) = 0.0494$
Experiment AB'	*Horse Snorts*	*Horse Meows*	*Bear Snorts*	*Bear Meows*
Probability	$\mu(A_1B'_1) = 0.7160$	$\mu(A_1B'_2) = 0.0494$	$\mu(A_2B'_1) = 0.2222$	$\mu(A_2B'_2) = 0.0123$
Experiment $A'B$	*Tiger Growls*	*Tiger Whinnies*	*Cat Growls*	*Cat Whinnies*
Probability	$\mu(A'_1B_1) = 0.7778$	$\mu(A'_1B_2) = 0.0864$	$\mu(A'_2B_1) = 0.0617$	$\mu(A'_2B_2) = 0.0741$
Experiment $A'B'$	*Tiger Snorts*	*Tiger Meows*	*Cat Snorts*	*Cat Meows*
Probability	$\mu(A'_1B'_1) = 0.0864$	$\mu(A'_1B'_2) = 0.0617$	$\mu(A'_2B'_1) = 0.0247$	$\mu(A'_2B'_2) = 0.8272$

limit, as the probability $\mu(A_iB_j)$, $\mu(A_iB'_j)$, $\mu(A'_iB_j)$ and $\mu(A'_iB_j)$ that the outcome A_iB_j, $A_iB'_j$, A'_iB_j and $A'_iB'_j, i, j = 1, 2$, respectively, is obtained in the corresponding experiment. Table 1 reports the judgement probabilities computed in this way. Referring to these probabilities, we can then calculate the expectation values, or correlation functions, of coincidence experiments AB, AB', $A'B$ and $A'B'$, as follows:

$$E(AB) = \mu(A_1B_1) - \mu(A_1B_2) - \mu(A_2B_1) + \mu(A_2B_2) = -0.8025 \quad (2)$$

$$E(AB') = \mu(A_1B'_1) - \mu(A_1B'_2) - \mu(A_2B'_1) + \mu(A_2B'_2) = 0.4568 \quad (3)$$

$$E(A'B) = \mu(A'_1B_1) - \mu(A'_1B_2) - \mu(A'_2B_1) + \mu(A'_2B_2) = 0.7037 \quad (4)$$

$$E(A'B') = \mu(A'_1B'_1) - \mu(A'_1B'_2) - \mu(A'_2B'_1) + \mu(A'_2B'_2) = 0.8765 \quad (5)$$

Inserting Eqs. (2)–(5) into Eq. (1), we get

$$\Delta_{CHSH} = E(A', B') + E(A', B) + E(A, B') - E(A, B) = 2.7901 \quad (6)$$

The numerical value $2.7901 \approx 2.79$ exceeds the classical limit imposed by the CHSH inequality and is slightly below the Cirel'son bound. However, while the deviation from the value 2 is statistically significant (p-value $1.44 * 10^{-5}$), the deviation from the value $2\sqrt{2}$ is not significant (p-value 0.4151), which entails that *The Animal Acts* data show a non-definite behaviour with respect to the Cirel'son bound. This empirical pattern confirms and strengthens the results obtained in [38] where a violation of the CHSH inequality was also observed, $\Delta_{CHSH} = 2.42$, see Sect. 2. The numerical difference between 2.79 and 2.42 in the two cognitive tests can be explained as due to a subtle, but important, difference in the introductory texts which were presented to the participants in the two tests. Indeed, the introductory text in [38] explicitly included the final sentence "Let all aspects of yourself play a role in the choice you make, ratio, but also imagination, feeling, emotion, and whatever." The individuals who took part to the 2011 *The Animal Acts* test paid some attention to this sentence, and this led some participants to make choices, not really expected a priori, e.g., some of them chose *The Bear Meows*, or *The Cat Growls*, or *The Tiger Meows*, because that was what they preferred when guided by the emotions and imagination that the test brought about to them. This final sentence was removed from the introductory text which was presented to the participants in the more recent *The Animal Acts* cognitive test. We believe that the above mentioned numerical difference depends on the fact that human minds are active entities, able to quickly adapt to the context of the questionnaire. Hence, the absence of a sentence explicitly suggesting to consider emotions and imaginations in their judgements led the participants in the more recent *The Animal Acts* test to make more concrete and natural, hence more expected a priori, choices, and this implied a higher numerical value of the violation of the CHSH inequality.

Coming back to *The Animal Acts* test presented above, the non-classical behaviour in it admits the presence of entanglement between the concepts *Animal* and *Acts* as a natural explanation, because *Animal* and *Acts* are connected by meaning and this connection gives rise to statistical correlations, those expressed by Eqs. (2)–(5) which cannot be represented in a classical Kolmogorovian probability framework.

Let us now come to *The Animal eats the Food* test. As mentioned in Sect. 2, we consider the concept *The Animal eats the Food* as a combination of the individual concepts *Animal* and *Food*. Let us, then, consider two pairs of items of *Animal*, namely, (*Cat*, *Cow*) and (*Horse*, *Squirrel*), and two pairs of items of *Food*, namely, (*Grass*, *Meat*) and (*Fish*, *Nuts*). The test can be illustrated as follows.

A sample of 81 individuals were presented in a 'within subjects design' a questionnaire which contained again the 4 coincidence experiments AB, AB', $A'B$ and $A'B'$ that are typical of a Bell-type test. More precisely, participants, after reading an introductory text on concepts and their combinations as in the first test, were asked to choose in each coincidence experiment one item in a list of 4 items, namely, the item that they judged as a good example of the conceptual combination *The Animal eats the Food*.

In coincidence experiment AB, participants had to choose among the 4 items:
(A_1B_1) *The Cat eats the Grass*
(A_2B_2) *The Cat eats the Meat*
(A_1B_2) *The Cow eats the Grass*
(A_2B_1) *The Cow eats the Meat*

If the response was (A_1B_1) or (A_2B_2), then experiment AB was attributed outcome $+1$; if the response was (A_1B_2) or (A_2B_1), then experiment AB was attributed outcome -1.

In coincidence experiment AB', participants had to choose among the 4 items:
($A_1B'_1$) *The Cat eats the Fish*
($A_1B'_2$) *The Cat eats the Nuts*
($A_2B'_1$) *The Cow eats the Fish*
($A_2B'_2$) *The Cow eats the Nuts*

If the response was ($A_1B'_1$) or ($A_2B'_1$), then experiment AB' was attributed outcome $+1$; if the response was ($A_1B'_2$) or ($A_2B'_1$), then experiment AB' was attributed outcome -1.

In coincidence experiment $A'B$, participants had to choose among the 4 items:
(A'_1B_1) *The Horse eats the Grass*
(A'_1B_2) *The Squirrel eats the Meat*
(A'_2B_1) *The Horse eats the Grass*

(A'_2B_2) *The Cow eats the Meat*

If the response was (A'_1B_1) or (A'_2B_2), then experiment $A'B$ was attributed outcome $+1$; if the response was (A'_1B_2) or (A'_2B_1), then experiment $A'B$ was attributed outcome -1.

Finally, in coincidence experiment $A'B'$, participants had to choose among the 4 items:

($A'_1B'_1$) *The Horse eats the Fish*
($A'_1B'_2$) *The Horse eats the Nuts*
($A'_2B'_1$) *The Squirrel eats the Fish*
($A'_2B'_2$) *The Squirrel eats the Nuts*

If the response was ($A'_1B'_1$) or ($A'_2B'_2$), then experiment $A'B'$ was attributed outcome $+1$; if the response was ($A'_1B'_2$) or ($A'_2B'_1$), then experiment $A'B'$ was attributed outcome -1.

As in the first test, after collecting relative frequencies of responses, we calculated in the large number limit the probability $\mu(X_iY_j)$ that the outcome X_iY_j is obtained in the coincidence experiment XY, $i, j = 1, 2$, $X = A, A'$, $Y = B, B'$. Table 2 reports the judgement probabilities computed in this way. The corresponding expectation values, or correlation functions, are then calculated as follows:

$$E(AB) = \mu(A_1B_1) - \mu(A_1B_2) - \mu(A_2B_1) + \mu(A_2B_2) = -0.7284 \quad (7)$$
$$E(AB') = \mu(A_1B'_1) - \mu(A_1B'_2) - \mu(A_2B'_1) + \mu(A_2B'_2) = 0.8025 \quad (8)$$
$$E(A'B) = \mu(A'_1B_1) - \mu(A'_1B_2) - \mu(A'_2B_1) + \mu(A'_2B_2) = 0.9012 \quad (9)$$
$$E(A'B') = \mu(A'_1B'_1) - \mu(A'_1B'_2) - \mu(A'_2B'_1) + \mu(A'_2B'_2) = 0.8519 \quad (10)$$

Inserting Eqs. (7)–(10) into Eq. (1), we get $\Delta_{CHSH} = 3.2840 \approx 3.28$.

The numerical value 3.28 violates both the CHSH inequality and the Cirel'son bound. In both cases, the deviation is statistically significant (p-values $2.81 * 10^{-11}$ and $4.40 * 10^{-2}$, respectively). Hence, this deviation confirms and strengthens the empirical patterns identified in document retrieval tests on the web in [32] and [46]. Also in this case, entanglement between the concepts *Animal* and *Food* is a natural

Table 2 The data collected in coincidence experiments on entanglement in the conceptual combination *The Animal eats the Food*

Experiment AB	*Cat eats Grass*	*Cat eats Meat*	*Cow eats Fish*	*Cow eats Meat*
Probability	$\mu(A_1B_1) = 0.1111$	$\mu(A_1B_2) = 0.3457$	$\mu(A_2B_1) = 0.5185$	$\mu(A_2B_2) = 0.0247$
Experiment AB'	*Cat eats Fish*	*Cat eats Nuts*	*Cow eats Fish*	*Cow eats Nuts*
Probability	$\mu(A_1B'_1) = 0.8765$	$\mu(A_1B'_2) = 0.0494$	$\mu(A_2B'_1) = 0.0494$	$\mu(A_2B'_2) = 0.0247$
Experiment $A'B$	*Horse eats Grass*	*Horse eats Meat*	*Squirrel eats Grass*	*Squirrel eats Meat*
Probability	$\mu(A'_1B_1) = 0.8889$	$\mu(A'_1B_2) = 0.0494$	$\mu(A'_2B_1) = 0$	$\mu(A'_2B_2) = 0.0617$
Experiment $A'B'$	*Horse eats Fish*	*Horse eats Nuts*	*Squirrel eats Fish*	*Squirrel eats Nuts*
Probability	$\mu(A'_1B'_1) = 0.0494$	$\mu(A'_1B'_2) = 0.1235$	$\mu(A'_2B'_1) = 0.0617$	$\mu(A'_2B'_2) = 0.8765$

candidate for explaining this non-classical behaviour, due to the meaning connections existing between *Animal* and *Food*.

The empirical findings above cannot generally be represented in a classical Kolmogorovian probability framework. We will show in the next two sections that these empirical findings can be represented in the Hilbert space formalism of quantum theory, independently of their behaviour with respect to the Cirel'son bound. However, we firstly need to present the essentials of a general quantum-theoretical framework that we have recently elaborated to model Bell-type situations in any empirical domain.

4 Construction of a General Hilbert Space Model

In this section we study how a theoretical framework can be constructed which models the two cognitive situations in Sect. 3 but is also general enough to cope with any domain in which Bell-type tests are performed. We will show that the violations of the CHSH inequality and the Cirel'son bound can be simultaneously explained by assuming that a strong form of 'quantum entanglement' is present in conceptual combinations. We refine and generalise the modelling scheme that we have elaborated in [41, 42] to identify the entanglement that occurs in Bell-type settings. The theoretical scheme has already been applied with success to model the web tests in [47] (see also Sect. 2). We refer to [48] for a detailed analysis and comparison of the entanglement situations occurring in physical and cognitive realms.

The general modelling scheme for Bell-type situations consists in the implementation of three main steps which we summarise in the following.

(i) One identifies in the empirical situation under investigation the composite conceptual entity and the individual conceptual entities composing it.

(ii) One recognises in the composite conceptual entity the states, measurements and outcome probabilities that are relevant to the phenomenon under study.

(iii) One represents entities, states, measurements and outcome probabilities using the Hilbert space representation of entities, states, measurements and outcome probabilities of quantum theory.

The application of steps (i)–(iii) to *The Animal Acts* situation has been presented in various papers [38, 41, 47]. Let us, then, focus here on how (i)–(iii) are applied to *The Animal eats the Food* situation. Let us start by coincidence experiment AB and its outcomes *The Cat eats the Grass*, *The The Cat eats the Meat*, *The Cow eats the Grass* and *The Cow eats the Meat*.

(i) The conceptual combination *The Animal eats the Food* can be considered as a composite conceptual entity made up of the individual conceptual entities *Animal* and *Food*.

(ii) Whenever an individual reads the introductory text which explains the details of the test and nature of the involved concepts, this set of instructions prepares the composite entity *The Animal eats the Food* in an initial state p which describes the general situation of an animal that eats food. This initial state is the unique conceptual state which all participants are confronted with in the test. In coincidence experiment AB, each respondent interacts with this uniquely prepared state p and operates as a measurement context e_{AB} for *The Animal eats the Food* which changes p into a generally different state. The latter state is not predetermined, as it depends on the concrete choice being made, which results as a consequence of this 'contextual interaction' between the respondent and the conceptual entity. More specifically, if the respondent chooses *The Cow eats the Grass*, that is, the outcome A_2B_1 is obtained in AB (see Sect. 3), the interaction between the entity *The Animal eats the Food* prepared in the state p and the (mind of the) respondent will determine a change of state of *The Animal eats the Food* from p to the state $p_{A_2B_1}$ which describes the more concrete situation of a cow that eats grass. When all responses are collected, a statistics of outcomes arises from this intrinsically indeterministic contextual process of state change. This outcome statistics is interpreted, in the large number limit, as the probability $P_p(A_2B_1)$ that the outcome A_2B_1 is obtained when the measurement e_{AB} is performed on the composite entity *The Animal eats the Food* in the initial state p.[2]

(iii) Let us finally come to the quantum mathematical representation. The entity *The Animal eats the Food* is associated with a Hilbert space and the state p is represented by a unit vector $|p\rangle$ of this Hilbert space. The measurement e_{AB} is instead represented by a self-adjoint operator or, equivalently, by a spectral family, on the Hilbert space whose eigenvectors represent the outcome states, or 'eigenstates', of e_{AB}, while outcome probabilities are obtained through the Born rule of quantum probability. This representation will become clear in Sect. 5.

The description above can be extended in a straightforward way to the other coincidence experiments of *The Animal eats the Food* situation. Hence, the complete identification of states, measurements, outcomes and outcome probabilities that are relevant to the situations in Sect. 3 is presented in the following.

For every $X = A, A'$, $Y = B, B'$, coincidence experiment XY corresponds to a measurement e_{XY} performed on the composite conceptual entity *The Animal Acts* (respectively, *The Animal eats the Food*) with 4 possible outcomes $X_iY_j, i, j = 1, 2$, where we choose $X_iY_j = +1$ if $i = j$ and $X_iY_j = -1$ if $i \neq j$, and 4 eigenstates $p_{X_iY_j}$ describing the state of *The Animal Acts* (*The Animal eats the Food*) after the outcome X_iY_j occurs in XY. Let us denote by $P_p(X_iY_j)$ the probability that the outcome X_iY_j is obtained when the measurement e_{XY} is performed on *The Animal Acts* (*The Animal eats the Food*) in the state p.

Once we have completed the identification of entities, initial state, measurements, eigenstates and outcome probabilities occurring in the *The Animal Acts* and *The Animal eats the Food* situations, we need to construct a general quantum framework in Hilbert space to model these conceptual situations. However, before doing so, we

---

[2] In [49]), we have called 'realistic' and 'operational' the above description of a measurement process on a conceptual entity, where the term 'realistic' refers to the preparation of the conceptual entity in a defined state and the term 'operational' refers to the measurements that are performed on the entity.

need to revisit a theoretical analysis that we have presented in detail in [27] (see also [41, 47]).

We preliminarily observe that in both *The Animal Acts* and *The Animal eats the Food* situations, all measurements e_{XY}, $X = A, A'$, $Y = B, B'$, have 4 outcomes $X_i Y_j, i, j = 1, 2$, which entails that both composite entities should be associated, as overall entities, with the complex Hilbert space $\mathbb{C}^4$ of all ordered 4-tuples of complex numbers. In addition, each state p of *The Animal Acts* (respectively, *The Animal eats the Food*) should be represented by a unit vector of $\mathbb{C}^4$ and each measurement on *The Animal Acts* (*The Animal eats the Food*) should be represented by a self-adjoint operator or, equivalently, by a spectral family, on $\mathbb{C}^4$. On the other hand, for every $i, j = 1, 2$, each outcome $X_i Y_j$ is obtained by juxtaposing the outcomes X_i and Y_j, e.g., *The Bear Growls* is obtained by syntactically juxtaposing the words "bear" and "growls". This operation defines a 2-outcome measurement e_X, $X = A, A'$ on the individual entity *Animal* (*Animal*) and a 2-outcome measurement e_Y, $Y = B, B'$ on the individual entity *Acts* (*Food*). Hence, each of these individual entities should be associated with the complex Hilbert space $\mathbb{C}^2$ of all ordered couples of complex numbers. But, then, the standard Hilbert space formalism prescribes that both composite entities *The Animal Acts* and *The Animal eats the Food* should be associated with the tensor product Hilbert space $\mathbb{C}^2 \otimes \mathbb{C}^2$. We stress that we are studying here an 'identification problem', that is, the problem of how the composite entity *The Animal Acts* (*The Animal eats the Food*) can be decomposed into the individual entities *Animal* (*Animal*) and *Acts* (*Food*)' in such a way that 'these individual entities can be recognised from measurements performed on the respective composite entities'. As such, we are doing an operation that is opposite to what one typically does in Bell-type situations in quantum physics, where one constructs or, better, 'composes', the measurements on the composite entity from measurements performed on individual entities.

From a mathematical point of view, the vector spaces $\mathbb{C}^4$ and $\mathbb{C}^2 \otimes \mathbb{C}^2$ are isomorphic, and any isomorphism can be expressed in terms of the relationship between the corresponding orthonormal (ON) bases. States of *The Animal Acts* are represented by unit vectors of $\mathbb{C}^4$, hence of $\mathbb{C}^2 \otimes \mathbb{C}^2$, which contains both vectors representing 'product states' and vectors representing 'entangled states'. Moreover, the vector space $L(\mathbb{C}^4)$ of all linear operators on $\mathbb{C}^4$ is isomorphic to the tensor product $L(\mathbb{C}^2) \otimes L(\mathbb{C}^2)$, where $L(\mathbb{C}^2)$ of all linear operators on $\mathbb{C}^2$. Analogously, the tensor product $L(\mathbb{C}^2) \otimes L(\mathbb{C}^2)$ contains both self-adjoint operators representing 'product measurements' and self-adjoint operators representing 'entangled measurements'.

Now, let $I : \mathbb{C}^4 \longrightarrow \mathbb{C}^2 \otimes \mathbb{C}^2$ be an isomorphism mapping a given ON basis of $\mathbb{C}^4$ onto a given ON basis of $\mathbb{C}^2 \otimes \mathbb{C}^2$. We say that a state p represented by the unit vector $|p\rangle \in \mathbb{C}^4$ is a 'product state' with respect to I, if two states p_A and p_B, represented by the unit vectors $|p_A\rangle \in \mathbb{C}^2$ and $|p_B\rangle \in \mathbb{C}^2$, respectively, exist such that $I|p\rangle = |p_A\rangle \otimes |p_B\rangle$. Otherwise, p is an 'entangled state' with respect to I. Next, we say that a measurement e represented by the self-adjoint operator $\mathscr{E}$ on $\mathbb{C}^4$ is a 'product measurement' with respect to I, if two measurements e_X and e_Y, represented by the self-adjoint operators $\mathscr{E}_X$ and $\mathscr{E}_Y$, respectively, on $\mathbb{C}^2$ exist such that $I\mathscr{E}I^{-1} = \mathscr{E}_X \otimes \mathscr{E}_Y$. Otherwise, e is an 'entangled measurement' with respect to

I. Hence, the notion of entanglement crucially depends on the 'isomorphism that is used to identify individual entities within a given composite entity'.

With reference to the Bell-type setting presented in this section, one can then prove that, if the measurements e_{XY} and $e_{XY'}$, $X = A, A', Y, Y' = B, B', Y' \neq Y$, are product measurements with respect to the isomorphism I, then, for every state p of the composed entity, the 'marginal law of Kolmogorovian probability' is satisfied, that is, for every $i = 1, 2$, $\sum_j P_p(X_i Y_j) = \sum_j P_p(X_i Y'_j)$. Analogously, if the measurements e_{XY} and $e_{X'Y}$, $X, X' = A, A', Y = B, B', X' \neq X$, are product measurements with respect to the isomorphism I, then, for every state p of the composed entity, the marginal law is satisfied, that is, for every $j = 1, 2$, $\sum_i P_p(X_i Y_j) = \sum_i P_p(X'_i Y_j)$ (see Theorem 2, [41]). In the case the marginal law is satisfied in all measurements, one can also prove that a unique isomorphism exists, which can be chosen to be the identity operator (see, Theorem 4, [41]).

It follows from the above that, if the marginal law is violated, as it occurs in both conceptual tests in Sect. 3,[3] then one cannot find a unique isomorphism between $\mathbb{C}^4$ and $\mathbb{C}^2 \otimes \mathbb{C}^2$ such that all measurements are product measurements with respect to this isomorphism. In this case, one cannot explain the violation of the CHSH inequality as due to the usual situation in quantum physics where all measurements are product measurements and only the initial, or pre-measurement, state is entangled. Furthermore, if the marginal law is systematically violated, 4 distinct isomorphisms I_{XY}, $X = A, A', Y = B, B'$, exist such that the measurement e_{XY} is a product measurement with respect to I_{XY}. As a consequence, there is no unique isomorphism allowing to identify individual entities of a given composite entity. Finally, if we consider a given isomorphism between $\mathbb{C}^4$ and $\mathbb{C}^2 \otimes \mathbb{C}^2$ with respect to which identifying individual entities of a composite entity in a given test, then it may happen that both the pre-measurement state and all measurements are entangled. This final remark suggests the following considerations.

Firstly, the non-classical connections which violate the CHSH inequality in both *The Animal Acts* and *The Animal eats the Food* situations can be reasonably attributed to the fact that 'the component individual concepts carry meaning and further meaning is created in the combination process'. Since the violation of the CHSH inequality indicates the presence of entanglement between the individual conceptual entities, then it is reasonable to assume that 'it is the quantum structure of entanglement which is able to theoretically capture the meaning connections that are created in these cases'. This suggests, in a quantum-theoretical perspective, that the initial state p of both composite entities *The Animal Acts* and *The Animal eats the Food* should be an entangled state. This entangled state would then capture the meaning connections that are created between *Animal* and *Acts* and also between *Animal* and *Food* when the respondent read the introductory text and start the questionnaire.

---

[3] The marginal law is systematically violated in *The Animal Acts* test. For example, $\mu(A_2 B_1) + \mu(A_2 B_2) = 0.8272 \neq 0.2345 = \mu(A_2 B'_1) + \mu(A_2 B'_2)$. Analogously, the marginal law is systematically violated in *The Animal eats the Food* test. For example, $\mu(A'_1 B_1) + \mu(A'_1 B_2) = 0.9383 \neq 0.1729 = \mu(A'_1 B'_1) + \mu(A'_1 B'_2)$.

Secondly, in both *The Animal Acts* and *The Animal eats the Food* situations, all measurements e_{XY}, $X = A, A', Y = B, B'$, violate the marginal law of Kolmogorovian probability. This suggests, again, in a quantum-theoretical perspective, that all these measurements should be entangled measurements. In addition, in each measurement, all outcomes $X_i Y_j$, $i, j = 1, 2$, correspond to combined concepts, e.g., *The Cat Meows*, *The Cat eats the Fish*, etc., which are thus in turn connected by meaning. This also suggests that all eigenstates $p_{X_i Y_j}$ should be entangled states.

The two considerations above applied to the Hilbert space representation of both *The Animal Acts* cognitive and web tests [41, 47]. We will see in the next section that these considerations also apply to the empirical tests in in Sect. 3.

5 Quantum Representation of Data

In this section, we work out a quantum representation in Hilbert space of the cognitive tests data in Sect. 3, following the methodology developed in [48] and already applied to previous tests [41, 47]. The considerations in Sect. 4 suggest the following quantum mathematical representation for both *The Animal Acts* and *The Animal eats the Food* situations.

The composite conceptual entity is associated with the Hilbert space $\mathbb{C}^4$ of all ordered 4-tuples of complex numbers. Let $(1, 0, 0, 0)$, $(0, 1, 0, 0)$, $(0, 0, 1, 0)$ and $(0, 0, 0, 1)\}$ be the unit vectors of the canonical ON basis of $\mathbb{C}^4$ and let us consider the isomorphism $I : \mathbb{C}^4 \longrightarrow \mathbb{C}^2 \otimes \mathbb{C}^2$ where this ON basis coincides with the ON basis of the tensor product Hilbert space $\mathbb{C}^2 \otimes \mathbb{C}^2$ made up of the unit vectors $(1, 0) \otimes (1, 0)$, $(1, 0) \otimes (0, 1)$, $(0, 1) \otimes (1, 0)$ and $(0, 1) \otimes (0, 1)$. In these ON bases, the initial state p of the composite entity is represented by the unit vector $|p\rangle = (ae^{i\alpha}, be^{i\beta}, ce^{i\gamma}, de^{i\delta})$, where $a, b, c, d \geq 0$, $a^2 + b^2 + c^2 + d^2 = 1$, α, β, γ, $\delta \in \Re$ and $\Re$ is the real line. One easily proves that $|p\rangle$ represents a product state if and only if the following condition is satisfied:

$$ade^{i(\alpha+\delta)} - bce^{i(\beta+\gamma)} = 0 \tag{11}$$

Otherwise, p represents an entangled state.

Let us now represent the measurements e_{AB}, $e_{AB'}$, $e_{A'B}$ and $e_{A'B'}$ in Sect. 4. Each measurement e_{XY}, $X = A, A'$, $Y = B, B'$, has 4 outcomes X_1Y_1, X_1Y_2, X_2Y_1 and X_2Y_2 and 4 eigenstates $p_{X_1Y_1}$, $p_{X_1Y_2}$, $p_{X_2Y_1}$ and $p_{X_2Y_2}$. As in Sects. 3 and 4, we set, for every $X = A, A', Y = B, B'$, $X_1Y_1 = X_2Y_2 = +1$ and $X_1Y_2 = X_2Y_1 = -1$. The measurement e_{XY} is represented by the self-adjoint operator $\mathscr{E}_{XY}$ on $\mathbb{C}^4$ or, equivalently, by the spectral family $\{|p_{X_1Y_1}\rangle\langle p_{X_1Y_1}|, |p_{X_1Y_2}\rangle\langle p_{X_1Y_2}|, |p_{X_2Y_1}\rangle\langle p_{X_2Y_1}|, |p_{X_2Y_2}\rangle \langle p_{X_2Y_2}|\}$, such that the eigenstates $p_{X_1Y_1}$, $p_{X_1Y_2}$, $p_{X_2Y_1}$ and $p_{X_2Y_2}$ are represented by the eigenvectors

$$|p_{11}\rangle = (a_{11}e^{i\alpha_{11}}, b_{11}e^{i\beta_{11}}, c_{11}e^{i\gamma_{11}}, d_{11}e^{i\delta_{11}}) \tag{12}$$
$$|p_{12}\rangle = (a_{12}e^{i\alpha_{12}}, b_{12}e^{i\beta_{12}}, c_{12}e^{i\gamma_{12}}, d_{12}e^{i\delta_{12}}) \tag{13}$$
$$|p_{21}\rangle = (a_{21}e^{i\alpha_{21}}, b_{21}e^{i\beta_{21}}, c_{21}e^{i\gamma_{21}}, d_{21}e^{i\delta_{21}}) \tag{14}$$
$$|p_{22}\rangle = (a_{22}e^{i\alpha_{22}}, b_{22}e^{i\beta_{22}}, c_{22}e^{i\gamma_{22}}, d_{22}e^{i\delta_{22}}) \tag{15}$$

of $\mathscr{E}_{XY}$, respectively. In Eqs. (12)–(15), coefficients are such that $a_{ij}, b_{ij}, c_{ij}, d_{ij} \geq 0$ and $\alpha_{ij}, \beta_{ij}, \gamma_{ij}, \delta_{ij} \in \Re$, $i, j = 1, 2$. For every $X = A, A'$, $Y = B, B'$, the self-adjoint operator $\mathscr{E}_{XY}$ can be expressed as a tensor product operator if and only if all unit vectors in Eqs. (12)–(15) represent product states. Otherwise, $\mathscr{E}_{XY}$ cannot be expressed as a tensor product operator, hence e_{XY} is an entangled measurement. For every $X = A, A'$, $Y = B, B'$, $i, j = 1, 2$, the probability $P_p(X_iY_j)$ of obtaining the outcome X_iY_j in a measurement of e_{XY} on the composite entity in the state p is then given by the Born rule of quantum probability, that is, $P_p(X_iY_j) = |\langle p_{ij}|p\rangle|^2$.

To find a quantum mathematical representation of the data in Sect. 3, for every measurement e_{XY}, the unit vectors in (12)–(15) have to satisfy the following three sets of conditions.

(i) Normalization. The vectors in Eqs. (12)–(15) are unitary, that is,

$$a_{11}^2 + b_{11}^2 + c_{11}^2 + d_{11}^2 = 1 \tag{16}$$
$$a_{12}^2 + b_{12}^2 + c_{12}^2 + d_{12}^2 = 1 \tag{17}$$
$$a_{21}^2 + b_{21}^2 + c_{21}^2 + d_{21}^2 = 1 \tag{18}$$
$$a_{22}^2 + b_{22}^2 + c_{22}^2 + d_{22}^2 = 1 \tag{19}$$

(ii) Orthogonality. The vectors in Eqs. (12)–(15) are mutually orthogonal, that is,

$$0 = \langle p_{11}|p_{12}\rangle = a_{11}a_{12}e^{i(\alpha_{12}-\alpha_{11})} + b_{11}b_{12}e^{i(\beta_{12}-\beta_{11})} + \\ + a_{11}c_{12}c^{i(\gamma_{12}-\gamma_{11})} + d_{11}d_{12}e^{i(\delta_{12}-\delta_{11})} \tag{20}$$
$$0 = \langle p_{11}|p_{21}\rangle = a_{11}a_{21}e^{i(\alpha_{21}-\alpha_{11})} + b_{11}b_{21}e^{i(\beta_{21}-\beta_{11})} + \\ + a_{11}c_{21}c^{i(\gamma_{21}-\gamma_{11})} + d_{11}d_{21}e^{i(\delta_{21}-\delta_{11})} \tag{21}$$
$$0 = \langle p_{11}|p_{22}\rangle = a_{11}a_{22}e^{i(\alpha_{22}-\alpha_{11})} + b_{11}b_{22}e^{i(\beta_{22}-\beta_{11})} + \\ + a_{11}c_{22}c^{i(\gamma_{22}-\gamma_{11})} + d_{11}d_{22}e^{i(\delta_{22}-\delta_{11})} \tag{22}$$
$$0 = \langle p_{12}|p_{21}\rangle = a_{12}a_{21}e^{i(\alpha_{21}-\alpha_{12})} + b_{12}b_{21}e^{i(\beta_{21}-\beta_{12})} + \\ + a_{12}c_{21}c^{i(\gamma_{21}-\gamma_{12})} + d_{12}d_{21}e^{i(\delta_{21}-\delta_{12})} \tag{23}$$
$$0 = \langle p_{12}|p_{22}\rangle = a_{12}a_{22}e^{i(\alpha_{22}-\alpha_{12})} + b_{12}b_{22}e^{i(\beta_{22}-\beta_{12})} + \\ + a_{12}c_{22}c^{i(\gamma_{22}-\gamma_{12})} + d_{12}d_{22}e^{i(\delta_{22}-\delta_{12})} \tag{24}$$
$$0 = \langle p_{21}|p_{22}\rangle = a_{21}a_{22}e^{i(\alpha_{22}-\alpha_{21})} + b_{21}b_{22}e^{i(\beta_{22}-\beta_{21})} + \\ + a_{21}c_{22}c^{i(\gamma_{22}-\gamma_{21})} + d_{21}d_{22}e^{i(\delta_{22}-\delta_{21})} \tag{25}$$

(iii) Probabilities. For every $X = A, A'$, $Y = B, B'$, $i, j = 1, 2$, the probability $P_p(X_iY_j)$ coincides with the empirical probability $\mu(X_iY_j)$ in Tables 1 and 2, that is,

$$\begin{aligned}\mu(X_1Y_1) = |\langle p_{11}|p\rangle|^2 = a^2a_{11}^2 + b^2b_{11}^2 + c^2c_{11}^2 + d^2d_{11}^2 + \\ + 2aba_{11}b_{11}\cos(\alpha - \alpha_{11} - \beta + \beta_{11}) + \\ + 2aca_{11}c_{11}\cos(\alpha - \alpha_{11} - \gamma + \gamma_{11}) + \\ + 2ada_{11}d_{11}\cos(\alpha - \alpha_{11} - \delta + \delta_{11}) + \\ + 2bcb_{11}c_{11}\cos(\beta - \beta_{11} - \gamma + \gamma_{11}) + \\ + 2bdb_{11}d_{11}\cos(\beta - \beta_{11} - \delta + \delta_{11}) + \\ + 2cdc_{11}d_{11}\cos(\gamma - \gamma_{11} - \delta + \delta_{11})\end{aligned} \tag{26}$$

$$\begin{aligned}\mu(X_1Y_2) = |\langle p_{12}|p\rangle|^2 = a^2a_{12}^2 + b^2b_{12}^2 + c^2c_{12}^2 + d^2d_{12}^2 + \\ + 2aba_{12}b_{12}\cos(\alpha - \alpha_{12} - \beta + \beta_{12}) + \\ + 2aca_{12}c_{12}\cos(\alpha - \alpha_{12} - \gamma + \gamma_{12}) + \\ + 2ada_{12}d_{12}\cos(\alpha - \alpha_{12} - \delta + \delta_{12}) + \\ + 2bcb_{12}c_{12}\cos(\beta - \beta_{12} - \gamma + \gamma_{12}) + \\ + 2bdb_{12}d_{12}\cos(\beta - \beta_{12} - \delta + \delta_{12}) + \\ + 2cdc_{12}d_{12}\cos(\gamma - \gamma_{12} - \delta + \delta_{12})\end{aligned} \tag{27}$$

$$\begin{aligned}\mu(X_2Y_1) = |\langle p_{21}|p\rangle|^2 = a^2a_{21}^2 + b^2b_{21}^2 + c^2c_{21}^2 + d^2d_{21}^2 + \\ + 2aba_{21}b_{21}\cos(\alpha - \alpha_{21} - \beta + \beta_{21}) + \\ + 2aca_{21}c_{21}\cos(\alpha - \alpha_{21} - \gamma + \gamma_{21}) + \\ + 2ada_{21}d_{21}\cos(\alpha - \alpha_{21} - \delta + \delta_{21}) + \\ + 2bcb_{21}c_{21}\cos(\beta - \beta_{21} - \gamma + \gamma_{21}) + \\ + 2bdb_{21}d_{21}\cos(\beta - \beta_{21} - \delta + \delta_{21}) + \\ + 2cdc_{21}d_{21}\cos(\gamma - \gamma_{21} - \delta + \delta_{21})\end{aligned} \tag{28}$$

$$\begin{aligned}\mu(X_2Y_2) = |\langle p_{22}|p\rangle|^2 = a^2a_{22}^2 + b^2b_{22}^2 + c^2c_{22}^2 + d^2d_{22}^2 + \\ + 2aba_{22}b_{22}\cos(\alpha - \alpha_{22} - \beta + \beta_{22}) + \\ + 2aca_{22}c_{22}\cos(\alpha - \alpha_{22} - \gamma + \gamma_{22}) + \\ + 2ada_{22}d_{22}\cos(\alpha - \alpha_{22} - \delta + \delta_{22}) + \\ + 2bcb_{22}c_{22}\cos(\beta - \beta_{22} - \gamma + \gamma_{22}) + \\ + 2bdb_{22}d_{22}\cos(\beta - \beta_{22} - \delta + \delta_{22}) + \\ + 2cdc_{22}d_{22}\cos(\gamma - \gamma_{22} - \delta + \delta_{22})\end{aligned} \tag{29}$$

Then, let us represent represent the initial state p of the composite conceptual entity by the unit vector

$$|p\rangle = \frac{1}{\sqrt{2}}(0, 1, -1, 0) \tag{30}$$

The vector in Eq. (30) represents a maximally entangled state and corresponds to the 'singlet spin state' that is used in typical Bell-type tests in quantum physics (see Sect. 1). There are several reasons for this choice. Firstly, our aim is to incorporate all possible entanglement of the state-measurement situation in the pre-measurement state, so that the ensuing measurements are as close as possible to product measurements. We are however aware that some measurements, if not all, are entangled, due to the violation of the marginal law (see Sect. 4). Secondly, it is known that the singlet spin state has specific symmetry properties, namely, it is always represented by a unit vector of the form in Eq. (30) independently of the ON basis in which the unit vector is expressed. This would intuitively correspond to the fact both *The Animal Acts* and *The Animal eats the Food* express more abstract concepts than the corresponding outcomes.

Thus, for every $X = A, A'$, $Y = B, B'$, conditions (i)–(iii) identify 20 equations which should be satisfied by the 32 real variables $a_{ij}, b_{ij}, c_{ij}, d_{ij}, \alpha_{ij}, \beta_{ij}, \gamma_{ij}, \delta_{ij}$, $i, j = 1, 2$. To simplify the calculation, we set, for every $i, j = 1, 2$, $\alpha_{ij} = \beta_{ij} = \gamma_{ij} = \delta_{ij} = \theta_{ij}$, where $\theta_{ij} \in \Re$. Thus, each unit vector in Eqs. (12)–(15) takes the form $|p_{ij}\rangle = e^{i\theta_{ij}}(a_{ij}, b_{ij}, c_{ij}, d_{ij})$, $i, j = 1, 2$. This reduces the total number of unknown variables to 20.

We are now ready to represent the empirical data in Sect. 3 in Hilbert space. We start by *The Animal Acts* test.

The eigenstates of the measurement e_{AB} are represented by the unit vectors

$$|p_{A_1B_1}\rangle = e^{i174.91°}(0.97, -0.10, 0.21, 0) \tag{31}$$

$$|p_{A_1B_2}\rangle = e^{i0.38°}(0, 0.88, 0.38, 0.29) \tag{32}$$

$$|p_{A_2B_1}\rangle = e^{i32.54°}(0.23, 0.35, -0.90, 0.12) \tag{33}$$

$$|p_{A_2B_2}\rangle = e^{i98.90°}(0.03, 0.32, 0, -0.95) \tag{34}$$

By applying the entanglement condition in Eq. (11), we can verify that all unit vectors are entangled, hence e_{AB} is an entangled measurement. However, one observes that the condition in Eq. (11) shows a relatively larger deviation from zero in the unit vector $|p_{A_2B_1}\rangle$. We can then say that the eigenstate $p_{A_2B_1}$, corresponding to *The Bear Growls*, is a 'relatively more entangled state'.

The eigenstates of the measurement $e_{AB'}$ are represented by the unit vectors

$$|p_{A_1B'_1}\rangle = e^{i0.07°}(0.14, 0.24, -0.96, 0.08) \tag{35}$$

$$|p_{A_1B'_2}\rangle = e^{i18.24°}(0.01, 0.31, 0, -0.95) \tag{36}$$

$$|p_{A_2B'_1}\rangle = e^{i64.68°}(-0.02, 0.92, 0.25, 0.30) \tag{37}$$

$$|p_{A_2B'_2}\rangle = e^{i132.68°}(0.99, -0.02, 0.14, 0) \tag{38}$$

Also in this case, all unit vectors are entangled, hence $e_{AB'}$ is an entangled measurement. The entanglement condition in Eq. (11) shows a relatively larger deviation from zero in the unit vector $|p_{A_1B'_1}\rangle$. We can then say that the eigenstate $p_{A_1B'_1}$, corresponding to *The Horse Snorts*, is a 'relatively more entangled state'.

The eigenstates of the measurement $e_{A'B}$ are represented by the unit vectors

$$|p_{A'_1B_1}\rangle = e^{i45.48^\circ}(0.25, 0.37, -0.88, 0.17) \tag{39}$$
$$|p_{A'_1B_2}\rangle = e^{i2.21^\circ}(0.01, 0.84, 0.42, 0.35) \tag{40}$$
$$|p_{A'_2B_1}\rangle = e^{i308.44^\circ}(0.97, -0.13, 0.22, 0) \tag{41}$$
$$|p_{A'_2B_2}\rangle = e^{i163.38^\circ}(0.05, 0.39, 0, -0.92) \tag{42}$$

All unit vectors are entangled, hence $e_{A'B}$ is an entangled measurement. The entanglement condition in Eq. (11) shows a relatively larger deviation from zero in the unit vector $|p_{A'_1B_1}\rangle$. We can then say that the eigenstate $p_{A'_1B_1}$, corresponding to *The Tiger Growls*, is a 'relatively more entangled state'.

Finally, the eigenstates of the measurement $e_{A'B'}$ are represented by the unit vectors

$$|p_{A'_1B'_1}\rangle = e^{i353.32^\circ}(0.96, -0.18, 0.23, 0) \tag{43}$$
$$|p_{A'_1B'_2}\rangle = e^{i143.07^\circ}(0.07, 0.35, 0, -0.93) \tag{44}$$
$$|p_{A'_2B'_1}\rangle = e^{i0.58^\circ}(0.01, 0.78, 0.55, 0.30) \tag{45}$$
$$|p_{A'_2B'_2}\rangle = e^{i41.95^\circ}(0.29, 0.49, -0.80, 0.20) \tag{46}$$

All unit vectors are entangled, hence $e_{A'B'}$ is an entangled measurement. The entanglement condition in Eq. (11) shows a relatively larger deviation from zero in the unit vector $|p_{A'_2B'_2}\rangle$. We can then say that the eigenstate $p_{A'_2B'_2}$, corresponding to *The Cat Meows* , is a 'relatively more entangled state'.

Let us now come to the representation of empirical data of *The Animal eats the Food* test. The eigenstates of the measurements e_{AB}, $e_{AB'}$, $e_{A'B}$ and $e_{A'B'}$ are respectively represented by the 4 unit vectors

$$|p_{A_1B_1}\rangle = e^{i156.99^\circ}(0.91, -0.06, 0.41, 0) \tag{47}$$
$$|p_{A_1B_2}\rangle = e^{i0.38^\circ}(0.01, 0.97, 0.13, 0.22) \tag{48}$$
$$|p_{A_2B_1}\rangle = e^{i117.49^\circ}(0.41, 0.11, -0.90, 0.03) \tag{49}$$
$$|p_{A_2B_2}\rangle = e^{i160.39^\circ}(0.02, 0.22, 0, -0.97) \tag{50}$$

$$|p_{A_1B'_1}\rangle = e^{i45.38^\circ}(0.22, 0.49, -0.84, 0.12) \tag{51}$$
$$|p_{A_1B'_2}\rangle = e^{i2.20^\circ}(0.01, 0.83, 0.52, 0.19) \tag{52}$$
$$|p_{A_2B'_1}\rangle = e^{i308.12^\circ}(0.97, -0.13, 0.18, 0) \tag{53}$$
$$|p_{A_2B'_2}\rangle = e^{i158.68^\circ}(0.03, 0.22, 0, -0.97) \tag{54}$$

$$|p_{A'_1B_1}\rangle = e^{i45.43^\circ}(0, 0.47, -0.87, 0.17) \quad (55)$$
$$|p_{A'_1B_2}\rangle = e^{i2.21^\circ}(0.02, 0.81, 0.50, 0.31) \quad (56)$$
$$|p_{A'_2B_1}\rangle = e^{i337.01^\circ}(0.99, -0.01, -0.01, 0) \quad (57)$$
$$|p_{A'_2B_2}\rangle = e^{i163.38^\circ}(0, 0.35, 0, -0.94) \quad (58)$$

$$|p_{A'_1B'_1}\rangle = e^{i353.52^\circ}(0.97, -0.13, 0.19, 0) \quad (59)$$
$$|p_{A'_1B'_2}\rangle = e^{i143.07^\circ}(0.02, 0.16, 0, -0.99) \quad (60)$$
$$|p_{A'_2B'_1}\rangle = e^{i0.58^\circ}(0.02, 0.85, 0.50, 0.14) \quad (61)$$
$$|p_{A'_2B'_2}\rangle = e^{i42.01^\circ}(0.22, 0.48, -0.84, 0.08) \quad (62)$$

Also in this case, referring to Eq. (11), one can show that all unit vectors in Eqs. (47)–(62) are entangled, hence 'all measurements e_{AB}, $e_{AB'}$, $e_{A'B}$ and $e_{A'B'}$ are entangled measurements'. In addition, as in *The Animal Acts* test, in each measurement, one eigenstate is represented by a unit vector which exhibits a larger deviation in the entanglement condition in Eq. (11). These 'relatively more entangled states' are $p_{A_2B_1}$, corresponding to *The Cow eats the Grass* in e_{AB}, $p_{A_1B'_1}$, corresponding to *The Cat eats the Fish* in $e_{AB'}$, $p_{A'_1B_1}$, corresponding to *The Horse eats the Grass* in $e_{A'B}$, and $p_{A'_2B'_2}$, corresponding to *The Squirrel eats the Nuts* in $e_{A'B'}$.

We have thus completed the quantum-mathematical representation of the data on *The Animal Acts* and *The Animal eats the Food* tests. This representation also suggests additional considerations, as follows.

We firstly observe that all measurements are entangled in the quantum representation in both *The Animal Acts* and *The Animal eats the Food* situations. This result is due to the violation of the marginal law of probability in both tests which forbids concentrating all the entanglement of the state-measurement situation into the state, as we have seen in Sect. 4. Moreover, in each measurement, all eigenstates are entangled. This result confirms with our suggestion in Sect. 4 that 'quantum entanglement theoretically captures the meaning connections between the individual concepts that form the composite conceptual entity, and that these meaning connections are not predetermined, but are created when the test is performed and each respondent interacts with the composite entity'.

Secondly, in both situations, there are some outcome eigenstates which exhibit a relatively higher degree of entanglement than others, which can exactly be explained with the fact that entanglement captures meaning connections, hence higher degrees of entanglement correspond to higher meaning connections. For example, in *The Animal Acts* situation, the eigenstate $p_{A_2B_1}$ of the measurement e_{AB}, corresponding to *The Bear Growls*, is relatively more entangled than the other eigenstates of e_{AB}, which can be naturally explained by the fact that the individual items *Bear* and *Growls*, which are concepts themselves, are relatively more connected by meaning. In fact, if we look at empirical probabilities (see Table 1, Sect. 3), we note that the outcome A_2B_1, corresponding to *The Bear Growls*, scores a high probability to be judged as a good example of the conceptual combination *The Animal Acts*. By contrast, items like *Bear Meows* and *Horse Growls* score a low probability and are less connected by meaning. And, indeed, the corresponding eigenstates are relatively less entangled. Analogously, in *The Animal eats the Food* situation, the eigenstate $p_{A'_2B'_2}$ of the measurement $e_{A'B'}$, corresponding to *The Squirrel eats the Nuts*, is relatively more entangled than the other outcome eigenstates of $e_{A'B'}$, which can be again explained by the fact that the individual items *Squirrel* and *Nuts*, which are concepts themselves, are relatively more connected by meaning. And, indeed, if we look at empirical probabilities (see Table 2, Sect. 3), we note that the outcome $A'_2B'_2$, corresponding to *The Squirrel eats the Nuts*, scores a high probability to be judged as a good example of the conceptual combination *The Animal eats the Food*. On the contrary, items like *The Cow eats the Meat*, *The Cat eats the Nuts* and *The Squirrel eats the Grass* score a low probability and are less meaning-connected. And, indeed, the corresponding eigenstates are relatively less entangled.

Thirdly, we observe that the quantum representation of the cognitive tests appears to be more complex and less symmetric than the quantum representation of the information retrieval tests on the web in [46] (see [47]), where all measurements were entangled but each spectral measure contained two product eigenstates and two entangled eigenstates. The reason is that many judgement probabilities were zero in [46], due to the fact that human minds are able to create additional meaning connections between concepts than corpuses of documents.

Fourthly, we observe that the violation of Cirel'son bound in *The Animal eats the Food* situation can still be explained in terms of quantum entanglement, in contrast to widespread beliefs. Hence, there is quantum entanglement also beyond Cirel'son bound, but this type of entanglement involves both states and measurements. This result is important and we will devote the next section to deepen it.

6 Quantum Entanglement Beyond Cirel'son Bound

The empirical results in the cognitive test on *The Animal eats the Food* presented in Sect. 3 show a significant violation of the Cirel'son bound and, as such, they confirm the results in the document retrieval tests on the web in [32] and [46]. As we have noticed throughout this paper, these results contrast with widespread beliefs in quantum physics. It is thus worth to carefully explain their meaning and implications.

The original question behind the determination of the Cirel'son bound was whether there is a fundamental limit to quantum nonlocality On the other side, in this paper and other papers, we have proved that it is possible to represent in Hilbert space empirical data collected in Bell-type tests which violate the CHSH inequality by an amount which also exceeds Cirel'son bound (see, e.g., [47, 48]). This means that the statistical correlations exhibited in these Bell-type tests, which are anyway non-classical and non-signalling, can be represented within quantum theory.

As we have seen in Sects. 4 and 5, the fundamental element which makes it possible to work out the above mentioned quantum-mathematical representation in Hilbert space is the suggestion that the measurements performed on the composite entity are such that (i) their outcomes are products of outcomes obtained in measurements separately performed on individual entities, but (ii) their eigenstates are generally entangled. Equivalently, the measurements performed on the composite entity are entangled measurements, hence they cannot generally be decomposed into product measurements. Cirel'son, and also those who investigated this bound after him, did not envisage such alternative. As a result, they implicitly assumed only product measurements in the derivation of the Cirel'son bound. If one also allows the possibility of using entangled measurements, then one can in principle violate the CHSH inequality in Eq. (1) by any amount within its mathematical limit, i.e. between -4 and $+4$ (see also [48]).

This result is relevant for various reasons. Firstly, the result follows from a careful investigation of the identification problem in which individual entities are identified through measurements that are performed on composite entities. Secondly, our investigation shows that a violation of Bell inequalities which also exceeds the Cirel'son bound can still be explained in terms of quantum entanglement. More important, this 'quantum entanglement beyond Cirel'son bound' is deeper and stronger than the entanglement that is generally believed to be responsible of the violation of Bell inequalities in quantum physics, that is, an entanglement that is only located in the state of the composite entity, or 'state entanglement', see Sect. 1. Indeed, a violation of the Cirel'son bound, in addition to a violation of Bell inequalities, can be explained by assuming the presence of entanglement in both the state and the measurements, or 'state-measurement entanglement'. Thirdly, together with the violation of the marginal law, the violation of the Cirel'son bound may have deep implications for the foundations of quantum physics, as it could shed new light on the problem of faster than light communication and signalling and could explain apparent deviations from quantum predictions in some Bell-type tests in physics. These deviations have been studied by proposing the hypothesis that they would be due to measurement errors [52, 53]. However, in [54] it is shown that they might well be of a fundamental nature and not be due to measurement errors. If the analysis put forward in [54] turns out to be true, then a whole new perspective is needed also on the violation of the Bell inequalities in physics, where some experimental tests have found a violation of the marginal law [10, 13]. We indeed believe that the type of signalling, that is generally believed to occur as a consequence of the violation of the marginal law in these tests, is 'not' the one imagined by physicists. Hence, due to its different nature, this type of signalling would not be able to travel faster than light, even if the coincidence events are space-like separated events [54].

Acknowledgements This work was supported by QUARTZ (Quantum Information Access and Retrieval Theory), the Marie Skłodowska-Curie Innovative Training Network 721321 of the European Union's Horizon 2020 research and innovation programme.

References

1. J. Bell, On the einstein podolsky rosen paradox. Phys. **1**, 195–200 (1964)
2. J.F. Clauser, M.A. Horne, A. Shimony, R.A. Holt, Proposed experiment to test local hidden-variable theories. Phys. Rev. Lett. **23**, 880–884 (1969)
3. I. Pitowsky, *Quantum Probability, Quantum Logic*, Lecture Notes in Physics, vol. 321, (Springer, Berlin Heidelberg, 1989)
4. A. Einstein, B. Podolsky, N. Rosen, Can quantum-mechanical description of physical reality be considered complete? Phys. Rev. **47**, 777–780 (1935)
5. D. Bohm, *Quantum Theory*, (Prentice-Hall, New-York, 1951)
6. B.S. Cirel'son, Quantum generalizations of bell's inequality. Lett. Math. Phys. **4**, 93–100 (1980)
7. B.S. Tsirelson, Some results and problems on quantum bell-type inequalities. Hadron. J. Suppl. **8**, 329–345 (1993)
8. M. Genovese, Research on hidden variable theories. A review of recent progresses. Phys. Rep. **413**, 319–396 (2005)
9. N. Brunner, D. Cavalcanti, S. Pironio, V. Scarani, S. Wehner, Bell nonlocality. Rev. Mod. Phys. **86**, 419 and 839 (2014)
10. A. Aspect, P. Grangier, G. Roger, Experimental realization of einstein-podolsky-rosen-bohm gedankenexperiment: A new violation of bell's inequalities. Phys. Rev. Lett. **49**, 91–94 (1982)
11. A. Aspect, J. Dalibard, G. Roger, Experimental test of bell's inequalities using time-varying analyzers. Phys. Rev. Lett. **49**, 1804–1807 (1982)
12. W. Tittel, J. Brendel, H. Zbinden, N. Gisin, Violation of bell's inequalities by photons more than 10 km apart. Phys. Rev. Lett. **81**, 3563–3566 (1998)
13. G. Weihs, T. Jennewein, C. Simon, H. Weinfurter, A. Zeilinger, Violation of bell's inequality under strict einstein locality condition. Phys. Rev. Lett. **81**, 5039–5043 (1998)
14. M. Giustina, A. Mech, S. Ramelow, B. Wittmann, J. Kofler, J. Beyer, A. Lita, B. Calkins, T. Gerrits, S. Woo Nam, R. Ursin, A. Zeilinger, Bell violation using entangled photons without the fair-sampling assumption. Nat. **497**, 227–230 (2013)
15. B.G. Christensen, K.T. McCusker, J. Altepeter, B. Calkins, T. Gerrits, A. Lita, A. Miller, L.K. Shalm, Y. Zhang, S.W. Nam, N. Brunner, C.C.W. Lim, N. Gisin, P.G. Kwiat, Detection-loophole-free test of quantum nonlocality, and applications. Phys. Rev. Lett. **111**, 1304–1306 (2013)
16. D. Aerts, Quantum structure in cognition. J. Math. Psychol. **53**, 314–348 (2009)
17. E. Pothos, J. Busemeyer, A quantum probability explanation for violations of 'rational' decision theory. Proc. R. Soc. B **276**, 2171–2178 (2009)
18. A. Khrennikov, *Ubiquitous Quantum Structure*, (Springer, Berlin, 2010)
19. J. Busemeyer, P. Bruza, *Quantum Models of Cognition and Decision*, (Cambridge University Press, Cambridge, 2012)
20. D. Aerts, J. Broekaert, L. Gabora, S. Sozzo, Quantum structure and human thought. Behav. Brain Sci. **36**, 274–276 (2013)
21. D. Aerts, L. Gabora, S. Sozzo, Concepts and their dynamics: A quantum theoretic modeling of human thought. Top. Cogn. Sci. **5**, 737–772 (2013)
22. E. Haven, A.Y. Khrennikov, *Quantum Social Science*, (Cambridge University Press, Cambridge, 2013)
23. P. Kvam, T. Pleskac, S. Yu, J. Busemeyer, Interference effects of choice on confidence. Proc. Natl. Acad. Sci. USA **112**, 10645–10650 (2015)
24. M.L. Dalla Chiara, R. Giuntini, E. Negri, A quantum approach to vagueness and to the semantics of music. Int. J. Theor. Phys. **54**, 4546–4556 (2015)
25. M.L. Dalla Chiara, R. Giuntini, R. Leporini, E. Negri, G. Sergioli, Quantum information, cognition, and music. Front. Psychol. **6**, 1583 (2015)
26. M. Melucci, *Introduction to Information Retrieval and Quantum Mechanics*, (Springer, Berlin Heidelberg, 2015)
27. D. Aerts, J. Aerts Arguëlles, L. Beltran, L. Beltran, I. Distrito, M. Sassoli de Bianchi, S. Sozzo, T. Veloz, Towards a quantum world wide web. Theor. Comput. Sci. **752**, 116–131 (2018)

28. D. Aerts, E. Haven, S. Sozzo, A proposal to extend expected utility in a quantum probabilistic framework. Econ. Theory **65**, 1079–1109 (2018)
29. J. Aerts Arguëlles, The heart of an image: Quantum superposition and entanglement in visual perception. Found. Sci. **23**, 757–778 (2018)
30. J. Aerts Arguëlles, S. Sozzo, How images combine meaning: Quantum entanglement in visual perception. Soft Comput. **24**, 10277–10286 (2020)
31. I. Schmitt, D. Zellhofer, A. Nurnberger, Towards quantum logic based multimedia retrieval, in *NAFIPS 2008 - 2008 Annual Meeting of the North American Fuzzy Information Processing Society*. https://doi.org/10.1109/NAFIPS.2008.4531329
32. D. Aerts, Quantum particles as conceptual entities: A possible explanatory framework for quantum theory. Found. Sci. **14**, 361–411 (2009)
33. P. Bruza, K. Kitto, D. Nelson, C. McEvoy, Is there something quantum-like about the human mental lexicon? J. Math. Psychol. **53**, 362–377 (2009)
34. P. Bruza, K. Kitto, D. Nelson, C. McEvoy, Quantum Interaction. QI 2009. in *Extracting spooky-activation-at-a-distance from considerations of entanglement*, ed. by P. Bruza, D. Sofge, W. Lawless, K. van Rijsbergen, M. Klusch, Lecture Notes in Computer Science, vol. 5494 (Springer, Berlin Heidelberg, 2009), pp. 71–83
35. B. Coecke, M. Sadrzadeh, S. Clark, Mathematical foundations for a compositional distributional model of meaning. Linguist. Anal. **36**, 345–384 (2010)
36. B. Piwowarski, I. Frommholz, M. Lalmas, K. van Rijsbergen, What can quantum theory bring to information retrieval. in *Proceedings of the 19th ACM international conference on Information and knowledge management*, (2010), pp. 59–68. https://doi.org/10.1145/1871437.1871450
37. D I. Frommholz, B. Larsen, B. Piwowarski, M. Lalmas, P. Ingwersen, Supporting polyrepresentation in a quantum-inspired geometrical retrieval framework. in *IIiX '10 Proceedings of the Third Symposium on Information Interaction in Context*, (2010), 115–124. https://doi.org/10.1145/1840784.1840802
38. D. Aerts, S. Sozzo, Quantum structure in cognition: Why and how concepts are entangled. quantum interaction. Lect. Notes Comput. Sci. **7052**, 116–127 (2011)
39. D. Zellhöfer, I. Frommholz, I. Schmitt, M. Lalmas, K. van Rijsbergen, Towards quantum-based DB+IR processing based on the principle of polyrepresentation, in *Advances in Information Retrieval, ECIR 2011, LNCS 6611*. ed. by P. Clough et al, (Springer, Berlin, Heidelberg, 2011), pp. 729–732
40. E. Di Buccio, M. Melucci, D. Song, Towards predicting relevance using a quantum-like framework. *Advances in Information Retrieval ECIR 2011. Lecture Notes in Computer Science 6611*, ed. by P. Clough et al, (Berlin Heidelberg, Springer , 2011), pp. 755–758
41. D. Aerts, S. Sozzo, Quantum entanglement in concept combinations. Int. J. Theor. Phys. **53**, 3587–3603 (2014)
42. D. Aerts, S. Sozzo, Entanglement Zoo I & II. ed. by H. Atmanspacher, E. Haven, K. Kitto, D. Raine, in *Quantum Interaction. QI 2013. Lecture Notes in Computer Science 8369*, (Springer, Berlin, Heidelberg, 2014), pp. 84-96, 97-109
43. P. Bruza, K. Kitto, B. Ramm, L. Sitbon, A probabilistic framework for analysing the compositionality of conceptual combinations. J. Math. Psychol. **67**, 26–38 (2015)
44. D. Aerts, J. Arguëlles, L. Beltran, L. Beltran, M. Sassoli de Bianchi, S. Sozzo, T. Veloz, Testing quantum models of conjunction fallacy on the World Wide Web. Int. J. Theor. Phys. **56**, 3744–3756 (2017)
45. G. Gronchi, E. Strambini, Quantum cognition and bell's inequality: A model for probabilistic judgment bias. J. Math. Psychol. **78**, 65–75 (2017)
46. L. Beltran, S. Geriente, Quantum entanglement in corpuses of documents. Found. Sci. **24**, 227–246 (2019)
47. D. Aerts, L. Beltran, S. Geriente, S. Sozzo, Quantum-theoretic modeling in computer science: A complex Hilbert space model for entangled concepts in corpuses of documents. Int. J. Theor. Phys. **60**, 710–726 (2021)
48. D. Aerts, J. Aerts Arguëlles, L. Beltran, S. Geriente, M. Sassoli de Bianchi, S. Sozzo, T. Veloz, Quantum entanglement in physical and cognitive systems: A conceptual analysis and a general representation. Eur. Phys. J. Plus **134**, 493 (2019)

49. D. Aerts, M. Sassoli de Bianchi, S. Sozzo, On the foundations of the brussels operational-realistic approach to cognition. Front. Phys. **4**, 17 (2016). https://doi.org/10.3389/fphy.2016.00017
50. D. Aerts, J. Aerts Arguëlles, L. Beltran, S. Geriente, M. Sassoli de Bianchi, S. Sozzo, T. Veloz, Spin and wind directions I: Identifying entanglement in nature and cognition. Found. Sci. **23**, 323–335 (2018)
51. D. Aerts, J. Aerts Arguëlles, L. Beltran, S. Geriente, M. Sassoli de Bianchi, S. Sozzo, T. Veloz, Spin and wind directions II: A bell state quantum model. Found. Sci. **23**, 337–365 (2018)
52. G. Adenier, A. Khrennikov, Is the fair sampling assumption supported by EPR experiments? J. Phys. B: At.Ic, Mol. Opt. Phys. **40**, 131–141 (2007)
53. G. Adenier, A. Khrennikov, Test of the no-signaling principle in the hensen loophole-free CHSH experiment. Fortschritte der Physik (Progress in Physics) **65**, 1600096 (2017)
54. D. Aerts, Quantum and concept combination, entangled measurements and prototype theory. Top. Cogn. Sci. **6**, 129–137 (2014)

Diederik Aerts is professor at the Vrije Universiteit Brussel (VUB) and founding director of the Leo Apostel Centre for Interdisciplinary Studies (CLEA). He is Editor-in-Chief of the international ISI and Springer Nature journal Foundations of Science (FOS) and a board member of the Worldviews group, founded by the philosopher Leo Apostel. He is president of the Centre for Quantum Social and Cognitive Science (IQSCS) and is a Fellow of the College of the International Institute for Advanced Studies in Systems Research and Cybernetics (IIAS). He was the scientific and artistic coordinator of the Einstein meets Magritte conference in 1995, where the world's leading scientists and artists gathered to reflect about science, nature, human action and society. CLEA was invited and participated in the World-Exhibition in Shanghai in 2010, with a follow up of this conference entitled 'Times of Entanglement'. Diederik Aerts started his research with a focus on the foundations of quantum physics and, during the last two decades, also on cognitive science. In this respect, he is considered a pioneer of the research domain called quantum cognition. Aerts received in 2020 the Prigogine Award for his work in quantum cognition.

Jonito Aerts Arguëlles studied Visual Art at the KASK school of Arts in Ghent Belgium, where he obtained his master's degree in 2017. During his studies he was exploring the interface between arts and science in many of his works. This made him engage in a broad range of interests leading through various disciplines and media, from photography and video to scientifically inspired installations. In the last year of his Master, he collaborated with an interdisciplinary group of scientists on different projects. Presently his interest for the ontology of an image has led him to his current Ph.D. research at the Center Leo Apostel at Brussels Free University, investigating how quantum structure can model the meaning content of visual entities.

Lester Beltran studied computer science at the Interface Computer College in Manila, the Philippines and mechanical engineering at Wrexham Glyndwr University, United Kingdom. He is affiliated with the Center Leo Apostel of Brussels Free University where he works on a interdisciplinary Ph.D. He was an Early Stage Researcher of QUARTZ (Quantum Information Access and Retrieval Theory), an Innovative Training Network (ITN) funded by the European Union's Horizon 2020 research and innovation programme under the Marie Skłodowska-Curie programme and now works on his Ph.D at the Center Leo Apostel. His research focuses on the investigation of quantum structures in computer science with applications to information retrieval and natural language processing.

Suzette Geriente studied psychology at the Philippine Normal University in Manila, the Philippines, and is affiliated with the Center Leo Apostel of Brussels Free University where she works on a interdisciplinary Ph.D. She was an Early Stage Researcher of QUARTZ (Quantum Information Access and Retrieval Theory), an Innovative Training Network (ITN) funded by the European

Union's Horizon 2020 research and innovation programme under the Marie Skłodowska-Curie programme and now works on her Ph.D. at the Center Leo Apostel. In her research she investigates quantum structures in cognition and how their appearance is related to quantum structures in computer science with a focus on entanglement.

Sandro Sozzo is associate professor at the University of Udine. His research interests mainly concern the foundations of natural and cognitive sciences. In particular, he has provided substantial and internationally recognised contributions in the development of a novel theoretical perspective for human judgment and decision-making. The new perspective has relevant applications in computer science. He is author of a hundred publications, he has organised several international conferences, managed special issues as guest editor, and was invited to give lectures in UK (Oxford), US, Canada, EU and China. He is the director of the "Centre for Quantum Social and Cognitive Science" (CQSCS), the secretary of the "International Quantum Structures Association", and the managing editor of the Springer Nature journal "Foundations of Science". He has recently been investigator in the consortium "Quantum Information Access and Retrieval Theory" (QUARTZ), involving seven European institutions, which has been funded by the EU for a Marie Curie Innovative Training Network. Finally, he has acted as peer-review expert on invitation by the EU COST Actions, the Economic & Social Research Council and Leverhulme Trust Fund, the US-Israel Bi-national Science Foundation, the Italian Ministry of University and Research, and the Swiss ETH Zurich.

Thinking and Deciding in a Complex World

Polina Khrennikova

Abstract Early 'paradoxical' findings in decision theory, such as Allais and Ellsberg paradoxes, revealed an existence of preference reversals and non-neutral attitudes to subjective risks that cannot be mathematically captured via the formula of total probability, thus showing that agents' decision making is at variance with the classical probabilistic information processing mode. In this paper, we discuss in more detail the decision making contexts pertaining to non-consequential reasoning, and how these instances of non-classical preferences can be described, and also quantified with the aid of alternative frameworks based on quantum probability theory. We consider the major challenges to a unified "quantum-like" framework of thinking and deciding under uncertainty and risk and elaborate on some successful solutions.

Keywords Contextual decision making · Disjunction effect · Quantum probability · Interference · Generalized observables

1 Introduction

The application of quantum probability-based models to the phenomena outside of quantum physics itself is no longer perceived as something exotic, despite the novelty of this interdisciplinary field. The domain of applications includes a wide range of decision-making problems in economics, cognitive science, politics and finance. Many of these findings focus on the violation of the classical probabilistic scheme and the axioms of Boolean logic. The best-known effects that violate the classical representation of observables as random variables in the joint probability space are the conjunction and disjunction errors. Non-additivity of beliefs and the conjunction errors exhibit themselves in violations of rational reasoning in the Savage Sure Thing Principle, [31] and mathematically, as a deviation from rules of classical probability theory, [26].

P. Khrennikova (✉)
Faculty of Behavioural, Management and Social Sciences, Financial Engineering Section, University of Twente, Drienerlolaan 5, 7522 NB Enschede, The Netherlands
e-mail: p.khrennikova@utwente.nl

A. Plotnitsky and E. Haven (eds.), *The Quantum-Like Revolution*,
https://doi.org/10.1007/978-3-031-12986-5_16

Decision theories under uncertainty and risk are the key building blocks in modern economic and finance models, as well as in public choice theory. In recent decades, starting from the advent of behavioural science and its entrance into the traditional domains of economics and finance, expected utility theories encountered a wave of experimental studies that challenged their axiomatic foundations. One of them is the rational mode of information processing and decision formation, grounded in the canonical formulation of Kolmogorovian probability theory.

Contextual behaviour and non-classical reasoning were initially explained with the aid of some qualitative accounts, by revealing contextual biases and external frames that bring a deviation from normative reasoning, (e.g., [13]). Some prominent models of preference formation, such as "Prospect Theory" due to [12] (see also [33]) were devised by relaxing the additivity of objective and subjective probabilities and allowing for state-dependent risk attitudes.

An alternative approach to human reasoning and decision making, inspired by the mathematics and probability theory of quantum physics, was adopted after more and more 'paradoxes' and biases were revealed in lab experiments and real-life decision-making data. Quantum probability applications to date have provided some accurate explanatory and descriptive accounts of non-consequential reasoning, with the potential to serve as a predictive instrument of human reasoning under uncertainty and risk.

Quantum probability models (often termed 'quantum-like, since they adopt the mathematics of quantum theory, but do not imply a correspondence between our cognitive states and quantum physics, such as that of our brains) relax the additivity and distributivity axioms of classical probability theory. Quantum-like models can capture non-additivity and contextuality in agents' prior beliefs across many uncertain states. Quantum probability also allows accommodating non-Bayesian formation of posterior beliefs. There is by now, a large body of literature that takes advantage of quantum-like models, to describe various decision making contexts in psychology, politics, economics and finance, and to accommodate the emergent probabilistic fallacies, cf., [1, 2, 5, 10, 11, 17, 18, 21, 29, 37]. See also some representative monographs that contain a broad-based evaluation of the empirical data from behavioural experiments, to devise quantum probability-based accounts that explicate these decision making patterns, [6, 10, 17].[1]

Many of these applications focus specifically on the decision-making situations, where *disjunction effect* is present, as it is a manifestation of non-classical information processing in 'uncertain' decision making states. The 'non-classicality' of human reasoning is well captured through the geometric properties of Hilbert space and the decision-making projectors therein.

In the following sections, we would like to discuss current advances of quantum probability-based models and consider the challenges in modeling human preferences and beliefs in various decision-making contexts. Finally, we evaluate possible avenues for future applications. In Sect. 2, we review the major findings on

[1] Some of the pioneering works on quantum(-like) probabilistic modeling of cognition, [14–16], were devised more than 20 years ago.

non-consequential reasoning and discuss—how the generalised quantum-like calculus can accommodate such decision-making patterns. In Sect. 2.2 we discuss how quantum-like models can serve as a remedy to preference formation in ambiguous states. In Sect. 3.2, we examine, how quantum models can describe contextual behaviour, most easily detected through 'order effects' in individual question-answers and task evaluations. We also take a more critical view on the applicability of the quantum formalism to human reasoning and decision making and consider some challenges with respect to the usage of the calculus of the conventional quantum theory. Some of the core problems can be characterized as an impossibility of using Hermitian projector-operators to capture all types of decision observables, and the limit to the usage of the mathematics of the Hilbert space to depict the rich domain of contextuality in human beliefs and preferences. Some proposed solutions are reviewed in Sect. 4. Finally, Sect. 5 offers some suggestions for the future refinement of the existing frameworks and a quantification of the sources of contextuality.

2 Non-consequential Reasoning and Its Manifestations

Nowadays, people are constantly faced with more complex decision-making tasks. The surrounding environment contains multiple layers of ambiguity regarding future economic, financial and political outcomes. Some of the recent rare events (some of these events were never observed before), such as BREXIT, the outbreak of COVID-19 and its impact upon economies and financial markets, have shown that it is increasingly challenging to contemplate consequences related to an uncertain future. One of the key factors behind non-consequential reasoning in economic, financial and routine day-to-day decisions is attributed to non-Bayesian processing of information, whereby decision-makers cannot form conditional preferences over outcomes. Consequentialism is grounded in people's ability to think through a hypothetical decision tree. It is at the heart of a celebrated principle of rational reasoning under uncertainty, the "Savage Sure Thing Principle," named after Leonard Savage, [31]. We shall discuss the essence of this principle next.

2.1 Savage Sure Thing Principle

The Sure Thing Principle (STP) is concerned with the elimination from consideration of the common outcomes in given states of the world. This principle assumes subjective uncertainty associated with these states and can be considered as an equivalent to the independence axiom of expected utility theory under objective risk. In a nutshell, this axiom allows establishing a choice preference in the condition of uncertainty, if the decision-maker knows that an act will produce a more desirable outcome in any state of the world. Savage exemplified how STP works in practice, with the aid of a widely cited story of a businessman, who contemplates purchasing a property before the Presidential elections in the US. The businessman is uncertain, whether the Republican Party or the Democrats will win the election campaign. He decides that he will buy the property in both states (considering his subjective utility associated with a purchase in each state of the world). STP can be considered as a manifestation of consequential reasoning, as individuals consider consequences of different acts given a set of states of the world and some subjective beliefs about these states. A central premise of consequential reasoning is that beliefs obey the axioms of classical probability theory (CP), including the usage of the Bayes formula for updating of these beliefs. It is also important to note, following [8], that the subjective probability estimates associated with some state B and a complimentary state, B' can be expressing an exogenous risk (the uncertainty about the state of nature, such as economic states), as well as a strategic risk, involving uncertainty about the choice of a strategic opponent, which is characteristic for game-theoretic settings (e.g., a competitor starts a price war). Both of these risks are subjective risks that require one to make personal estimates based on their prior beliefs.

In many decision making experiments, entailing both monetary and non-monetary outcomes, the collected decision making statistics did not obey the above stated principle (e.g., [6, 8, 21, 22, 29, 32, 35]) and recently, [11]. It was shown that a choice A probability (frequency), without any certain information given on an event B, and its negation, B', was very often below the conditional choice frequencies $A|B$ and $A|B'$, as well as below the total probability (frequency) of A, expressed via the disjunction of these conditional choice outcomes, given by the formula of total probability, [26].[2]

A common feature of the above findings on "disjunction effect" is that an inaction is usually preferred to an action when the states of the world are unknown. These findings apply to experiments, such as "Two-Stage Gambling Task" and its variations, where probabilities are given exogenously [11, 35], but also to decision making tasks, where no probabilities of the states of the world are known, see, e.g., "Hawaii Vacation" experiment in [32]. In "Prisoner's Dilemma" type experiments, a lack of evaluation of counterparty's actions leads to a higher rate of cooperation, despite of this strategy being irrational from a game-theoretic perspective, see [32], and a further analysis in [8, 29]. In the "Hawaii Vacation" experiment the willingness to purchase a ticket to Hawaii was very low among participants, who were in the condition of uncertainty with respect to the outcome of an exam that they took. Only 7% of the participants were willing to purchase the vacation package on the spot, when exam outcome (pass, fail) was unknown, versus an average of 55.5%, who were keen to go on a vacation, both after a pass and a fail grade. We can see that a lack of information about state probabilities makes the decision-making process even more 'non-classical' from the viewpoint of CP. This phenomenon can be handled with the aid of quantum-like models and specifically, via the use of a more general calculus of QP, as will be discussed in Sect. 3.

[2] This means that in the formula of total probability, there is a super-additivity of disjunctions, where $p(A) < p(A \cap B) \cup p(A \cap B')$.

2.2 *Thinking and Deciding with an Ambiguous Future*

In more complex decision-making settings, with uncertain economic states, following the tenets of consequential reasoning becomes even more challenging. One has to make decisions, before acquiring information about the future economic and financial environment and the corresponding short, or long-term outcomes. Nevertheless, some choices must be made here and now (e.g., to buy or to sell shares, take a mortgage, or possibly invest in education). So, the key question that could be posed is: *How do we make decisions, when our beliefs are vague and the exact probabilities of events in question are not attainable?*

We note that standard finance frameworks operate under the assumptions that agents are forming homogeneous subjective beliefs, following the rules of CP theory. This is the essence of preference formation under Subjective Expected Utility Theory (SEUT) conceived by Savage, [31]. The subjective beliefs of agents are not observable per se but can be inferred from their preferences. Some studies aim to elicit these beliefs to ascertain, whether they obey the *linearity* condition of classical probability theory see, e.g., [34]. While SEUT assumes that both objective and subjective probabilities obey laws of probability theory, Knight, [25], distinguishes these two concepts in his early work: *"Uncertainty must be taken radically distinct from the familiar notion of Risk [...] It will appear that a measurable uncertainty or 'risk' proper, as we shall use the term, is so far from an unmeasurable one that is not in effect an uncertainty at all"*. (p. 19). Empirical evidence in behavioural finance studies and psychological experiments show that individuals usually dislike, vague information, or ambiguity, and prefer known risks. This behavioural pattern is known as "Ellsberg-type" ambiguity aversion [9]. Attitudes towards ambiguity are not universal, and people can have heterogeneous ambiguity attitudes, for instance, when the experienced events shift their attitude towards some kinds of uncertainties. We can mention the study, [30], that detects a more rare phenomenon of 'ambiguity seeking', because of a shift in an individual reference point, caused by experienced gains or losses. There are also other contextual factors, such as emotions, [4], that can come into play showing that the agents' ambiguity attitude is not stable and can deviate from CP-based rules of information processing under subjective risk. In this regard, it becomes of a great significance to further explore the settings that can induce instances of non-classical information processing resulting in non-consequential reasoning.[3]

---

[3] There are also some successful frameworks in behavioural economics that capture the non-additivity of beliefs over objective and subjective risks, cf., [12, 34]. As noted in [6, 37] and recently in [3], models based on quantum probability have a core distinguishing feature in describing and also quantifying 'vague' beliefs.

3 Quantum-Like Modeling of Co-Existing Beliefs and State Dependent Choice

Due to its more general axiomatics, QP has gained recognition in the modeling of contextual beliefs and preferences in various fields of decision making. There is now an established field of quantum-like models that can deal with ambiguous attitudes and their evolution, see [1, 2] as well as how these attitudes lead to non-consequential reasoning and contextual preferences, just to mention some representative works, [5, 6, 10, 17, 22, 29, 37, 39]. What is the key to quantifying the emergent contextuality in beliefs and preferences? Quantum-like models allow to depict preferences and beliefs that are vague and indeterminate providing a measure for this indeterminacy with the aid of so called *superposition states* and *interference* of uncertain probabilities, or more formally, their amplitudes.

3.1 *Interference in Quantum Probability*

In quantum probability (QP) calculus an introduction of non-commuting observables enables to capture non-jointly distributed random variables. One can speak about the impossibility of joint measurement of these types of observables. Instead of a classical probability measure with respect to some random variables A and B one considers the interference of their complex probability amplitudes.[4] Non-consequential reasoning can be explained mathematically, as a manifestation of the non-additivity and non-commutativity of probabilistic beliefs and preferences. One can capture this non-additivity with the aid of a generalized formula of total probability, [36].

$$P(A=\alpha) = \sum_{\beta} P(B=\beta)P(A=\alpha|B=\beta)$$

$$+2\sum_{\beta_1<\beta_2} \cos\theta_{\beta_1\beta_2}\sqrt{P(B=\beta_1)P(A=\alpha|B=\beta_1)P(B=\beta_2)P(A=\alpha|B=\beta_2)} \tag{1}$$

The interference term on the right-hand side can serve as a parameter, quantifying the deviation of beliefs from objective probabilities (prior and posterior). The size of $\cos\theta$ can vary depending on, the (i) initial cognitive state of the decision-maker; (ii) type of a random variable (observable) that is measured. QP can account for the complementarity of beliefs concerning future states of the world in economic and financial decisions. Complementarity in beliefs suggests an impossibility to

[4] Mathematically, complex probability amplitudes are given as Hilbert space vectors. In difference to the Bayes rule of probability update, quantum conditional probability is obtained following Born rule that describes an update of the quantum state. We refer the interested reader to the monographs [6, 10].

simultaneously posses firm estimates about the realisation of different states of the world. Works [28, 38] offer a deeper treatment of the notion of complementarity and its meaning in quantum physics and psychology.

3.2 *State Dependence and Order Effects*

Another deviation from the commutativity principle of preferences and beliefs, widely known in psychology and opinion poll studies, is known as the order effect when several question-observables are measured after each other. The order effect is related to specific features of human cognition, whereby new information or question reveals some preferences that the decision-maker was unable to establish in a condition of uncertainty. Formally, for the random variables corresponding to questions A, B, joint probability distribution does not hold, when $P(A = \alpha, B = \beta) \neq P(B = \beta, A = \alpha)$. Such types of contextual preferences have been successfully modelled via non-commuting observables, corresponding to the random variables, A and B, see, (e.g., [6, 39]). In the later contribution, authors considered decision making data from a large-scale organisational survey and modeled the obtained response statistics with the aid of the quantum probability framework.

The introduction of quantum-like models was not free from paradoxes, since not all psychological effects could be nicely accommodated with the aid of 'standard' quantum probability models. In other words, decision making statistics exhibits neither quantum nor classical features as shown in [6, 17, 29]. Hence, a generalization using the mathematics from quantum information theory was proposed in order to come as close as possible to a 'unified' framework of human contextual beliefs and preferences.

4 Contextual Beliefs and Preferences: What are the Key 'Paradoxes' of Quantum-Like Models?

4.1 *Non-double Stochasticity Problem*

One of the major caveats with the above reviewed experimental findings on disjunction effect is the following. While the cognitive statistics is violating the rules of classical probability theory, it is also not obeying a so-called *double-stochasticity* constraint. Double-stochasticity implies a unit additivity of the transition probabilities in the transition probability matrix, [17, 29]. This feature of cognitive statistics constraints the application of standard quantum physical formalism based on the orthogonal Hermitian projectors.

To be able to fully accommodate the decision making tasks in a two-dimensional Hilbert space, a class of more general projectors, relaxing the orthogonality constraint of projector-operators, "POVMs" (Positive operator valued measures) can be used following their application in quantum information theory, [7]. The authors in [11, 23] employ such types of POVMs, to accommodate preference reversals in voting statistics and in investment choice.[5]

4.2 *Order Effects and Response Replicability*

We have already discussed in Sect. 3.2 that order effects can be faithfully represented with the aid of non-commuting observables that are given by conventional Hermitian operators. At the same time, there is another feature of individual responses or preferences, coined "response replicability". The same response would be usually repeated by an individual if being asked the same question twice. This is knowns as the "A-A" type response replicability. Also, question surveys show that answers and preferences would be naturally stable over short time horizons and in the absence of new information. Hence, if questions are asked in an order A-B-A, the answer to the question A should not change. This brings a caveat in using conventional quantum probability to capture this second type of response replicability, and at the same time to accommodate order effects of the type "A-B" and "B-A". One can say that there exists a 'paradox', pertaining to the usage of non-commuting Hermitian operators to capture preference reversals. This is due to the impossibility to jointly accommodate both psychological effects as firstly highlighted in [19]. This 'no-go' result was at variance with the key feature of quantum probability, namely, the principle of non-commutativity. This problem was further researched in [20], and a more inclusive probabilistic framework inspired by a generalized quantum measurement theory (based on usage of non-projective operators) was recently proposed in [27].[6] Of course, these generalisations add some complexity and additional free parameters to the quantum framework, which is the cost one has to pay for an "all-inclusive" model of contextual decision making.

4.3 *What if the Interference Magnitude Is Too High?*

Finally, we would like to mention another possible problem related to the usage of conventional quantum probability framework. This problem emerges when the decision-making statistics exhibits non-additivity of a very high magnitude and

[5] Some authors approached non-additivity of transition probabilities, by increasing the dimension of the state-space representation of the observables, see [29]. In these models degenerate spectra can be used to represent eigenvalues by corresponding bases that are given as subspaces of larger dimensions.

[6] The authors in [27] employ a slightly different mathematics, to accommodate both of these psychological effects in a single operational framework of contextual preferences. In a nutshell, the distinguishing feature of this approach allows for an additional influence of a 'mental environment' upon the cognitive state of a decision-maker.

hence, cannot be described with the aid of complex probability amplitudes in a Hilbert space. For instance, if the parameter, $\cos\theta$, defined in Eq. (1) is not bounded by ± 1, the usage of complex numbers is not possible. Authors in [17] proposed a solution to this problem, by introducing a hyperbolic analogue of the interference coefficient, where the probability amplitudes are given by hyperbolic, rather than complex numbers. An interference of a very high magnitude ($\cos\theta > 3$) was detected in the survey statistics on voting preferences and a hyperbolic state space was introduced to capture this effect, [21].

5 Future Directions and Concluding Remarks

Quantum probability models are a good analogue of classical probability-based decision frameworks and modifications thereof. The key advantage is that quantum-like models are built upon an already existing axiomatic and operational framework, used to describe the uncertainty associated with a systems' behaviour in quantum physics. Hence, the quantum probability is a complete axiomatic framework. We can see that the richness of contextuality of human beliefs and preferences might often go beyond the axiomatics of standard quantum formalism. We have reviewed in this work the key caveats related to a straightforward application of quantum probability and the solutions offered so far, by relaxing some additional constraints (Hilbert space representation, usage of a small dimensional state space, and usage of orthogonal projectors). We can also witness that the non-classicality of beliefs can depend on various objective factors (such as the number of random variables to consider, familiarity with the decision making task, previous decisions, etc.), as well as on the mental environment shaped by our emotions and mood. There are many studies in behavioural economics and psychology showing the deep impact of experienced emotions upon beliefs and risk attitudes, e.g., [4, 24]. It would be valuable, to capture the impact of emotions and other mental factors in the construction of quantum decision observables. Other factors that could also be considered in advancing this unified framework are, for instance, biological and cognitive factors. More experiments dealing with the causes behind our sensitivity to internal and external context would be promising to better quantify and predict the impact of these contextual settings upon human behaviour.

References

1. D. Aerts, S. Sozzo, J. Tapia, Identifying quantum structures in the Ellsberg paradox. Int. J. Theor. Phys. **53**, 3666–3682 (2014)
2. D. Aerts, E. Haven, S. Sozzo, A proposal to extend expected utility in a quantum probabilistic framework. Econ. Theory **65**, 1079–1109 (2018)

3. H. Atmanspacher, I. Basieva, J.R. Busemeyer, A. Khrennikov, E.M. Pothos, R. Shiffrin, Z. Wang, What are the appropriate axioms of rationality for reasoning under uncertainty and resource constrained systems?. Behav. Brain Sci. **43**, e2 (2020)
4. A. Baillon, P.D. Koellinger, T. Treffers, Sadder but wiser: the effects of emotional states on ambiguity attitudes. J. Econ. Psychol. **53**, 67–82 (2016)
5. J.R. Busemeyer, E.M. Pothos, R. Franco, J.S. Trueblood, A quantum theoretical explanation for probability judgement errors. Psychol. Rev. **118**(2), 193–218 (2011)
6. J.R. Busemeyer, P.D. Bruza, *Quantum Models of Cognition and Decision* (Cambridge University Press, Cambridge, 2012)
7. P. Busch, M. Grabowski, P. Lahti, *Operational Quantum Physics* (Springer, Berlin, 1995)
8. R. Croson, The disjunction effect and reason-based choice. Organ. Behav. Hum. Decis. Process. **80**(2), 118–133 (1999)
9. D. Ellsberg, Risk, ambiguity and the Savage axioms. Q. J. Econ. **75**, 643–669 (1961)
10. E. Haven, A. Khrennikov, *Quantum Social Science* (Cambridge University Press, Cambridge, 2013)
11. E. Haven, P. Khrennikova, A quantum-probabilistic paradigm: non-consequential reasoning and state dependence in investment choice. J. Math. Econ. **78**, 186–197 (2018)
12. D. Kahneman, A. Tversky, Prospect theory: an analysis of decision under risk. Econometrica **47**, 263–291 (1979)
13. D. Kahneman, *Thinking, Fast and Slow* (Penguin Books, 2011)
14. A. Khrennikov, Classical and quantum mechanics on information spaces with applications to cognitive, psychological, social and anomalous phenomena. Found. Phys. **29**, 1065–1098 (1999)
15. A. Khrennikov, On quantum-like probabilistic structure of mental information. Open Syst. Inf. Dyn. **11**(3), 267–275 (2004)
16. A. Khrennikov, Information dynamics in cognitive, psychological, social, and anomalous phenomena, *Fundamental Theories of Physics* (Kluwer, Dordrecht, 2004)
17. A. Khrennikov, *Ubiquitous Quantum Structure* (Springer, Berlin, 2010)
18. A. Khrennikov, E. Haven, Quantum mechanics and violations of the sure-thing principle: the use of probability interference and other concepts. J. Math. Psychol. **53**, 378–388 (2009)
19. A. Khrennikov, I. Basieva, E.N. Dzhafarov, J.R. Busemeyer, Quantum models for psychological measurements: an unsolved problem. PLoS ONE **9**(10) (2014)
20. A. Khrennikov, I. Basieva, Quantum model for psychological measurement: from the projection postulate to interference of mental observables represented as positive operator valued measures. NeuroQuantology **12**(3), 324–336 (2014)
21. P. Khrennikova, Quantum-like modelling of the non-separability of voters' preferences in the U.S. political system. Lect. Notes Comput. Sci. **8951**, 196–209 (2015)
22. P. Khrennikova, E. Haven, Instability of political preferences and the role of mass-media: a dynamical representation in a quantum framework. Philos. Trans. R. Soc. A **374**, 20150106 (2015)
23. P. Khrennikova, E. Haven, Quantum generalized observables framework for psychological data: a case of preference reversals in US elections. Philos. Trans. R. Soc. A **375**, 20160391 (2017)
24. D. Kliger, O. Levy, Mood-induced variation in risk preferences. J. Econ. Behav. Organ. **52**(4), 573–584 (2003)
25. F.H. Knight, *Risk, Uncertainty, and Profit* (Hart, Schaffner and Marx, Houghton Mifflin Company, Boston, 1921)
26. A.N. Kolmogorov, *Grundbegriffe der Warscheinlichkeitsrechnung* (Springer, Berlin, 1933) (English translation: *Foundations of the Probability Theory*. Chelsea Publishing Company, New York, 1956)
27. M. Ozawa, A. Khrennikov, Application of theory of quantum instruments to psychology: combination of question order effect with response replicability effect. Entropy **22**(1), 37 (2019). https://doi.org/10.3390/e22010037

28. A. Plotnitsky, What is complementarity?: Niels Bohr and the architecture of quantum theory. Phys. Scr. **2014**(T163), 014002 (2014)
29. E. Pothos, J. Busemeyer, A quantum probability explanation for violations of rational decision theory. Proc. R. Soc. Lond. B Biol. **276**, 2171–2178 (2009)
30. M. Roca, A.J. Hogarth, A.J. Maule, Ambiguity seeking as a result of the status quo bias. J. Risk Uncertain. **32**, 175–194 (2006)
31. L.J. Savage, *The Foundations of Statistics* (Wiley, 1954)
32. E. Shafir, A. Tversky, Thinking through uncertainty: non-consequential reasoning and choice. Cogn. Psychol. **24**, 449–474 (1992)
33. E. Shafir, Uncertainty and the difficulty of thinking through disjunctions. Cognition **49**, 11–36 (1994)
34. A. Tversky, G. Fox, Weighting risk and uncertainty. Psychol. Rev. **102**(2), 269–283 (1995)
35. A. Tversky, E. Shafir, The disjunction effect in choice under uncertainty. Psychol. Sci. **3**(5), 305–309 (1992)
36. J. von Neumann, *Mathematische Grundlagen der Quantenmechanik* (Springer, Berlin-Heidelberg-New York, 1932). (English translation: *Mathematical Foundations of Quantum Mechanics*. Princeton University Press, Princeton, 1955)
37. Z. Wang, J. Busemeyer, H. Atmanspacher, E. Pothos, The potential for using quantum theory to build models of cognition. Top. Cogn. Sci. **5**, 672–688 (2013)
38. Z. Wang, J. Busemeyer, Reintroducing the concept of complementarity into psychology. Front. Psychol. **6**, 1822 (2015). https://doi.org/10.3389/fpsyg.2015.01822
39. L.C. White, E.M. Pothos, The cost of asking: how evaluations bias subsequent judgements. Decision **7**(14), 259–286 (2020)

Polina Khrennikova is Assistant Professor in the Financial Engineering section, within the Faculty of Behavioural, Management and Social Sciences at the University of Twente in the Netherlands. Before joining the University of Twente, Polina held a position of Assistant Professor in Finance at the University of Leicester in the UK. She holds a MSc in International Business Studies and Economics from the Department of Economics, Linnaeus University in Sweden and a PhD in Finance from the School of Business, University of Leicester. She also studied Philosophy of AI Foundations at the University of Umea in Sweden. Polina's research embraces human decision-making and information processing under uncertainty and risk. Her current research interests are in the fields of behavioural finance and economics. In her latest publications, she explores how individuals process information and deal with uncertainty and risk and uses multidisciplinary methods to devise novel mathematical models of these processes. Her works appeared, among others, in the Journal of Mathematical Psychology, Philosophical Transactions of the Royal Society A, Physica A: Statistical Mechanics and Its Applications, and Journal of Mathematical Economics.

Contextuality and Random Variables

Ehtibar Dzhafarov

Abstract Contextuality is a property of systems of random variables. The identity of a random variable in a system is determined by its joint distribution with all other random variables in the same context. When context changes, a variable measuring some property is instantly replaced by another random variable measuring the same property, or instantly disappears if this property is not measured in the new context. This replacement/disappearance requires no action, signaling, or disturbance, although it does not exclude them. The difference between two random variables measuring the same property in different contexts is measured by their maximal coupling, and the system is noncontextual if one of its overall couplings has these maximal couplings as its marginals.

1 Preamble

Quantum physicists like telling people (perhaps even themselves) that their field is strange and counterintuitive. Contextuality, especially when it takes the form of nonlocality, is one of its strange and counterintuitive notions. Indications of contextuality, such as violations of the Bell-type inequalities, are sometimes referred to as paradoxes. Like everything else in quantum physics, contextuality involves probabilities, hence random variables. And these, unlike the physical issues described by them, are usually taken to be clear and well-known: nothing strange or counterintuitive about random variables, they are merely mathematical tools, on a par with derivatives and integrals.

However, if I had any propensity to mystify my readers, I would argue that random variables are very strange objects. A random variable is a pure potentiality until it is realized, i.e., until it "collapses" into a single value being observed. Why is it less intriguing that the measurement problem in quantum physics (the wonderment

E. Dzhafarov (✉)
Psychological Sciences, Purdue University, West Lafayette, IN, USA
e-mail: ehtibar@purdue.edu

A. Plotnitsky and E. Haven (eds.), *The Quantum-Like Revolution*,
https://doi.org/10.1007/978-3-031-12986-5_17

at why the Schrödinger wave, which is essentially a special way of describing a random variable with a spatiotemporal distribution, collapses into a specific value being observed)? The textbook definition says that a random variable is a function from a probability space to a measurable space, but one would look in vain for any utilization of this fact in the quantum physical literature. What are these probability spaces on which the random variables are defined? How does one know that two observations belong to a single random variable rather than two different variables? Can one speak of values of a random variable counterfactually, in terms of what its value might have been, had it not been what it was observed to be? Questions and wonderments like this can be multiplied. I am not, however, into mystifying my readers. All these questions have clear answers, but this clarity is not of an evident variety, one cannot achieve it without nontrivial conceptual work. And once one achieves clarity about random variables, I will argue, this clarity is imparted on the substantive issues they describe, contextuality including.

One researcher who forcefully argued that contextuality and nonlocality are primarily matters of probability theory rather than physics was Andrei Khrennikov [1, 2]. He does not seem to maintain this position currently, and his arguments when he maintained it were different from those presented in this paper. Nevertheless, I think my views are close to Khrennikov's former views in spirit.

2 Random Variables Within a System

In 1989 David Mermin published a popular-level discussion of the nonlocal form of contextuality [3] (based on his 1981 work, added to [3] as an appendix). I will present Mermin's reasoning in a modified form. Consider the well-known Alice-Bob scenario, in which Alice chooses between two settings, denoted 1 and 2, and Bob chooses between his two settings, also denoted 1 and 2. The outcomes of Alice's measurements at either setting can be $+1$ or -1, and the same is true for Bob's measurements. Because of the way the experiment is set up (e.g., with Alice's and Bob's measurements being spacelike separated), Alice's choice of a setting cannot influence Bob's measurements, and vice versa. Let us use the term "context ij" to describe the situation in which Alice chooses setting $i \in \{1, 2\}$ and Bob chooses setting $j \in \{1, 2\}$. Suppose that the outcomes of the two measurements in the four contexts have the following probabilities:

context 22	Bob's outcome $= +1$	Bob's outcome $= -1$	
Alice's outcome $= +1$	0	1/2	1/2
Alice's outcome $= -1$	1/2	0	1/2
	1/2	1/2	

contexts 11, 12, 21	Bob's outcome $= +1$	Bob's outcome $= -1$	
Alice's outcome $= +1$	1/2	0	1/2
Alice's outcome $= -1$	0	1/2	1/2
	1/2	1/2	

. (1)

This describes what is commonly referred to as a *PR-box* [4], a highly contextual system by all reasonable measures [5], and one that violates the CHSH inequalities [6] to the maximal extent algebraically possible (I will explain this in Sect. 3). Mermin's reasoning is aimed at showing that there is something paradoxical ("extremely perplexing," he says) about such a system of random variables.[1]

We begin with context 11, and denote the two random variables representing the outcomes of Alice's and Bob's measurements as follows:

$$\text{context } 11 \left[\begin{array}{r|l} \text{Alice's setting } 1 & \text{random variable } A_1 \\ & \\ \text{Bob's setting } 1 & \text{random variable } B_1 \end{array} \right. \quad \Pr\left[A_1 = B_1\right] = 1. \qquad (2)$$

That the probability of $A_1 = B_1$ is 1 follows from the joint distribution (1) for context 11.

Now, Mermin proposes what he calls the *Strong Baseball Principle* (SBP),[2] that I will formulate as follows:

(SBP) if Alice's random variable at her setting i can in no way be influenced by Bob's choice of his setting, then the same random variable can represent her measurement outcomes in both context $i1$ and context $i2$; analogously, Bob's measurement outcomes at his setting j can be represented by the same random variable in both context $1j$ and context $2j$.

In particular, in context 12, Alice's measurement outcomes can be represented by the same A_1 as they are in context 11, and in context 21, Bob's measurement outcomes can be represented by the same B_1 as they are in context 11:

[1] Mermin does mention the distributions (1), with perfect correlations and anticorrelations, but for a detailed reasoning he uses the distributions in which the joint probabilities 1/2 and 0 are replaced with $(1 + \cos \frac{\pi}{4})/4$ and $(1 - \cos \frac{\pi}{4})/4$, respectively. The difference is not significant for my presentation.

[2] So dubbed because of the simile with the belief that watching a baseball game on one's TV cannot affect the game's outcome.

$$\begin{array}{ll}
\text{context 12} & \left[\begin{array}{r|l} \text{Bob's setting 2} & ? \\ \text{Alice's setting 1} & A_1 \end{array}\right] \\
\\
\text{context 11} & \left[\begin{array}{r|l} \text{Alice's setting 1} & A_1 \\ \\ \text{Bob's setting 1} & B_1 \end{array}\right] \quad \Pr[A_1 = B_1] = 1\,. \\
\\
\text{context 21} & \left[\begin{array}{r|l} \text{Bob's setting 1} & B_1 \\ \text{Alice's setting 2} & ? \end{array}\right]
\end{array} \tag{3}$$

We can easily fill in the places held by the question marks. We know from (1) that the two random variables in context 12 are perfectly correlated, so once we have determined one of them, the other must copy it (and the same holds for context 21):

$$\begin{array}{ll}
\text{context 12} & \left[\begin{array}{r|l} \text{Bob's setting 2} & B_2 \\ \\ \text{Alice's setting 1} & A_1 \end{array}\right] \quad \Pr[A_1 = B_2] = 1 \\
\\
\text{context 11} & \left[\begin{array}{r|l} \text{Alice's setting 1} & A_1 \\ \\ \text{Bob's setting 1} & B_1 \end{array}\right] \quad \Pr[A_1 = B_1] = 1\ . \\
\\
\text{context 21} & \left[\begin{array}{r|l} \text{Bob's setting 1} & B_1 \\ \\ \text{Alice's setting 2} & A_2 \end{array}\right] \quad \Pr[A_2 = B_1] = 1
\end{array} \tag{4}$$

Then we apply SBP once again, and conclude that in context 22 the measurements by Alice and Bob can be represented by the same random variables A_2 and B_2 as they are in contexts 21 and 12, respectively:

$$\text{context 22} \quad \left[\begin{array}{r|l} \text{Alice's setting 2} & A_2 \\ \\ \text{Bob's setting 2} & B_2 \end{array}\right] \quad \Pr[A_2 = B_2] = 1. \tag{5}$$

That the probability of $A_2 = B_2$ is 1 follows from the chain

$$\Pr[A_2 = B_1] = 1, \Pr[B_1 = A_1] = 1, \Pr[A_1 = B_2] = 1. \tag{6}$$

But we know from (1) for context 22 that the probability of $A_2 = B_2$ is zero, not 1. We have run into a contradiction.

A contradiction always means that some of the assumptions made in the process of reasoning, perhaps unawares, are wrong. The first impulse one might have is to declare that the random variables with the distributions (1) are impossible, but this can easily be dismissed. One way to do this is to note that the same reasoning with the same contradiction at the end can be obtained with distributions that are empirically observed and codified by a well-established theory. This is the argument chosen by Mermin, who uses distributions that are predicted by quantum mechanics for a certain choice of the four settings (directions in which spins are measured) in the standard EPR/Bohm experiment with spin-1/2 particles [6] (see footnote 1). Of course, one can always challenge the validity of quantum mechanics and the veracity of the experiments corroborating its predictions (which one would have to do if the contradiction we arrived at could not be dissolved by any other means). A much better argument therefore would be to simply note that the random variables with distributions (1) exist mathematically, as appropriately chosen functions on certain probability spaces. We will get to this later, however.

Assuming we are satisfied there is nothing wrong with our distributions, where else can one seek the cause of the contradiction we derived? One might try to deny the possibility that Bob's setting have no influence on Alice's measurements (which probably remains the most commonly held interpretation of nonlocality among non-physicists). However, this would be wrong ("disquieting," Mermin says) in view of what physics says about information propagation: e.g., it is ruled out if Alice's and Bob's measurements are spacelike separated. Note that Alice has no means to infer Bob's setting because in contexts $i1$ and $i2$ the random variables representing the outcomes of her measurements have indistinguishable distributions ($+1$ and -1 occurring with equal probabilities). Therefore the hypothetical ways in which Bob's setting would influence Alice's measurements would have to be contrived to remain hidden, in addition to contradicting physical theory.

Mermin too dismisses the "hidden action at a distance" resolution of the contradiction he derives, and he suggests that the culprit is SBP.

> Many people want to conclude from this [the contradiction – E.D.] that what happens at A does depend on how the switch is set at B, which is disquieting in view of the absence of any connections between the detectors. The conclusion can be avoided, if one renounces the Strong Baseball Principle, maintaining that indeed what happens at A does not depend on how the switch is set at B, but that this is only to be understood in its statistical sense, and most emphatically cannot be applied to individual runs of the experiment. To me this alternative conclusion is every bit as wonderful as the assertion of mysterious actions at a distance. I find it quite exquisite that, setting quantum metaphysics entirely aside, one can demonstrate directly from the data and the assumption that there are no mysterious actions at a distance, that there is no conceivable way consistently to apply the Baseball Principle to individual events (p. 49).

While it is quite obvious to me that SBP is wrong, I do not think Mermin's explanation is sufficiently transparent. Let us try to understand it. Mermin says "individual runs" because in his exposition he does not even mention random variables, speaking instead of very long sequences of realizations thereof. If treated informally, this only obfuscates analysis, and if treated rigorously, complicates it. A sequence of realizations of a random variables is a random process, an indexed set of identically distributed random variables. I suggest therefore that Mermin's transplantation of a sequence of realizations from one context to another should simply be understood as placing in these contexts one and the same random variable, the way we have done this in (4) and (5). And this is what must not be done, Mermin says based on the contradiction this led us into, and I think there is no reasonable way to disagree with this prohibition. Consider any single context, say, 11 in (2). The realizations of A_1 and B_1 there come in pairs, they have therefore a well-defined joint distribution. In particular, one has an opportunity to decide, by looking at a long enough sequence of the paired realizations, whether they are perfectly correlated. By contrast, in (4), if one looks at A_1 in context 11 and A_1 in context 12, their realizations cannot co-occur, because context 11 and 12 are mutually exclusive. One has no non-arbitrary way of pairing a value of A_1 in context 11 with a value of A_1 in context 12. There is no meaningful sense of asking whether they are correlated, perfectly or otherwise. But then it means that we have made a mistake by denoting them by the same symbol: A_1 is a single random variable, even if mentioned many times, and $A_1 = A_1$ holds with probability 1. In contexts 11 and 12 therefore we have two different random variables with one and the same distribution. This is, I suggest, how one could understand Mermin's assertion "that indeed what happens at A does not depend on how the switch is set at B, but that this is only to be understood in its statistical sense."

How should we amend the representations (4) and (5) to avoid contradiction? The answer is simple: we should use different symbols for A_1 in context 11 and A_1 in context 12, e.g., denote them by A_1^{11} and A_1^{12}, respectively (and analogously for other pairs of random variables transplanted from one context to another by SBP).[3] One can write $A_1^{11} \sim A_1^{12}$, where $\sim$ means "is distributed as", but statements like $A_1^{11} = A_1^{12}$, $A_1^{11} \neq A_1^{12}$, $A_1^{11} + A_1^{12} = 2$, etc. are all void of meaning because A_1^{11} and A_1^{12} posses no joint distribution. The system of random variables in our example can now be presented as follows:

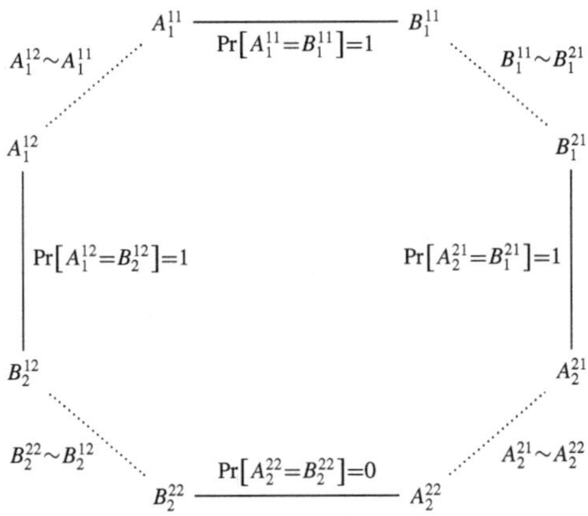

(7)

---

[3] This notation for the random variables has redundancy in it: the subscripts can be recovered from the superscripts (contexts) and the symbol (A or B) used for the random variable. Thus, A_1^{11} could simply be written as A^{11}, and B_2^{12} as B^{12}. I leave the notation redundant by a deliberate choice, however, to make the structure of the random variables maximally transparent.

We are no longer driven into contradiction: the value of $\Pr\left[A_2^{22} = B_2^{22}\right]$ can in no way be inferred from other components of this diagram, because none of them contains the variables A_2^{22} and B_2^{22}. If my interpretation of Mermin's conclusion is deemed reasonable, his was a valuable observation for the 1980s, although I doubt it could be well understood the way it was formulated.

To generalize, a random variable within a system of random variables is identified not only by what it measures (e.g., A_1^{11} and A_1^{12} both measure the same property, say, the spin of a particle in Alice's direction 1) but also by the context in which it is recorded (here, by the directions chosen by both Alice and Bob for their measurements). Let me dispel two possible objections to this general statement.

One is that we still can write A_1 for both A_1^{11} and A_1^{12} but keep in mind that we deal with two different sequences of realizations of A_1. Indeed, one might argue, there is nothing wrong in saying that one records a random variable today, and then the same random variable is recorded tomorrow. The response to this argument is that it is acceptable only if one is allowed to be informal, hoping this will not lead to confusion. A rigorous treatment of random variables requires that whenever one attaches different contexts to them (in this example, day of the measurement, today or tomorrow), one deals with different random variables. After all, to say "A_1 today" and "A_1 tomorrow" means to denote them differently, albeit sloppily. They are (perhaps) identically distributed, but they are distinct and have no joint distribution. Recall that the rigorous definition of a sequence of realizations of a random variable R (a sample of its values) is the sequence of different random variables, $R_1, R_2, \ldots$, each of which is distributed as R. The fact that in most applications they are also assumed to be independent is more subtle, and its mathematical meaning is captured through the notion of couplings that we will discuss later.

Another, often heard objection is that by saying that A_1^{11} and A_1^{12} are different random variables, one somehow admits that something in the contexts *causes* A_1^{11} to *transform* into A_1^{12} as context 11 is replaced with context 12. So in the EPR/Bohm scenario, one might argue, we still have some kind of an action at a distance. This objection is merely a play on the words "causes" and "transforms." If A_1^{11} and A_1^{12} are identically distributed, Alice has no means to distinguish them. Which means that no information, no action is transferred from Bob's setting to Alice's measurements. A_1^{11} and A_1^{12} are different random variables only for someone who, like Mermin, gets information from both Alice and Bob, both about their settings chosen and outcomes obtained. What changes when context 11 is replaced with context 12 is the relation between Alice's measurements and Bob's measurements, and this, because of the fundamentally relational nature of random variables (as explained below), means that Mermin knows that in contexts 11 and 12 Alice deals with two different random variables.

Having dealt with these objections, can we say that with my interpretation of Mermin's analysis we, at least informally (because a more formal treatment is to follow), have explained the strangeness of contextuality in terms of random variables? The answer is, we have not. In fact, unexpected as this might come, Mermin's conclusion in my interpretation equally applies to any and all systems of random variables, contextual or not. Consider, e.g., the following modification of the distributions (1):

contexts 11, 12, 21, 22	Bob's outcome $= +1$	Bob's outcome $= -1$	
Alice's outcome $= +1$	$1/2$	0	$1/2$
Alice's outcome $= -1$	0	$1/2$	$1/2$
	$1/2$	$1/2$	

. (8)

This system is clearly noncontextual. It can be viewed as describing a single pair of perfectly correlated random variables, with setting choices being fake. It is still true, however, that the contexts 11 and 12 are mutually exclusive, and insofar as these settings are not ignored, Alice's measurements in these contexts must be represented by different (though identically distributed) random variables that have no joint distribution. The diagram representing this situation is identical to (7), except $\Pr\left[A_2^{22} = B_2^{22}\right]$ is now 1 rather than zero. This makes no difference for how one treats the random variables because in both cases $\Pr\left[A_2^{22} = B_2^{22}\right]$ is completely unrelated to other elements of the diagrams.

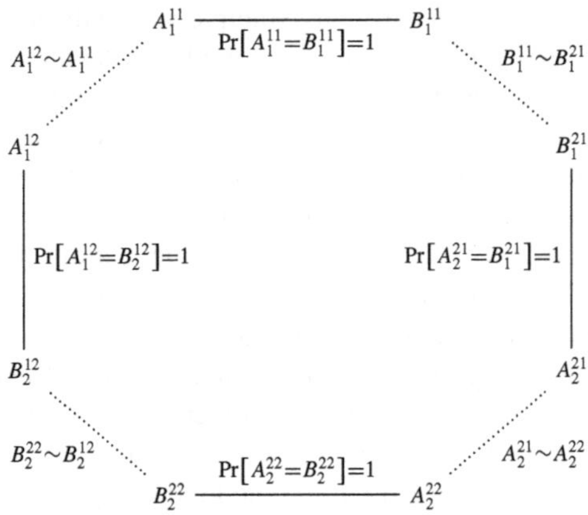

(9)

Somewhat paradoxically, therefore, having resolved the contradiction brought in by SBP, Mermin (or at least my interpretation of his analysis) loses the distinction between contextual and noncontextual systems. There is no way, e.g., to derive Bell-type inequalities for systems like (7) and (9), because the joint distributions involved of the four pairs of random variables

$$\left(A_1^{11}, B_1^{11}\right), \left(A_2^{21}, B_1^{21}\right), \left(A_2^{22}, B_2^{22}\right), \left(A_1^{12}, B_2^{12}\right), \tag{10}$$

are logically unrelated to each other.

Some researchers derive from this that the notion of contextuality is flawed. In particular, Bell-type inequalities, according to this view, are simply invalid, based on the mistake of following SBP. It seems that this is also Andrei Khrennikov's view [7], although his implementation of the context-dependence of random variables is different from the one presented here [8].

While one is free not to make distinctions one finds uninteresting, I find this position less than constructive. It is true that the context-indexing of random variables precludes the naive notion of contextuality, but what one better get rid of is the naivety rather than the notion. As it turns out, there is a conceptual and mathematical tool that enables us to readily distinguish systems like (7) from those like (9). As a bonus, this tool, while incompatible with SBP, justifies and formalizes the counterfactual reasoning on which SBP is based. The questions like "what would the outcome of one's measurement be if it were made in a context other than the one in which it is made" translate into rigorous and non-controversial mathematical problems.

3 Sample Spaces and Couplings

The mathematical tool in question is (probabilistic) *coupling*. Before introducing it, however, let us make sure we understand why two random variables in different contexts do not have a joint distribution (from which it follows also that they can never be one and the same random variable).

All random variables in this paper are assumed to be *dichotomous*,[4] because of which a random variable is defined as a function $X : S \to \{-1, 1\}$, with the following properties:

1. S belongs to a probability space (S, Σ, μ), where Σ is a sigma algebra of subsets of S, and μ a probability measure $\Sigma \to [0, 1]$;
2. $X^{-1}(\{1\}) \in \Sigma$, and $\Pr[X = 1] = \mu\left(X^{-1}(\{1\})\right)$; $X^{-1}(\{-1\}) \in \Sigma$, and $\Pr[X = -1] = \mu\left(X^{-1}(\{-1\})\right)$;
3. $\Pr[X = 1] + \Pr[X = -1] = 1$.

The set S is often called a *sample space*, but I prefer to use this term for the probability space (S, Σ, μ). Random variables X and Y are jointly distributed if and only if they are functions on the same sample space. If they are, then

$$\Pr[X = 1, Y = 1] = \mu\left(X^{-1}(\{1\}) \cap Y^{-1}(\{1\})\right). \tag{11}$$

Realizations of X and Y are then defined in pairs. If they are not on the same sample space, $\Pr[X = 1, Y = 1]$ is undefined, and no pairing scheme for their realizations exists.

We see that, by definition, to construct a set of jointly distributed random variables means to specify a sample space and define these random variables as functions on this sample space. What usually remains unclear to a student of these textbook definitions is the nature of a sample space. What is it and how can it be (re)constructed? The answer to this question is so simple that it can be surprising. Let us discuss this

[4] In my favorite theory of contextuality, confining the consideration to dichotomous variables is not a loss of generality, because all random variables within a system have to be replaced by sets of jointly distributed dichotomous variables before the system can be subjected to contextuality analysis [9, 10].

answer in detail, using a system of the same format as above, but this time with more arbitrary joint distributions in the four contexts: for $i \in \{1, 2\}$ and $j \in \{1, 2\}$,

$$
\begin{array}{c|c|c|c}
\text{context } ij & B_j^{ij} = +1 & B_j^{ij} = -1 & \\
\hline
A_i^{ij} = +1 & r_{ij} & p_i - r_{ij} & p_i \\
\hline
A_i^{ij} = -1 & q_j - r_{ij} & 1 - p_i - q_j + r_{ij} & 1 - p_i \\
\hline
 & q_j & 1 - q_j &
\end{array}. \quad (12)
$$

The only constraint imposed by my notation here is that the distribution of A_i^{ij} is the same for $j = 1$ and $j = 2$ (in both cases $\Pr\left[A_i^{ij} = 1\right] = p_i$), and analogously for the distribution of B_j^{ij}. This property of a system of random variables is called *consistent connectedness*, or *no-disturbance*. This is precisely the property that guarantees that Alice has no way of distinguishing A_i^{i1} and A_i^{i2}, the difference being only available to someone who receives information about both Alice's and Bob's settings and outcomes.

We assume, of course, that the distribution in (12) is well-defined, that is, all probabilities shown are numbers between 0 and 1. As it turns out, this is all one needs to say that the random variables A_i^{ij} and B_j^{ij} with this joint distribution exist as mathematical objects. Indeed, consider the following probability space (S, Σ, μ_{ij}) for context ij: $S = \{a, b, c, d\}$, $\Sigma = 2^S$, and μ_{ij} is defined by the probability mass function

$$
\begin{array}{c||c|c|c|c}
x = & a & b & c & d \\
\hline
\mu_{ij}(\{x\}) = & r_{ij} & p_i - r_{ij} & q_j - r_{ij} & 1 - p_i - q_j + r_{ij}
\end{array}. \quad (13)
$$

The random variables are now defined as the functions

$$
A_i^{ij}(x) = \begin{cases} +1 \text{ if } x \in \{a, b\} \\ -1 \text{ if } x \in \{c, d\} \end{cases}, \; B_j^{ij}(x) = \begin{cases} +1 \text{ if } x \in \{a, c\} \\ -1 \text{ if } x \in \{b, d\} \end{cases}. \quad (14)
$$

The resulting system of random variables can be presented in the form of the following *content-context matrix*:

A_1^{11}		B_1^{11}		context 11
A_1^{12}			B_2^{12}	context 12
	A_2^{21}	B_1^{21}		context 21
	A_2^{22}		B_2^{22}	context 22
Alice's 1	Alice's 2	Bob's 1	Bob's 2	Alice-Bob system

(15)

The term *content* (of a random variable) refers to that which the random variable is measuring, or settings from which the measured property can be deduced. In the matrix above, the contents are listed at the bottom.

Clearly, up to the labeling of the values of S, this construction is unique. Moreover, any other sample space $\left(S'_{ij}, \Sigma'_{ij}, \mu'_{ij}\right)$ on which A_i^{ij} and B_j^{ij} can be defined is reducible to this $\left(S, \Sigma, \mu_{ij}\right)$, in the following sense: denoting by X_a the pre-image of $\left[A_i^{ij} = +1, B_j^{ij} = +1\right]$ in S'_{ij}, by X_b the pre-image of $\left[A_i^{ij} = +1, B_j^{ij} = -1\right]$, etc., one can map X_a into a, X_b into b, and so on, to define A_i^{ij} and B_j^{ij} as functions on $\left(S, \Sigma, \mu_{ij}\right)$. The latter therefore is the most economic sample space possible. The general logic of the construction should be clear. Whenever a joint distribution of hypothetical random variables is well-defined, these random variables exist as functions defined on a sample space, and the most economic version of the latter can be uniquely constructed from the joint distribution. There is never a situation in which one can say that random variables with a given joint distribution do not exist (provided no a priori constraints are imposed on their sample space).

It is also clear from this construction why A^{i1} and A^{i2} are distinct random variables even if they are identically distributed. They are defined on different sample spaces: even if one chooses to denote the elements of the sample set in the same way, $\{a, b, c, d\}$, the respective measures μ_{i1} and μ_{i2} are as distinct as are the joint distributions in contexts $i1$ and $i2$ in (12).

In the previous section I mentioned "the fundamentally relational nature of random variables," because of which the identity of random variables cannot be the same in different contexts. One can see now that this expression has a precise mathematical meaning: the sample space on which a given random variable is defined is determined by its joint distribution with all other random variables in the same context. A reasonable analogy is provided by a set of points in a metric space without coordinates. Each point is characterized by its distances to the rest of the points, so moving even one of the latter changes the point in question instantly. No spooky transfer of information is involved, these are changes that occur by definition. Of course, like all analogies, this one also has its drawbacks. In particular, it is possible to say that "this point" (one pointed at) changes its identity when other points change their positions. In a system of random variables one can only say that a random variable in one context is different from a random variable that measures the same thing in a different context. However, it seems to me that the analogy with distances does a very good job in dispelling remnants of mystery in the term "nonlocality."

Let me now introduce the conceptual tool that would allow us to speak of contextual and noncontextual systems. A *coupling* of several random variables $X_1, \ldots, X_n$ is a set of jointly distributed variables $Y_1, \ldots, Y_n$ such that $Y_i \sim X_i$ for $i = 1, \ldots, n$. Note that $X_1, \ldots, X_n$ need not be jointly distributed, and in fact in all applications we are interested in, they are not. In other words, each of X_i is defined on its own sample space, whereas all Y's are defined on yet another sample space. Using the term "*stochastically unrelated*" for random variables no two of which possess a joint distribution, $X_1, \ldots, X_n$ and $(Y_1, \ldots, Y_n)$ viewed as a single random variable, are stochastically unrelated. In relation to our discussion of Mermin's SBP, couplings can be thought of as answers to the counterfactual question "How could these random variables be jointly distributed if they were jointly distributed?"

Any set $X_1, \ldots, X_n$ of random variables has a coupling, and generally it has an infinity of couplings, i.e., infinity of $(Y_1, \ldots, Y_n)$ with different joint distributions (but the same marginal distributions, because by definition, $Y_i \sim X_i$ for $i = 1, \ldots, n$).[5] Therefore, to use couplings as a means for categorizing the sets $X_1, \ldots, X_n$, one should only be interested in whether a set $X_1, \ldots, X_n$ has a coupling *subject to some specified constraints* [11], as we will see shortly.

In the Contextuality-by-Default (CbD) theory, the notion of a coupling is applied to a system of random variables in two ways. We first construct couplings for all pairs of the same-content random variables. We have four of them in system (15):

$$\left\{A_1^{11}, A_1^{12}\right\}, \left\{A_2^{21}, A_2^{22}\right\}, \left\{B_1^{11}, B_1^{21}\right\}, \left\{B_2^{12}, B_2^{22}\right\}. \tag{16}$$

Denoting a coupling of $\left\{A_1^{11}, A_1^{12}\right\}$ by $\left(\tilde{A}_1^{11}, \tilde{A}_1^{12}\right)$, we look among these couplings for one in which the probability of $\tilde{A}_1^{11} = \tilde{A}_1^{12}$ is as large as possible, given the individual distribution of $\tilde{A}_1^{11} \sim A_1^{11}$ and $\tilde{A}_1^{12} \sim A_1^{12}$. The reason one is interested in such *maximal couplings* is that the maximal probability in question is a natural measure of similarity between A_1^{11} and A_1^{12}, when they are taken in isolation from their respective contexts. Put counterfactually, "if they were jointly distributed and no other random variables existed," they could coincide as often as this maximal probability. Suppose we have computed these maximal probabilities, and

$$\begin{aligned} \max \Pr\left[\tilde{A}_1^{11} = \tilde{A}_1^{12}\right] &= \omega_1, \ \max \Pr\left[\tilde{A}_2^{21} = \tilde{A}_2^{22}\right] = \omega_2, \\ \max \Pr\left[\tilde{B}_1^{11} = \tilde{B}_1^{21}\right] &= \omega_3, \ \max \Pr\left[\tilde{B}_2^{12} = \tilde{B}_2^{22}\right] = \omega_4, \end{aligned} \tag{17}$$

where each of the maxima is taken over all possible couplings of the corresponding pair in (15).

We next construct a coupling for the entire system (15), or more precisely, a coupling of the four stochastically unrelated random variables

$$X_1 = \left(A_1^{11}, B_1^{11}\right), X_2 = \left(A_1^{12}, B_2^{12}\right), X_3 = \left(A_2^{21}, B_1^{21}\right), X_4 = \left(A_2^{22}, B_2^{22}\right). \tag{18}$$

Such a coupling consists of the jointly distributed random variables

$$Y_1 = \left(\bar{A}_1^{11}, \bar{B}_1^{11}\right), Y_2 = \left(\bar{A}_1^{12}, \bar{B}_2^{12}\right), Y_3 = \left(\bar{A}_2^{21}, \bar{B}_1^{21}\right), Y_4 = \left(\bar{A}_2^{22}, \bar{B}_2^{22}\right), \tag{19}$$

with $Y_i \sim X_i$ $(i = 1, \ldots, 4)$. There are generally an infinity of such couplings, and in each of them we can compute

[5] One could, obviously, create many copies of $(Y_1, \ldots, Y_n)$, identically distributed but defined on different sample spaces. We should agree therefore that we make no distinction between them: a coupling is entirely identified by its distribution.

$$\Pr\left[\bar{A}_1^{11} = \bar{A}_1^{12}\right] = \omega_1',\ \Pr\left[\bar{A}_2^{21} = \bar{A}_2^{22}\right] = \omega_2',$$
$$\Pr\left[\bar{B}_1^{11} = \bar{B}_1^{21}\right] = \omega_3',\ \Pr\left[\bar{B}_2^{12} = \bar{B}_2^{22}\right] = \omega_4'. \tag{20}$$

Obviously,

$$\omega_1' \le \omega_1, \omega_2' \le \omega_2, \omega_3' \le \omega_3, \omega_4' \le \omega_4. \tag{21}$$

What we need to determine is whether there is a coupling in which all these inequalities become equalities, i.e.,

$$\Pr\left[\bar{A}_1^{11} = \bar{A}_1^{12}\right] = \omega_1,\ \Pr\left[\bar{A}_2^{21} = \bar{A}_2^{22}\right] = \omega_2,$$
$$\Pr\left[\bar{B}_1^{11} = \bar{B}_1^{21}\right] = \omega_3,\ \Pr\left[\bar{B}_2^{12} = \bar{B}_2^{22}\right] = \omega_4. \tag{22}$$

In other words, we seek couplings of system (15) that preserve both the distributions within the contexts (as they should by the definition of a coupling) and the similarity values (17) between the content-sharing variables. If no such couplings exist, one can say that the contexts make the content-sharing random variables more dissimilar than they are when they were taken in isolation. Such a system is called *contextual*. Otherwise, if such a coupling exists (generally not uniquely), the system is *noncontextual*.

This simple and, I would argue, highly intuitive logic of (non)contextuality is sufficient to restore all the contextuality results obtained in the literature and then go much further, doing so without compromising the rigorous mathematics of random variables and, in particular, without falling into the SBP trap. For a consistently connected system, as in (12), the elements of each pair in (16) are identically distributed, and it is easy to see that in this case

$$\omega_1 = \omega_2 = \omega_3 = \omega_4 = 1. \tag{23}$$

Indeed, the maximal coupling $\left(\tilde{A}_1^{11}, \tilde{A}_1^{12}\right)$ of $\left\{A_1^{11}, A_1^{12}\right\}$, e.g., has the distribution

content 1 of Alice	$\tilde{A}_1^{12} = +1$	$\tilde{A}_1^{12} = -1$	
$\tilde{A}_1^{11} = +1$	p_1	0	p_1
$\tilde{A}_1^{11}, = -1$	0	$1 - p_1$	$1 - p_1$
	p_1	$1 - p_1$	

(24)

It follows that in this case we seek couplings (Y_1, Y_2, Y_3, Y_4) in (19) such that $\Pr\left[\bar{A}_1^{11} = \bar{A}_1^{12}\right] = 1$, $\Pr\left[\bar{B}_1^{11} = \bar{B}_1^{21}\right] = 1$, and so on. Equivalently, we seek a *reduced coupling* [12] in which the variables $\bar{A}_1^{11}$ and $\bar{A}_1^{12}$ can be replaced with a single $\bar{A}_1$, the variables $\bar{B}_1^{11}$ and $\bar{B}_1^{21}$ can be replaced with a single $\bar{B}_1$, etc.

With this reformulation, one comes as close to the intuition underlying SBP as it is possible without committing the logical error of the original SBP. Thus, the system in our opening example, with distributions (1), is contextual because otherwise it would have to have a coupling in which

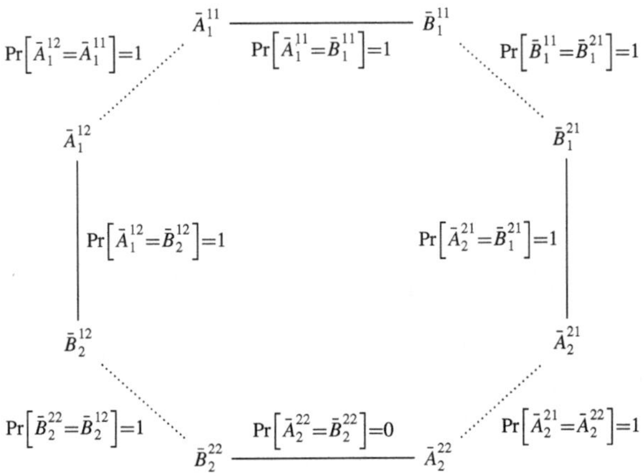

(25)

A chain of equalities it contains,

$$\Pr\left[\bar{A}_1^{11} = \bar{B}_1^{11}\right] = 1, \ldots, \Pr\left[\bar{A}_2^{22} = \bar{B}_2^{22}\right] = 0, \ldots, \Pr\left[\bar{A}_1^{12} = \bar{A}_1^{11}\right] = 1, \quad (26)$$

is obviously impossible.

As shown in detail in [13], the language of probabilistic couplings, when applied to consistently connected systems, allows one to formalize both the counterfactual formulation of contextuality and its formulation in terms of the hidden variable models with noncontextual mapping. I will not be repeating this discussion here. The Bell-type inequalities for consistently connected systems are derived in essentially the same way as they are derived traditionally. For instance, system (15) with distributions (12) can be shown to be contextual (i.e., not to have a coupling with the stipulated properties) if and only if

$$\max\left(\pm\left\langle A_1^{11} B_1^{11}\right\rangle \pm \left\langle A_1^{12} B_2^{12}\right\rangle \pm \left\langle A_2^{21} B_1^{21}\right\rangle \pm \left\langle A_2^{22} B_2^{22}\right\rangle\right) > 2, \quad (27)$$

where the maximum is taken over all choices between + and − in front of each expected value $\langle \ldots \rangle$ such that the number of the minus signs is odd [14, 15]. This is the well-known CHSH inequality, except it is commonly written as

$$\max\left(\pm\langle A_1 B_1\rangle \pm \langle A_1 B_2\rangle \pm \langle A_2 B_1\rangle \pm \langle A_2 B_2\rangle\right) > 2. \quad (28)$$

The latter form, however, is logically flawed, as it employs SBP and places the same random variable in different contexts. In fact, this inequality simply cannot be satisfied, because the variables A_1, A_2, B_1, B_2 in it must be jointly distributed (defined on the same sample space) by the following diagram:

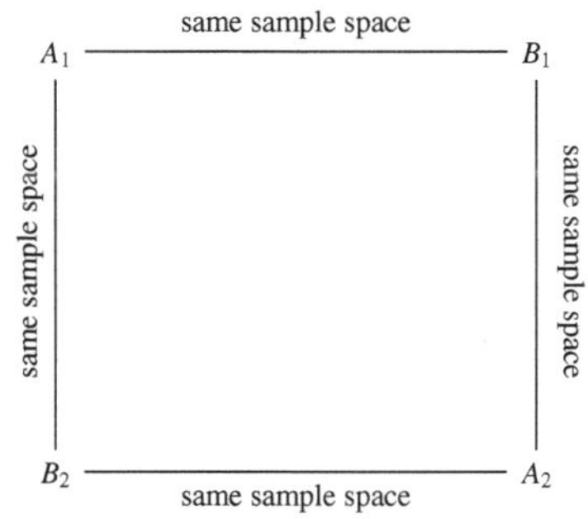

(29)

4 Overt Influences Versus Contextuality (or on the Magic of Words)

There is one situation in which SBP cannot even be considered. It is the case of *inconsistently connected systems*, or systems with disturbance. Using again the system (15) as an example, suppose that the distributions of the random variables are as follows: for $i \in \{1, 2\}$, $j \in \{1, 2\}$,

context ij	$B_j^{ij} = +1$	$B_j^{ij} = -1$	
$A_i^{ij} = +1$	r_{ij}	$p_{ij} - r_{ij}$	p_{ij}
$A_i^{ij} = -1$	$q_{ij} - r_{ij}$	$1 - p_{ij} - q_{ij} + r_{ij}$	$1 - p_{ij}$
	q_{ij}	$1 - q_{ij}$	

. (30)

The difference between this and (12) is in the marginal distributions: they are no longer necessarily the same for A_i^{i1} and A_i^{i2} (generally, $p_{i1} \neq p_{i2}$), nor are they necessarily the same for B_j^{1j} and B_j^{2j} (generally, $q_{1j} \neq q_{2j}$).[6] Suppose, e.g., that in the EPR/Bohm experiment, Alice's and Bob's measurements are timelike separated, i.e., transmission of information from settings of one of them to measurement outcomes of another is possible. Say, Bob sends certain π-rays of frequency 1 when he chooses setting 1, and he sends π-rays of frequency 2 when he chooses setting 2, so that the outcomes of Alice's measurements, A_i^{i1} and A_i^{i2}, can be affected by these rays differently. Clearly, we have here dependence of Alice's measurements not only on

[6] The use of "not necessarily" here is to indicate that consistently connected systems, with distributions (12), are merely a special case of the inconsistently connected ones, with distributions (30).

her choice of a setting i but on the entire context ij. By definition, one can speak of *context-dependence* here.

But does this context-dependence necessarily mean that the system (15) with distributions (30) is contextual? Some researchers think that the answer to this question must be affirmative, unless the distributions are reduced to (12), in which case a system may be contextual or noncontextual. However, unless some unknown to me laws compel the meaning of the word "contextuality" to be derived from the way it sounds, or its closeness to "context-dependence," this is not the only possible answer. A more constructive approach would be to consider contextuality as a *form* of context-dependence, and to ask whether it can be separated from and studied together with inconsistent connectedness, viewed as another form of context-dependence, on the level of marginal distributions.

The definition of contextuality given in the previous section is in fact formulated for (generally) inconsistently connected systems. The values of $\omega_1, \omega_2, \omega_3, \omega_4$ defined in (17) generally are not all equal to 1. Thus, the maximal couplings of the content-sharing pair $\{A_1^{11}, A_1^{12}\}$ now has the distribution

content 1 of Alice	$\tilde{A}_1^{12} = +1$	$\tilde{A}_1^{12} = -1$	
$\tilde{A}_1^{11} = +1$	$\min(p_{11}, p_{12})$	$p_{11} - \min(p_{11}, p_{12})$	p_{11}
$\tilde{A}_1^{11}, = -1$	$p_{12} - \min(p_{11}, p_{12})$	$1 - p_{11} - p_{12} + \min(p_{11}, p_{12})$	$1 - p_{11}$
	p_{12}	$1 - p_{12}$	

, (31)

whence

$$\omega_1 = \max \Pr\left[\tilde{A}_1^{11} = \tilde{A}_1^{12}\right] = 1 - |p_{11} - p_{12}|. \tag{32}$$

Similar formulas hold for other content-sharing pairs. In all other respects, however, the logic of contextuality remains precisely as previously described: one seeks an overall coupling of the system subject to the constraints (22), and the system is contextual if and only if no such couplings exist. The interpretation of contextuality also remains the same as it was for consistently connected systems: contextuality means that the content-sharing random variable within their respective contexts (i.e., considered jointly distributed with other variables) are more dissimilar than when they are isolated from their contexts. An inconsistently connected system can be contextual or noncontextual, by precisely the same logic as in the special case when the system is consistently connected.

This approach is more constructive than simply declaring any inconsistently connected system contextual, because it provides greater differentiation among systems of random variables, while properly reducing to special cases when more restricted definitions apply. One can offer specific arguments in favor of our definition of contextuality [5, 11, 16–20], of which I will mention one. First, observe that some contextual systems are more contextual than others, with respect to the following, intuitively plausible way of measuring contextuality. Consider the value

$$\omega' = \max\left(\omega_1' + \omega_2' + \omega_3' + \omega_4'\right)$$
$$= \max\left(\Pr\left[\bar{A}_1^{11} = \bar{A}_1^{12}\right] + \Pr\left[\bar{A}_2^{21} = \bar{A}_2^{22}\right] + \Pr\left[\bar{B}_1^{11} = \bar{B}_1^{21}\right] + \Pr\left[\bar{B}_2^{12} = \bar{B}_2^{22}\right]\right) \tag{33}$$

with the maximum taken over all possible couplings of system (15). This value cannot exceed

$$\omega_1 + \omega_2 + \omega_3 + \omega_4 = 4 - |p_{11} - p_{12}| - |p_{21} - p_{22}| - |q_{11} - q_{21}| - |q_{12} - q_{22}|, \tag{34}$$

because of which the nonnegative quantity

$$\mathsf{CNTX} = \omega_1 + \omega_2 + \omega_3 + \omega_4 - \omega' \tag{35}$$

can be taken for a measure of contextuality. A system is noncontextual if this quantity is zero. In this paper's opening example, system (15) with distributions (1) is consistently connected, so $\omega_1 = \omega_2 = \omega_3 = \omega_4 = 1$. This system is contextual because, as we have seen, it is not possible for a coupling to satisfy the chain of equalities in (25). The value of CNTX for this system can be shown to be 1, and this can be shown to be the highest possible value of CNTX across all systems of format (15) [5, 15, 21].

Now, let us introduce a small disturbance in our example, making the distributions

context 22	$B_2^{22} = +1$	$B_2^{22} = -1$	
$A_2^{22} = +1$	0	$1/2 + \varepsilon$	$1/2 + \varepsilon$
$A_2^{22} = -1$	$1/2 - \varepsilon$	0	$1/2 - \varepsilon$
	$1/2 - \varepsilon$	$1/2 + \varepsilon$	

contexts $ij = 11, 12, 21$	$B_j^{ij} = +1$	$B_j^{ij} = -1$	
$A_i^{ij} = +1$	$1/2$	0	$1/2$
$A_i^{ij} = -1$	0	$1/2$	$1/2$
	$1/2$	$1/2$	

. (36)

Intuition tells us that the degree of contextuality in this system should be only slightly different from the value of CNTX in the previous case, for $\varepsilon = 0$. And indeed, the degree of contextuality here is

$$\mathsf{CNTX} = 1 - 2\varepsilon. \tag{37}$$

We would not have such a smooth change of the degree of contextuality with $\varepsilon \to 0$ if we based the contextuality of the system with $\varepsilon > 0$ on the difference of the marginal probabilities alone.

On the other hand, in our second example, system (15) with distributions (8) is noncontextual, i.e., $\mathsf{CNTX} = 0$. If we introduce the same small perturbation as above, the distributions will be

context 22	$B_2^{22}=+1$	$B_2^{22}=-1$	
$A_2^{22}=+1$	$1/2+\varepsilon$	0	$1/2+\varepsilon$
$A_2^{22}=-1$	0	$1/2-\varepsilon$	$1/2-\varepsilon$
	$1/2-\varepsilon$	$1/2+\varepsilon$	

(38)

contexts $ij=11,12,21$	$B_j^{ij}=+1$	$B_j^{ij}=-1$	
$A_i^{ij}=+1$	$1/2$	0	$1/2$
$A_i^{ij}=-1$	0	$1/2$	$1/2$
	$1/2$	$1/2$	

It can be shown that this system remains noncontextual, $\mathsf{CNTX}=0$, as ε increases from 0 to $1/2$. Again, this is what one should expect based on the definition of contextuality. Zero CNTX at $\varepsilon=0$ means that the system has a coupling in which the value of ω' reaches $\omega_1+\omega_2+\omega_3+\omega_4$, which in this case has the maximal possible value, 4. Clearly, this is even easier to achieve if $\omega_1+\omega_2+\omega_3+\omega_4$ has a smaller value, $4-2\varepsilon$.

As we have seen, an inconsistently connected system can be contextual or noncontextual, and this lays the ground for a richer classification of systems than the indiscriminate notion of context-dependence. It seems reasonable to maintain that being able to make finer differentiations is always desirable, provided it is done in a principled way. Nevertheless some researchers keep coming up with the revelatory insight that it is possible to present both contextuality and inconsistent connectedness as context-dependence and to refuse to distinguish them. Sometimes this is presented as the only position consistent with the "ontological" (or "ontic") models, in which (continuing to use our example) A_i^{ij} and B_j^{ij} are presented as functions

$$A_i^{ij}=f(i,j,\lambda)\,,\ B_j^{ij}=g(i,j,\lambda)\,, \tag{39}$$

where λ is some "hidden" variable. The term "ontological/ontic" is supposed to hint at something happening in reality, as opposed to purely mathematical descriptions. However, as a general approach, (39) is purely descriptive rather than explanatory, because it is trivially applicable to any system of random variables. It is in fact nothing more than a mathematically lax version of constructing an unconstrained overall coupling for system (15). We know that this is always possible. Recall, that to make $\left(A_i^{ij},B_j^{ij}\right)$ for $i,j\in\{1,2\}$ jointly distributed they have to be presented as functions on the same sample space. The random variable λ is nothing but the identity function on this sample space. More rigorously, of course, one has to write

$$\bar{A}_i^{ij}=f(i,j,\lambda)\,,\ \bar{B}_j^{ij}=g(i,j,\lambda)\,, \tag{40}$$

or

$$\left(A_i^{ij}, B_j^{ij}\right) \sim (f(i, j, \lambda), g(i, j, \lambda)), \tag{41}$$

because $\left(A_i^{ij}, B_j^{ij}\right)$ for different i, j are stochastically unrelated.

One source of misunderstanding leading some to considering (39) as an alternative to CbD is the suggestive terminology I and my colleagues coined for inconsistent connectedness: we called it (the manifestation of) *direct influences*, as opposed to contextual influences [22–25]. For instance, the distribution of A_i^{i1} may be different from that of A_i^{i2} because Bob sends his π-rays that affect the outcomes of Alice's measurements. This intuition leads some to point out, as if this were a discovery of a flaw in CbD, that the π-rays can also account for contextuality. One needs nothing but the π-rays, according to this reasoning. It is simply that some effects of the π-rays are overt, and are reflected in the differences of marginal distributions, while other effects of π-rays are hidden, and we call them (mistakenly, according to this criticism) contextuality [26]. This assertion is being justified, not surprisingly, by the very same possibility of representing a system by (39). One can construct various toy examples to demonstrate this, but the fact remains that (39) is applicable universally. As I have mentioned, it is simply a restatement of the possibility to construct an unconstrained overall coupling of any system.

To see that all of this is completely off target, it would suffice to replace the term *direct influence* with *overt influence*. In retrospect, this would have been a better term, and I intend to use it in the future. The criticism in question then would look like this: CbD distinguishes overt effects (observable on the level of marginal distributions) and contextual effects in systems of random variables, while we (the critics) say that some context-dependence in such systems can be overt and some hidden. This is no more than a terminological quibble, provided the hidden influences are to be revealed by means of the CbD-based contextuality analysis. However, the criticism in question seems to lead its proponents to simply lump together all context-dependence for systems that are not known to be consistently connected. If one accepts this position, in physics, contextuality analysis will be reserved to situations when no physical transfer of information from Bob's settings to Alice's measurements (and vice versa) is allowed by laws of physics. If tomorrow the physicists concluded that superluminal transmission is possible after all, the EPR/Bohm contextuality would have to be suspect. In systems like the original Kochen-Specker one [27] or KCBS system [28], where the measurements in each context are made on the same particle, contextuality is inherently suspect, as there it hinges on the fact that the current quantum mechanical accounts of these systems involve no forces or other forms of interference. In CbD, however, contextuality does not depend on the state of substantive theories: e.g., the EPR/Bohm system with certain choices of directions by Alice and Bob is contextual in both contemporary quantum theory and in Bohmian mechanics, where hidden superluminal transmission is built in. All of this is discussed and explained in our earlier publications, e.g. [17]. Quoting from the latter work,

> [...] to defend a definition is a difficult task. A good definition of a term should be intuitively plausible (although sometimes one's intuition itself should be "educated" to make it plausible), it should include as special cases all examples and situations that are traditionally considered to fall within the scope of the term, it should lead to productive development (to allow one to prove nontrivial theorems), and have a growing set of applications. I believe contextuality in the sense of CbD satisfies all these desiderata (p. 14).

To summarize:

1. (Non)contextuality is a property of systems of random variables. It is a special form of context-dependence, the other form of context-dependence being inconsistent connectedness.
2. Being a purely mathematical property, (non)contextuality of a system does not depend on substantive theories of the empirical situations represented by the system.
3. The identity of a random variable in a system is determined by its joint distribution with all other random variables in the same context. When context changes, a variable measuring some property is instantly replaced by another random variable measuring the same property (in the language of CbD, having the same content), or it instantly disappears (if the property is not measured in the new context).
4. In particular, if the measurements described by the random variables in each contexts are separated by spacelike intervals, then the disappearance or replacement of a random variable by another random variable with the same content occurs instantly in response to spacelike separated changes in the context. No action at a distance is involved.
5. The difference between two random variables having the same content in different contexts is measured by their maximal coupling, and the system is noncontextual if one of its overall couplings has these maximal couplings as its marginals.
6. A contextual system, by contrast, makes the content-sharing random variables in different contexts more dissimilar than they are in isolation.
7. A system can be contextual or noncontextual irrespective of whether it is consistently connected.

References

1. A. Khrennikov, Bell-Boole inequality: nonlocality or probabilistic incompatibility of random variables? Entropy **10**, 19–32 (2008)
2. A. Khrennikov, Bell's inequality: physics meets Probability. Inf. Sci. **179**, 492–504 (2009)
3. N. Mermin, Can you help your team tonight by watching on TV? More experimental metaphysics from Einstein, Podolsky, and Rosen, in *Philosophical Consequences of Quantum Theory: Reflections on Bell's Theorem*, ed. by J.T. Cushin, E. McMullin (University of Notre Dame Press, 1989), pp. 38–59
4. S. Popescu, D. Rohrlich, Quantum nonlocality as an axiom. Found. Phys. **24**, 379–385 (1994)

5. J.V. Kujala, E.N. Dzhafarov, Measures of contextuality and noncontextuality. Philos. Trans. R. Soc. A **377**, 20190149 (2019)
6. J.F. Clauser, M.A. Horne, A. Shimony, R.A. Holt, Proposed experiment to test local hidden-variable theories. Phys. Rev. Lett. **23**, 880–884 (1969)
7. D. Avis, P. Fischer, A. Hilbert, A. Khrennikov, Single, complete, probability spaces consistent with EPR-Bohm-Bell experimental data, in *Foundations of Probability and Physics-5, AIP Conference Proceedings*, ed. by A. Khrennikov, vol. 750 (AIP, Melville, New York, 2009), pp. 294–301
8. E.N. Dzhafarov, J.V. Kujala, Embedding quantum into classical: contextualization vs conditionalization. PLoS One **9**(3), e92818 (2014). https://doi.org/10.1371/journal.pone.0092818
9. E.N. Dzhafarov, J.V. Kujala, Contextuality-by-default 2.0: systems with binary random variables, in *Lecture Notes in Computer Science*, ed. by J.A. de Barros, B. Coecke, E. Pothos, vol. 10106 (2017), pp. 16–32
10. E.N. Dzhafarov, V.H. Cervantes, J.V. Kujala, Contextuality in canonical systems of random variables. Philos. Trans. R. Soc. A **375**, 20160389 (2017)
11. E.N. Dzhafarov, Replacing nothing with something special: contextuality-by-default and dummy measurements, in *Quantum Foundations, Probability and Information*, ed. by A. Khrennikov, T. Bourama (Springer, Berlin, 2018), pp. 39–44
12. E.N. Dzhafarov, J.V. Kujala, Probability, random variables, and selectivity, in New *Handbook of Mathematical Psychology*, ed. by W. Batchelder et al. (Cambridge University Press, 2016), pp. 85–150
13. E.N. Dzhafarov, On joint distributions, counterfactual values, and hidden variables in understanding contextuality. Philos. Trans. R. Soc. A **377**, 20190144 (2019)
14. E.N. Dzhafarov, J.V. Kujala, J.-Å. Larsson, Contextuality in three types of quantum-mechanical systems. Found. Phys. **7**, 762–782 (2015)
15. J.V. Kujala, E.N. Dzhafarov, Proof of a conjecture on contextuality in cyclic systems with binary variables. Found. Phys. **46**, 282–299 (2016)
16. J.V. Kujala, E.N. Dzhafarov, J.-Å. Larsson, Necessary and sufficient conditions for extended noncontextuality in a broad class of quantum mechanical systems. Phys. Rev. Lett. **115**, 150401 (2015)
17. E.N. Dzhafarov, J.V. Kujala, Conversations on contextuality, in *Contextuality from Quantum Physics to Psychology*, ed. by E.N. Dzhafarov, S. Jordan, R. Zhang, V.H. Cervantes (World Scientific, New Jersey, 2015), pp. 1–22
18. E.N. Dzhafarov, J.V. Kujala, Context-content systems of random variables: the contextuality-by-default theory. J. Math. Psychol. **74**, 11–33 (2016)
19. E.N. Dzhafarov, J.V. Kujala, Contextuality analysis of the double slit experiment (with a glimpse into three slits). Entropy **20**, 278 (2018). https://doi.org/10.3390/e20040278
20. E.N. Dzhafarov, Contents, contexts, and basics of contextuality, in *From Electrons to Elephants and Elections: Saga of Content and Context*, ed. by S. Wuppuluri, I. Stewart (To be published by Springer - The Frontiers Collection, 2021)
21. E.N. Dzhafarov, J.V. Kujala, V.H. Cervantes, Contextuality and noncontextuality measures and generalized Bell inequalities for cyclic systems. Phys. Rev. A **101**, 042119 (2020). (available as arXiv:1907.03328.) + Erratum: Contextuality and noncontextuality measures and generalized Bell inequalities for cyclic systems [Phys. Rev. A **101**, 042119 (2020)]. Phys. Rev. A **101**, 069902
22. V.H. Cervantes, E.N. Dzhafarov, Snow Queen is evil and beautiful: experimental evidence for probabilistic contextuality in human choices. Decision **5**, 193–204 (2018)
23. V.H. Cervantes, E.N. Dzhafarov, True contextuality in a psychophysical experiment. J. Math. Psychol. **91**, 119–127 (2019)
24. I. Basieva, V.H. Cervantes, E.N. Dzhafarov, A. Khrennikov, True contextuality beats direct influences in human decision making. J. Exp. Psychol.: Gen. **148**, 1925–1937 (2019)
25. E.N. Dzhafarov, The contextuality-by-default view of the sheaf-theoretic approach to contextuality. To be published in *Samson Abramsky on Logic and Structure in Computer Science and Beyond (Outstanding Contributions to Logic)*, ed. by A. Palmigiano, M. Sadrzadeh (Springer Nature, 2021)

26. H. Atmanspacher, T. Filk, Contextuality revisited: signaling may differ from communicating, in *Quanta and Mind. Synthese Library (Studies in Epistemology, Logic, Methodology, and Philosophy of Science)*, ed. by J. de Barros, C. Montemayor, vol. 414 (Springer, Cham, 2019)
27. S. Kochen, E.P. Specker, The problem of hidden variables in quantum mechanics. J. Math. Mech. **17**, 59–87 (1967)
28. A.A. Klyachko, M.A. Can, S. Binicioglu, A.S. Shumovsky, A simple test for hidden variables in spin-1 system. Phys. Rev. Lett. **101**, 020403 (2008)

Ehtibar N. Dzhafarov is professor of psychological sciences at Purdue University, with PhD degree from Moscow University, 1979. His work has been primarily in mathematical psychology, psychophysics, and foundations of quantum theory (theory of contextuality). His research interests include behavioral "sorites paradox," abstract psychophysics of discrimination, response time analysis, and abstract theory of measurement. He is the principal author of the Generalized (or Universal) Fechnerian Scaling theory in psychophysics, and the interdisciplinary Contextuality-by-Default theory. Further information can be found in http://www.psych.purdue.edu/~ehtibar/.

Can Quantum Non-identity Exist in Social Phenomena?

J. Acacio de Barros

Abstract In this chapter, I discuss indistinguishability and identity in quantum mechanics. I then argue that, similarly to quantum particles, digital money has indistinguishable aspects. This indistinguishability suggests that one can apply quantum-like statistical mechanics to financial systems where digital money is used.

1 Introduction

Quantum mechanics has many peculiar characteristics that are hard to reconcile with a classical view of the world. For instance, quantum properties are contextual because one cannot consistently assign properties without reference to a measurement process [43]. Since a property depends on an observer's measurement, this aspect has been used to argue that specific interpretations of quantum mechanics are idealistic. What "something is" may depend on the choices made by an observer.

Another peculiarity of the quantum world is nonlocality [27]. Nonlocality is a form of contextuality-at-a-distance, where the properties of an entangled quantum system seem to be instantly connected to those of another far-away system. This connection is not explainable by properties set at a common source in the system's past lightcone [4, 5].

However, perhaps among the strangest aspects of quantum theory, the indistinguishability of objects is arguably one of the most puzzling. It is a property of quantum objects that, for certain circumstances, they are absolutely indistinguishable in a most fundamental way. Schrödinger, for instance, argues that indistinguishability goes beyond the purely epistemic aspects implied by the word, i.e., that we cannot distinguish between two particles because we lack information about them [53]. Instead, he argues that quantum particles are indistinguishable because they lack identity. They do not seem to have an explicit ontology that can be thought about

J. Acacio de Barros (✉)
School of Humanities and Liberal Studies, San Francisco State University,
San Francisco, CA, USA
e-mail: barros@sfsu.edu

A. Plotnitsky and E. Haven (eds.), *The Quantum-Like Revolution*,
https://doi.org/10.1007/978-3-031-12986-5_18

in classical terms. This lack of identity seems to have broad consequences in our description of the quantum world [19–22].

Recent research has also suggested that some strange aspects of quantum physics have equivalents in the social sciences. For example, researchers have used quantum-like formulations to model order effects in decision-making [57],[1] contextuality in semantics [1], and pricing options [29], to name a few applications. The application of quantum formalisms to the social sciences has become a burgeoning area, with many scientific papers and books published.

Though it is unclear why the quantum formalism adequately describes social phenomena, some researchers proposed plausible explanations for this effectiveness. For example, in decision-making, the contextuality of outcomes could be the consequence of a neural-oscillator Stimulus-Response-like model of how the brain works [15]. In this neural-oscillator network, the monotonicity of classical probability theory breaks down because of wave-like interference between oscillators. Another possibility could be that human decision-makers use intuitionistic logic. If this is the case, then the quantum algebra of orthomodular quantum lattices approximates the intuitionistic lattice [50]. Finally, quantum-like phenomena may be simply a manifestation of contextual observations [26]. In this situation, quantum mechanics provides a natural probability calculus consistent with context-dependency.[2] Common to those approaches is that the underlying reasons for using a quantum formalism are all "classical," i.e., they do not rely on any actual quantum effects. Thus the common choice of referring to them as "quantum-like."

Among the puzzling aspects of quantum physics, the one that is mainly discussed and accepted in the quantum-like social sciences' literature is contextuality. The reason is that even if nonlocality existed for social sciences, observing it would be impossible with current technology [26]. So, even though nonlocality is sometimes considered, it is always understood differently from the way it is in physics. In physics, nonlocality requires correlations that cannot be explained by causal mechanisms that act in a faster-than-light fashion. In the literature of quantum-like social sciences, nonlocality is simply the violation of probability rules explainable by context dependency. However, this author is unaware of any discussions about the lack of identity and indistinguishability in this literature. This chapter aims to discuss indistinguishability and identity in social sciences and speculate about possible applications of ideas from quantum statistical mechanics in this realm.

This chapter is organized as follows. First, in Sect. 2, I discuss the concepts of identity and indistinguishability, both in classical physics and in quantum mechanics. I then, in Sect. 3, take the strong metaphysical stance that the ontology of quantum particles implies their non-identity, and I examine some of its consequences. In Sect. 4, I examine some proposals that use non-classical statistics outside of physics, suggesting the utility of indistinguishability in the social sciences. In Sect. 5, I argue

---

[1] A classical, i.e., non-quantum, explanation may be found in [48].

[2] Albeit one that is severely constrained [14]. We argued elsewhere that if only contextuality is at play, the quantum probability calculus restricts too much the realm of possibilities of social science experiments [17, 25].

that economics is another possible field where indistinguishability may be applied, and I speculate about how one could expect a quantum-like lack of identity to affect our descriptions of digital money in economics. Finally, in Sect. 6, I end with some remarks and future perspectives.

I met Professor Andrei Khrennikov for a brief chat in 2007 during the first Quantum Interaction conference at Stanford University. In 2012 Andrei invited me to give a talk at his legendary annual conference on the foundations of quantum mechanics, to which I became a regular participant. During those conferences (and others), I had a better opportunity to interact with Andrei. I was struck by his enthusiasm about quantum physics and quantum-like social sciences. So much so that he (together with Emmanuel Haven, who co-edited this book with Arkady Plotnitsky) motivated me to continue investigating the connection between quantum probability and the social sciences. I believe this chapter fits the spirit of Andrei's work on quantum-like phenomena. It is a privilege to dedicate this paper to him and have it as part of a Festschrift in his honor.

2 Identity and Indistinguishability

This section examines the intimately related concepts of identity and indistinguishability. Let us start with the idea of identity. Intuitively, we have a sense of identity because we identify with ourselves. We then extend this idea of identity to other people. Andrei Krhrennikov has his identity, which is different from mine or this chapter's readers (assuming the reader is not Khrennikov). We can further broaden this idea of identity to animals, such as my pet fish or the squirrel who eats my fruits. For example, my fish is different from other fishes of the same species who live in the river. He may be very similar to all other fishes, but my fish is the one who is in my fish tank. We can also add objects to the universe of entities that have identity. This pen on my desk is different from another pen in my drawer, and they are not the same, despite being almost identical (i.e., the same model). Finally, since we added objects, why stop there? We can confer identity to the atoms that make my pen or the elementary particles that make those atoms. There is nothing wrong with the idea of "things" having an identity, and it even seems intuitive to assume so. In this sense, an object can only be identical to itself, identity being the property of sameness.

The intuitive idea of identity is fine for casual conversations, but what do we mean by having an identity? Let us examine the pen on my desk and the pen in my drawer. If I put each side by side, they look the same. If I am not careful to label them, I can quickly get confused as to which is which. Does that mean they are the same? Probably not. One may have a minor scratch, or may have less ink, or have some manufacturing differences, and so on. There are tiny details that, if carefully examined, would tell us that indeed they are different pens. The idea here is that even though they look the same, the pens are different because they have different properties. Since there will always be a property that two objects do not share, this can be considered a property view of identity. In this view, for an object to be identical to something else, they need to be the same. Them being the same means having all the same properties. So, even if my pens were identical, i.e., if all their properties were the same, they would still be located in different places: one is on my desk and the other in my drawer.

So, it seems that properties offer a reasonable way to define identity: there cannot be two different things having all the same properties. This idea was represented in Leibniz's Principle of the Identity of Indiscernibles (PII). This principle states that x is equal to y if and only if all properties of x are the same as y's. Formally, PII is often expressed as

$$\forall P\,(P(x) \iff P(y)) \iff x = y,$$

where P's are properties of x and y. We emphasize that this representation of Leibniz's PII equates identity with the logical symbol of equality.

Leibniz's law is at the core of classical logic and, consequently, the vast majority of the mathematics used in the natural and social sciences. For example, let us consider the Zermelo–Fraenkel (ZF) axioms that formalize naive set theory [46]. In ZF, the only objects are sets, and the only properties of objects are given by the binary predicate "$\in$." Identity shows up as the binary predicate $x = y$, codifying the idea that x and y are the same sets, i.e., the same objects in ZF. This equality comes from first-order logic, which undergirds ZF set theory. For our purposes, in first-order logic, equality satisfies the following axioms:

Reflexivity: $\forall x\,(x = x)$

Substitutivity: $\forall x \forall y\,(\varphi(x) \rightarrow \varphi(y))$ for any formula φ where x is a free variable and $\varphi(y)$ as obtained by replacing any occurrences of x with y.

In addition to the logic axioms of equality, ZF adds the axiom of extensionality. The axiom of extensionality states that two sets x and y are the same if and only if they have precisely the same elements. Formally, this can be written as

$$\forall x \forall y\,[\forall z\,(z \in x \iff z \in y) \iff (x = y)]\,.$$

So, identity appears to be related to an ontological aspect of the objects, which are sets in the universe of ZF. First, using the reflexivity axiom from logic, we assume that an object can only be identical to itself. Second, if two sets share all properties, the extensionality axiom implies they are not two objects but one and the same. So, the concept of identity is baked into the foundations of mathematics, with first-order logic and set theory dealing with objects that satisfy Leibinz's PII.

There is one common objection to the ideas expressed above. When talking about objects' properties and equating them with identity, we effectively conflate the epistemic notion of indistinguishability with the ontological notion of identity. Identity, one can argue, is an inherent property of objects: they either have an identity or not. Indistinguishability, on the other hand, seems to be an epistemic concept. It seems to come from our empirical inability to distinguish two things that, in principle, are not the same.

Let us take, for instance, two "identical" twin brothers. They may seem qualitatively identical, in the sense that one may not, at least at first glance, distinguish one from the other, except that one sees a brother to one's right and another to one's left. However, each brother has their own lived experience, peculiarities, and individual (albeit non-perceptible to some people) differences. So, it would be strange to say that they are identical. Perhaps they may differ at the molecular level; perhaps one's synapses reacted differently to slightly different proximal stimuli. Nevertheless, if one were to obtain further information, such as a scar or a memory, one could, *in principle*, distinguish between them. In other words, they seem indistinguishable because one lacks further information about them. They are each an individual because they have their own lived lives, even if everything else was the same.

To summarize our discussions in this section, we saw that identity is an ontological principle. We assign identity and individuality to objects because it seems natural to do so. The pen in my drawer is different from the pen on my desk. They have, in this sense, an identity. Even if I cannot distinguish between them, i.e., even if they are indistinguishable, I can still talk and think about two different pens, one in my drawer and another on my desk. However, attaching identity to objects is a metaphysical move: when we do so, we are stating that there are two pens, and they are different things, even if they are indistinguishable. Furthermore, this ontological principle is reflected in the mathematics used to describe nature. For instance, most modern standard mathematics is formulated within ZF set theory, which uses the concept of identity through Leibniz's principle of identity of indiscernibles. In the next section, we shall see that some results in quantum physics question this ontological assumption that objects are individuals.

3 On the Ontology of Quantum Physics

As we mentioned above, one of the most puzzling aspects of quantum theory is that particles seem to lack individuality. This section will examine the issue of indistinguishability of particles and how it seems to imply this lack of individuality of particles.

Let us start with the origins of this discussion in physics. In the early 20th century, physicists quickly realized that the statistics of quantum particles seem to be different from the classical statistics of Boltzman. It all started with a letter that Satyendra Bose wrote to Albert Einstein. This letter included his rejected paper, in which he detailed how, based on the black-body radiation laws discovered by Plank, quantum statistics was different from Boltzman's statistics. Einstein quickly translated the article to German and had it published in *Zeitschrift für Physik*, under Bose's request. The critical idea in Bose's article is that, since two photons of the same wavelength are indistinguishable, then their mathematical description had to accommodate this indistinguishability and be reflected in the statistics.

Let us start with the mathematical description. Assume that we have two particles (say, two photons), and we describe each as $\psi(\mathbf{r}_1)$ and $\phi(\mathbf{r}_2)$, where the r's subscripts refer to a position measurement of particle one or two, and where ψ and ϕ are two distinct wavefunctions. Since the particles are indistinguishable, we cannot know which particle is in the state ϕ and which is in the state ψ. Therefore, the wavefunction for the entire system, composed of particles one and two, needs to be invariant under permutations of the particles. Otherwise, we would distinguish particle one from two. There are two possible forms for this wavefunction, namely

$$\Psi_s = \frac{1}{\sqrt{2}}\left[\psi(\mathbf{r}_1)\phi(\mathbf{r}_2) + \psi(\mathbf{r}_2)\phi(\mathbf{r}_1)\right]$$

or

$$\Psi_a = \frac{1}{\sqrt{2}}\left[\psi(\mathbf{r}_1)\phi(\mathbf{r}_2) - \psi(\mathbf{r}_2)\phi(\mathbf{r}_1)\right].$$

The first form, Ψ_s, is symmetric under permutations of one and two, whereas the second form, Ψ_a is anti-symmetric. The symmetric form corresponds to particles with integer spin, as the case for photons, and are, aptly, called bosons (in honor of Bose). The anti-symmetric wavefunction describes particles with fractional spin, such as electrons, called fermions (after the physicist Enrico Fermi, who formulated the quantum statistical mechanics for electrons).

Interestingly, the above symmetries or anti-symmetries of the wavefunction are based on indistinguishability, an epistemic concept, as we saw, but have real-world consequences. For bosons, this symmetry leads to two bosons occupying the same quantum state. We can have as many bosons as we wish occupying the same state, a phenomenon called Bose-Einstein Condensation. But this phenomenon happens because we cannot count two bosons as two individuals. Instead, we count them as two (indistinguishable) non-individuals.

To illustrate the above point, imagine we have three particles, A, B, and C. We want to distribute them through two possible states, X and Y. Classically, there are eight possibilities: (I) A, B, and C are in X; (II) A and B are in X, but C is in Y; and so on. Since we are talking about statistics, we need to compute the probability of finding a particle or two in X or Y. To do so, we pick one of the particles, say A, and randomly distribute it to either X or Y. Then we repeat the process with B and C. The possible choices are shown in Fig. 1. From the tree diagram, we can conclude that the probability of having all three particles in state X is $1/8$, two in state X is $3/8$, and so on.

The counting is different if A, B, and C are non-individuals. As shown in Fig. 2, if we cannot distinguish A, B, and C, we cannot notice a difference between A and B in state X and C in state Y from, say, A being in state Y and B and C in state X. Thus, the probability of all three particles being on state X is not $1/8$ anymore, but the much higher value $1/4$. This probability is $1/4$ because states with two particles in X or two particles in Y are indistinguishable in a fundamental way, as we cannot follow the tree diagram for particles who lack individuality.

Here we find ourselves in an exciting situation. Bose argued that the indistinguishability of quantum particles results in a statistical mechanics different from

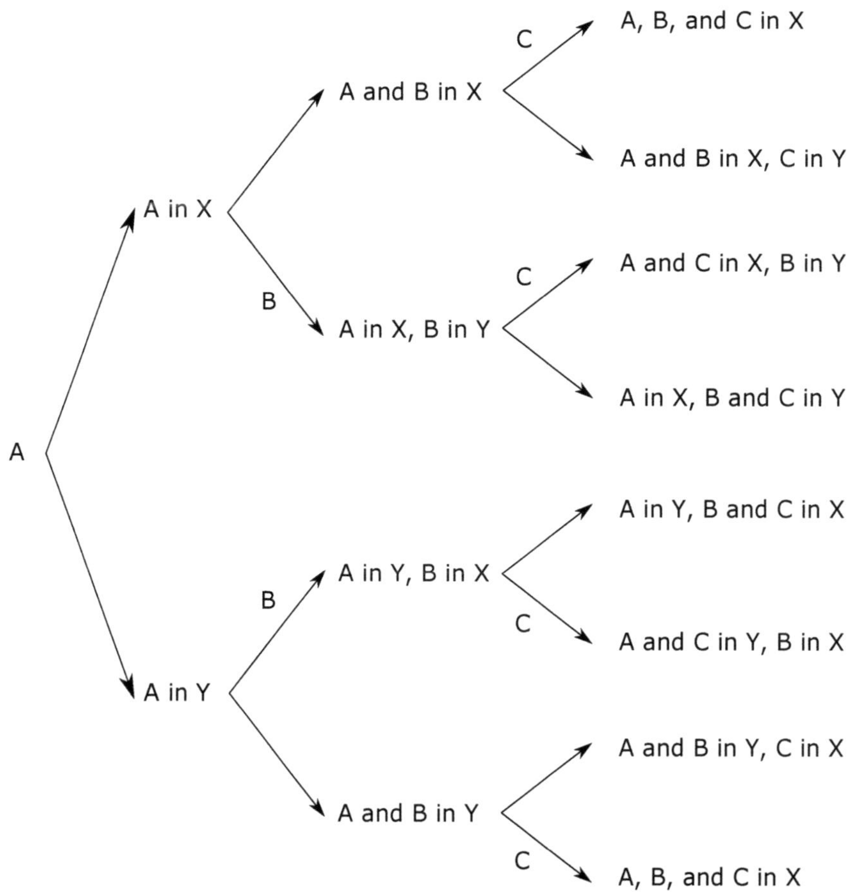

Fig. 1 Probability tree diagram for three particles A, B, and C distributed between two states, X or Y

Boltzman's classical one. However, as we saw above, indistinguishability is an epistemic concept: just because we cannot distinguish particle A from B does not mean they are not individuals. However, the counting shown in Fig. 2 suggests that we cannot treat them as individuals. It is not just that we do not know how to distinguish them, but that we cannot because they lack identity. This suggests that the issue is not epistemic but ontological.

To emphasize this, let us look at the example given by Hermann Weyl. Let us consider a quantum system composed of two "identical" particles (e.g., two electrons) that can be in one of two energy states, E_1 or E_2, $E_1 \neq E_2$. Weyl points out that for such systems, the wavefunction needs to be either symmetric or anti-symmetric. If anti-symmetric, both "individuals" cannot be in the same quantum state, i.e., the only constraint is that the system has total energy $E_1 + E_2$. Because $E_1 + E_2$ is

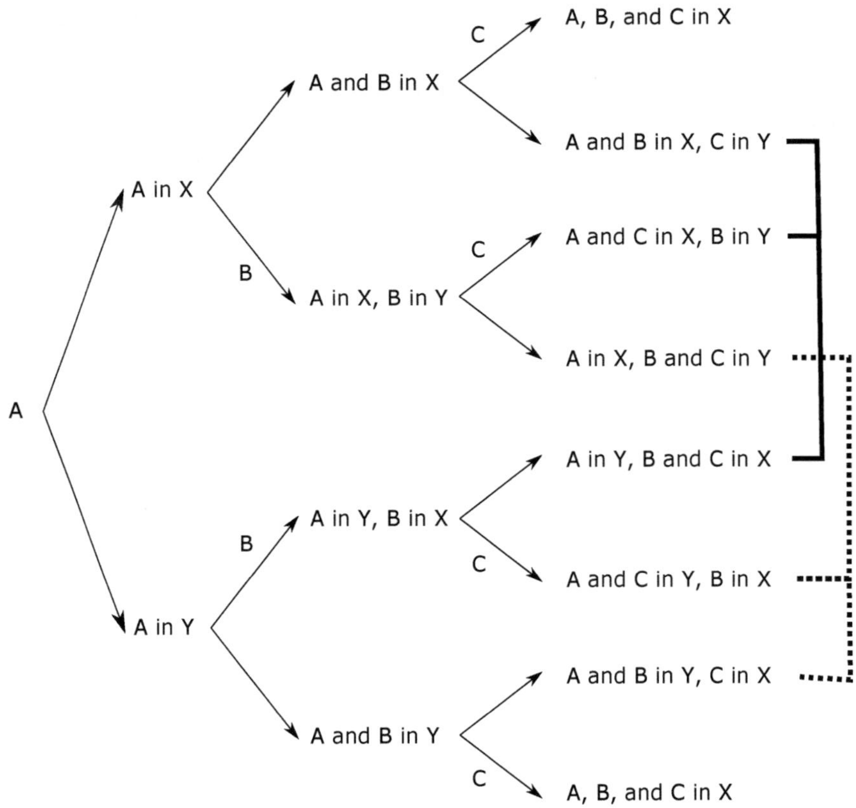

Fig. 2 Probability tree diagram for three particles A, B, and C distributed between two states, X or Y. The solid line to the right shows indistinguishable states with two particles in X and one in Y. In contrast, the dashed lines show states with one in X and two in Y. Since they are indistinguishable at a fundamental level, they cannot be counted as different, and they are equiprobable to three particles in either X or Y

the only possibility, and because the question of which particle has energy E_1 and which has energy E_2 is meaningless, Weyl points out that we count the permutation of 1 and 2 as the same. This symmetry in the permutation of particles should be contrasted with the way of counting possible ways of reaching $E_1 + E_2$ classically. If we have two classical particles, I and M, one can have energy E_1 and the other E_2. Therefore, we may *in principle* figure out which is which, and use this to count two possibilities: (1) I has energy E_1 and M has E_2, or (2) I has energy E_2 and M has E_1. Thus, classically we have two counts, not just one. This difference in statistics, which results in measurable effects (such as Pauli's exclusion principle for anti-symmetric particles), led Weyl to explain this quantum characteristic the following way:

> That $E_1 + E_2$ occurs only once in [the symmetric case...] means: the possibility that one of the identical twins Mike and Ike is in the quantum state E_1 and the other in the quantum state E_2 does not include two differentiable cases which are permuted on permuting Mike and Ike; it is impossible for either of these individuals to retain his identity so that one of them will always be able to say "I'm Mike" and the other "I'm Ike." Even in principle one cannot demand an alibi of an electron! [58]

Unlike Mike and Ike, quantum particles not only have no identity card, but they seem not to be even distinct individuals. For a quantum system of two identical particles, talking of particle A and B is nothing more than a *façon de parler*, as they only make sense as part of a whole that cannot itself be split or even accounted for as made of two objects with individual properties.

We emphasize that this lack of individuality does not imply that we do not have two "things." For example, one can measure one electron at point A and another at B, and we may be tempted to say that we distinguish them by their location. However, we cannot say that electron "Ike" is in A and "Mike" is in B. All we can say is that one electron is in A and another is in B, but not which one. This impossibility to say which electron is where is reflected in the formal description of the system, as we need to use an anti-symmetric wave function to describe it. So, we have two things, but they lack identity.

Is this non-individuality necessary? Perhaps not. Let us examine what happens to the concept of individuals with one of the most researched realistic interpretations of quantum mechanics: the de Broglie-Bohm pilot-wave interpretation, usually referred to as Bohm's interpretation. Bohm's interpretation uses a classical-like ontology where the individuality of particles is at its core. Though physicists often complain of Bohm's theory as a return to the classical viewpoints of the 19th century, it includes contextuality at its core, through the boundary conditions of the wavefunction solutions. Furthermore, since the Bohmian quantum potential changes instantly, such boundary conditions instantaneously affect the state of a system. Thus, we could hardly say that Bohm's theory is a return to the classical view of the world unless we ignored the development of classical physics in the 19th century.

The effects of indistinguishability of fermions are present as quantum potential effects that keep them apart.[3] In Bohm's theory, as in quantum theory, indistinguishability is imposed ad hoc through the wave function's symmetrization (or antisymmetrization). This assumption about the wave function, introduced by an indistinguishability argument, affects the particles through the quantum potential. We then find ourselves in a weird situation in Bohm's theory: we assume indistinguishability as a *fundamental* aspect of the quantum world affecting the quantum potential, and we then predict real effects, only to deny the reality of indistinguishability through an ontology associated with the physical model. This difficulty is not present with Bohm's theory on contextuality or nonlocality, where the ontology is clearly compatible.

Furthermore, Bohmian mechanics has individuals because, in the theory, one can follow the trajectories of each particle and use those trajectories as an "identity

---

[3] The Bohmian picture is more challenging for bosons [35].

card." These identity cards are obtained within Bohm's interpretation. However, such trajectories cannot be observed. In the lab, subatomic particles do not exhibit such trajectories, and instead show up as discrete clicks in spacetime on the screen of a detector. This observed behavior is different from the classical, as we cannot ensure (unless we believe in Bohm's interpretation) that a particle that we detected here is the same detected there. As Schrödinger puts it,

> [w]hen a familiar object reenters our ken, it is usually recognized as a continuation of previous appearances, as being the same thing. The relative permanence of individual pieces of matter is the most momentous feature of both everyday life and scientific experience. If a familiar article, say an earthenware jug, disappears from your room, you are quite sure that somebody must have taken it away. If after a time it reappears, you may doubt whether it really is the same one – breakable objects in such circumstances are often not. You may not be able to decide the issue, but you will have no doubt that the doubtful sameness has an indisputable meaning – that there is an unambiguous answer to your query. So firm is our belief in the continuity of the unobserved parts of the string! [53, pp. 190]

This is different from what happens with quantum particles, as Weyl pointed out.

We end this section with a comment on our own view on this subject. As we discussed in other papers [19–22], we believe that an ontology of non-individuals is part of the microscopic world. We believe that such an ontology even explains some of the puzzles of quantum mechanics, such as the Bell-EPR apparent paradox if we consider a realistic interpretation. The inconsistency that comes out of a Bell-type inequality has as underlying assumption the individuality of particles [22].

4 Indistinguishability Outside of Physics

In this section, we will speculate whether quantum indistinguishability and its relative, quantum non-identity, have anything to contribute to the modeling of social phenomena. We begin with some discussions on the use of quantum formalisms in social sciences. We then talk about possible applications of quantum indistinguishability.

First, let us distinguish between quantum mechanics and its formalism. Most readers will find this discussion elementary, but, unfortunately, the conflation of those two ideas is a source of much confusion, particularly for those unfamiliar with the literature on quantum social sciences. Quantum mechanics is a fundamental theory in physics that describes microscopic (and some macroscopic) phenomena incredibly accurately. As it was developed, it was meant to describe physical phenomena. Since it is a fundamental theory, it is widely relied on and used outside of physics. For example, one cannot imagine modern chemistry doing without at least some knowledge of quantum mechanics, which is essential to our understanding of the periodic table. To a lesser degree, quantum mechanics also enters the realm of biology, indirectly through chemistry and directly through processes that rely on non-classical dynamics. More directly, the study of quantum processes in biology forms a field of its own, quantum biology. Prominent phenomena studied in quantum biology are magnetoreception, olfaction, photosynthesis, and vision [47].

In the early years of quantum mechanics, some prominent physicists hypothesized that quantum processes might be relevant not only to chemistry and biology but perhaps to understand consciousness and the brain. Notably, [55] connected the measurement problem with the mind of an observer (for more details, see [54]). This connection was later on deepened by [45] and [59]. It is not, therefore, surprising that many are still exploring such connections. However, even though some celebrated physicists proposed this idea, most physicists see this connection as unlikely. This attitude is also true for other social quantum phenomena. Though many physicists would find it credible that quantum mechanics may appear in chemical or biological processes, most would view the appearance of quantum phenomena in the social sciences with intense skepticism. The main reason for such skepticism is that social phenomena involve the interaction of very complex systems (humans) in a thermal bath (the environment). In such situations, quantum decoherence would make any quantum effects quickly disappear, and classical descriptions would be possible [56]. However, some prominent researchers, such as Physics Nobel laureates Roger [51] and Brian [36], still believe there is a connection between quantum mechanics and psychology through consciousness. For a discussion on the different views concerning quantum mechanics and the mind, we direct the reader to [24].

Differently from the above view, in the mid-2000s, William Lawless and Peter Bruza discussed the possibility of organizing a conference in social interactions and quantum mechanics. Their idea was that social interactions might be context-dependent in the same way that quantum observables are. In 2007, Lawless and Bruza, together with C. J. van Rijsbergen and Don Sofge, organized the first Quantum Interactions conference as part of the 2007 AAAI Spring Symposium Series held at Stanford University.[4] This conference was the beginning of several other meetings. It resulted in the creation of a vibrant community whose goal was not only to seek quantum-like phenomena in social interactions but also to investigate if the contextual probability calculus developed by the founders of quantum mechanics could also be useful outside of physics. Its scope covered areas as diverse as computer science, linguistics, econophysics, cognition, and decision-making. According to this view, quantum mechanics should not be seen as a theory of microscopic phenomena applied to the social sciences. Instead, it should be thought of as a contextual calculus whose application transcends the physics of the very small. From the 2007 Stanford conference, a burgeoning field of research emerged, with hundreds of articles and several books published on the subject (see, for example, [9, 32, 34] and references therein). Now, quantum-like models exist in many distinct areas, such as econophysics [29–31], cognition [15, 39, 40, 49, 52], decision-making [10, 12, 13, 33, 37], political sciences [38, 41, 42], and linguistics [2, 7, 8], to name a few.

The use of the quantum formalism in social sciences is not far-fetched. First, in this context, the quantum formalism has nothing to do with the quantum mechanics of microscopic systems but instead with particular aspects of social phenomena, espe-

[4] See the Preface to [18] for a brief historical account of this conference series.

cially contextuality [26]. In fact, it is possible to create classical models of cognitive processes that exhibit quantum-like characteristics, such as violations of monotonicity in probability theory [11, 15]. As such, we can think of the quantum formalism as one of the many different ways to represent complex contextual phenomena.[5] It presents the advantage that it is well developed and studied in the context of quantum physics. Therefore, we use this perspective to think about indistinguishability outside of physics, i.e., that the quantum formalism may offer insights into the social sciences.

Now, attempts to use indistinguishability are not new outside of physics. For example, Bose-Einstein distributions were used in information retrieval [3]. The idea here is that the indistinguishability of "particles," in this case, linguistic elements, comes from the inability to attach identity to them. This lack of identity comes from the fact that language is contextual, a case that resembles the indistinguishability of quantum properties examined by [21]. Therefore, cases of linguistic contextuality would be reflected on the deviation of classical statistics toward a Bose-Einstein one. Another use of Bose-Einstein statistics appears in the work of [6]. Their paper uses the Bose-Einstein statistics to examine complex networks such as the Internet and show that they can form Bose-Einstein condensate states. Finally, another example of using indistinguishability in social sciences is Andrei Khrennikov's "social laser" [38]. In his paper, Khrennikov uses an analogy between the inverted population in lasers, where the majority of the atoms are in an excited state, and social energy. He hypothesizes that when the population in a state of high social energy reaches a critical rate, we expect the "stimulated emission" of social energy to surpass the spontaneous emissions, leading to an action amplification by the social agents. Indistinguishability enters into Khrennikov's model the same way that it shows up in Einstein's model of Plank's radiation.

However, where else should we expect indistinguishability to play a role outside of quantum physics. In the next section, we will examine a possible application of indistinguishability in economic theory.

5 Indistinguishability in Economics: Perspectives

A prime candidate for using indistinguishability is economics. The goal of economics is, among others, to study social interactions that involve value. Classic examples are economic models of supply and demand that describe the relation between prices of goods or services as their availability changes. The medium of exchange is, in most industrialized societies, fiat money. So, let us focus on money.

First, let us start with a brief and oversimplified history of money (the interested reader is referred to some classic texts on the subject, such as [28]). In less complex

[5] We have argued elsewhere that perhaps a more general approach, such as extended probability theories, might be better suited than Hilbert-space models [16, 17, 23]. However, this chapter will focus on the quantum route and not examine other extended probabilities.

societies, exchanging of goods or services happens through bartering, i.e., through direct exchanges. For example, someone who wants rum may have a small vegetable farm that produces food. Since their vegetable farm does not produce rum, but their neighbor does, they may agree to a direct exchange of some vegetables for a bottle of rum. Of course, this is a very cumbersome and, sometimes, impractical process. Therefore, some societies developed the concept of a medium exchange for bartering, usually in the form of something widely considered valuable (such as, say, salt or gold). So, instead of our two farmers negotiating every time how many vegetables in exchange for a bottle, they may choose to establish the exchange in terms of, say, grams of salt. This way, even if the vegetable farmer has no vegetables, i.e., in winter, they may still access rum through its exchange with salt. This type of medium of exchange is called *commodity money*, gold coins being one of its best known examples. Commodity money was later replaced with *representative* or *commodity-backed money*, essentially a token for the actual commodity. For example, instead of having a coin made out of gold, one may carry a piece of paper money that has the backing of an authority (usually a government) stating that this particular piece of paper can be, if requested, exchanged by the equivalent amount in gold. This form of money gave way to *fiat money,* as many governments, who usually backed their currency with gold, silver, or platinum, moved from commodity money. In the US, the gold standard was abandoned by Nixon in 1971. As such, fiat money has no value except within a social contract. Finally, we get to *digital money*, the form that most money takes in the US nowadays. Instead of issuing actual bills, as it has done in the past, first attached to a commodity (e.g., gold) and later as fiat money, nowadays central banks issue money electronically in databases.

Commodities, representative, and even fiat money are individuals. Once a bill is printed, it has a serial number. Though two dollar bills have the same intrinsic value, they are not the same, in the sense that they constitute classical individuals. They are exchangeable in the same way that the three classical particles are exchangeable in Fig. 1, meaning that nothing changes if we replace one with another. Nevertheless, the way we count is that of individuals.

Digital money, on the other hand, is not constituted of individuals. One cannot track each digital dollar, but only how many dollars there are at a specific time in a particular account. In this sense, digital money is countable, as it has cardinality. However, unlike print money, one cannot lay down each bill side by side and map them to the set of integers, as this process requires to keep count of the order in which we are mapping. This lack of ordinality means that it does not make sense to think of them as standard sets: they are not representable by standard mathematics [44]. Perhaps the proper way to treat such systems mathematically is to use a theory grounded on quasi-sets.

Quasi-set theory is an extension of set theory to include non-individuals. Set theory can be formalized in first-order logic through Zermelo-Fraenkel's axioms (ZF). Most of modern mathematics is done under ZF set theory, sometimes adding extra axioms, such as the Continuum Hypothesis or the Axiom of Choice [46]. In ZF set theory, the only elements are sets (e.g., the empty set $\emptyset$, the set containing the empty set $\{\emptyset\}$, and so on). Quasi-set theory includes ZF plus urelement, i.e., objects that are not sets, m and M. The intuitive idea is that the urelements M correspond to individuals whereas m to objects that are non-individuals. In quasi-set theory, we may have $\{a, a\}$ as different from $\{a\}$ if a is an m-object; the first set has cardinality two whereas the second has cardinality one. However, we cannot talk about ordinality for the quasi-set $\{a, a\}$; how would we distinguish the "first" a from the other?

Another possibility is to use the tricks developed in quantum theory to create a probabilistic theory of indistinguishables. As we saw, in quantum theory, when two particles are indistinguishable, we impose symmetries in their Hilbert space description. A possibility would be to construct Hilbert space models for money and then impose symmetries in their description. As we mentioned above, whether this symmetry is one of bosons or fermions is an empirical question. Nevertheless, since digital money lacks identity, either quasi-sets of a Hilbert space formalism seem reasonable to model it. Another approach is to look at the consequences of non-individuals statistics, as Khrennikov has done for the social laser.

6 Final Remarks

In this chapter, we examined the concept of indistinguishability and identity in physics and then argued that money has some of the characteristics of indistinguishable particles. Furthermore, by construction, digital money has cardinality but not ordinality. Therefore, it constitutes a prime candidate for being treated as indistinguishable not only because we cannot know but because it actually lacks identity. This situation is similar to quasi-set theory, where sets with no ordinality but with definite cardinality exist.

Because digital money does not have an identity, they share the same statistics as elementary particles: a statistics of indistinguishability. Whether digital money behaves like bosons or fermions is a question we cannot answer at this point. However, one possible consequence of this indistinguishability of digital money is that models such as Khrennikov's social laser could probably be developed for it. For instance, the fact that massive quantitative easing happened in the US without any increase in inflation suggests a form of coherent "money emission." We believe those are interesting questions for further investigation.

Acknowledgements I would like to thank Professors Emmanuel Haven and Arkady Plotnitsky for inviting me to contribute this chapter to the Festschrift in honor of Professor Khrennikov and for their useful comments and suggestions. I also want to thank Professor Haven for his discussions about indistinguishability and economics and finance.

References

1. D. Aerts, Quantum structure in cognition. J. Math. Psychol. **53**(5), 314–348 (2009)
2. D. Aerts, J. Broekaert, L. Gabora, T. Veloz, The Guppy Effect as Interference, in *Quantum Interaction*, ed. by J.R. Busemeyer, F. Dubois, A. Lambert-Mogiliansky, M. Melucci. Lecture Notes in Computer Science, vol. 7620 (Springer, Berlin, 2012), pp. 36–47
3. G. Amati, C.J. Van Rijsbergen, Probabilistic models of information retrieval based on measuring the divergence from randomness. ACM Trans. Inf. Syst. **20**(4), 357–389 (2002)
4. J. Bell, On the Einstein-Podolsky-Rosen paradox. Physics **1**(3), 195–200 (1964)
5. J. Bell, On the problem of hidden variables in quantum mechanics. Rev. Mod. Phys. **38**(3), 447–452 (1966)
6. G. Bianconi, A.L. Barabási, Bose-Einstein condensation in complex networks. Phys. Rev. Lett. **86**(24), 5632–5635 (2001). (American Physical Society)
7. P. Bruza, K. Kitto, D. Nelson, C. McEvoy, Is there something quantum-like about the human mental lexicon? J. Math. Psychol. **53**(5), 362–377 (2009)
8. P. Bruza, K. Kitto, B.J. Ramm, L. Sitbon, A probabilistic framework for analysing the compositionality of conceptual combinations. J. Math. Psychol. **67**, 26–38 (2015)
9. J. Busemeyer, P. Bruza, *Quantum Models of Cognition and Decision* (Cambridge University Press, Cambridge, 2012)
10. J. Busemeyer, Z. Wang, A. Lambert-Mogiliansky, Empirical comparison of Markov and quantum models of decision making. J. Math. Psychol. **53**(5), 423–433 (2009)
11. J.R. Busemeyer, P. Fakhari, P. Kvam, Neural implementation of operations used in quantum cognition. Prog. Biophys. Mol. Biol. (2017)
12. J.R. Busemeyer, Z. Wang, A. Khrennikov, I. Basieva, Applying quantum principles to psychology. Phys. Scripta **2014**(T163), 014007 (2014)
13. J.R. Busemeyer, Z. Wang, J.T. Townsend, Quantum dynamics of human decision-making. J. Math. Psychol. **50**(3), 220–241 (2006)
14. J.A. de Barros, Joint probabilities and quantum cognition, in *AIP Conference Proceedings*, vol. 1508, ed. by A. Khrennikov, A.L. Migdall, S. Polyakov, H. Atmanspacher (American Institute of Physics, Vaxjo, Sweden, 2012), pp. 98–107
15. J.A. de Barros, Quantum-like model of behavioral response computation using neural oscillators. Biosystems **110**(3), 171–182 (2012)
16. J.A. de Barros, *Decision Making for Inconsistent Expert Judgments Using Negative Probabilities* Lecture Notes in Computer Science. (Springer, Berlin, 2014), pp.257–269
17. J.A. de Barros, Beyond the quantum formalism: consequences of a neural-oscillator model to quantum cognition, in *Advances in Cognitive Neurodynamics (IV)*. ed. by H. Liljenström. Advances in Cognitive Neurodynamics. (Springer, Netherlands, 2015), pp. 401–404
18. J.A. de Barros, B. Coecke, E. Pothos, in *Quantum Interaction: 10th International Conference, QI 2016, San Francisco, CA, USA, July 20-22, 2016, Revised Selected Papers*. Springer. Google-Books-ID: Chb_DQAAQBAJ (2017)
19. J.A. de Barros, F. Holik, Indistinguishability and negative probabilities. Entropy **22**(8), 829 (2020). (Number: 8 Publisher: Multidisciplinary Digital Publishing Institute)
20. J.A. de Barros, F. Holik, D. Krause, Contextualityy and indistinguishability. Entropy **19**(9), 435 (2017)
21. J.A. de Barros, F. Holik, D. Krause, *Indistinguishability and the Origins of Contextuality in Physics* (Philosophical Transactions of the Royal Society A, Mathematical, Physical and Engineering Sciences, 2019)
22. J.A. de Barros, J.P. Jorge, F. Holik, On the assumptions underlying KS-like contradictions (2021). arXiv: 2103.06830 [quant-ph]
23. J.A. de Barros, J.V. Kujala, G. Oas, Negative probabilities and contextuality. J. Math. Psychol. **74**, 34–45 (2016)
24. J.A. de Barros, C. Montemayor (eds.), *Quanta and Mind: Essays on the Connection Between Quantum Mechanics and Consciousness*, Synthese Library, vol. 414. (Springer, Cham, 2019)

25. J.A. de Barros, G. Oas, Some examples of contextuality in physics: implications to quantum cognition, in *Contextuality, from Quantum Physics to Psychology*. ed. by E.N. Dzhafarov, R. Zhang, S.M. Jordan (World Scientific, 2015), pp. 153–184
26. J.A. de Barros, P. Suppes, Quantum mechanics, interference, and the brain. J. Math. Psychol. **53**(5), 306–313 (2009)
27. A. Einstein, B. Podolsky, N. Rosen, Can quantum-mechanical description of physical reality be considered complete? Phys. Rev. **47**(10), 777–780 (1935)
28. J.K. Galbraith, *Money. Princeton University Press* (Money, Publication Title, 2017)
29. E. Haven, A discussion on embedding the Black-Scholes option pricing model in a quantum physics setting. Phys. A **304**(3–4), 507–524 (2002)
30. E. Haven, The wave-equivalent of the Black-Scholes option price: an interpretation. Phys. A **344**(1–2), 142–145 (2004)
31. E. Haven, Pilot-wave theory and financial option pricing. Int. J. Theor. Phys. **44**(11), 1957–1962 (2005)
32. E. Haven, A. Khrennikov, *Quantum Social Science* (Cambridge University Press, Cambridge, 2013)
33. E. Haven, A. Khrennikov, Quantum probability and the mathematical modelling of decision-making. Philos. Trans. Roy. Soc. A: Math. Phys. Eng. Sci. **374**(2058), 20150105 (2016). (Royal Society)
34. E. Haven, A. Khrennikov, *The Palgrave Handbook of Quantum Models in Social Science* (Springer, 2017)
35. P.R. Holland, *The Quantum Theory of Motion: An Account of the de Broglie-Bohm Causal Interpretation of Quantum Mechanics* (Cambridge University Press, 1995)
36. B.D. Josephson, The physics of mind and thought. Act. Nerv. Super. **61**(1), 86–90 (2019)
37. A. Khrennikov, Quantum-like model of cognitive decision making and information processing. Biosystems **95**(3), 179–187 (2009)
38. A. Khrennikov, 'Social Laser': action amplification by stimulated emission of social energy. Philos. Trans. Roy. Soc. A: Math. Phys. Eng. Sci. **374**(2058), 20150094 (2016). (Royal Society)
39. A. Khrennikov, I. Basieva, E.N. Dzhafarov, J.R. Busemeyer, Quantum models for psychological measurements: an unsolved problem. PLoS ONE **9**(10), e110909 (2014)
40. A.Y. Khrennikov, E. Haven, Quantum mechanics and violations of the sure-thing principle: the use of probability interference and other concepts. J. Math. Psychol. **53**(5), 378–388 (2009)
41. P. Khrennikova, E. Haven, A. Khrennikov, An application of the theory of open quantum systems to model the dynamics of party governance in the us political system. Int. J. Theor. Phys. **53**(4), 1346–1360 (2014)
42. P. Khrennikova, A. Khrennikov, F. Haven, The quantum-like description of the dynamics of party governance in the US political system (2012). arXiv:1206.2888
43. S. Kochen, E.P. Specker, The problem of hidden variables in quantum mechanics. J. Math. Mech. **17**, 59–87 (1967)
44. D. Krause, On a quasi-set theory. Notre Dame J. Formal Logic **33**(3), 402–411 (1992)
45. F. London, E. Bauer, La théorie de l'observation en mécanique quantique (1939)
46. Y.I. Manin, *A Course in Mathematical Logic for Mathematicians* (Springer Science & Business Media, 2009). (Google-Books-ID: ftK9pUj8SaMC)
47. A. Marais, B. Adams, A.K. Ringsmuth, M. Ferretti, J.M. Gruber, R. Hendrikx, M. Schuld, S.L. Smith, I. Sinayskiy, T.P.J. Krüger, F. Petruccione, R. van Grondelle, The future of quantum biology. J. Roy. Soc. Interface **15**(148), 20180640 (2018). (Royal Society)
48. C. Moreira, J.A. de Barros, Order effects in Bayesian updates. Proc. Annu. Meeting Cognit. Sci. Soc. **43**(43) (2021)
49. C. Moreira, A. Wichert, Quantum probabilistic models revisited: the case of disjunction effects in cognition. Front. Phys. **4** (2016)
50. L. Narens, *Probabilistic Lattices: With Applications to Psychology* (World Scientific, 2014). Google-Books-ID: Bh23CgAAQBAJ
51. R. Penrose, *The Emperor's New Mind: Concerning Computers, Minds, and the Laws of Physics* (Oxford University Press, Oxford, NY, 1989)

52. E.M. Pothos, J.R. Busemeyer, Can quantum probability provide a new direction for cognitive modeling? Behav. Brain Sci. **36**(03), 255–274 (2013)
53. E. Schrödinger, What is an elementary particle? *Annual Report of the Board of Regents of The Smithsonian Institution* (1950), pp. 183–196
54. P. Suppes, J.A. de Barros, Quantum Mechanics and the Brain. In *Quantum Interaction: Papers from the AAAI Spring Symposium*, Technical Report SS-07-08 Menlo Park, CA. (AAAI Press, 2007), pp. 75–82
55. J. von Neumann, *Mathematical Foundations of Quantum Mechanics; Translated by Robert T. Beyer from the 1932 German*. (Princeton University Press, Princeton, NJ, 1955)
56. D. Waldner, Schrödinger's cat and the dog that didn't bark: Why quantum mechanics is (probably) irrelevant to the social sciences. Crit. Rev. **29**(2), 199–233 (2017). Taylor & Francis
57. Z. Wang, T. Solloway, R.M. Shiffrin, J.R. Busemeyer, Context effects produced by question orders reveal quantum nature of human judgments. Proc. Natl. Acad. Sci. **111**(26), 9431–9436 (2014)
58. H. Weyl, *The Theory of Groups and Quantum Mechanics* (Courier Corporation, 1950). Google-Books-ID: jQbEcDDqGb8C
59. E. Wigner, Remarks on the mind-body problem. *The Scientist Speculates* (1961), pp. 284–302

Author Biography

Jose Acacio de Barros is a physicist and Professor of Liberal Studies at San Francisco State University. Before going to SFSU, he was at the physics department of the Federal University in Juiz de Fora, Brazil. He earned a Ph.D. and MS in Physics from the Brazilian Center for Research in Physics under the supervision of Francisco Doria. Most of his research is interdisciplinary, ranging from the foundations and philosophy of physics and mathematics to psychology and physics education. He has published numerous academic articles in prestigious physics, psychology, mathematics, and philosophy journals. His current research focuses on quantum ontology and its relation to contextuality and the hard problem of consciousness. He is also interested in how the mathematical formalism of quantum mechanics informs us of issues outside of physics. Dr. de Barros has held visiting faculty positions at the Institute for Mathematical Studies in the Social Sciences (Stanford University), the Center for the Studies of Language and Information (Stanford University), and COPPE (Federal University of Rio de Janeiro). At Stanford, he was a long-term collaborator of Patrick Suppes. Dr. de Barros is a Corresponding Member of the Brazilian Academy of Philosophy.

Index

A
Agent, 15, 67, 69, 72, 73, 77–87, 145, 158, 282, 283, 285, 327, 328, 331, 372
Amplitudes, 19, 34, 35, 110, 186, 235, 236, 243, 332, 335
Asymmetries, 33, 290, 291
Axiomatization (of quantum mechanics), 19, 191, 193, 210

B
Bayesianism, 62, 79–81
Bayesian probability, 62, 73, 78, 285
Belief, 3, 21, 51, 62, 63, 68–70, 82, 85, 87, 158, 256, 285, 299, 303, 306, 321, 327–333, 335, 341, 370
Bell state, 121, 251–253, 255
Bell theorem, 2, 5, 6, 8, 9
Bohmian mechanics, Bohmian theory, 12, 14, 97, 110, 357, 369
Bohr-Einstein debate, 2, 4, 5
Boolean logic, 327
Born rule, 72, 83, 94, 108, 113, 118, 120, 235, 236, 243, 312, 316, 332
Bose-Einstein condensate, 372

C
Category, 30–32, 107, 248, 249, 255, 256, 281, 291
Category theory, 248, 249, 255
Causality, 30, 32, 33
CHSH inequality, 21, 283, 299–301, 303–306, 308–311, 314, 321, 322, 341, 352
Cirel'son bound, 21, 299, 301, 303, 304, 306, 308, 310, 311, 321, 322
Classical, 3, 6, 8–13, 15, 18–20, 26, 28, 30, 32, 34, 48, 49, 52, 55, 71, 73, 77, 82, 98–100, 103–105, 110, 116–119, 125, 126, 130, 132, 136, 148, 152, 181, 183, 184, 186, 187, 215, 216, 219, 225, 227, 235–238, 240, 241, 244, 253, 260, 281, 285, 288, 289, 293, 300, 308, 309, 311, 327, 330–333, 335, 361, 362, 364, 365, 367–373
Classical concepts, 31
Classical probability theory, 287, 288, 327, 328, 330, 331, 333, 362
Clifford reduction, 168–170, 177, 178
Communication, 26, 28, 33–35, 37, 38, 51, 52, 94, 121, 322
Complementarity, 2, 5, 9, 10, 31, 34, 38, 40, 45, 131, 236, 248, 332, 333
Complete compositionality, 20, 247, 256, 258, 261
Completeness, 5–7, 72, 257, 303
Composite systems, 128, 141, 143, 166, 195, 207, 208
Compositionality, 20, 247–254, 256, 257, 259, 261, 262
Conjugate logic, 18, 19, 157–162, 164, 171, 172
Conjunction fallacy, 288, 289, 292
Consistent connectedness, 348, 354, 356–358
Context, 5, 9, 11, 13, 15, 43, 51, 52, 72, 85, 96, 100, 104, 119, 139, 172, 181, 184–188, 193, 247, 250, 251, 258, 291, 293,

A. Plotnitsky and E. Haven (eds.), *The Quantum-Like Revolution*,
https://doi.org/10.1007/978-3-031-12986-5

309, 312, 327, 328, 335, 339–354, 356–358, 362, 371, 372
Contextuality, 9, 21, 119, 158, 171, 172, 184, 187, 237, 289, 291, 300, 301, 328, 329, 332, 335, 339, 340, 345–347, 351–358, 361, 362, 369, 372
Contextuality-by-Default (CbD) theory, 350
Contextual quantization, 184, 186, 188
Continuous variables, 215, 223, 230
Correlations, 2, 3, 5–10, 18, 114, 135, 137, 138, 145, 157, 159, 165, 166, 168, 170–172, 240, 244, 300, 304, 305, 308–310, 321, 341, 362
Couplings, 21, 31, 47, 108, 111–115, 117, 118, 120, 339, 345, 347, 349–352, 354–358
Cut, 104, 283

D
Data compression, 18, 136
Decoherence, 78, 110, 111, 114, 118, 119, 371
Degree of belief, 70
Delayed-choice experiment, 31, 40–42, 44, 47
Density matrix, 99, 101–103, 105, 107–109, 112–114, 116–119, 140, 150, 152, 188, 216, 217, 223, 224, 226–230, 238, 260
Density operator, 19, 136, 137, 140, 194, 196, 197, 199, 207, 215–217, 222, 224, 229, 230
Determinism, 32, 75
Discreteness, 8, 9
Disjunction effect, 327, 328, 330, 333
Double-slit experiment, 97, 109, 163

E
Entanglement, 2, 5, 6, 12, 18, 20, 21, 43, 72, 126, 130–132, 135, 137, 138, 143, 146, 147, 185, 236, 252, 282, 283, 285, 299–306, 308–311, 314, 318–322
Entropy, 13, 18, 99, 100, 106, 148, 152, 237, 335, 358
Epistemic logic, 18, 157, 158, 164
EPR experiment, 5–8
Event, 3, 8, 15, 30, 32, 33, 36, 38–40, 42, 43, 47, 48, 51–54, 78, 79, 83, 86, 87, 103, 110, 119–121, 125, 126, 132, 185, 187, 196, 201, 202, 243, 281, 282, 287, 288, 292, 322, 329–331, 343
Experimental device, 30

F
Fermionic quantum information (theory), 138
Fermionic systems, 18, 135, 136, 138–140, 142, 148, 151
Fock states, 215, 219, 221, 225, 226
Formalism, 3, 9–11, 13–15, 19–21, 30, 37, 71, 74, 82, 85, 95, 96, 99, 103, 104, 116, 119, 120, 136, 159, 165, 169, 171, 181–185, 188, 215, 217, 230, 244, 248, 253, 261, 302, 303, 306, 311, 313, 329, 333, 335, 362, 370–372, 374
Frames, 20, 65, 74, 236–238, 328
Frequency interpretation, 116

G
Gleason theorem, 185, 187, 188

H
Hadamard transformation, 165, 172
Hermitian operators, 334
Hilbert space, 18, 20, 21, 74, 94–96, 126, 127, 137, 139–143, 158, 159, 172, 187, 191, 193, 194, 196, 205, 207, 216, 217, 222, 235–237, 243, 244, 254, 273, 282, 299, 301–303, 305, 306, 311–313, 315, 318, 321, 328, 329, 332, 333, 335, 374

I
Indeterminacy, 36, 332
Indirect measurement, 15, 206, 207, 209, 210, 212
Indistinguishability, 21, 22, 301, 361–367, 369, 370, 372, 374
Information, 2, 6, 9–12, 14, 16–19, 25–27, 29, 30, 33–39, 43–45, 48–50, 53–57, 62, 64, 72, 73, 100, 105, 106, 114, 117, 121, 135–139, 143–145, 148, 150–152, 157, 159, 181, 230, 236, 237, 270, 284–287, 301, 303–305, 321, 322, 327–331, 333, 334, 343, 345, 348, 349, 353, 357, 361, 365, 372
Interpretation, 3–10, 12, 14, 30, 35, 38, 43, 44, 49, 57, 63, 64, 74, 75, 91–94, 96, 98, 99, 106–109, 111, 112, 114, 116, 117, 119, 120, 126, 130, 132, 136, 142, 144, 149, 158, 182, 197, 199, 235, 244, 260, 261, 283, 343, 345, 346, 354, 361, 369, 370

K

Knowledge, 13–15, 18, 34, 35, 37, 48–51, 54, 57, 65, 68–70, 83, 99–101, 105–107, 117, 119, 130, 144, 157, 158, 164, 169, 171, 282, 290, 291, 322, 370
Kochen-Specker theorem, 5–7, 283

L

Language, 3, 5, 30, 37, 39, 51, 56, 66, 72, 74, 76, 83, 95, 119, 127, 158, 159, 172, 185, 248, 250–253, 255, 258, 260, 284, 301, 304, 305, 352, 358, 372
Law, 9–11, 26, 27, 30–33, 39, 44, 53, 55, 72, 73, 76, 78, 83, 86, 96, 106, 113, 120, 121, 160, 162–164, 167, 173–175, 185, 186, 196, 197, 199, 204, 206, 208, 210, 301, 302, 305, 314, 315, 318, 320, 322, 331, 354, 357, 364, 365
Linguistics, 83, 84, 106, 110, 249–251, 258, 259, 371, 372
Local discriminability, 130, 138
Locality, 7, 8, 138, 283

M

Meaning circuit hypothesis (Wheeler), 16, 57
Measurement, 8–10, 14, 15, 18–20, 26–31, 34–40, 42, 43, 45, 48–55, 78–81, 85, 92, 96, 100–121, 138, 151, 157–160, 162–172, 183–187, 191–193, 196–199, 201–208, 210–212, 216, 248, 253, 271–273, 275–278, 299–303, 305, 306, 311–316, 318–322, 332, 339–343, 345–347, 353, 357, 358, 361, 366, 371
Measuring apparatus, 196, 197, 203–206, 208
Measuring instrument, 9–11, 19, 203, 286
Mereological holism, 17, 125, 126, 129–132
Modality, 19, 119, 184–188

N

Nebit (negative bit), 20, 235, 237–244
Negative probability, 19, 20, 216, 235–238, 243, 244
'No-go' result, 334
Noncommutativity, 332, 334
Noncontextuality, 171
Nonlocality, 2, 3, 5–9, 12, 57, 63, 151, 186, 237, 253, 300, 321, 339, 340, 343, 349, 358, 361, 362, 369
Nonrealism, 5

O

Object, 9, 18, 28, 31, 43, 48, 52, 53, 67, 70, 76, 77, 79, 84, 86, 96, 101, 105, 108, 119, 125, 126, 128, 130, 132, 181–183, 185, 192, 196, 201, 203, 205, 207, 208, 243, 258, 290, 339, 348, 361, 363–365, 369, 370, 373, 374
Observer, 28–37, 43, 44, 46–51, 54, 55, 57, 66, 72, 80, 81, 100, 101, 201, 208, 282, 287, 361, 371
Ontology, 26, 45, 50, 56, 63, 64, 72–74, 78, 82–84, 97, 99, 103, 109, 110, 119, 120, 238, 361, 362, 365, 369, 370
Order effects, 15, 20, 269–271, 278, 329, 333–335, 362

P

Participatory universe, 11, 16, 28, 36, 52
Phenomenology, 83
Phenomenon, 9, 10, 20, 26, 30, 31, 33–36, 38, 39, 41–43, 46, 48, 52, 54, 57, 72, 75, 84, 87, 103, 171, 182, 183, 185, 204, 205, 216, 240, 300, 311, 330, 331, 366
Physical object, 18, 50, 57, 125, 132
Physical reality, 2, 5, 6, 11, 28, 30, 31, 33, 39, 51, 56, 182
Positive Operator Value Measure (POVM), 19, 80, 86, 147, 191, 193, 196, 197, 200, 201, 208, 211, 238, 334
Potentiality, 35, 39, 52, 53, 339
Pragmatism, 61, 66, 67, 69, 83
Probability, 3, 6, 7, 12, 14, 15, 18, 19, 34, 62, 63, 77–80, 82, 86, 87, 91, 107–109, 116, 118, 120, 136, 137, 144, 145, 164, 173, 183–188, 193, 196–201, 204, 208, 210, 215–217, 222–224, 226–230, 235–244, 270, 272, 276, 277, 281–283, 285, 287–289, 293, 300–302, 304, 305, 308–312, 314–317, 320, 321, 327, 328, 330–335, 339–344, 347, 348, 350, 355, 362, 366–368, 371–373
Probability density, 110, 215, 216, 224, 225
Probability distribution, 12, 19, 149, 185, 192, 194, 196–198, 200, 201, 203, 206, 208, 211, 215, 217, 218, 222, 223, 225–227, 230, 237–240, 293, 333

Probability representation, 19, 215, 217–219, 222, 223, 226, 229, 230
Projector, 21, 126–129, 131, 143, 149–151, 185, 187, 188, 229, 230, 270, 271, 274–276, 282, 328, 329, 333–335
Property, 11, 18, 19, 21, 26, 31, 36, 37, 39, 41, 43, 44, 48–51, 57, 73, 101, 103, 104, 112, 118, 125–132, 138, 143, 146, 150, 151, 157, 162, 164, 167, 172, 181–183, 186–188, 196, 199, 205, 207, 210, 211, 216, 222, 225, 230, 239, 279, 282, 284, 286, 301, 302, 318, 328, 330, 339, 345, 347, 348, 352, 358, 361, 363, 364, 369, 372
Property, 282, 284, 286
Purification, 102, 103, 107, 137, 142, 143, 146, 147

Q

QQ equality, 15, 20, 21, 269–271, 276, 278
Quantum algebra, 362
Quantum Bayesianism (QBism), 16, 17, 19, 62–65, 78, 80–82, 84–86, 116
Quantum cognition, 20, 21, 269, 278, 282
Quantum computing, 12, 186, 236, 248, 253
Quantum foundations, 1, 2, 5–9, 12, 14, 16, 18, 20, 25, 61, 96, 98, 116, 136, 247, 248, 262
Quantum holism, 17, 20, 126, 128, 130
Quantum information theory, 16
Quantum-like models, 328–330, 332, 333, 335, 371
Quantum logic, 15, 18, 126, 127, 158, 172
Quantum measurement, 11, 15, 17, 19–21, 36, 41, 42, 48, 55, 72, 79, 80, 86, 87, 91, 93, 96, 98, 99, 104, 106, 108–112, 119, 191, 193, 204, 205, 212, 269–271, 273, 278, 334
Quantum object, 7, 9–11, 105, 361
Quantum phenomena, 3, 4, 6–9, 11, 17, 19, 27, 34, 35, 38–40, 42, 46, 49, 50, 54, 57, 371
Quantum probability, 3, 12, 14, 15, 21, 35, 62, 185, 225, 312, 316, 327, 328, 331–335, 362, 363
Quantum state, 19, 35, 44, 53, 62, 63, 74, 77, 81, 85, 100, 106, 138, 146, 147, 159, 185, 192, 193, 196, 198, 201, 202, 206, 208, 215–219, 223, 227, 229, 230, 235, 237, 243, 332, 366, 367, 369
Quantum systems, 3, 6, 8, 14, 18, 19, 37, 41, 49, 51, 54, 74, 76, 77, 90, 98, 99, 105, 117, 119, 125, 126, 128–130, 132, 138, 157, 161, 164, 181, 184, 186, 192, 194, 196, 205, 207, 210, 215–218, 230, 248, 361, 367, 369
Quantum world, 31, 47, 73–75, 181, 182, 186, 361, 362, 369
Quantum probability, 281, 282
Qubit, 135, 137, 139–141, 143, 152, 153, 188, 215, 217, 226–230, 238, 257

R

Random field, 12
Random walk, 98, 235, 238–241
Rationalism, 61, 69, 75, 83
Realism, 2–5, 7, 19, 35, 36
Reality, 26, 27, 29–31, 61–63, 66–70, 73, 74, 79–81, 83, 85, 282
Repeatability hypothesis (von Neumann's), 19, 192, 202
Response replicability, 15, 271, 334, 335

S

Savage sure-thing principle, 327, 329
Schrödinger compositionality, 247, 253, 256, 261
Schrödinger equation, 95, 96, 216–218, 220, 227, 230
Shannon entropy, 136, 148, 152, 237
State space, 19, 191, 193–195, 207, 210, 211, 335
Stern-Gerlach, 105, 271, 272
Strong Baseball Principle, 341, 343
Super-correlations, 282, 283
Superposition, 44, 72, 106, 137, 141, 149, 185, 215–217, 221, 229, 230, 274, 282, 283, 285, 301, 332
Symmetry, 26, 96, 162, 167, 188, 302, 318, 366, 368, 374
Symplectic tomography, 217, 223

T

Tensor product, 100, 102, 140, 150, 195, 272, 273, 301, 302, 305, 313, 315, 316
Theory of measurement, 192
Tomography, 18, 138, 143, 151, 217, 222, 223

U

Uncertainty, 52, 84, 116, 121, 136, 158, 161, 164, 171, 172, 282, 327–331, 333, 335

Uncertainty relations, 9, 10, 164, 216
Unitary transformation, 129, 185, 187, 236, 272

V
Von Neumann entropy, 106, 107, 130, 135–138, 148, 151, 154

W
Wave function, 19, 29, 94, 96, 100, 101, 107–110, 119, 215–222, 224, 225, 230, 236, 369
Wigner function, 216, 224, 225, 236
Wigner's friend, 78, 79, 80, 81, 119

MIX
Papier aus verantwortungsvollen Quellen
Paper from responsible sources
FSC® C105338

If you have any concerns about our products,
you can contact us on
ProductSafety@springernature.com

In case Publisher is established outside the EU,
the EU authorized representative is:
Springer Nature Customer Service Center GmbH
Europaplatz 3, 69115 Heidelberg, Germany

Printed by Libri Plureos GmbH
in Hamburg, Germany